WATER IN MINERAL PROCESSING

Edited by Jaroslaw Drelich

Contributing Editors:
Jiann-Yang Hwang
Jack Adams
D.R. Nagaraj
Xiaowei Sun
Zhenghe Xu

PROCEEDINGS OF THE FIRST INTERNATIONAL SYMPOSIUM

Published by
Society for Mining, Metallurgy, and Exploration

Society for Mining, Metallurgy, and Exploration, Inc. (SME)
12999 E. Adam Aircraft Circle
Englewood, Colorado, USA 80112
(303) 948-4200 / (800) 763-3132
www.smenet.org

SME advances the worldwide mining and minerals community through information exchange and professional development. With members in more than 70 countries, SME is the world's largest association of mining and minerals professionals.

ISBN 978-0-87335-349-6
Ebook 978-0-87335-356-4

Library of Congress Cataloging-in-Publication Data

Water in mineral processing / edited by Jaroslaw Drelich ; contributing editors, Jiann-Yang Hwang ... [et al.]. -- 1st ed.
p. cm.
Includes bibliographical references and index.
ISBN 978-0-87335-349-6 (print) -- ISBN 978-0-87335-356-4 (ebook)
1. Mineral industries--Water-supply. 2. Mineral industries--Waste disposal--Environmental aspects. 3. Water reuse. 4. Dredging spoil--Environmental aspects. I. Drelich, J. (Jaroslaw) II. Hwang, Jiann-Yang.
TD428.D74W38 2012
622.028'6--dc23

2011047136

Contents

Preface

One of the major challenges that the mining and minerals processing industry faces in the 21st century is dealing with the management of ever decreasing water resources. More than 80 countries experience water shortages that threaten industrial activities and the health of their citizens. In the United States, water shortages are also of concern, not only in dry, western states, but also in Alabama, Florida, Hawaii, Illinois, Louisiana, Maryland, Missouri, South Carolina, and Virginia. The solutions to the problem include reduced consumption of water, development and use of water-free technologies, treatment and recycling of industrial and municipal water streams, and desalting of brines and water from oceans and seas.

To conserve water resources, water from industrial activities must be recycled or discharged after treatment to meet the standards of freshwater. Although removal of suspended and floating particles has improved over the years, the search for better and more economical technologies continues. More challenging is the removal of dissolved solids and organics. Current technologies in this area are insufficient and, consequently, even continuous recycling of industrial water contributes to a decline in its quality, particularly through increases in salinity and hardness of the water and increased content of dissolved organics, which affect mineral processing operations, corrosion of equipment, water use, and reclamation efforts.

In recognition of these challenges, the SME Mineral and Metallurgical Processing Division organized this First International Symposium on Water in Mineral Processing during the 2012 SME Annual Meeting in Seattle, Washington. Through this symposium, and similar symposia to be organized in the future, we intend to promote innovative water-use and water-purification technologies for mineral processing applications. Because most mineral processing operations require high water use to sustain the parameters for selective separation of minerals, the symposium intends to address the needs for sustainable technologies with reduced water consumption and reduced discharge of process-affected water. Strong emphasis is also given to fundamental aspects of the effects of process water quality on recovery and grades of valuable minerals products.

The program comprises a plenary session with keynote addresses by Robert Dunne from Newmont Mining Corporation and Ronald S. Oremland of the U.S. Geological Survey, and four regular sessions on the following topics: Processing with Sea Water and Saline Solutions, Water Treatment and Biological Methods, Effect of Water Quality on Minerals Processing, and Water and Tailings Management, with two additional keynote presentations by Sergio Castro of the Universidad de Concepción (Chile) and Enrico Drioli from the University of Calabria (Italy). We thank all of the keynote speakers for accepting our invitation and sharing their expertise and time.

All keynote and regular presentations from the symposium are collected in this book. We sincerely thank the authors for their contributions, and we thank all the reviewers for their time and effort. We also acknowledge the help and support received from the publisher. We hope that you will find this book timely and useful.

Water Water Everywhere and Not a Drop to Drink, Nor Do I Know Its Whereabouts

Robert Dunne
Newmont Mining Corporation, Englewood, CO, USA

ABSTRACT

The demand for water is driven primarily by population and concomitant economic growth. Water requirements are predicted to grow considerably in the next decades while supplies will remain relatively constant or decline due to over pumping of aquifers, changing weather patterns and increased water pollution and contamination.

Mining activities are often located in remote, arid environments, with limited access to high-quality water. The water used in mining operations comes from a variety of sources and the sources and quality of the water varies from operation to operation. Mining impacts on water quantity and quality are among the most contentious aspects of mining development. The main problem for the mining industry is to generate confidence in developing a responsible, sustainable and transparent water management strategy that is recognized as such by all stakeholders. This paper provides an overview of water in the wider global arena and compares this to how the mining industry has dealt with water stewardship over the last couple of decades.

INTRODUCTION

The demand for water is driven primarily by population and concomitant economic growth. Water requirements are predicted to grow considerably in the next decades while supplies will remain relatively constant or decline due to over pumping of aquifers, changing weather patterns and increased water pollution and contamination. While all regions will experience water scarcity to some degree, for some countries it will become more acute (Vaux 2011). It is estimated that by 2025, as water scarcity intensifies, more countries will be severely short of water. Water shortages appear to have started in earnest around 1900, when around 2% of the world's populations were subjected to acute water shortages (Oelkers et al. 2011). This increased to 9% by 1960 and since then the number of people subject to chronic water shortages have increased rapidly to around 35 percent by 2005 (Kummu et al. 2010).

Overall, some 66% of the water withdrawn from the environment is used in agriculture, 23% by industry (includes mining) and 7% by households (Horrigan et al. 2002). Annually, about 42,500 cubic km^3 of water is available for human use, and 2600 km^3 is consumed (primarily for agriculture). Globally, water supplies are virtually fully allocated and water use is between 50 and 85% of freshwater supply. In total, approximately 4000 km^3/y of water is withdrawn from surface water and groundwater sources worldwide to meet human needs. Of this roughly 20% is taken from groundwater (Boswinkel 2000). It is estimated that the "safe" consumption of runoff is essentially already allocated in that it will be needed to feed a growing global human population (Barrett 2011).

GLOBAL WATER RESOURCE

The total amount of water in the hydrosphere consists of "free" water in liquid, solid or gaseous states in the atmosphere, on the Earth's surface, and in the crust down to a depth of

Table 1. Major earth surface water reservoirs (based on Gleick, 1996)

Water Source	Percent of Total Water	Mean Residence Time of Water
Oceans, seas, bays	96.5	2500 years
Ice caps, glaciers	1.74	9700 years
Groundwater	1.69	1400 years
Fresh	0.76	—
Saline	0.93	—
Ground ice and permafrost	0.022	10,000 years
Lakes	0.013	17 years
Fresh	0.007	—
Saline	0.007	—
Atmosphere	0.001	8 days
Swamps	0.0008	5 years
Rivers	0.0002	16 days
Biological water	0.0001	4 hours
Totals	100	

approximately 2000 m. Approximate estimates of the Earth's water resources are shown in Table 1 (Gleick 1996). Some 97.5% is saline water largely in the oceans, so that only 2.5% is fresh water. The greater proportion of the fresh water is in the form of ice and permanent snow cover in polar and mountainous regions (Trenberth et al. 2007). The other two sources are surface water (rivers and lakes) and atmospheric water (via rainfall) and store no more than 0.0012% of global water reserves. These sources are the most accessible for economic needs and are of vital importance to water ecosystems. However, a significant fraction of this water flows in isolated rivers (e.g., in South America and the Arctic) and from a practical viewpoint is unavailable. Besides this these resources are rapidly replenished by natural processes. Almost all of the remaining water is contained in underground aquifers. Water storage in rivers and lakes is small, some two orders of magnitude less than groundwater volume. The mean residence time of water in rivers and the atmosphere is estimated to be 16 and 8 days, respectively. Although lakes store more freshwater than either rivers or the atmosphere, they take on average 17 years to refill.

WATER AVAILABILITY

Rainfall and Surface Storage

Surface water and groundwater sources have up until now, ensured the growth of our society. Surface water can be very important locally however, there are limitations in the supply as the availability of surface water can fluctuate markedly according to rainfall (monsoon) or snowmelt cycles (Vaux 2011). As rainfall supplies approximately 80% of the water used for agricultural production worldwide these fluctuations can have serious impacts on agricultural production.

The prevailing practices of water resource management during the last century placed heavy reliance on the construction of reservoirs/dams and canals to store and distribute water (Vaux 2011). Constructed reservoirs/dams are able to accommodate seasonal changes, but often at the cost of large evaporative losses and expensive infrastructure. Surface waters are also vulnerable to cycles of drought. The naturally small storage of surface waters means that declines in precipitation quickly translate into reductions in stream flows. As water demand continues to grow so water available for development becomes increasingly scarce. The costs of water storage infrastructure have risen disproportionately because low-cost sites and water supplies have already been developed and because the costs of civil works have increased faster than the rate of inflation (Vaux 2011).

Half the world's people now live in cities and are heavily dependent on their water-supply systems. Examples of these are: Los Angeles uses of local groundwater but import the bulk of its supply from surface waters in the Owens Valley, the Sacramento-Dam Joaquin Delta, and the Colorado River, hundreds of kilometers away. Beijing, which has relied historically on local supplies, is now tapping the Yangtze River 1000 km to the south. In less than a century, sociological tools needed to grow their cities seemingly without regard for natural limits on the availability of local water supplies (Schwartz and Ibaraki 2011).

Groundwater

Groundwater is by far the most readily available water resource of freshwater. The huge volumes of water stored in aquifers together with slow rates of "natural" depletion make groundwater a resilient source of water (Vaux 2011). Groundwater is commonly extracted from aquifers big and small and the rate of "external" withdrawal of groundwater around the world has increased substantial however, the rates in many instances far exceed the natural replenishment (Schwartz and Ibaraki 2011). Moreover, increasing populations, with their associated agricultural and industrial activities, create sources of pollution that further limit supplies of water. Furthermore, intensive extraction of groundwater can create unintended impacts on surface waters. The inherent coupling of groundwater and surface waters means that groundwater consumed for irrigation reduces the flow in streams that normally receive natural inflows of groundwater (e.g., springs) (Schwartz and Ibaraki 2011).

In the United States, two-thirds of piped water (mostly in large cities) comes from surface water, but in rural areas, drinking water is frequently not obtained from public water suppliers and nearly all domestic supply is reliant on groundwater (Johnston et al. 2011; Kenny et al. 2009).

Lakes

Lakes also are a "large" source of freshwater and well as "quality" water for industrial usage. The latest survey undertaken by the International Lake Environment group, in co-operation with the United Nations Environment Program, of 217 lakes worldwide found that 54% of lakes in Asia are eutrophic (oxygen depleted by nutrients), compared to 41% in South America and 28% in Africa (Jørgensen 2001). The greatest water-quality problems in lakes are found in countries with large populations and with scarce financial resources. The regulation of nutrients to protect drinking water supplies is urgently needed (Johnston et al. 2011)

Aquifer Storage

A process known as aquifer storage and recovery *(*ASR*)*, has emerged as an increasingly attractive water-management strategy in the last two decades. ASR and variants such as aquifer storage transport and recovery (Rinck-Pfeiffer et al. 2005) involves the storage of water in sub-surface aquifers when water is plentiful and recovered during times of peak demand or water stress (Pyne 1995). Peak storm water flows and wastewater streams can be harvested, treated passively (e.g., constructed wetlands) or actively (e.g., dissolved air flotation/filtration) and injected into confined aquifers for subsequent recovery for non-potable purposes (Pyne 1995; Pavelic et al. 2006a; Swierc et al. 2005). ASR has been carried out for municipal, industry and agriculture use. In Texas, there are two operating ASR facilities-one was developed in Kerrville for the Upper Guadalupe River Authority in the early 1990s, and the other the San Antonio Water System in south Bexar county and this came online in June 2004. The first agriculture ASR wells were put into service in Oregon in the autumn of 2006 injecting spring flood waters into the aquifer (www.ecy.wa.gov). The shallow recharged water is then recovered as potable water and injected into the deep basalt aquifer.

Wastewater

Today, wastewater treatment technologies have advanced to the point where household and industrial wastewater can be cleaned so as to meet the standards prescribed for household use. For example, the Orange County Water District, located in Southern California, produces significant quantities of water from wastewater, and this recycled water is recharged to local aquifers from which it is ultimately extracted to serve household needs (Vaux 2011). The district employs several technologies, of which the most advanced feature artificial membranes that are used in a reverse osmosis process to clean the water prior to direct injection into the underlying aquifer. These technologies are relatively costly and can be employed economically only under certain conditions,

WATER QUALITY

The deterioration of water quality can be either natural (geogenic) or human induced (anthropogenic).

Geogenic Deterioration

Geogenic contaminants in groundwater are related not only to the local and regional geology but also to the conditions that facilitate contaminant release from aquifer sediments (Smedley and Kinniburgh 2002).

The chemistry of most surface water and shallow groundwater is the result of water-rock-gas interactions. The evaporation of water concentrates solutes, in certain circumstances resulting in the crystallization of minerals such as calcite and gypsum. The combination of evaporation and mineral precipitation can lead to high sodium, chloride, and/or sulfate concentrations. Soil gases commonly have carbon dioxide because of root respiration and organic matter decay. Therefore percolation of water through organic-rich soil can further lower pH, leading to the dissolution of acid soluble minerals. Weathering reactions in the soil zone include the dissolution of unstable silicates, ion exchange reactions are fast, and water commonly establishes equilibrium with these minerals and mineral surfaces. In contrast, silicate mineral dissolution in nature is controlled by slow kinetics, and the precipitation of secondary clay minerals is an even more sluggish process (Zhu and Schwartz 2011).

The two most important geogenic contaminants from a health aspect (Hopenhayn 2006) are arsenic (Ravenscroft et al. 2009) and fluoride (Fawell et al. 2006). Hundreds of millions of individuals, mostly in Asia, are exposed to these geogenic contaminants at levels 10 to 100 times greater than drinking water standards.

Anthropogenic Deterioration

Anthropogenic contaminants are chemical contaminants that have been introduced to the environment by the activity of man. These contaminants include industrial chemicals inadvertently released into the environment, as well as those derived from land use activities such as oils and grease flushed off roadways and agricultural chemicals. Chemical plants, manufacturing facilities, gas stations, repair shops, landfills and mining activities all have the potential to release contaminants into the environment.

Modern agriculture is capable of contaminating shallow aquifers with fertilizer, pesticides and agricultural waste (e.g., manure). In the High Pains Aquifer (central USA), enhanced recharge in irrigated regions has mobilized natural chloride and nitrate from the unsaturated zoned and transported trace quantities of pesticides into shallow groundwater (Gurdak et al. 2009). For this reason efforts are in place and growing to protect groundwater from contamination from agricultural, mining and industrial activities, and improper disposal of solid waste and domestic wastewater (Zhu and Schwartz 2011). In recent years it has been recognized that submarine groundwater discharges into coastal oceans contributing 5–10% of freshwater inflows to oceans, as well as being a significant source of nutrients and pollutants for coastal ecosystems (Johnston et al. 2011).

As rainwater falls through the atmosphere it is chemically altered; for example it dissolves and incorporates sea salt aerosols, sulfate, ammonium, and nitrate from natural and human sources such as power plants, automobile exhaust, smelters, and agriculture (Zhu and Schwartz 2011). Rainwater also equilibrates with atmospheric carbon dioxide giving unpolluted rainwater a pH of around 5–6. Furthermore, this rainwater water may be chemically altered if it is able to react with dry particulate aerosols on the ground accumulated between precipitation events.

Anthropogenic changes in water flows to the oceans may also affect the global cycles of some elements (Oelkers et al. 2011). The flow of water down rivers transports more than 7.6×10^{13} kg/y of elements in dissolved form and within suspended materials to the oceans (Viers et al. 2009). This process is reported to play a major role in the global cycles of most elements, including phosphorus and nitrogen (Meybeck 1982; Vitousek et al. 1997; Filippelli 2008). Similarly the transport of calcium by rivers from the continents to the oceans and its subsequent precipitation as carbonate minerals is an important component of the long-term carbon cycle (Berner et al. 1983; Gislason et al. 2006).

WATER TREATMENT

In an attempt to address the growing need for water and to attenuate environmental damage, desalination treatment is growing rapidly (Shannon et al., 2008). A range of desalination processes are used worldwide and the most common is reverse osmosis (Fritzman et al. 2007). It is estimated that there are around 21,000 desalination plants in use today in more than 120 countries around the world (Oelkers et al. 2011), almost 50% use seawater to produce fresh water and the rest use brackish water. Technical improvements have enabled the cost to decrease substantially over the past three decades; the cost to desalinate 1 m^3 of seawater can now be US$1.00 or less (Yermiyahu et al. 2007). A large part of the cost is associated with energy consumption and therefore the overall cost can escalate rapidly and be vastly different on a regional basis (Andrianne and Alardi 2002). The combination of lower cost and an increasing need for fresh water has led to substantial increase in desalination treatment worldwide. Between 1994 and 2004 desalination capacity has reportedly increased from 6.3 to

13 km^3/year. Desalination currently supplies less than 0.3% of global water needs and thus contributions a small amount only to global requirements.

WATER IN THE MINING INDUSTRY

Mining activities are often located in remote, arid environments, with limited access to high-quality water. The water used in mining operations comes from a variety of sources, including groundwater, surface water, seawater, water from dams on site or sourced from water treatment plants. The source and quality of the water varies from operation to operation. Mining impacts on water quantity and quality are among the most contentious aspects of mining development. Negative environmental impacts of past mining activity generally cause local and downstream populations to worry that existing and new mining activity will adversely affect their water supply and quality (Brown 2003; Bebbington and Williams 2008). The main problem for the mining industry is to generate confidence in developing a responsible, sustainable and transparent water management strategy that is recognized as such by all stakeholders.

WATER USE IN THE MINING INDUSTRY

Water is a critical input for mining, mineral processing smelting, refining and petroleum businesses. Proportionately most of the water in mineral processing is for flotation and to a lesser degree for heap leaching of copper and gold ores. Other water uses include dust suppression (McIntosh and Croni 2003), equipment cooling, slurry conveyance and human needs at mine sites. In recent times many sites are replacing high quality (potable) water with poorer quality water to help conserve local water supplies. From a mining aspect, water, especially high quality water, is an increasingly scarce and valued resource around the world.

WATER RESOURCES AND QUALITY IN THE MINING INDUSTRY

There are a multitude of water resources available to the mining industry and Kaufmann and Nadler provided an excellent diagrammatic representation of these in their paper (Kaufmann and Nadler 1965) that is still valid today. The diagram, which has been modified, is shown in Figure 1 and the water resources available are summarized as follows.

Groundwater

As mentioned previously mining activities are often located in remote and arid environments with limited surface water supply, for this reason groundwater is the main source of water for mining and processing activities. Due to the need of good quality for human consumption and farming the quality of underground water for mining is usually "brackish." As an example in Western Australia water supply is mainly from bore field sources and this water is hyper-saline. Measured salinities are in excess of 100,000 ppm total dissolved solids and, in some cases, in excess of 200,000 ppm, six times that of sea water.

Sea Water

Some historic and current examples of sea water application are the Texada Mill in British Columbia where sea water was used in the 1970s to produce both high-grade sinter iron concentrate and a by-product copper-gold-silver flotation concentrate (Haig-Smillie 1974). The Black Angel lead-zinc mine in Greenland operated from 1973 to 1990 used sea water in the grinding and flotation circuits (Poling and Ellis 1995). The Batu Hijau Copper Mine

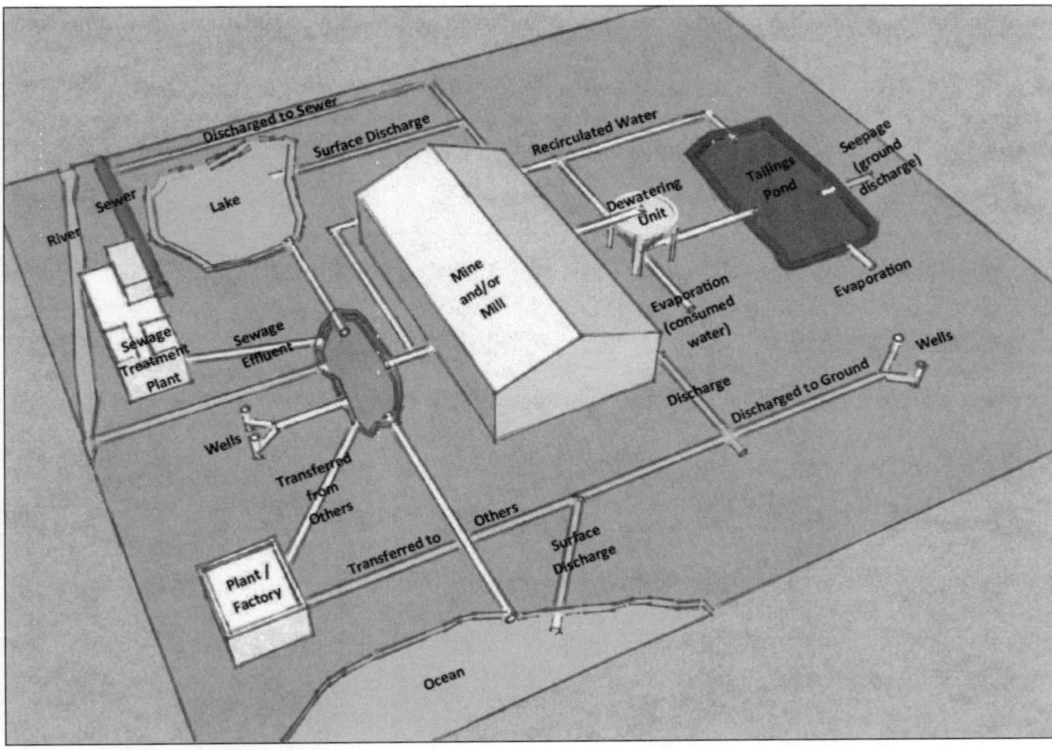

Figure 1. Water resources available to the mining industry (based on Kaufmann and Nadler 1975)

on the island of Sumbawa, Indonesia, initially intended to use fresh water from surface runoff for the processing plant. However, the lack of sufficient storage capacity and the large variations in seasonal rainfall led to the decision to use sea water. Sea water is used for 7 to 8 months of the year, the balance of the time water is a blend of seawater and mine water runoff from stockpile, waste and mine dewatering (Wirfiyata and McCaffery 2010). In Chile, the Michilla Copper Mine located near the coast and 100 km north of Antofagasta, has been using sea water since the 1990s for copper leaching (Wiertz 2009). More recently, Esperanza, a copper mining project under construction and located north east of Antofagasta decided to exclusively use sea water for flotation (Wiertz 2006; Chadwick 2009). Interestingly about half of Rio Tinto's annual water withdrawal is sea water (600 billion litres) and is used for cooling water at Rio's owned power stations (Rio Tinto Sustainability Report 2010).

Desalinated Sea Water and Groundwater

In Chile, several mine sites located in the arid mountains of the Andes are currently looking at using reverse osmosis (RO) plant located on the seashore to supply their operation with desalinated sea water (Wiertz 2009). The Escondida mine filed an environmental impact assessment in late 2009 for sea water desalination plant using reverse osmosis (RO) to provide water for a proposed capacity increase at the mine. The Candelaria Copper Mine has also looked at reverse osmosis (RO) options to provide fresh water to alleviating impacts on the mine site impact on the local aquifers (Levay et al. 2006). BHP Billiton has for some time been evaluating seawater desalination on the coast of South Australia and pumping it inland to supply the Olympic Dam mine with fresh water to cater for the mine's proposed expansion (Levay et al. 2006; BHP Billiton Sustainability Report 2010).

Reverse osmosis (RO) is the key membrane-based filtration/desalination technology used to treat saline groundwater at many mines in the Australian outback to provide potable water and high quality water for metallurgical purposes (e.g., elution of gold loaded activated carbon).

Sewage Water

The simplest, and perhaps most common synergies in wastewater application are those where one operation takes the wastewater from another as all or part of its input. Many examples exist where treated effluent is used in this way. Examples are the Levack concentrator in Sudbury, Canada (Browne and Butler 1970), Cadia Gold Mine, New South Wales, Australia (Levay et al. 2001), Queensland Alumina Limited (QAL) in Queensland (Stegink et al. 2003) and the Port Kembla steel works in South Australia (Hird 2006).

MINIMIZING WATER CONSUMPTION IN THE MINING INDUSTRY

The incentive for recycling water on a mine is to lower the demand for externally supplied water. It is generally accepted that any recycling of water is an environmentally desirable practice. Recycling can also lead to lower operating costs as less water needs to be purchased and/or pumped. Furthermore any residual reagents can be returned in the recycle water, thereby lowering reagent consumptions (Coleman and Wallace 1978). There also are positive social implications from water recycling, especially if there is competition for scarce surface water resources in semi-arid or arid regions, or if extraction of underground water impacts local users of the same groundwater resource. One of the earliest references on "good water recycle practice" stated that at the Copper Cliff mill, Sudbury, Canada, recycled water increased from 50% to 72% during 1941 (Browne and Butler 1970). Records in the USA showed that on average about 69% of mine site water was recycled in the 1970s (Bailey 1970; Joe 1973; Pickett and Joe 1974; Hyatt 1976). By the 1980s the recycle had increased up to 73% (Watt et al. 1982). Information provide in recent sustainability reports (2009 and 2010) for large global mining companies show that water recycled on a company-wide basis to be in the range of 50 to 75%.

There are a number of factors driving the industry to increase "dewatering" (solid liquid separation) in the mining industry. These include limited water availability and reduced chemical consumption (Klein and Hallbom 2009). The "dewatering" process separates water from solids and provides the opportunity to recycle water. There are many dewatering process that achieve this including screening (Doerffer and Heinrich 2009; Mathewson et al. 2006), conventional and paste thickening (Chambers et al. 2003), filtration (Mathewson et al. 2006) and sedimentation in the tailings dam. To enhance the separation of water from the solids in thickening and sedimentation selected chemicals are used for coagulation and flocculation (Levy et al. 2006; de Krester et al. 2009). Thickener equipment technology also has advanced greatly over the last two decades to produce higher underflow densities and especially for the treatment of "difficult to dewater" ores (Schoenbrunn et al. 2009). Enhanced dewatering of colloidal hydrophilic mineral dispersions, via flocculation and gravity-assisted thickening, remains an important and technically challenging problem. This is especially so for ores with high clay contents. Prudent selection and mode of addition of different polymer types and structure can promote synergistic effects between cationic and anionic flocculants leading to significantly improvements in the dewatering of heterogeneous mineral tailings (Addai-Mensah and Ralston 2005; Levy et al. 2006).

For thickened tailings disposal, dry stacking and paste fill, the rheology must be manipulated and exploited (Sofra and Boger 2000; Levy et al. 2006) to ensure maximum efficiency of the entire disposal process including pumping (Simms et al. 2008; Cooke 2008). Recently novel dewatering aids have been shown to substantially reduce the cycle time for filtration and the filter cake moistures content (Asmatulu et al. 2005). Advances in dewatering equipment over the last two decades, especially the development of large capacity vacuum and pressure filter equipment (Mathewson et al. 2006) has presented an opportunity for storing tailings in an "unsaturated" state. The filtered tailings can then be transported to the storage area by conveyor or truck and placed, spread and compacted to form an unsaturated, dense and stable tailings "stack" (often termed a "dry stack").

Most of the world's mineral processing plants store tailings in engineered impoundments to insure structural integrity. These structures must be designed and constructed to hold and manage large quantities of water. Water losses (evaporation and leakage into subsurface) from tailings impoundments are receiving increased scrutiny from regulators and the public, who now insisting that viable alternatives be considered for any new mining development (Pinto and Barrera 2008; AMEC 2008). Water losses by evaporation can be reduced by minimizing the water surface area, reducing the water temperature and/or minimizing the exchange of water vapor with dry air above the water. Evaporation reduction methods can be grouped into either structural (i.e., wind breaks, alternative dam design or cover techniques). Several companies currently manufacture floating balls, floating covers, chemical mono-layers and shade structures to minimize evaporation from storage dams. A novel "floating modular" cover to reduce evaporation has been developed by Rio Tinto (Takos et al. 2006).

WATER TREATMENT IN THE MINING INDUSTRY

The practice of water treatment is to improve water quality either before recycle or prior to discharge. As far back as the 1960s it was recognized that stricter requirements on the quality of discharge water would be imposed by regulatory authorities, thus necessitating the treatment of virtually all discharged water (Kaufman 1967). At that time it was predicted that mineral treatment plants would recycle substantially more water and incur additional costs associated with chemically treating discharge water or in recirculating the water.

Mine water treatment technologies have evolved substantially during the past twenty years, driven by a sustained research effort and by the need of mining companies to address their long-term liabilities associated with contaminated mine water (van Niekerk et al. 2006; Kuyucak 2006). A number of reliable and proven neutralization and desalination technologies are now available and shown in Figure 2. Lime is still the principal alkali pH modifier for flocculation to improve water clarity and/or for alkalinity to promote metal precipitation as hydroxides (Kemmer and Beardsley 1971; Kuyucak 2006; van Niekerk et al. 2006). The next preferred option is treatment by reverse osmosis (RO). Queensland Nickel uses RO to treat and recycle contaminated water; Ranger Uranium Mine and the Ulan Coal Mine use it to treat excess water prior to discharge into the environment (Levy et al. 2006). During the last two decade biological treatment of wastewaters has receive a lot of attention. The biological sulfate reduction (BSR) process is one of the more successful technologies and in today's terms it is considered established and proven (van Niekerk et al. 2006; Kuyucak 2006). Nowadays contaminated mine water is not always considered a liability, but may be exploited as a valuable water resource in many water scarce catchments (Levy et al. 2006).

Brine solutions generated from RO treatment of mine and process waters normally need further treatment to remove and stabilize the precipitated metals prior to disposal into land

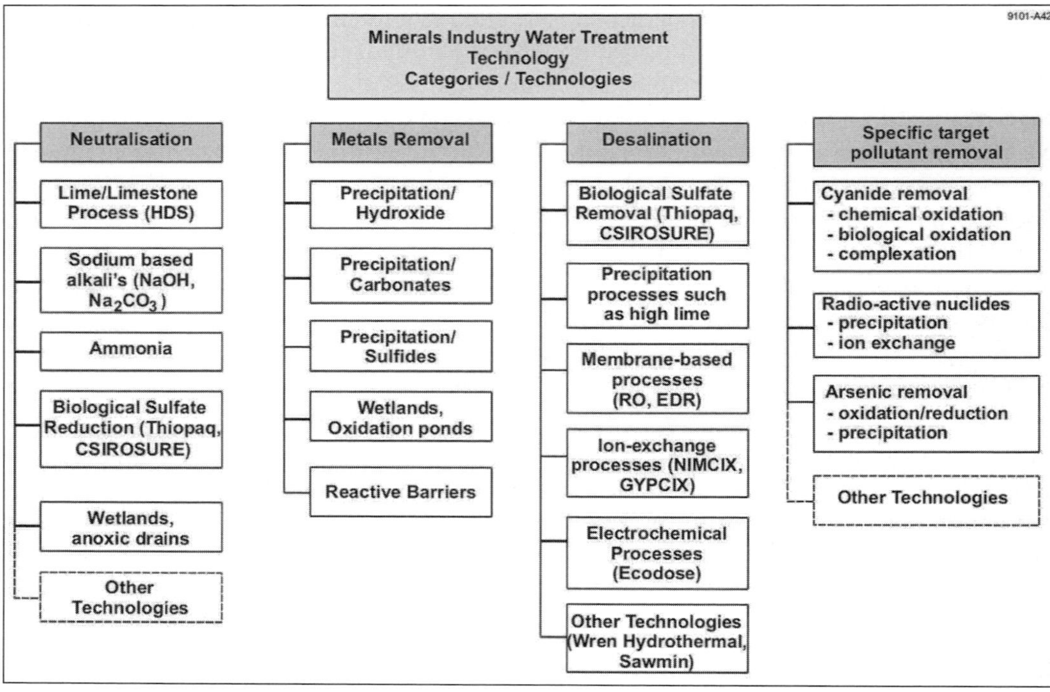

Figure 2. Commonly available water treatment technologies (van Niekerk et al., 2006)

fill. The process is complex involving complex chemistry to achieve the extremely low metal levels required by statutory organizations (Madin 2006). The subject matter on all the available treatment options is extensive and a few recent references on the subject are provided (Fisher et al. 1999; O'Leary 1996; Kuyucak 2006; Levy et al. 2006; McCloskey et al. 2008; Gusek et al. 2008; McCloskey et al. 2009; Sibrell et al. 2008; Croall 2009; Hollis 2009).

Sludge generated from water treatment, such as ARD precipitation or treatment of RO brine solutions, is either stored in plastic lined facilities on site or sent to external permitted land filled sites. Disposal of sludge often constitutes a significant portion of the operating cost of an AMD treatment plant.

WATER TREATMENT IN THE OIL INDUSTRY

In the oil industry the treatment of water containing more than 40,000 mg/L TDS using RO is considered not to be economically feasible even at favorable energy prices (Gregory et al. 2011; Cline et al. 2009). For high-TDS waters, vibratory shear-enhanced processing (VSEP) has been applied to membrane technologies (Jaffrin 2008). In VSEP, flat membranes are arranged as parallel discs separated by gaskets. Shear is created by vibrating a leaf element tangent to the membrane surface and this lifts solids and fouling material off the membrane surface, thereby reducing colloidal fouling and polarization of the membrane (New Logic Research 2004). VSEP technology has been used successfully in the treatment of water from offshore oil production (Fakhru'l-Razi et al. 2009).

Several other technologies that have been or are being developed in the oil industry for treating TDS waters include ion exchange and capacitive deionization (Jurenka 2007), which are limited to the treatment of low-TDS water; freeze-thaw evaporation, which is restricted to cold climates; evaporation ponds, which are restricted to arid climates; and artificial wetlands and agricultural reuse (Veil et al. 2004), which are greatly limited by the salinity tolerance of plant and animal life. Distillation and crystallization are mature technologies that rely on evaporating the wastewater to separate the water from its dissolved constituents (Doran and Leong 2000). Distillation removes up to 99.5% of dissolved solids and has been estimated to reduce treatment and disposal costs by as much as 75% for produced water from shale oil development (ALL Consulting 2003). However, distillation is a very energy-intensive process.

STRATEGIC AND MINE SITE WATER MANAGEMENT PLANNING

A strategic water management framework utilizes a system-level approach to the accounting of water in the mine water circuit and surrounding region. It identifies the operational tasks on-site associated with water, such as dewatering of ore bodies, grinding, flotation, dust suppression, water quality treatment, and the storing and discharge of water. It explicitly represents the flows of water between tasks within sites and the flows of water into and out of sites within a region. A systems approach allows benchmarking of mine performance against industry best practice, and examination of how operations interact with surrounding environment, community and other stakeholders,

Water management plans are required for most mining operations and include requirements to implement controls to mitigate water use and discharge impacts. A dynamic water balance that takes account of cyclic variations (i.e., wet and dry seasons) is a major part of any mine site water management plan. The quality of water is also an important consideration, as it can effect production or increase operation costs. As a result, each mining operation will have its own unique water strategy.

WATER DISPOSAL IN MINING

The way that effluent mine water is treated depends on the characteristics of the effluent, the nature of the receiving water (such as rivers and lakes), the possibility for reusing the water, and the specific legal requirements of the local country authority. Obviously before making any decision on what treatment system to use for a particular effluent, it is essential to consider its physical, chemical and biological characteristics in order to asses the potential for reducing its concentration and volume, in accordance with the principle of cleaner production.

CONCLUSIONS

Key considerations in today's world of mining are the availability of water for ore treatment and the amount of water that can be replenished or that needs to be discharged. All stakeholders associated with the mining industry are demanding that the industry continually looks at ways to reduce external water requirements and implement best practice water management strategies. Where mines discharge water into the environment there will be ever increasing requirements to improve water quality. Increased water recycle, as a means to decrease external water demand will become a day to day focus at mines of the future. This implies that dewatering practices and water treatment unit operations will be of greater importance in the day to day operation at a mine site.

ACKNOWLEDGMENTS

My sincere thanks to Charlene Glendill for sourcing references, Andrea Bensch for assistance with the diagram and Christie Kelsay for document formatting.

REFERENCES

Addai-Mensah, J., and Ralston, J. 2005. Investigation of the role of interfacial chemistry on particle interactions, sedimentation and electro-osmotic dewatering of model kaolinite dispersions. *Powder Technology,* 35–39.

ALL Consulting. 2003. *Handbook on Coal Bed Methane Produced Water: Management and Beneficial Use Alternatives.* United States Department of Energy, National Energy Technology Laboratory. www.gwpc.org/elibrary/documents/general/Coalbed%20Methane%20Produced%20Waterr%20Management%20and%20Beneficial%20Use%20Alternatives.pdf, 322.

AMEC Earth and Environmental, Inc. 2008. *Rosemont Copper Company Filtered Tailings Dry Stacks; Current State of Practice.* Final report.

Andrianne, J., and Alardin, F. 2002. Thermal and membrane process economics: Optimized selection for seawater desalination. *Desalination*, 153–311.

Asmatulu, R., Zhand, J., Yoon, R.H., Hull, C., Kerr, M., Khan, S., Lampinen, P., Brodin, P., and Bolin, N. 2005. Novel dewatering aids for mineral concentrates and coal. *Proceedings CMP*, 665–681.

Barrett, D. 2011. Strategic water management in the mining industry. *AusIMM Bulletin*, Feb., 56–77.

Bailey, R.P. 1970. Case for reclamation of mineral processing water. *CMJ*, 87–92.

Bebbington, A., and Williams, M. 2008. Water and mining conflicts in Peru. *Mountain Research and Development*, 190–195.

Berner, R.A., Lasaga, A.C., and Garrels, R.M. 1983. The carbonate-silicate geochemical cycle and its effects on atmospheric carbon dioxide over the past 100 million year. *American Journal of Science,* 283, 641–683.

Blumenstein, E.P., and Gusek, J.J. 2008. Designing a biochemical reactor for selenium and thallium removal from bench scale testing through pilot construction. *Proceedings Sixth International Symposium Hydrometallurgy 2008.* Edited by C.A. Young, P.R. Taylor, C.G. Anderson, and Y. Choi. Littleton, CO: SME, 117–129.

Boswinkel, J.A. 2000. Information Note. International Groundwater Resources Assessment Centre (IGRAC). Netherlands Institute of Applied Geoscience, Netherlands.

Brown, E.T. 2003. Water for a sustainable minerals industry—A review. *Proceedings Water in Mining, AusIMM*, 3–12.

Browne, R.N., and Butler, H.R. 1970. Water reuse at Inco's Sudbury Mills. *CMJ*, 83–87.

Chadwick, J. 2009. Copper recovery. *Mining Magazine*, 23–30.

Chambers, R., Plewes, H., Pottie, J., Murray, L., and Burgess, A. 2003. Water recovery from a mine in the Atacama desert. *Proceedings Water in Mining*, 261–270.

Cline, J.T., Kimball, B.J., Klinko, K.A., and Nolen, C.H. 2009. Advances in Water Treatment Technology and Potential Affect on Application of USDW. GWPC: 2009, Underground Injection Control Conference, Antonio, TX, January 26–29, 2009.

Coleman, R.L., and Wallace, B.P. 1978. Tailings disposal in Canada, milling practice in Canada. *CIMM*, 16, 13–20.

Cooke, R. 2008. Pipeline design for paste and thickened tailings systems. *Proceedings Tailings and Mine Waste*, 95–100.

Croall, J. 2009. Using reverse osmosis technology to treat gold tailings supernatant or how to make a superbug. SME Annual Meeting, 1–6.

Doerffer, M., and Heinrich, R. 2009. Efficient dewatering solutions on vibrating screens. *Proceedings Tailings and Mine Waste*, 101–112.

De Kretser, R.G., Powell, M., Scales, P.J., and Lim, J. 2009. The water efficient plant of the future—Towards a holistic process chain approach. *Proceedings Water in Mining, AusIMM*, 65–70.

Doran, G., and Leong, L.Y.C. 2000. *Developing a Cost Effective Solution for Produced Water and Creating a 'New' Water Resource.* United States Department of Energy, National Energy Technology Laboratory. DOE/MT/95008-4.

Fakhru'l-Razi, A., Pendashteh, A., Abdullah, L.C., Biak, D.R.A., Madaeni, S.S., and Abidin, Z.Z. 2009. Review of technologies for oil and gas produced water treatment. *Journal of Hazardous Materials,* 170: 530–551.

Fawell, J., Bailey, K., Chilton, J., Dahi, E., Fewtrell, L., and Magara, Y. 2006. *Fluoride in Drinking-Water.* WHO Drinking-Water Quality Series. London: IWA Publishing. pp. 134. www.who.int/water_sanitation_health/publications/fluoride_drinking_water_full.pdf.

Filippelli, G.M. 2008. The global phosphorus cycle: Past, present, and future. *Elements,* 4: 89–95.

Fritzmann, C., Löwenberg J., Wintgens, T., and Melin, T. 2007. State-of-the-art of reverse osmosis desalination. *Desalination,* 216: 1–76.

Gislason, S.R., Oelkers, E.H., and Snorrason, A. 2006. Role of river-suspended material in the global carbon cycle. *Geology,* 34: 49–52.

Gleick, P.H. 1996. Water Resources. In: *Encyclopedia of Climate and Weather.* Edited by S.H. Schneider. New York: Oxford University Press, 2: 817–823.

Gregory, K.B., Vidic, R.D., and Dzombak, D.A. 2011. Water management challenges associated with the production of shale gas by hydraulic fracturing. *Elements,* 7: 181–186.

Gurdak, J.J., McMahon, P.B., Dennehy, K., and Qi, S.L. 2009. Water Quality in the High Plains Aquifer, Colorado, Kansas, Nebraska, New Mexico, Oklahoma, South Dakota, Texas, and Wyoming, 1999–2004. *U.S. Geological Survey Circular* 1337, 63.

Gusek, J., Conroy, K., and Rutkowski, T. 2008. Past, present and future for treating selenium-impacted water. *Proceedings Tailings and Mine Waste,* 281–290.

Haig-Smillie, L.D. 1974. Sea water flotation. *Proceedings Annual CMP,* 263–267.

Hollis, K.M. 2009. A $10 million reverse osmosis plant, and it is all water under the bridge: recent challenges in site water management at Newmont, Waihi Gold, Aus IMM, NZ branch 2009, 1–15.

Hopenhayn, C. 2006. Arsenic in drinking water: Impact on human health. *Elements,* 2: 103–107.

Horrigan, L., Lawrence, R.S., and Walker, P. 2002. How sustainable agriculture can address the environmental and human health harms of Industrial agriculture. *Environmental Health Perspectives,* 110: 445–456.

Hyatt, D.E. 1976. Effluent guideline compliance—Technology and cost elements. Western Mining Conference.

Jaffrin, M.Y. 2008. Dynamic shear-enhanced membrane filtration: A review of rotating disks, rotating membranes and vibrating systems. *Journal of Membrane Science,* 324: 7–25.

Joe, E.G. 1973. Summary of recent trends in Canadian mineral processing. *CIMM Bulletin,* 107–109.

Johnston, R.B., Berg, M., Johnson, A., Tilley, E., and Hering, F.G. 2011. Water and sanitation in developing countries: Geochemical aspects of quality and treatment. *Elements,* 7: 163–168.

Jørgensen, S.E. 2001. The impact of eutrophication. *Lakes and Reservoirs. Volume 3.* Water Quality: International Lake Environment Committee and United Nations Environmental Programme, 125.

Jurenka, B. 2007. *Electrodialysis (ED), and Electrodialysis Reversal (EDR).* United States Department of Interior, Bureau of Reclamation. www.usbr.gov/pmts/water/publications/reportpdfs/Primer%20Files/07%20-%20Electrodialysis.pdf.

Kaufman, A., and Nadler, M. 1965. *Water Use in the Mineral Industry.* USBM IC 8285, 1–37.

Kaufman, A. 1967. Water use in the mineral industry. *Trans. SME-AIME,* March, 83–90.

Kemmer, F.N., and Beardsley, J.A. 1971. Chemical treatment of wastewater from mining and mineral processing. *EMJ,* 92–97.

Kenny, J.F., Barber, N.L., Hutson, S.S., Linsey, K.S., Lovelace, J.K., and Maupin, M.A. 2009. *Estimated Use of Water in the United States in 2005.* United States Geological Survey, Reston, VA, 52. pubs.usgs.gov/circ/1344/pdf/c1344.pdf.

Klein, B., and Hallbom, D. 2009. Rheology in mineral processing. In: *Recent Advances in Mineral Processing Plant Design.* Edited by M. Deepak, P.R. Taylor, E. Spiller, M. LeVier. Littleton, CO: SME, 406–418.

Kummu, M., Ward, P.J., de Moel, H., and Varis, O. 2010. Is physical water scarcity a new phenomenon? Global assessment of water shortage over the last two millennia, *Environmental Research Letters*, 5: 034006.

Kuyucak, N. 2006. Selecting suitable methods for treating mining effluents. *Proceedings Water in Mining, AusIMM*, 267–276.

Levay, G., Smart, R.St.C., and Skinner, W.M. 2001. The impact of water quality on flotation performance. *J SAIMM*, 101, 69–75.

Levay, G., and Schumann, R. 2006. A systematic approach to water quality in the mineral processing industry. *Water in Mining, AusIMM*, 277–289.

Levy, V., Fabre, R., Goebel, B., and Hertle, C. 2006. Water use in the mining industry—Threats and opportunities. *Proceedings Water in Mining, AusIMM*, 289–295.

Little, M.G., and Jackson, R.B. 2010. Potential impacts of leakage from deep CO_2 geosequestration on overlying fresh-water aquifers. *Environmental Science and Technology*, 44, 9225–9232.

Madin, C. 2006. Water treatment options, the mining applications, processes and case studies. *Proceedings Water in Mining Conference*, 297–305.

Mathewson, D., Norris, R., and Dunne, M. 2006, Cost effective screening, dewatering and water treatment. *Proceedings Green Processing Conference*, 125–132.

McCloskey, J., Twidwell, L., Park, B., and Fallon, M. 2008. Removal of selenium oxyanions from industrial scrubber waters utilizing elemental iron. *Proceedings Sixth International Symposium Hydrometallurgy 2008.* Edited by C.A. Young, P.R. Taylor, C.G. Anderson, and Y. Choi. LIttleton, CO: SME, 140–148.

McCloskey, J., Twidwell, L., Hsin-Hsinng, H., and Goldstein, L. 2009. Fundamental procedures to evaluate and design industrial waste water metals treatment systems, case study discussions. In: *Recent Advances in Mineral Processing Plant Design.* Edited by M. Deepak, P.R. Taylor, E. Spiller, and M. LeVier. Littleton, CO: SME, 268–279.

McIntosh, K.S., and Cronin, D.J. 2003. Bauxite mining in water supply catchments—Water conservation and quality protection, Brisbane. *Proceedings Water in Mining*, AusIMM, 203–210.

Meybeck, M. 1982. Carbon, nitrogen, and phosphorous transport by world rivers. *American Journal of Science*, 282: 401–450.

New Logic Reseach, 2004. Using VSEP to treat produced water: An effective and economical solution. New Logic Research, Inc. www.vsep.com/pdf/Produced Water.pdf.

Oelkers, E.H., Hering, J.G., and Zhu, C. June 2011. Water: Is there a global crisis? *Elements*, 7: 157–162.

O'Leary, W. 1996. Wastewater recycling and environmental constraints at a base metal mine and process facilities. *Water Science Technology*, 33(10-11): 371–379.

Pavelic, P., Dillon, P., and Barry, K. 2006a. Hydraulic evaluation of aquifer storage and recovery (ASR) with urban storm water in a brackish limestone aquifer. *Hydrogeology Journal*, 14, 1544–1555.

Pickett, D.E., and Joe, E.G. 1974. Water recycling experience in Canadian mills. *Trans. SME-AIME*, 230–235.

Pinto, M., and Barrera, S. 2008. Tailings impoundments-water balance-key variables. *Proceedings Water in Mining WIM 2008*, 233–241.

Poling, G.W., and Ellis, D.V. 1995. Importance of geochemistry, the Black Angel Lead-Zinc Mine, Greenland. *Marine Georesources and Geotechnology*, 13, 101–118.

Pyne, R.D.G. 1995. *Groundwater Recharge and Well: A Guide to Aquifer Storage and Recovery.* Boca Raton, FL: Lewis.

Ravenscroft, P., Brammer, H., and Richards, K. 2009. *Arsenic Pollution: A Global Synthesis.* Chichester: Wiley-Blackwell, 588.

Rinck-Pfeiffer, S., Pitman, C., and Dillon, P. 2005. Stormwater ASR in practice and ASTR (Aquifer Storage Transfer and Recovery) under investigation in Salisbury, South Australia. *Proc. ISMAR5, Berlin*, June, 151–159. unesdoc.unesco.org/images/0014/001492/149210E.pdf.

Schoenbrunn, F., Niederhauser, M., and Baczek, F. 2009. Paste thickening of tailings: Process and equipment design fundamentals relative to deposition goals. *Recent Advances in Mineral Processing Plant Design*. Edited by M. Deepak, P.R. Taylor, E. Spiller, and M. LeVier. Littleton, CO: SME, 455–465.

Schwartz, F.W., and Ibaraki, M. June 2011. Groundwater: A resource in decline. *Elements*, 7, 175–179.

Shannon, M.A., Bohn, P.W., Elimelech, M., Georgiadis, J.G., Mariñas, B.J., and Mayes, A.M. 2008. Science and technology for water purification in the coming decades. *Nature*, 452, 301–310.

Sibrell, P.L, Tucker, T.W., Kehler, T., and Fletcher, J.W. 2008. Removal of phosphorus from wastewater using ferroxysorb sorption media produced from AMD sludge. *Proceedings Sixth International Symposium Hydrometallurgy 2008*. Edited by C.A. Young, P.R. Taylor, C.G. Anderson, and Y. Choi. Littleton, CO: SME, 86–92.

Simms, P., Fisseha, B., Henriquez, J., and Bryan, R. 2008. Desiccation and rheology in cyclic surface deposition of gold paste tailings. *Proceedings Tailings and Mine Waste*, 269–280.

Smedley, P.L., and Kinniburgh, D.G. 2002. A review of the source, behaviour and distribution of arsenic in natural waters. *Applied Geochemistry*, 1, 517–568.

Sofra, F., and Boger, D. 2000. Exploiting the rheology of mine tailings for dry disposal. *Proceedings Tailings and Mine Waste*, 169–180.

Stegink, H.D.J., Lane, J., Barker, D.J., and Pei, B. 2003. Water usage reductions at Queensland Alumina. *Proceedings Water in Mining, AusIMM*, 293–299.

Swierc, J., Page, D., Van Leeuwen, J., and Dillon, P. 2005 *Preliminary Hazard Analysis and Critical Control Points Plan (HACCP)—Salisbury Stormwater to Drinking Water Aquifer Storage Transfer and Recovery (ASTAR) Project*. CSIRO Technical report 20/05, Adelaide.

Takos, J.N., Vagias, N., Shaw, R., Coghill, M., and Easman, M. 2006 Rio Tinto floating module. *Proceedings Water in Mining, AusIMM*, 369–372.

Trenberth, K.E., Smith, L., Qian, T., Dai, A., and Fasullo, J. 2007. Estimates of the global water budget and its annual cycle using observational and model data. *Journal of Hydrometeorology–Special Section*, 8: 758–769.

Van Niekerk, A.M., Wurster, A., and Cohen, D. 2006.Technology advance in mine water treatment in Southern Africa over the past 20 years. *Proceedings Water in Mining*, 373–378.

Vaux, H., Jr. 2011. Water conservation, efficiency, and reuse. *Elements*, 7: 187–191.

Veil, J.A., Puder, M.G., Elcock, D., and Redweik, R.J., Jr. 2004. A white paper describing produced water from production of crude oil, natural gas, and coal bed methane. United States Department of Energy, Argonne National Laboratory W-31-109-Eng-38, p. 79. www.evs.anl.gov/pub/doc/ProducedWatersWP0401.pdf.

Viers, J., Dupré, B., and Gaillardet, J. 2009. Chemical composition of suspended sediments in world rivers: New insights from a new database. *Science of the Total Environment*, 407, 853–868.

Vitousek, P.M., Aber, J.D., Howarth, R.W., Likens, G.E., Matson, P.A., Schindler, D.W., Schlesinger, W.H., and Tilman, D.G. 1997. Human alteration of the global nitrogen cycle: Sources and consequences. *Ecological Applications*, 7, 737–750.

Ward, J., Dillon, P., and Grandgirand, A. 2008. Applying robust separation to compare the opportunities for market base policy approaches for aquifer storage and recovery in Australia and France CEMAGREF.

Watt, J.G., and Horton, R.C. 1982. *The Domestic Supply of Critical Minerals*. USBM.

Wiertz, J.V. 2006. When best water use efficiency is not enough, what can the mining industry do? *Proceedings Water in Mining, AusIMM*, 13–18.

Yermiyahu, U., Tal, A., Ben-Gal, A., Bar-Tal, A., Tarchitzky, J., and Lahav, O. 2007. Rethinking desalinated water quality and agriculture. *Science*, 318: 920–921.

Zhu, C., and Schwartz, F.W. June 2011. Hydrogeochemical processes and controls on water quality and water management. *Elements*, 7: 169–174.

Arsenic and Life: Bacterial Redox Reactions Associated with Arsenic Oxyanions

Ronald S. Oremland
U.S. Geological Survey, Menlo Park, CA, USA

ABSTRACT

Oxyanions of the Group 15 element arsenic (As), namely arsenite [As(III)] and arsenate [As(V)], have been known for millennia to be potent poisons. In more recent times it was discovered that certain microorganisms, in contrast to human, have evolved specific resistance mechanisms guarding them against the presence of these oxyanions in their surroundings, thereby allowing for their grow in As-contaminated environments. Subsequent discoveries have revealed that biologically-mediated redox reactions occurring in the cell envelope of some prokaryotes can either oxidize As(III) to As(V), or reduce As(V) to As(III). These redox reactions are highly exergonic, and result in biological energy conservation that sustains growth. Hence, arsenic oxyanions can be thought of as an energy source ripe for microbiological exploitation, in other words as a food source. In some cases these redox reactions can be linked to chemo- or photo-autotrophy. Since arsenic often co-occurs in the ores of precious metals, the exploitation of these microbiological redox reactions for the purpose of site remediation or of waste sequestration (via immobilization) holds promise for future investigation.

INTRODUCTION

In contrast to the Group 15 elements nitrogen and phosphorus, the lower atomic weight neighbors of arsenic in the Periodic Table, the latter is infamous for its ability to end life rather than to sustain it. The biochemical mechanisms for acute arsenic toxicity appear to be its action as an inhibitor of oxidative phosphorylation [when in the form of As(V)] acting as an analog of phosphate, and by its binding to the sulf-hydryl groups of key respiratory enzymes [when occurring as As(III)] (Hughes, 2002; Stolz et al., 2010). Arsenic became popular for the specific purpose of people poisoning (aka: murder) because not only was it effective but its soluble oxyanions were generally colorless, odorless and tasteless when added to the food or beverage of the unsuspecting victim (Emsley, 2005). Its nefarious employment in this capacity became quite popular until the development of the forensic Marsh Test in the 19th century, which was devised as a means to detect the presence of arsenic in biological tissues. Today an even more insidious poisoning phenomenon is evident. There is growing recognition that naturally-occurring arsenic found as solid phase minerals embedded in drinking water aquifer matrices can become mobilized into the aqueous phase, thereby posing dire threats to human health. The problem will certainly become more acute as the human population expands in numbers and water resources become further constrained. Arsenic poisoning ("arsenicosis") is most severe and most highly visible in Asia (Fendorf et al., 2010). In Bangladesh alone some 40–50 million people drink water that is well above the WHO recommended of 10 ppb, and arsenicosis is endemic to the population, with many thousands of life-threatening cases and deaths occurring annually (Nordstrom, 2002; Smith et al., 2002; Oremland and Stolz, 2005). Arsenic mobilization in the environment is a complicated interdisciplinary phenomenon, including important components of geochemistry, hydrochemistry, hydrology, and microbiology. There is a broad literature dealing with various aspects of this subject that is well beyond the scope of this brief chapter to recount, but significant reviews have been published on various aspects of the topic (e.g., Smedley and Kinniburgh, 2002). Since this chapter's focus is upon the microbiological perspective, the reader is referred to the reviews of Mukhopadhay et al. (2002), Rosen, (2002), Oremland and Stolz (2003), Silver and Phung (2005), Stolz et al., (2006), Oremland et al., (2009), Páez-Espino et al., 2009), and Stolz et al., (2010).

ARSENIC RESISTANCE

Arsenic oxyanions can enter the cytoplasm of microbial cells by transiting the cell membrane. The mechanisms by which this is achieved depend on the chemical species under scrutiny, with As(V) mimicking phosphate and actively brought inside by phosphate transporters. Arsenite freely enters via aqua-glycerol porins, at least this occurs at circum-neutral pH where the As(III) ion has no charge. Once inside the cell arsenic oxyanions still hold the capacity to cause biochemical mayhem and hence must be pumped outward. In the case of As(V), this is first achieved via its reduction to As(III) through the action of a low molecular weight, soluble arsenate-reductase located in the cytoplasm (Mukhopadhyay and Rosen, 2002). In prokaryotes like *E. coli* this is achieved by the ArsC system, while in unicellular eukaryotes like *S. cerevisiae* by an analogous protein termed Arr2p (Simon & Phung, 2005). The arsenic oxyanions are then transported out of the cell in a membrane-associated enzyme complex that also includes the action of an ATPase. Hence, a key point to bear in mind is that removal of arsenic oxyanions from the internal milieu of the cell is an energy-requiring process. It was initially thought that expression of As-resistance genes were responsible for the observed speciation of arsenic in the environment, especially for the presence of As(III) in anoxic sediments and waters. However, the discovery of dissimilatory As(V) reduction (i.e., arsenate respiration) changed the paradigm away from resistance mechanisms being the primary cause of observed arsenic speciation (Ahmann et al., 1994; Laverman et al., 1995; Newman et al., 1997). Nonetheless, As(V) respiring anaerobic bacteria like *Shewanella* strain ANA-3 also have a full cassette of As-resistance genes to compliment their respiratory As(V) reductases (Saltikov and Newman, 2003). This enables the organism to not only generate energy via As(V) reduction, but to do so in the presence of high concentrations of this toxicant.

The notion that arsenic is a universal toxicant for all forms of life unless specifically adapted for its resistivity has recently been challenged. Strain GFAJ, an otherwise "common" halophilic bacterium of the γ-Proteobacteria isolated from Mono Lake, California was found to be capable of using As(V) in lieu of phosphate to sustain its growth (Wolfe-Simon et al., 2011). The results suggested that this organism was capable of inserting As into key biological macromolecules where P should be normally be present, most sensationally into the phosphodiester linkages of nucleic acids (e.g., DNA). This provocative report has stirred up much commentary and criticism, especially in the "blog-o-sphere" but also in reputable reviews (i.e., Rosen et al., 2011; Silver and Phung, 2011). Not all the commentary has been negative, and two papers support the idea that As-DNA molecules are stable based on the authors' quantum mechanical calculations (Denning and MacKerrell Jr, 2011; Mládek et al., 2011). If confirmed, the work has profound importance with regard to what specific major elements other than C, N, P, S, O, and H can actually constitute life as we currently understand its compositional base to be restricted. The idea that As could substitute for P in nucleotides was first shown to be possible over 45 years ago in ^{74}As(V)-radiolabel experiments conducted with mouse cancer cells (Kay, 1965). The initial incorporation of the radiolabel into nuclear DNA was followed by its transcription into cytoplasmic RNA. At the time the goal of this work was to better understand the role of arsenic in promoting lung cancer in eukaryotes rather than its substitution for phosphorus as a possible basis for life (E. Kay, personal communication, 2011).

ARSENOTROPHY

Arsenotrophy is here defined as the "redox" reactions involving either As(V)-reduction or As(III) oxidation that couple to ATP generation, that is they conserve energy for the concomitant growth of the microorganism (Hoeft et al., 2010). In the case of As(III) oxidation, excluded are the incidental reactions not directly coupled to growth that may be otherwise be involved in detoxification or immobilization of the As-oxyanion. Hence, if As(III) acts as a the primary electron donor (energy source) the microbe therefore obtains its carbon from reduction of CO_2, and is a chemo-autotroph. Most of the chemo-autotrophic As(III) oxidizers previously described are obligate aerobes and couple their oxidation of As(III) to the reduction of O_2 (Oremland and Stolz, 2003; Garcia-Dominguez et al., 2008). However, more recently some novel As(III) oxidizers were found to be able to employ nitrate in lieu of O_2 in this capacity (Oremland et al., 2002; Hoeft et al., 2007; Rhine et al., 2006,; 2007; Sun et al., 2008; 2009). Indeed, oxidation of arsenite can bypass the need to couple to strong biological oxidants (e.g., O_2, NO_3^-) entirely. Anoxygenic photosynthesis carried out by bacteria like *Ectothiorhodospira* can employ As(III) as its electron donor in the light, much as they do for sulfide, but in this case resulting in the formation of As(V) along with the conservation of energy for growth (Buddinhoff and Hollibaugh, 2008; Kulp et al., 2008).

Running in the counter direction to the above-described process, arsenate respiration employs the As(V) oxyanion as a respiratory electron acceptor. This enables anaerobes to conserve energy for growth from their biochemical oxidation of organic matter (or of H_2 or sulfide) resulting in the formation of As(III) and oxidized substances like acetate, CO_2, or sulfate in the case of sulfide (Oremland and Stolz, 2003; Hoeft et al., 2004). Because the product As(III) is more toxic and more hydrologically mobile than is As(V), the process of As(V) respiration is of great environmental significance. In certain arsenic-rich environments, such as Mono Lake, California, bacterial As(V) respiration represents an important pathway of carbon re-mineralization, shuttling about 10% of annual water column primary productivity via this respiratory route (Oremland et al., 2000; Hollibaugh et al., 2005).

The enzymes that control these reactions, namely As(III) oxidase and respiratory As(V) reductase are both members of the broad family of Mo-containing DMSO-type reductases (McEwan et al., 2002). Both are hetero-dimers with the larger, Mo-containing reaction center subunit (~87–88 kDa) associated with a smaller iron-sulfur cluster containing subunit of 14–24 kDa (Silver and Phung, 2005). Despite their outward similarities, they are controlled by two distinctly different sets of genes, designated *aoxBA* for the arsenite-oxidase complex and *arrAB* for the arsenate reductase complex (Oremland et al., 2009). Until quite recently these enzymes were thought to function uni-directionally *in vivo*. However, the full genome annotation for *Alkalilimnicola erhlichii* altered that perception because while it clearly lacked a homolog of *aox*, it contained two homologs of the reductase gene *arrA* but was incapable of reducing As(V) (Hoeft et al., 2007). Subsequent biochemical studies revealed that this organism's respiratory As(V) reductase was actually reversible, running in the oxidative mode when coupled with nitrate (Richey et al., 2009). An investigation using "knockout" mutants of *A. ehrlichii* clearly demonstrated that only one of its two *arrA* homologs was involved in As(III) oxidation. However, the incapacitation of this section of its genome (protein designated 0216) rendered the *A. ehrlichii* mutant incapable of growth on As(III) and nitrate, or of As(III) oxidation and nitrate reduction (Zargar et al., 2010). This reverse-running, anaerobic As(V) reductase gene was therefore designated as "*arxA*" and was phylogenetically aligned with As(V) reductases (*arrA*), suggesting a common ancestor distinct from that of aerobic As(III) oxidases (*aoxB*). In a similar vein, *Ectothiorhodospira* strain PHS-1, another member of the γ-Proteobacteria isolated from Mono Lake also lacked an *aox*-type of As(III) oxidase gene, but did have an *arx*-type homolog closely aligned to that of *A. ehrlichii* (Kulp et al., 2008). These observations suggest that this clade of anaerobic As(III) oxidases may be widespread in nature.

ARSENIC AND ARSENOTROPHS IN THE ENVIRONMENT

Arsenic ranks as 20th on the list of elemental abundance within the Earth's crust, lying well below oxygen (#1), silicon (#2), and for the purposes of this section aluminum (#3) and iron (#4). The latter two elements are significant here because of the interactions they have with arsenic oxyanions that govern their hydrologic mobility. In general, As(V) adsorbs more strongly with and to a greater diversity of minerals than does As(III). In particular, As(V) adsorbs to aluminum oxides while As(III) does not (Zobrist et al., 2000), and this is of great significance considering the widespread abundance of clays and alumino-silicates. Adsorbed As(V) on an aluminum oxide surfaces can undergo direct dissmilatory reduction by *Sulfurospirillum barnesii* which results in the net accumulation of As(III) in the aqueous phase (Zobrist et al., 2000). The interaction of arsenic species and iron minerals is far more complex as both As(V) and As(III) adsorb with differential strengths to these surfaces, and the adsorptive capacity tends to diminish with increasing degree of crystallization of the minerals (Dixit and Hering, 2003). Direct bacterial reduction of Fe(III) contained in ferrihydrite can liberate adsorbed As(V) into solution (Langner and Inskeep, 2000), but this process is retarded with time and microbial activity as the amorphous ferrihydrite is converted to more crystalline forms like goethite and siderite (Campbell et al., 2006; Tufano et al., 2008). These minerals also have a tendency to bind any As(III) formed from As(V) reduction, further complicating interpretation of results (Zobrist et al., 2000). Nonetheless, the above-cited laboratory studies are necessary to aid our understanding of the question of arsenic mobility in sub-surface aquifers as this interdisciplinary phenomenon has great implications for human health. These complex interactions between dominant mineral phases, indigenous bacterial species, hydrology and hydrochemistry all have a bearing on the question as has been noted, for example, with investigations conducted in Bangladesh (Harvey et al., 2002; Islam et al., 2004; Akai et al., 2004; 2008) as well by continued pursuit in laboratory investigations with sediments (Fakih et al., 2009; Pearcy et al., 2011). Moreover, microbial weathering of minerals containing key nutrients can also result in the aqueous mobilization of arsenic from the solid phase. Hence, phosphorus-starved cells of *Burkholderia fungorum* were able to break down the matrix of the mineral apatite [$Ca_5(PO_4)_3$(F, Cl, H)] to liberate phosphate, and in doing so released associated arsenic into solution (Mallioux et al., 2009). Ironically, *in situ* aquifer bioremediation efforts, such as enhancement of reductive dechlorination of solvents via addition of electron donors (i.e., lactate) can have a secondary effect in promoting arsenic mobility via dissimilatory reduction of solid phase Fe(III) and As(V) (He et al., 2010).

A number of investigations have harnessed the discoveries made with respect to arsenotrophic enzymes to detect the presence of these genes in DNA extracted from bacterial populations that reside in the environment. Thus, after first identifying the gene encoding the arsenate-reductase of *Shewanella* strain ANA-3 (Saltikov and Newman, 2003), Malasarn et al. (2004) were the first to report successful primer amplification of *arrA* amplicons from DNA taken from the sediments of Hiawee Reservoir in California. Similar success was also achieved with enrichments taken from Chesapeake Bay mud (Song et al., 2009). The arsenic-rich extreme environments of Mono Lake and Searles Lake have also proven amenable to amplification of these genes (Kulp et al., 2006; 2007; Hoeft et al., 2010). There has been considerable recent research conducted on the physiology, biochemistry, genetics, and environmental microbiology/biogeochemistry of aerobic arsenite-oxidizing bacteria (e.g., Wilkie and Hering, 1998; Santini et al., 2000; 2004; 2007; Gihring et al., 2001; Langner et al., 2001; Muller et al., 2003; Salmassi et al., 2006; D'Imperio et al., 2007; Quéméneur et al., 2008). However, there has been only a limited amount published to date on attempts to amplify and characterize *aox*-sequences from natural habitats, and these were primarily from hot spring environments located in Yellowstone National Park (Inskeep et al., 2007; Clingenpeel et al., 2009). Recently, however, a paper emerged reporting a high diversity of environmental *aoxB* sequences in DNA extracted from sediments located downstream of the Gabes-Gottes mine located in France (Heinrich-Salmeron et al., 2011).

ARSENIC SEQUESTRATION

Arsenic can be removed from the aqueous phase and thereby immobilized by a number of strategies collectively termed sequestration. The goal of sequestration can be to prevent its hydrologic movement from a point source, thereby foreclosing down-gradient arsenic contamination. Alternatively, sequestration can be employed as a means to remove arsenic from contaminated water so that it can be safely employed for human consumption or agricultural purposes. In practicality, sequestration can be achieved either by adsorption onto an appropriate solid surface (or exchange resin) or via precipitation by de novo formation of a complex insoluble mineral phase. In both processes there can be significant involvement of As-metabolizing bacteria. Hence, since As(V) adsorbs more strongly and to a greater variety of solid matrices than does As(III), the efficacy of adsorptive removal is dependent on the arsenic chemical species in solution. Bacterial oxidation of As(III) to As(V) is therefore a key primary step in any adsorptive removal strategy, and aerobic biofilms have been investigated with this purpose in mind (Michel et al., 2007; Andrianisa et al., 2008). Straight-forward photochemical methods have also been pursued (Neppolian et al., 2008) as well as purely chemical adsorptive mechanisms that are capable of sequestering both As(III) and As(V) for small community-level water purification purposes (Sarkar et al., 2008). The possibility that *in situ* arsenic sequestration within subsurface aquifers could be accomplished by such means is appealing, but the limited solubility of oxygen in water makes such an approach impractical. Aqueous subsurface injection of an alternative to oxygen, namely a strong oxidant like nitrate is actually quite feasible and was shown to immobilize dissolved arsenic transport in groundwater experiments in Bangladesh (Harvey et al., 2002). Sun et al. (2008) showed that anoxic biological As(III) oxidation in freshwater systems can be achieved with nitrate reduction, and then went on to demonstrate that combined As(III) and Fe(II) oxidation with nitrate could be a practical strategy for constraining sub-surface arsenic mobility as it not only increases the amount of As(V) but also produces more Fe(III)-rich adsorptive sites for this oxyanion to bind with (Sun et al., 2009). While this approach may possibly be useful in temporarily arresting arsenic hydrologic migration, one must continually monitor such sites and if necessary continuously supply them with a source of nitrate. The problem of arsenic contamination could then become one of nitrate contamination of groundwater aquifers, which would still render the water un-potable for human and livestock consumption.

Another means of sequestration involves precipitation reactions. These primarily involve reaction of soluble arsenic oxyanions with free sulfide, and when formed in combination with ferrous ions can yield arsenopyrites (Corkhill and Vaughan, 2009). A word of caution is in order as there is complicated chemistry involved, some of it focused on the stability of thio-arsenic compounds in sulfidic waters as well as the ability of sulfide to drive the oxidation of As(III) to As(V) (Holz and Tossell, 2008).The biological precipitation of arsenic trisulfide, essentially as the yellow-colored mineral orpiment, was first reported in cultures of the sulfate-reducing bacteria *Desulfotomaculum auripigmentum*, which was also capable of As(V)-respiration (Newman et al., 1997 a & b). Natural deposits of arsenic sulfides attributable to these microbiological reactions have been noted in salt playas located in the Andes Mountains (Demergasso et al., 2007), as well as along Yellowstone National Park hot springs. The latter occur as sequential chemical reactions along a spring outflow gradient (Planer-Friedrich et al., 2007). However, biological precipitation reactions cannot be entirely ruled out, as some anaerobic thermophiles can precipitate realgar by employing As(V)-resistance enzymes (Ledbetter et al., 2007). Under highly alkaline conditions thio-arsenic compounds are quite soluble. Thus, although sulfate-reduction and As(V)-respiration take place in situ in the anoxic water column of Mono Lake, California (pH = 9.8), the resulting thioarsenites remain dissolved under anoxia (Oremland et al., 2000; Hoeft et al., 2004; Hollibaugh et al., 2005) but can be oxidized back to As(V) and sulfate by bacterial nitrate-reduction (Oremland et al., 2002; Hoeft et al., 2002) or by sulfur-oxidizing bacteria (Fisher et al., 2008). This contrasts with freshwater lakes where thioarsenites precipitate into the sediments during anoxic stratification. Their dissolution requires either water column aeration upon turnover or the presence of nitrate to break-up the precipitate and release As(V) into the water column (Senn and Hemond, 2002).

Sequestration of initially mobile arsenic oxyanions can be achieved *in situ* by naturally-occurring microbial processes. Thus arsenic oxyanions released from the site of a former pesticide manufacturing plant on the shores of a tidal estuary (San Francisco Bay) was immobilized down-gradient by formation of realgar and orpiment phases when encountering anoxic, sulfidic regions having active sulfate-reduction (Root et al., 2009). Several laboratory investigations have pursued studying the effects of sulfate-reduction on arsenic speciation and mobility in systems containing a solid-phase matrix such as ferrihydrite columns (Kocar et al., 2010), landfill materials (Keimowitz et al., 2007), and ferrihydrite-slurries (Saalfield et al., 2009). These studies further reinforced the concept that overall, sulfate-reduction can significantly contribute to arsenic sequestration in complex aquifer systems containing Fe(III), As(V), and sulfate, and that organic amendments can enhance this process. Nonetheless, a high degree of complexity is associated with these processes, and under certain conditions biogenic sulfide can actually cause the reverse to occur, namely mobilization of solid phase arsenic entrapped in minerals (i.e., arsenopyrite). Zhu et al. (2008) found that under conditions of redox disequilibrium, where sulfide plumes encounter oxic or hypoxic natural waters, an exchange of solid phase As with aqueous free sulfide occurs. These counterintuitive observations of complexity have an important cautionary tale when devising strategies for arsenic sequestration employing sulfate reduction. Meso-scale investigations of arsenic sequestration by sulfate-reduction and its release under subsequent aeration using incubated aquifer materials over long term treatments revealed insights into practical *in situ* management of contaminated systems (Onstott et al., 2011).

SUMMARY

In this brief review I have attempted to highlight some aspects of the behavior of microorganisms that alters our understanding of how they mediate the chemical speciation of arsenic in the environment. It is clear that a number of diverse microbes, both prokaryotes and eukaryotes, have evolved effective strategies which enable them to survive in the presence of high arsenic oxyanion concentrations. Recognition that the redox changes associated with either bacterial As(III) oxidation or As(V) respiration couples with energy conservation came later on, but at the time they were first announced it was a surprising observation. Those discoveries prompted further investigations concerning microbial redox process that influenced the mobility of arsenic oxyanions in nature, a realm once thought to be governed entirely by chemical and physical processes. This area of research has important implications for human health and the well-being of broad segments of humanity world-wide. Exploitation of these fundamental microbial processes holds out the hope that they can be put to use in removing or sequestering arsenic from aquifers so that it does not impact communities down-gradient or at the well-head. Finally, microorganisms that respire As(V) and also reduce sulfur species have the ability to form nano-scaled minerals. Some of these arsenic-sulfides may have applications in the realm of emerging nano-technologies, such as "nano-photonics" (Lee et al., 2007). The employment of microorganisms in this realm rather than harsh chemicals offers the possibility of a "green" synthesis approach to form novel materials (Pearce et al., 2011).

REFERENCES

Adrianisa, H.A., Ito, A., Sasaki, A., Aizawa, J., and Umita, T. 2008. Biotransformation of arsenic species by activated sludge and removal of bio-oxidized arsenate from wastewater by coalgulation with ferric chloride. *Water Res.* 42: 4809–4817.

Akai, J., Kanekiyo, A., Hishida, N., Ogawa, M., Naganuma, T., Fukuhara, H., and Anawar, H.N. 2008. Biogeochemical characterization of bacterial assemblages in relation to release of arsenic from South East Asia (Bangladesh) sediments. *Appl. Geochem.* 23: 3177–3186.

Akai, J., Izumi, K., Fukuhara, H., Masuda, H., Nakano, S., Yoshimura, T., Ohfuji, H., Anawar, H.M., and Akai, K. 2004. Mineralogical and geomicrobiological investigations on groundwater arsenic enrichment in Bangladesh. *Appl. Geochem.* 19: 215–230.

Budinhoff, C.R., and Hollibaugh, J.T. 2008. Arsenite-dependent photoautotrophy by *Ectothiorhodospira*-dominated consortium. *ISME J.* 2: 340–343.

Campbell, K.M., Malassarn, D., Saltikov, C.W., Newman, D.K., and Hering, J.G. 2006. Simultaneous microbial reduction of iron(III) and arsenic(V) in suspensions of hydrous ferric oxide. *Environ. Sci. Technol.* 40: 5950–5955.

Clingenpeel, S.R., D'Imperio, S., Oduro, H., Drischel, G.K., and McDermott, T.R. 2009. Cloning and *in situ* expression studies of the *Hydrogenobaculum* arsenite oxidase genes. *Appl. Environ. Microbiol.* 75: 3362–3365.

Corkhill, C.L., and Vaughan, D.J. 2009. Arsenopyrite oxidation—A review. *Appl. Geochem.* 24: 2342–2361.

Demergasso, C.S., Chong, G.D., Escudero, L.G., Mur, J.J.P., and Pedrós-Alió, C. 2007. Microbial precipitation of arsenic sulfides in Andean salt flats. *Geomicrobiol. J.* 24: 1–14.

Denning, E., and MacKerell Jr., A. 2011. Impact of arsenic/phosphorus substitution on the intrinsic conformational properties of the phosophodiester backbone of DNA investigated using *ab initio* quantum mechanical calculations. *J. Amer. Chem. Soc.* 133: 5770–5772.

D'Imperio, S., C.R. Lehr, M. Breary, and T.R. McDermott. 2007. Autoecology of an arsenite chemolithotroph: sulfide constraints on function and distribution in a geothermal stream. *Appl. Environ. Microbiol,* 73: 7067–7074.

Emsley, J.M. 2005. *The Elements of Murder: A History of Poison.* Oxford: Oxford University Press.

Fakih, M., Davranche, M., Dia, A., Nowack, B., Morin, G., Petijean, P., Châtellier, X., and Gruau, G. 2009. Environmental impact of As(V)-Fe oxyhydroxide reductive dissolution: An experimental insight. *Chem. Geol.* 259: 290–303.

Fendorf, S., Michael, H.A., and van Geen, A. 2010. Spatial and temporal variations of groundwater arsenic in south and south east Asia. *Science* 328: 1123–1127.

Fischer, J.C., Wallschläger, D., Planer-Friedrich, B., and Hollibaugh, J.T. 2008. A new role for sulfur in arsenic cycling. *Environ. Sci. Technol.* 42: 81–85.

Garcia-Dominguez, E., Munford, A., Rhine, E.D., Paschal, A., and Young, L.Y. 2008. Novel autotrophic arsenite-oxidizing bacteria isolated from soil and sediments. *FEMS Microbiol. Ecol.* 66: 401–410.

Gihring, T.M., Druschel, G.K., McCleskey, R.B., Hammers, R.J., and Banfield, J.F. 2001. Rapid arsenite oxidation by *Thermus aquaticus* and *Thermus thermophilus*: Field and laboratory investigations. *Environ. Sci. Technol.* 35: 3857–3862.

Harvey, C.F., Swartz, C.H., Badruzzaman, A.B.M., Keon-Blute, N., Yu, W., Ali, M.A., Jay, J., Beckie, R., Niedan, V., Brabander, D., Oates, P.M., Ashfaque, K.N., Islam, S., Hemond, H.F., and Ahmed, M.F. 2002. Arsenic mobility and groundwater extraction in Bangladesh. *Science* 298: 1602–1606.

He, Y.T., Fitzmaurice, A.G., Bilgin, A., Choi, S., O'Day, P., Horst, J., Harrington, J., Reisinger, H.J., Burris, D.R., and Hering, J.G. 2010. Geochemical processes controlling arsenic mobility in groundwater: A case study of arsenic mobilization and natural attenuation. *Appl. Geochem.* 25: 69–80.

Helz, G.R., and Tossell, J.A. 2008. Thermodynamic model for arsenic speciation in sulfidic waters: A novel use of *ab initio* computations. *Geochim. Cosmochim. Acta* 72: 4457–4468.

Heinrich-Salmeron, A., Cordi, A., Brochier-Armanet, C., Halter, D., Pagnout, C., Abbaszadeh-Fard, E., Montaut, D., Seby, F., Bertin, P.N., Bauda, P., and Arsène-Ploetze, F. 2011. Unsuspected diversity of arsenite-oxidizng bacteria as revealed by widespread distribution of the *aoxB* gene in prokaryotes. *Appl. Environ. Microbiol.* 77: 4685–4692.

Hoeft, S.E., Kulp, T.R., Han, S., Lanoil, B., and Oremland, R.S. 2010. Coupled arsenotrophy in a photosynthetic hot spring biofilm from Mono Lake, California. *Appl. Environ. Microbiol.* 76: 4633–4639.

Hoeft, S.E., Switzer Blum, J., Stolz, J.F. Tabita, F.R., Witte, B., King, G.M., Santini, J.M., and Oremland, R.S. 2007. *Alkalilimnicola ehrlichii*, sp. nov., a novel, arsenite-oxidizing haloalkaliphilic γ-Proteobacterium capable of chemoautotrophic or hetereotrophic growth with nitrate or oxygen as the electron acceptor. *Int. J. System. Evol. Microbiol.* 57: 504–512.

Hoeft, S.E., Kulp, T.R., Stolz, J.F., Hollibaugh, J.T., and Oremland, R.S. 2004. Dissimilatory arsenate reduction with sulfide as the electron acceptor: Experiments with Mono Lake water and isolation of strain MLMS-1, a chemoautotrophic arsenate-respirer. *Appl. Environ. Microbiol.* 70: 2741–2747.

Hoeft, S.E., Lucas, F., Hollibaugh, J.T., and Oremland, R.S. 2002. Characterization of microbial arsenate reduction in the anoxic bottom waters of Mono Lake, California. *Geomicrobiol. J.* 19: 23–40.

Hollibaugh, J.T., Carini, S., Girley, U.H., Jellison, R., Joye, S.B., LeCleir, G., Meile, C., Vasquez, L., and Wallschlaeger, K. 2005. Arsenic speciation in Mono Lake, California: Response to seasonal stratification and anoxia. *Geochim. Cosmochim. Acta* 69: 1925–1937.

Hughes, M.F. 2002. Arsenic toxicity and potential mechanisms of action. *Toxicol. Lett.* 133: 1–16.

Inskeep, W.P., Macur, R.E., Mamamura, N., Warelow, T.P., Ward, S.A., and Santini, J.M. 2007. Detection, diversity and expression of aerobic bacterial arsenite oxidase genes. *Environ. Microbiol.* 9: 934–943.

Islam, F.S., Gault, A.G., Boothman, C., Polya, D., Charnock, J.M., Chaterjee, D., and Lloyd, J.R. 2004. Role of metal-reducing bacteria in arsenic release from Bengal delta sediments. *Nature* 430: 68–71.

Kay, E.R.M. 1965. Incorporation of radioarsenate into proteins and nucleic acids of the Ehrlich Lettré ascites carcinoma *in vitro*. *Nature* 206: 371–373.

Keimowitz, A.R., Mailloux, B.J., Cole, P., Stute, M., Simpson, H.J., and Chillrud, S.N. 2007. Laboratory investigations of enhanced sulfate reduction as a groundwater arsenic remediation strategy. *Environ. Sci. Technol.* 41: 6718–6724.

Kocar, B.D., Borch, T., and Fendorf, S. 2010. Arsenic repartitioning during biogenic sulfidization and transformation of ferrihydrite. *Geochim. Cosmochim. Acta* 74: 980–994.

Kulp, T.R., Hoeft, S.E., Asao, M., Madigan, M.T., Hollibaugh, J.T., Fischer, J.C., Stolz, J.F., Culbertson, C.W., and Oremland, R.S. 2008. Arsenic(III) fuels anoxygenic photosynthesis in hot spring biofilms from Mono Lake, California. *Science* 321: 967–970.

Kulp, T.R., Han, S., Saltikov, C.W., Lanoil, B.D., Zargar, K., and Oremland, R.S. 2007. Effects of imposed salinity gradients on dissimilatory arsenate reduction, sulfate reduction, and other microbial processes in sediments from two California soda lakes. *Appl. Environ. Microbiol.* 73: 5130–5137.

Kulp, T.R., Hoeft, S.E., Miller, L.G., Saltikov, C.W., Murphy, J.N., Han, S., Lanoil, B., and Oremland, R.S. 2006. Dissimilatory arsenate and sulfate reduction in sediments of two hypersaline, arsenic-rich soda lakes: Mono and Searles Lakes, California. *Appl. Environ. Microbiol.* 72: 6514–6526.

Langner, H.W., Jackson, C.R., McDermott, T.R., and Inskeep, W.P. 2001. Rapid oxidation of arsenite in a hot spring ecosystem, Yellowstone National Park. *Environ. Sci. Technol.* 35: 3302–3309.

Langner, H.W., and Inskeep. W.P. 2000. Microbial reduction of arsenate in the presence of ferrihydrite. *Environ. Sci. Technol.* 34: 3131–3136.

Ledbetter, R.N., Connon, S.A., Neal, A.L., Dohnalkova, A., and Magnuson, T.S. 2007. Biogenic mineral production by a novel arsenic-metabolizing thermophilic bacterium from the Alvord Basin, Oregon. *Appl. Environ. Microbiol.* 73: 5928–5936.

Lee, J.H., Kim, M.G., Yoo, B.Y., Myung, N.V., Maeng, J.S., Lee, T., Dohnalkova, A.C., Frederickson, J.K., Sadowsky, M.J., and Hur, H.G. 2007. Biogenic formation of photoactive arsenic-sulfide nanotubes by *Shewanella* sp. strain HN-41. *Proc. Nat'l Acad. Sci. USA* 104: 20410–20415.

Mallioux, B.J., Alexandrova, E., Keimowitz, A.R., Wovkulick, K., Freyer, G.A., Herron, M., Stolz, J.F., Kenna, T.C., Pichler, T., Polizzotto, M.L., Dong, H., Bishop, M., and Knappett, P.S. 2009. Microbial mineral weathering for nutrient acquisition releases arsenic. *Appl. Environ. Microbiol.* 75: 2558–2565.

Michel, C., Jean, M., Dictor, M.-C., Delorme, F., Morin, D., and Garrido, F. 2007. Biofilms of As(III)-oxidising bacteria: Formation and activity studies for bioremediation process development. *Environ. Biotechnol.* 77: 457–467.

McEwan, A.G., Ridge, J.P., McDevitt, C.A., and Hugenholtz, P. 2002. The DMSO reductase family of microbial molybdenum enzymes: Molecular properties and role in dissimilatory reduction of toxic elements. *Geomicrobiol. J.* 19: 3–22.

Mládek, A., Šponer, J., Sumpter, B.G., Fuentes-Cabrera, M., and Šponer, J.E. 2011. On the geometry and electronic structure of the As-DNA backbone. *J. Phys. Chem. Lett.* 2: 389–392.

Mukhopadhyay, R., and Rosen, B.P. 2002. Arsenate reductases in prokaryotes and eukaryotes. *Environ. Health Perspect.* 110: 745–748.

Mukhopadhyay, R., Rosen, B.P., Phung, L.T., and Silver, S. 2002. Microbial arsenic: From geocycles to genes and enzymes. *FEMS Microbiol. Rev.* 26: 311–325.

Muller, D., Lièvremont, D., Simeonova, D.D., Hubert, J-C., and Lett, M-C. 2003. Arsenite oxidase aox genes from a metal-resistant B-Proteobacterium. *J. Bacteriol.* 185: 135–141.

Neppolian, B., Celik, E., and Choi, H. 2008. Photochemical oxidation of arsenic(III) to arsenic(V) using peroxydisulfate ions as an oxidizing agent. *Environ. Sci. Technol.* 42: 6179–6184.

Newman, D.K., Beveridge, T.J., and Morel, F.M.M. 1997a. Precipitation of arsenic trisulfide by *Desulfotomaculum auripigmentum*. *Appl. Environ. Microbiol.* 63: 2022–2028.

Newman, D.K., Kennedy, E.K., Coates, J.D., Ahmann, D., Ellis, D.J., Lovley, D.R., and Morel, F.M.M. 1997b. Dissimilatory arsenate and sulfate reduction in *Desulfotomaculum auripigmentum* sp. nov., *Arch. Microbiol.* 168: 380–388.

Nordstrom, D.K. 2002. Worldwide occurrences of arsenic in groundwater. *Science* 296: 2143–2144.

Onstott, T.C., Chan, E., Polizzotto, M.L., Lanzon, J., and DeFlaun, M.F. 2011. Precipitation of arsenic under sulfate reducing conditions and subsequent leaching under aerobic conditions. *Appl. Geochem.* 26: 269–285.

Oremland, R.S., Wolfe-Simon, F., Saltikov, C.W., and Stolz, J.F. 2009. Arsenic in the evolution of earth and extraterrestrial ecosystems. *Geomicrobiol. J.* 26: 522–536.

Oremland, R.S., and Stolz, J.F. 2005. Arsenic, microbes, and contaminated aquifers. *Trends in Microbiol.* 13: 45–49.

Oremland, R.S., and Stolz, J.F. 2003. The ecology of arsenic. *Science* 299: 939–944.

Oremland, R.S., Hoeft, S.E., Bano, N., Hollibaugh, R.A., and Hollibaugh, J.T. 2002. Aanerobic oxidation of arsenite in Mono Lake water and by a facultative arsenite-oxidizing chemoautotroph, strain MLHE-1. *Appl. Environ. Microbiol.* 68: 4795–4802.

Oremland, R.S., Dowdle, P.R., Hoeft, S.E., Sharp, J.O., Schaefer, J.K., Miller, L.G., Switzer Blum, J., Smith, R.L., Bloom, N.S., and Wallschlaeger, D. 2000. Bacterial dissimilatory reduction of arsenate and sulfate in meromictic Mono Lake, California. *Geochim. Cosmochim. Acta* 64: 3073–3084.

Páez-Espino, D., Tamaames, J., de Lorenzo, V., and Cánovas, D. 2009. Microbial responses to environmental arsenic. *Biometals* 22: 117–130.

Pearce, C.I., Baesman, S.M., Switzer Blum, J., Fellowes, J.W., and Oremland, R.S. 2011. Nanoparticles formed from microbial oxyanion reduction of Group 15 and Group 16 metalloids. In *Microbial Metal and Metalloid Metabolism: Advances and Applications.* Edited by Stolz, J.F., and Oremland, R.S. Washington, DC: American Society for Microbiology Press. p. 297–319.

Pearcy, C.A., Chevis, D.A., Huag, T.J., Jeffries, H.A., Yang, N., Tang, J., Grimm, D.A., and Johannesson, K.H. 2011. Evidence of microbially mediated arsenic mobilization from sediments of the Aquia aquifer, Maryland, USA. *Appl. Geochem.* 26: 575–586.

Planer-Friedrich, B., London, J., McMlsekey, R.B., Nordstron, D.K., and Wallschläger, D. 2007. Thioarsenates in geothermal waters of Yellowstone National Park: Determination, preservation, and geochemical importance. *Environ. Sci. Technol.* 41: 5245–5251.

Quéméneur, M., Heinrich-Salmeron, A., Muller, D., Lièvremont, D., Jauzein, M., Bertin, P.N., Garrido, F., and Joulian, C. 2008. Diversity surveys and evolutionary relationships of *aoxB* genes in aerobic arsenite-oxidizing bacteria. *Appl. Environ. Microbiol.* 74: 4567–4573.

Rhine, E.D., Garcia-Dominguez, E., Phelps, C.D., and Young, L.Y. 2005. Environmental microbes can speciate and cycle arsenic. *Environ. Sci. Technol.* 39: 9569–9573.

Rhine, E.D., Phelps, C.D., and Young, L.Y. 2006. Anaerobic arsenite oxidation by novel denitrifying isolates. *Environ. Microbiol.* 8: 899–908.

Root, R.A., Vlassopoulos, D., Rivera, N.A., Rafferty, M.T., Andrews, C., and O'Day, P.A. 2009. Speciation and natural attenuation in a tidally influenced shallow aquifer. *Geochim. Cosmochim. Acta* 73: 5528–5553.

Rosen, B. 2002. Biochemistry of arsenic detoxification. *FEBS Lett.* 529: 86–92.

Rosen, B.P., Abdul, A., and McDermott, T.R. 2011. Life and death with arsenic. *Bioessays* 33: 350–357.

Saalfield, S.L., and Bostick, B.C. 2009. Changes in iron, sulfur, and arsenic speciation associated with bacterial sulfate reduction in ferrihydrite-rich systems. *Environ. Sci. Technol.* 43: 8787–8793.

Salmassi, T.M., Walker, J.J., Newman, D.K., Leadbetter, J.R., Pace, N.R., and Hering, J.G. 2006. Community and cultivation analysis of arsenite-oxidizing biofilms at Hot Creek. *Environ. Microbiol.* 8: 50–59.

Saltikov, C.W., and Newman, D.K. Genetic identification of a respiratory arsenate reductase. *Proc. Nat'l Acad. Sci. USA.* 100: 10983–10988.

Santini, J.M., Kappler, U., Ward, S.A., Honeychurch, M.J., vanden Hoven, R.N., and Bernhardt, P.V. 2007. The NT-26 cytochrome c_{552} and its role in arsenite oxidation. *Biochim. Biopphys. Acta* 1767: 189–196.

Santini, J.M., and vanden Hoven, R.N. 2004. Molybdenum-containing arsenite oxidase of the chemolithoautotrophic arsenite oxidizer NT-26. *J. Bacteriol.* 186: 1614–1619.

Santini, J.M., Sly, L.I., Schnagl, R.D., and Macy, J.M. 2000. A new chemolithoautotrophic arsenite-oxidizing bacterium isolated from a gold mine: Phylogenetic, physiological, and preliminary biochemical studies. *Appl. Environ. Microbiol.* 92–97.

Sarkar, S., Blaney, L.M., Gupta, A., Ghosh, D., and Sengupta, A.K. 2008. Arsenic removal from groundwater and its safe containment in a rural environment: Validation of a sustainable approach. *Environ. Sci. Technol.* 42: 4268–4273.

Senn, D.B., and Hemond, H.F. 2002. Nitrate controls on iron and arsenic in an urban lake. *Science* 296: 2372–2376.

Silver, S., and Phung, L.T. 2005. Genes and enzymes involved in bacterial oxidation and reduction of inorganic arsenic. *Appl. Environ. Microbiol.* 71: 599–608.

Silver, S., and Phung, L.T. 2011. Novel expansion of living chemistry or just a serious mistake? *FEMS Microbiol. Lett.* 315: 79–80.

Smith, A.H., Lopipero, P.A., Bates, M.N., and Steinmaus, C.M. 2002. Arsenic epidemiology and drinking water standards. *Science* 296: 2145–2146.

Song, B., Chyun, E., Jaffè, P.R., and Ward, B.B. Molecular methods to detect and monitor dissimilatory arsenate-respiring bacteria (DARB) in sediments. *FEMS Microbiol. Ecol.* 68: 108–117.

Stolz, J.F., Basu, P., Santini, J.M., and Oremland, R.S. 2006. Arsenic and selenium in microbial metabolism. *Ann. Rev. Microbiol.* 60: 107–130.

Stolz, J.F., Basu, P., and Oremland, R.S. 2010. Microbial arsenic metabolism: New twists on an old poison. *Microbe* 5: 53–59.

Sun, W., Sierra, R., and Field, J.A. 2008. Anoxic oxidation of arsenite linked to denitrification in sludges and sediments. *Water Res.* 42: 4569–4577.

Sun, W., Sierra-Alvarez, R., Milner, L., Oremland, R.S,, and Field, J.A. 2009. Arsenite and ferrous iron oxidation linked to chemolithotrophic denitrification for the immobilization of arsenic in anoxic environments. *Environ. Sci. Technol.* 43: 6585–6591.

Tufano, K.J., Reyes, C., Saltikov, C.W., and Fendorf, S. 2008. Reductive processes controlling arsenic retention: Revealing the relative importance of iron and arsenic reduction. *Environ. Sci. Technol.* 42: 8283–8289.

Wilkie, J.A., and Hering, J.G. 1998. Rapid oxidation of geothermal As(III) in streamwaters of the eastern Sierra Nevada. *Environ. Sci. Technol.* 32: 657–662.

Wolfe-Simon, F., Switzer Blum, J., Kulp, T.R., Gordon, G.W., Hoeft, S.E., Stolz, J.F., Webb, S.M., Davies, R.C.W., Anbar, A.D., and Oremland, R.S. 2011. A bacterium that can grow by using arsenic instead of phosphorous. *Science* (in press).

Zargar, K., Hoeft, S., Oremland, R., and Saltikov, C. 2010. Genetic identification of a novel arsenite oxidase, *arxA*, in the haloalkaliphilic, arsenite-oxidizing bacterium *Alkalilimnicola ehrlichii* strain MLHE-1. *J. Bacteriol.* 192: 4633–4639.

Zhu, W., Young, L.Y., Yee, N., Serges, M., Rhine, E.D., and Reinfelder, J.R. 2008. Sulfide-driven arsenic mobilization from arsenopyrite and black shale pyrite. *Geochim. Cosmochim. Acta* 72: 5243–5250.

Zobrist, J., Dowdle, P.R., Davis, J.A., and Oremland, R.S. 2000. Mobilization of arsenite by dissimilatory reduction of adsorbed arsenate. *Environ. Sci. Technol.* 34: 4747–4753.

Challenges in Flotation of Cu-Mo Sulfide Ores in Sea Water

Sergio Castro
Department of Metallurgical Engineering, University of Concepción, Chile

ABSTRACT

Sea water and saline water are usually considered detrimental for the flotation of copper-molybdenum ores. A desalination stage is assumed necessary to improve the quality of water for the use in flotation process. However, the effect of sodium chloride and other electrolytes improve the floatability of natural hydrophobic minerals. Hence, salinity by itself is not a limitation in mineral flotation. However, the chemistry of sea water is much more complex that a simple sodium chloride solution. Chemical factors such as the buffering effect of sea water, which increases lime consumption; and the precipitation of hydrophilic colloidal particles in alkaline media, which may exhibit depressing effect on some minerals, must be taken into account. It was found that when Cu-Mo sulfide ores are floated in sea water, molybdenite (MoS_2) is strongly depressed at pH>9.5. Upgrading of copper concentrates takes place by pyrite depression, but when pH is increased up to the values at which molybdenite is depressed, non-conventional pyrite depressants are needed to avoid Mo losses. Therefore, the use of sea water requires innovative solutions to these specific problems. This paper attempts to describe the advances in the Cu-Mo sulfide ore flotation in sea water, and to identify the most important and challenging issues.

INTRODUCTION

Mining companies in different countries are facing water scarcity, and as a result they are competing with other water users such as farmers and local communities. The use of sea water and underground saline water has a high potential to satisfy these water demands, but they are usually considered of poor metallurgical quality.

Reverse osmosis offers an adequate technology to transform sea water into drinking water, being the most widely used desalination technology (Greenlee, et al., 2009). However, it has two major problems in mineral processing: the high cost, and the production of concentrated brines that can strongly affect the environment (Barrera and Cerna, 2009; Randall et al., 2011; Lee et al., 2011). Hence, the major challenge in copper flotation is the successful use of sea water without desalination.

The question is, is it necessary to desalinate sea water before its use in flotation? To answer this question some factors need to be take into account. First, the floatability of natural hydrophobic minerals is significantly improved in concentrated electrolyte solutions (Klassen and Mokrousov, 1963). Hence, salinity is not an obstacle but a positive aspect for particle/bubble attachment in flotation, as it has been discussed elsewhere (Laskowski and Castro, 2008; Castro and Laskowski, 2011). Additionally, electrolytes prevent bubble coalescence as fothers do, and improve foamability of frother solutions, being both positive factors for flotation (Castro et al., 2010; 2012a; 2012b). The problem is the complex chemistry of sea water because it contains not only simple electrolytes but also some secondary ions (additional to NaCl). These ions may interfere with the flotation of some minerals. Hence, the flotation behaviour of Cu, Mo and Fe sulfides in sea water is very different from that in NaCl solutions.

One of the first references on flotation of copper ores in sea water in Chile dated back to 1930 (Burn). Later, Rey and Raffinot (1966) reported that sea water flotation technology was very successful in a number of mills and it could be used extensively. Successful pilot plant flotation tests were reported for the Andacollo deposit in Chile (Cu-Mo-Au-Ag ore), pH 9.5 was recommended for the rougher circuit—by using thionocarbamate as collector and pine oil as frother—and pH between 9.5 and 10.5 for cleaner stages (Morales, 1975).

Basic studies showed the floatability of chalcocite is strongly affected in sea water. Chalcocite exhibits only a narrow peak of floatability around pH 9. Pyrite is also strongly depressed in sea water, even at pH below 7. The floatability of chalcopyrite is not affected in NaCl solutions, but the information on the effect of sea water is not available (Alvarez and Castro, 1977; Laskowski and Castro, 2008). On the other hand, Lekki and Laskowski, (1972) emphasized the significant role of the frother (α-terpineol) on the flotation recovery of chalcocite when copper ores are floated with saline mine water at laboratory scale. These results are in good agreement with the frothing properties reported for α-terpineol in sea water (Castro et al., 2012b). On the other hand, galvanic effects that arise by direct contact between sulfide particles also play a role in sea water flotation, for example the loss of floatability of chalcocite in the presence of pyrite by Cu ions released to the solution by electrochemical mechanisms (Alvarez and Castro, 1977; Parraguez et al., 2009).

The most important recent advance involving the use of sea water for Cu-Mo-Au flotation, at large mining scale, is the Esperanza concentrator at Sierra Gorda in Chile (Antofagasta Minerals S.A AMSA). A chalcopyrite Cu-Au ore (0.57% Cu and 0.23 g Au/ton) is processed in a 95,000 tpd capacity flotation plant, by using sea water without any pre-treatment. The operation just started in the January–March 2011 period. Sea water is pumped 145 Km from the Pacific Ocean to a 60,000 m^3 pool at the mine site located at an altitude of 2,300 m above sea level. It is planning to produce around 200,000 tons of Cu (as copper concentrate) and around 285,000 Au ounces per year in concentrate (Parraguez et al., 2009; Thiele, 2011). Other similar but small flotation plant is the Las Luces concentrator at Taltal in Chile (Minera Las cenizas) which has been operating during the last 17 years with sea water (chalcocite Cu-Au ore). Sea water is pumped from the Pacific Ocean (2,200 m^3/day) to the concentrator site located at 7 Km distance and there is stored in a 2,600 m^3 pool. Recycled sea water from the tailing dam constitutes around 65%, while only 35% is fresh sea water. Cu recovery is usually in the range of 81–85%, and Au recovery around 71% (Monardes, 2009). Other important case is the Batu Hijau concentrator (Newmont, operating from 2000), which uses sea water for processing a gold-rich phorphyry copper ore (Chalcopyrite-bornite), located at the Indonesian island of Sumbawa, with an annual production in 2005 of 325,500 ton Cu; and 719,000 oz Au.

The aim of this work is to analyse and discuss the recent developments in the Cu-Mo sulfide ores flotation in sea water and to identify most urging challenges.

EXPERIMENTAL

This study was conducted in the laboratories of the University of Concepción using real copper ore samples provided by two Chilean mining companies. In both cases standard laboratory flotation tests were carried out (standard reagents, pH, etc.). In the tests carried out with sea water this water was employed in both grinding and flotation stages.

Table 1. Standard flotation conditions employed for rougher and cleaner tests in fresh water and sea water (sample 1)

	Flotation Reagents, g/ton				Time, min		
Stage	Sascol-95	Matcol TC-123	Matfroth-355	Diesel	Grinding	Conditioning	Flotation
Grinding	—	—	—	10	3' 30"	—	—
Rougher	11	22	10	—	—	5	7
Cleaner	—	—	—	—	—	3	4

Table 2. Standard flotation conditions employed in the rougher and cleaner tests in fresh water and sea water (sample 2)

	Flotation Reagents, g/ton				Time, min		
Stage	Matcol D-101	IsobX	X-133/ Pine oil	Diesel	Grinding	Conditioning	Flotation
Grinding	—	—	—	20	7' 05"	—	—
Rougher flotation	35	5	10	—	—	4	8

Note: As activator 100 g/ton NaSH.

Flotation Set-up and Procedure

With the exception of results presented in Figure 3, all laboratory flotation tests were carried out with the same copper ore (sample 1) (chalcopyrite-bornite): 0.56%Cu, 5.76%Fe and 0.017%Mo. The experimental conditions were:1,204.4 g of sample for rougher flotation tests; 35% solids (by wt.); pH as indicated in figures; grinding size analysis: 29.3% + 100 Tyler mesh (P_{80} = 210 μm). For rougher flotation tests a 2.7 L Agitair LA-500 flotation cell was employed; and a 1.5L for cleaner flotation tests. All tests were carried out at 900 rpm impeller speed rate and air flow rate of 10 L/min. For open cycle tests, two rougher flotation concentrates were employed to carry out one cleaner flotation test (without re-grinding); and pH 9 was employed for rougher flotation in sea water and pH 10.5 in fresh water. The general scheme of reagents and time for rougher tests is given in Table 1.

Only for the flotation tests showed in Figure 3, a different copper ore (sample 2) (chalcocite-chalcopyrite) was employed: 0.99%Cu, 1.7%Fe and 0.007%Mo. The experimental conditions were: 1,015.7 g; impeller speed, 900 rpm; air flow rate 7 L/min; 30% solids (by wt.); grinding size analysis: 30% +100 Tyler mesh (P_{80} = 198.3 μm). Details are given in Table 2.

All flotation reagents were provided by the mining companies and were employed as received. Sea water was a local sample taken from Concepción coast (*Bellavista-Tomé*); and fresh water was Concepción tap water.

RESULTS AND DISCUSSION

Copper Flotation Technology in Fresh Water

The current Cu-Mo flotation technology in fresh water usually includes a rougher flotation stage at pH around 10.0–10.5, and one or two cleaner stages with one scavenger circuit operating at pH 11.5 to 12.0. Lime is used as pyrite depressant and a minimum of 28–30%Cu

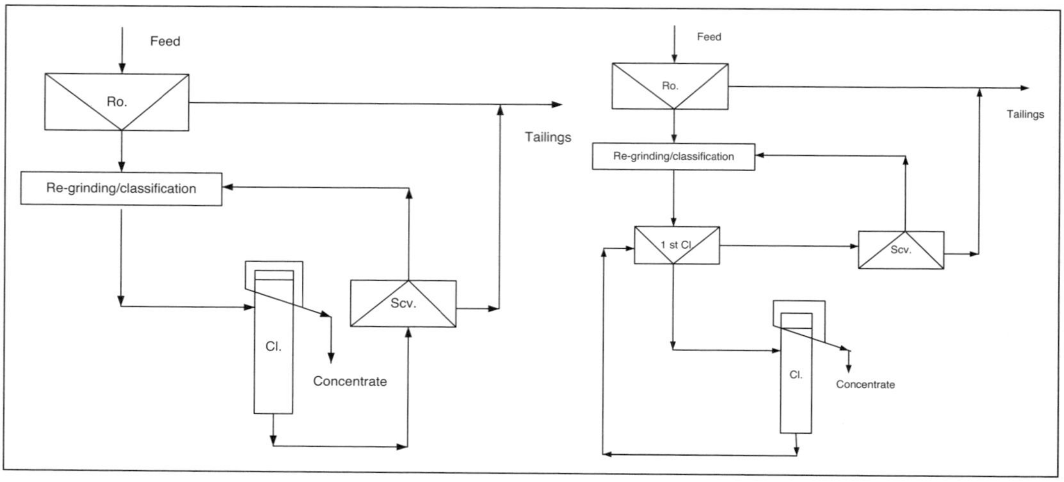

Figure 1. Flowsheets employed for copper ores flotation

in the final concentrate is achieved. With this technology a high pyrite ore can be processed with high Mo and Au recovery. The main flowsheets are illustrated in Figure 1.

According to the plant practice experience, the same flowsheets employed for the flotation of Cu-Mo sulfides in fresh water are useful with sea water. The flowsheet including two cleaner stages seems to be more convenient to obtain high final concentrate grade in sea water, particularly for low Cu grade ores.

Chemical Factors in Sea Water

Two most important chemical factors of sea water that may affect the flotation process are: (a) the buffering effect of sea water, which significantly increases lime consumption in flotation pulps; and (b) the precipitation of secondary ions at alkaline pH with potential depressing effect on the flotation of some sulfide minerals.

Buffering Effect of Sea Water

The buffering effect of sea water increases significantly lime consumption when pH is adjusted to values traditionally used for copper sulfide flotation in fresh water. Sea water has a natural pH of around 7.8–8.2, which depends on salinity and concentration of the couples carbonate/bicarbonate ions (HCO_3^-/CO_3^{2-}) and boric acid/borate ions ($B(OH)_3/B(OH)_4^-$). These ions are responsible for the buffering effect of sea water (Pytkowicz and Atlas, 1975).

Figure 2 shows a comparison between the consumption of lime in fresh water and sea water for a typical copper ore under usual rougher flotation conditions. It was observed that in sea water lime consumption increases exponentially with pH.

Alkalinity of sea water and the majority of ground waters are caused by dissolved bicarbonate ions (HCO_3^-). The most important reactions for maintaining acid-base balance in sea water are the carbonic acid/bicarbonate ion buffer.

The simultaneous equilibrium reactions of interest—in distilled water—are:

$$CO_2 + H_2O \rightleftharpoons H_2CO_3 \tag{1}$$

The equilibrium constant is 1.70×10^{-3}: hence, the majority of the carbon dioxide is not converted into carbonic acid, remaining as CO_2 molecules.

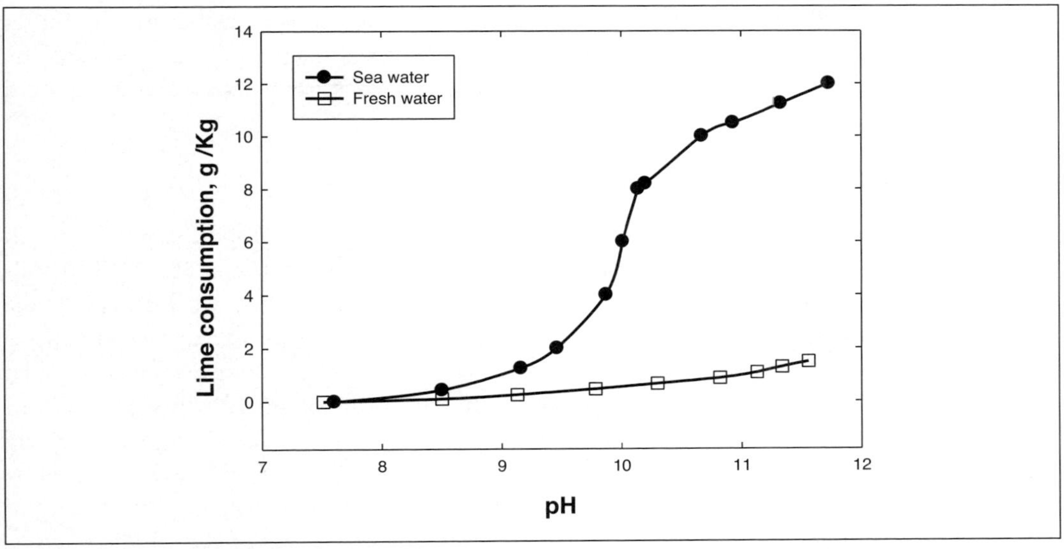

Figure 2. Lime consumption in rougher flotation as a function of pH adjusted by lime for a copper ore floated in fresh water and sea water (sample 1)

$$H_2CO_3 \rightleftharpoons H^+ + HCO_3^- \quad (2)$$
$$K_{a1} = 4.45 \times 10^{-7};\ pK_{a1} = 6.352 \text{ at } 25°C$$

$$HCO_3^- \rightleftharpoons H^+ + CO_3^{2-} \quad (3)$$
$$K_{a2} = 4.69 \times 10^{-11};\ pK_{a2} = 10.329 \text{ at } 25°C$$

Therefore, at 25°C and pH 6.35 the concentrations of H_2CO_3 and HCO_3^- species are the same; and at pH 10.33 the concentration of ions HCO_3^- and CO_3^{2-} are equivalent. However, the values of the constants for the dissociation of the carbonic acid in sea water need a correction due to the effect of salinity (Millero et al., 2006).

Sea water contains around 0.027 g/kg of borate ions (BO_3^{3-}). Although the boric acid can appear in solution as polyborates, the main reaction in aqueous solution is the conversion to borate ions, and depending on the borate anion considered, according to Raposo et al. (2003) this reaction could be expressed as:

$$B(OH)_4^- + H^+ \rightleftharpoons B(OH)_3 + H_2O \quad (4)$$

$$H_2BO_3^- + H^+ \rightleftharpoons B(OH)_3 \quad (5)$$

$$BO_2^- + H^+ + H_2O \rightleftharpoons B(OH)_3 \quad (6)$$

$$B(OH)_3 + H_2O \rightleftharpoons B(OH)_4^- + H^+ \qquad pKa \sim 8.55 \quad (7)$$

The ortho-boric acid is a week acid with the following dissociation constants:

$$H_3BO_3 \rightleftharpoons H^+ + H_2BO_3^- \qquad K_1 = 5.8 \times 10^{-10} \quad (8)$$

$$H_2BO_3^- \rightleftharpoons H^+ + HBO_3^{2-} \qquad K_2 = 1.8 \times 10^{-13} \quad (9)$$

$$HBO_3^{2-} \rightleftharpoons H^+ + BO_3^{3-} \qquad K_3 = 1.6 \times 10^{-14} \quad (10)$$

Therefore, ions dissociated from carbonic and boric acid are responsible for the buffer effect observed in sea water and in some saline waters. However, magnesium, calcium, sulfate, and bicarbonate ions are also very important in defining the chemistry of sea water in flotation systems, because they participate in side precipitation reactions.

Colloidal Precipitation of Secondary Ions

Salinity of sea water is typically 35‰, and NaCl concentration is around 0.5–0.6M, with important secondary ions such as, sulfate ions (2.7 g/kg); magnesium ions (1.29 g/Kg); calcium ions (0.41 g/Kg); bicarbonate ions (0.145 g/Kg); borate ions (0.027 g/Kg); etc. These ions are not indifferent for the flotation process. They may influence the surface chemistry and thereby the floatability of some minerals. Some of these secondary ions can form colloidal precipitates at pH>10. Calcium and magnesium ions can form colloidal hydroxides, carbonates and sulfates. These precipitates can depress some mineral species like molybdenite. There may be several reasons for depression of molybdenite, including hydrophilic surface coating by some of these colloidal precipitates. The main precipitation reactions between sea water and lime under flotation conditions are:

$$Ca(OH)_2 + Mg^{2+} \rightleftarrows Mg(OH)_2 + Ca^{2+} \tag{11}$$

$$Ca^{2+} + 2OH^- \rightleftarrows Ca(OH)_2 \tag{12}$$

$$Ca(OH)_2 + SO_4^{2-} + 2H_2O \rightleftarrows CaSO_4{\cdot}2H_2O + 2OH^- \tag{13}$$

$$Mg(OH)_2 + SO_4^{2-} \rightleftarrows MgSO_4 + 2OH^- \tag{14}$$

$$Ca(OH)_2 + CO_3^{2-} \rightleftarrows CaCO_3 + 2OH^- \tag{15}$$

$$Mg(OH)_2 + CO_3^{2-} \rightleftarrows MgCO_3 + 2OH^- \tag{16}$$

$$K_{ps}\ Mg(OH)_2 = 1{,}2 \times 10^{-11} \tag{17}$$

$$K_{ps}\ Ca(OH)_2 = 4{,}1 \times 10^{-6} \tag{18}$$

$$K_{ps}\ CaCO_3 = 6{,}7 \times 10^{-7} \tag{19}$$

Flotation of Cu-Mo Ores in Sea Water

Two samples of porphyry Cu ores were evaluated through rougher laboratory flotation tests in sea water, by using the flotation reagents corresponding to the current standard conditions applied in commercial plants. The purpose of the first tests was to compare the Cu rougher recovery in sea water and fresh water as a function of pH. It was found that the flotation of copper ores in sea water depends on mineralogy of copper sulfides. For porphyry copper ores with a predominance of chalcopyrite ($CuFeS_2$), Cu recovery is not sensitive to pH. Figure 3 shows the effect of pH on the rougher Cu recovery in sea water and fresh water (under the same conditions) for a chalcopyrite ore.

However, when chalcocite (Cu_2S) and covellite (CuS) are the predominating species a different flotation pattern was found and in this case Cu recovery exhibits a peak about pH 10, as is shown in Figure 4. These results are in good agreement with micro-flotation tests carried

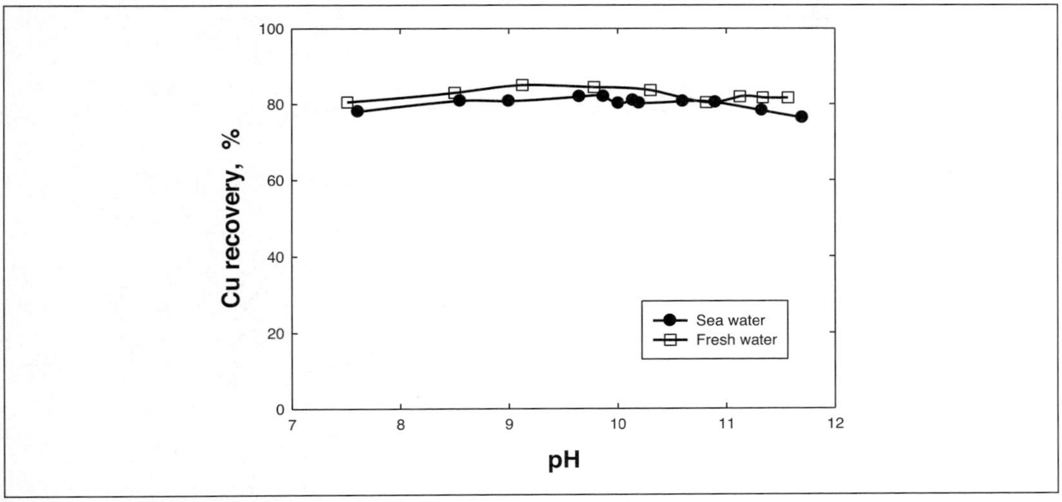

Figure 3. Rougher Cu recovery as a function of pH adjusted by lime for a copper ore floated in fresh water and sea water (sample 1)

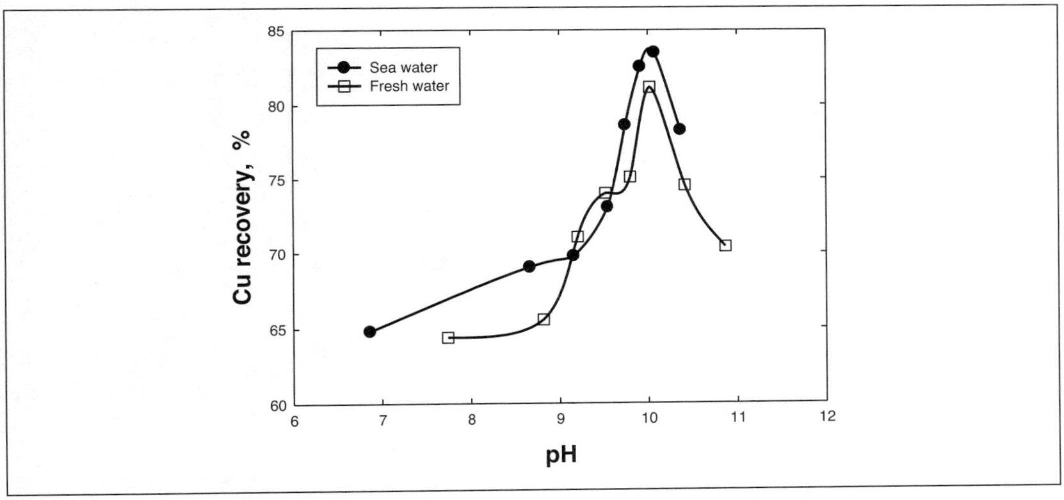

Figure 4. Rougher Cu recovery as a function of pH adjusted by lime for a different copper ore (sample 2) floated in fresh water and sea water

out with a pure chalcocite sample floated in sea water (Alvarez and Castro, 1977; Laskowski and Castro, 2008).

Therefore, for the rougher flotation of Cu-Mo sulfide ores, a reasonable high Cu recovery is feasible to be obtained in sea water, taking into account mineralogy of Cu sulfides and flotation pH. Figure 5 illustrates the effect of pH on Cu, Mo and Fe recovery in sea water. It must be noted that strong depression of Mo was observed in alkaline media. It was found that Mo and Fe recovery were strongly decreased at pH higher than 9.5, showing that the floatability of molybdenite (MoS_2) and pyrite (FeS_2) are very sensitive to pH in sea water. Along the same line, Figure 6 shows that Fe recovery in sea water is lower than in fresh water at the same pH, probably because the amount of lime in the system is much higher. Therefore, copper ores with

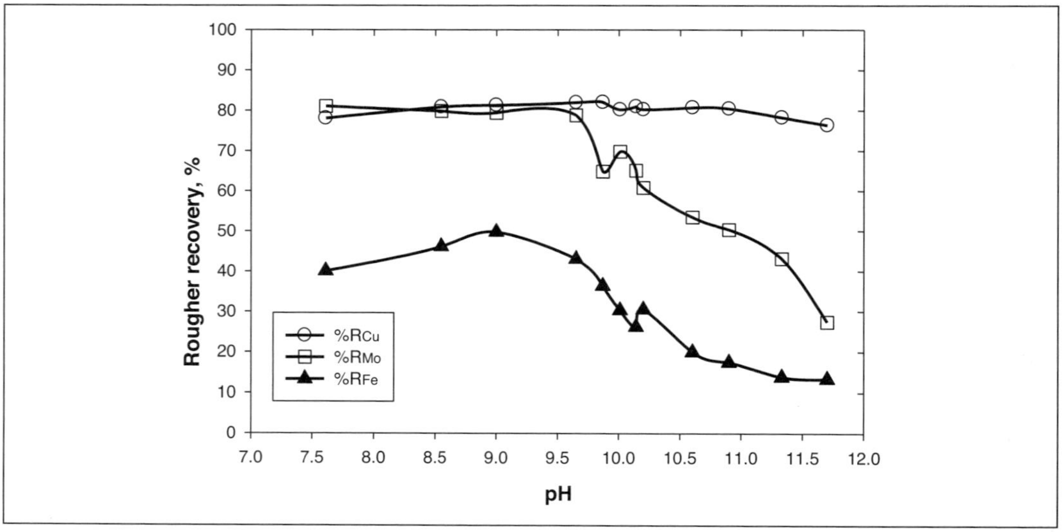

Figure 5. Rougher Cu, Mo, and Fe recovery as a function of pH adjusted by lime for a copper ore floated in sea water (sample 1)

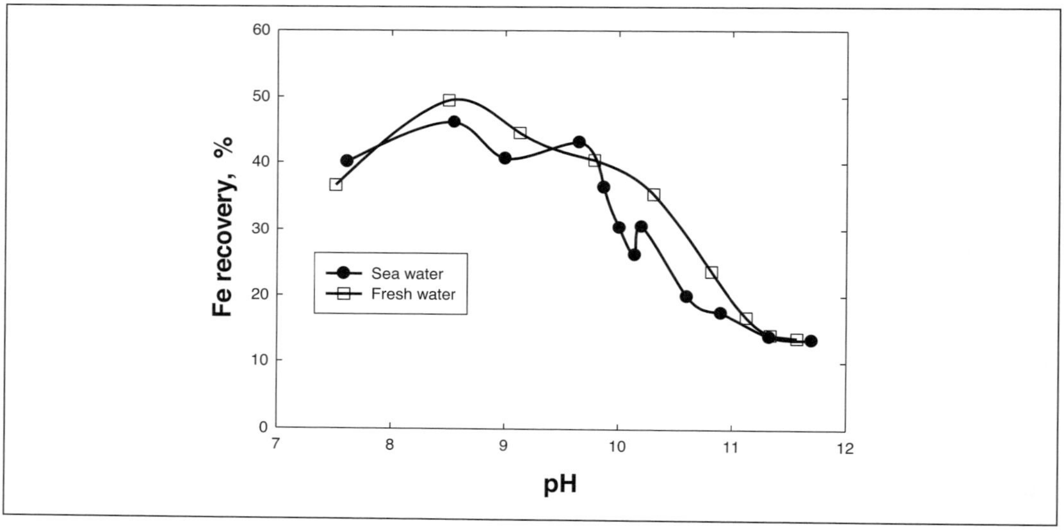

Figure 6. Rougher Fe recovery as a function of pH adjusted by lime for a copper ore floated in fresh water and sea water (sample 1)

Mo as the main by-product face a serious problem: how to depress pyrite without depressing Mo. This issue will be more important for high pyrite copper ores.

Depression of Molybdenite in Sea Water

Mo recovery in sea water is highly affected by pH, because the floatability of molybdenite is strongly depressed at pH>9.5. Figure 7 shows the differences in Mo rougher recovery between sea water and fresh water as a function of pH. It indicates that Mo is dramatically depressed when pH is increased over pH 9.5 in sea water, but not in fresh water, suggesting that secondary ions may play a depressing role in molybdenite sea water flotation. Taking into

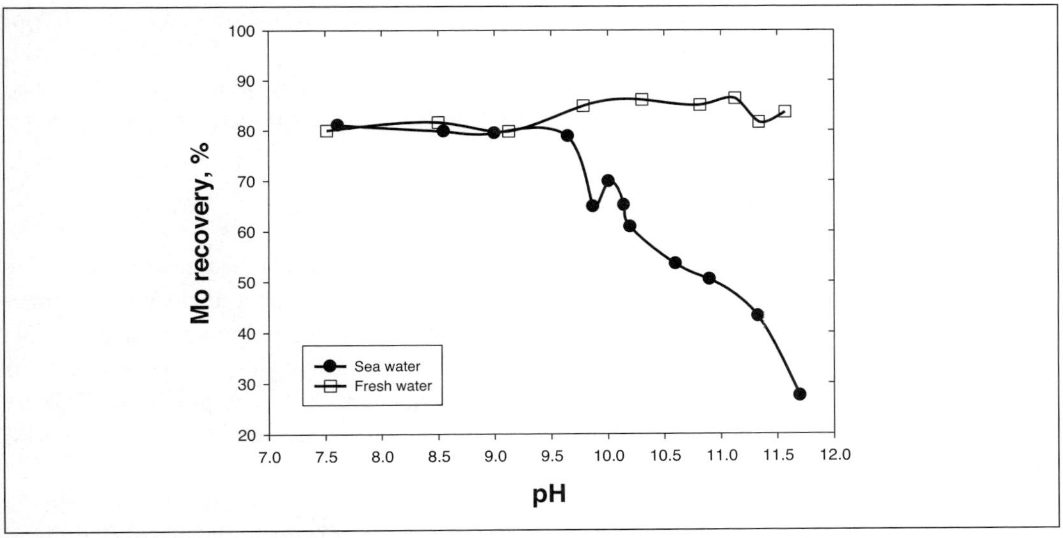

Figure 7. Rougher Mo recovery as a function of pH adjusted by lime for a copper ore floated in fresh water and sea water (sample 1)

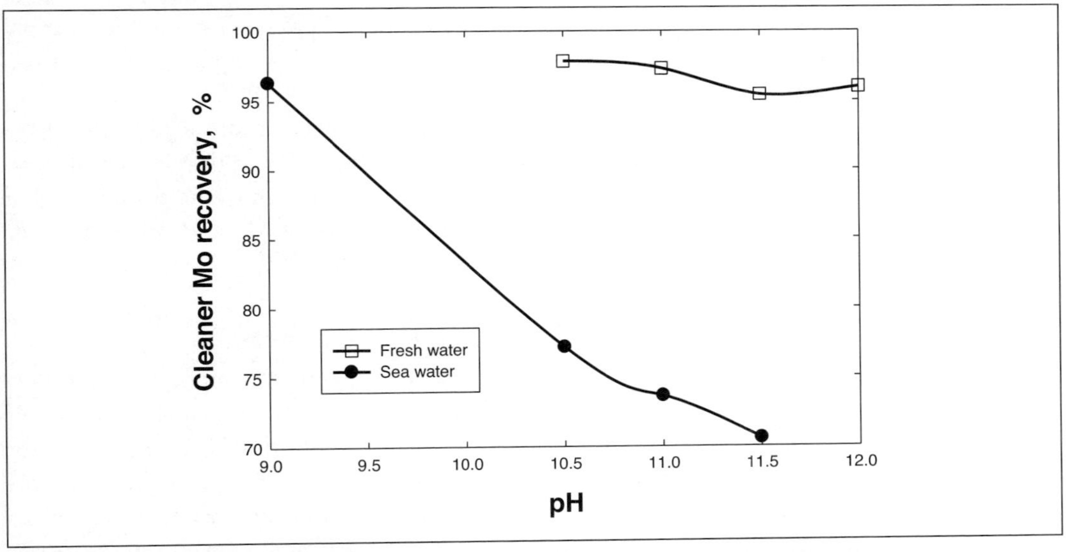

Figure 8. Mo recovery in cleaner flotation (first cleaner) as a function of pH adjusted by lime for a copper ore floated in fresh water and sea water (rougher: pH 9 in sea water; and pH 10.5 in fresh water) (sample 1)

account that molybdenite is easily depressed and that high lime doses are needed to adjust pH in rougher flotation, values below pH 9.5 are recommended. This is in agreement with some other data which suggest a pH range from 8.5 to 9.5 for copper sulfides sea water flotation for rougher circuits (Parraguez et al., 2009; Morales, 1975).

The current technology in fresh water for cleaner circuits needs the addition of lime up to pH 11.5–12 to depress pyrite. However, in sea water a strong Mo depression may take place in the first cleaner stage, as can be seen in Figure 8. These results confirm the incompatibility between Mo recovery and pyrite depression when lime is used as the only pyrite depressing agent.

Therefore, for high pyrite Cu-Mo ores the conventional technology based on the use of lime to depress pyrite in cleaner circuits is not suitable. Depending on rougher and cleaner pH, the choice are two fold: (a) to obtain a high Cu concentrate grade by effective pyrite depression, but with low Mo recovery; or (b) to obtain high Mo recovery operating at low pH but obtaining low Cu concentrate grade (high pyrite content).

Nonconventional Pyrite Depressants Alternative to Lime

The reagent used traditionally to depress pyrite in Cu cleaning stages is lime. However due to the buffer effect of sea water the consumption of lime in the rougher and cleaner-scavenger stages is very high and if the same pH scheme is followed colloidal precipitates which are able to depress molybdenite (and gold) are formed. Therefore, the use of lime to depress pyrite is not convenient in sea water and new pyrite depressants that can act at low pH (< pH 9.5) are needed. However, this should be an inorganic compound since polymers usually also depress molybdenite flotation (Castro and Laskowski, 1997).

Therefore, in order to obtain 28%–30% Cu grade in the final concentrate a depressing agent alternative to lime is required. It has been reported that in sea water the use of cyanide in small doses (6 g/ton, regarding feed to the plant) makes it possible to obtain a 28%Cu grade in the final concentrate (Parraguez et al., 2009). However, cyanide is strongly toxic and is not environmentally friendly. The difficulty of upgrading Cu concentrates in cleaner-scavenger circuits is a function of pyrite content in the ore. For low pyrite content it is feasible to apply lime in cleaner stages up to pH 10.2; however for higher pyrite content an additional depressant agent is necessary.

Finally, when sea water is used, the rougher stage must be carried out at pH 8.5–9.5 to avoid high lime consumption and losses of Cu by-products (Mo, Au), and the cleaners-scavenger circuits at a pH not higher than 10.2 However, for copper ores with pyrite content higher than 2%, and Mo as the main by-product, the probability of losing Mo and Au recovery or producing low Cu final concentrate is very high (Parraguez et al., 2009).

CONCLUSIONS

The rougher and cleaner pHs when flotation is carried out in sea water must be different from those traditionally used in fresh water. The flotation of molybdenite in sea water is strongly depressed in rougher and cleaner stages at pH>9.5, limiting the use of lime for pyrite depression, particularly in cleaner stages. Therefore, the following are the most important aspects for Cu-Mo sulfide ores flotation in sea water: High Mo recovery and strong pyrite depression in cleaner circuits, in order to produce a 28–30% Cu final concentrate grade. As shown by selected examples in this paper, the main challenges facing the use of sea water in Cu-Mo sulfide flotation are:

- High recovery not only for copper but also of its valuable by-products (Mo and Au).
- Pyrite depression in sea water at lower pH (pH<9.5) from that traditionally used in fresh water (pH 11.5–12.0)
- Reducing the excess of lime consumption in sea water (into the range 0.7–1.0 Kg/ton).
- Development and application of new pyrite depressants in cleaner circuits able to operate at moderate pH in sea water (pH<9.5).
- Obtaining high Cu concentrate grade (28–30%) for high pyrite copper ores (<3% Py).

ACKNOWLEDGMENT

Funding for this project was provided by the CORFO-INNOVA CHILE (Project 08CM01-18), with industrial sponsorship from: Antofagasta Minerals (*Minera Esperanza*); BHP Billiton (*Minera Escondida Ltd.*); Anglo American Chile (*Mantos Blancos*); and Teck (*Carmen de Andacollo*), through AMIRA-AUSTRALIA operating Research Grant P968.

REFERENCES

Alvarez, A., and Castro, S. 1977. Flotation of chalcocite and chalcopyrite in seawater and concentrated saline water. *Proc. IV Encontro Nacional de tratamento de Minerios*. São Jose dos Campos, Brazil, 1:39–44 (Spanish text).

Barrera, J., and Cerna, M. 2009. Comparative analysis between desalinated and non-desalinated seawater for a concentrator. *Proc. PROCEMIN* 2009, Santiago-Chile, pp. 267–276.

Burn, A.K. 1930. The flotation of chalcopyrite in seawater. *Bulletin Institution of Mining and Metallurgy*, No. 314.

Castro, S., and Laskowski J.S. 1997. The effect of hydrophilic and hydrophobic polymers on molybdenite flotation. *Proc. 5th Southern Hemisphere Meeting on Mineral Technology*, Buenos Aires-Argentina, Intemin, pp. 117–120.

Castro, S., Venegas, I., Landero, A., and Laskowski, J.S. 2010. Frothing in seawater flotation systems. *Proc. XXV Int. Mineral Processing Congress*, Brisbane, pp. 4039–4047.

Castro, S., and Laskowski J.S. 2011. *Froth Flotation in Saline Water, KONA* (forthcoming).

Castro, S., Toledo, P., and Laskowski, J.S. 2012a. Foaming properties of flotation frothers at high electrolyte concentrations (Proc. this symposium).

Castro, S., Ramos, O., Cancino, J.P., and Laskowski, J.S. 2012b. Frothing in the flotation of copper sulfide ores in sea water (Proc. this symposium).

Greenlee, L.F., Lawler, D.F., Freeman, B.D., Marrot, B., and Moulin, P. 2009. Reverse osmosis desalination: Water source, technology, and today's challenges. *Water Research* 43:2317–2348.

Klassen, V.I., and Mokrousov, V.A. 1963. *An Introduction to the Theory of Flotation*. London: Butterworths.

Laskowski, J.S., and Castro, S. 2008. Flotation in concentrated aqueous electrolyte solutions. *Proc. 11th International Mineral Processing Symposium*, Belek-Antalya, Turkey, pp. 281–290.

Lee, K.P., Arnot, T.C., and Mattia, D. 2011. A review of reverse osmosis membrane materials for desalination—Development to date and future potential. *Journal of Membrane Science*, 370:1–22.

Lekki, J., and Laskowski, J.S. 1972. Influence of NaCl on the flotation of copper sulfide ores. *Minerales*, 27(118):3–7. (Spanish text)

Millero, F.J., Graham, T.B., Huang, F., Bustos-Serrano, H., and Pierrot, D. 2006. Dissociation constants of carbonic acid in seawater as a function of salinity and temperature. *Marine Chemistry* 100:80–94.

Monardes, A. 2009. Use of seawater in grinding-flotation operations and tailing disposal. *Proc. XI Symposium on Mineral Processing Moly-Cop 2009.* (Spanish text)

Morales, J.E. 1975. Flotation of the Andacollo's ore in pilot plant by using seawater. *Minerales*, XXX(130):16–22. (Spanish text).

Parraguez, L., Bernal, L., and Cartagena, G. 2009. Chemical study for selectivity and recovery of metals sulphides by flotation using seawater. *Proc. PROCEMIN* 2009, pp. 323–333.

Pytkowicz, R.M., and Atlas, E. 1975. Buffer intensity of seawater. *Limnology and Oceanography*, 20(2):222–229.

Randall, D.G., Nathoo, J., and Lewis, A.E. 2011. A case study for treating a reverse osmosis brine using eutectic freeze crystallization—Approaching a zero waste process. *Desalination*, 266:256–262.

Raposo, J.C., Zuloaga, O., Olazabal, M.A., and Madariaga, J.M. 2003. Development of a modified Bromley methodology for the estimation of ionic media effects on solution equilibria. Part 5. The chemical model of boric acid in aqueous solution at 25°C and comparison with arseniuous acid. *Fluid Phase Equilibria*, 207:81–95.

Rey, M., and Raffinot, P. 1966. Flotation of ore in sea water: high frothing, soluble xanthate collecting. *World Mining*, (June):18.

Thiele, C. 2011. Esperanza mine enters in production. *Nueva Minería*, 3(25):46–51 (Spanish text).

Correlation of Graphite Flotation and Gas Holdup in Saline Solutions

S. Alexander, J. Quinn, J.E. van der Spuy, and J.A. Finch
Department of Mining and Materials Engineering, McGill University, Montreal, Quebec, Canada

ABSTRACT

Two-phase (solution-air) batch gas holdup tests were undertaken in a laboratory bubble column with six inorganic salts (KCl, NaCl, Na_2SO_4, $MgCl_2$, $CaCl_2$, $MgSO_4$). The electrolytes grouped according to the gas holdup: salts containing divalent ions (1-2, 2-1 and 2-2 cation-anion pair) increased gas holdup at lower molar concentrations than salts containing monovalent ions (1-1 salts) and a dependence on ionic strength was demonstrated. The gas holdup was shown to correlate with the graphite flotation results of Pugh et al. (1997). In the presence of coalescence inhibiting salts, gas dispersion appears to play a significant role in dictating flotation performance.

INTRODUCTION

Over the past decades, water use in mineral processing has come under in½creased scrutiny. With fresh water being scarce in many mining locales, the use of process water with high inorganic salt content is increasingly common. This ranges from use of sea or bore water to water recycling which can lead to build-up of soluble salts. Table 1 gives examples of flotation operations which utilize(d) saline process water and typical water composition (Texada data from Haig-Smillie (1972), Raglan data from Quinn et al. (2007), Mt. Keith data from Peng and Seaman (2011) and George (1996), Leinster data from Nesset et al. (2007), potash flotation data from Laskowski et al. (2003)). The extreme case is potash processing where flotation occurs in a saturated brine solution.

INORGANIC SALTS AND GAS DISPERSION

There have been several studies on the effect of inorganic salts on bubble coalescence inhibition and gas dispersion (e.g., bubble size) (Lessard and Zieminski 1971; Craig et al. 1993; Zahradnik et al. 1999). Relatively high levels of certain inorganic salts have been shown to affect gas dispersion in a similar manner to conventional flotation frothers. For example, Quinn et al. (2007) showed that 0.4 M NaCl solution not only produced similar gas dispersion (bubble size and gas holdup) but yielded similar solids concentrate rate in flotation tests on Brunswick ore (Xstrata Zinc) as 8–10 ppm Methyl Iso-Butyl Carbinol (MIBC), a typical frother and dosage. In comparison to frothers which inhibit bubble coalescence at low concentrations (mM), inorganic salts are only effective in relatively high concentrations (0.05–1 M) (Lessard and Zieminski 1971; Craig et al. 1993; Zahradnik et al. 1999).

Unlike flotation frothers, inorganic salts (typically) slightly increase surface tension, ions tending towards the bulk solution (as most ions are hydrated) rather than the water/air interface. Foulk and Miller (1931) discussed the concept of positive and negative adsorption by

Table 1. Examples of flotation operations that utilize saline process water

Operation	Company	Location	Water Type	TDS (ppm)	Major Elements
Raglan	Xstrata	Quebec	Recycle	~30 000	Na^+, Ca^{2+} SO_4^{2-}, $S_2O_3^{2-}$
Texada	Texada Mines	British Columbia	Sea	~35 000	Na^+, Cl^-
Mt. Keith	BHP Billiton	Australia	Bore	60 000 – 80 000	Na^+, K^+, Ca^{2+}, Mg^{2+}, Cl^-, SO_4^{2-}
Leinster	BHP Billiton	Australia	Bore	~120 000	N/A
Potash	Various	Canada	Brine	Saturated	Na^+, K^+, Cl^-

surfactants (frothers) and inorganic salts, respectively, at the water/air interface and how both could result in stable film formation (inhibition of bubble coalescence).

Various researchers have identified the ability of certain inorganic salts to inhibit bubble coalescence (Marrucci and Nicodemo 1967; Lessard and Zieminski 1971; Craig et al. 1993; Hofmeier et al. 1995; Laskowski et al. 2003; Craig 2004). Much of the work involved contacting bubble pairs at adjacent capillaries and determining the percentage that coalesce. A transition concentration at which salts inhibit bubble coalescence has been proposed (Lessard and Zieminski 1971; Craig et al. 1993; Zahradnik et al. 1999). Zieminski and Whittemore (1971) showed that salts containing multi-valence ions have lower transition concentrations compared to monovalent ions. Craig et al. (1993) developed a combining rule which indicates whether a specific cation-anion pair (inorganic salt) will inhibit bubble coalescence.

There has been considerable research into the effects of salts in coal flotation (Klassen and Vlasova 1967; Yoon and Sabey 1989; Laskowski 1986; Kurniawan et al. 2011) but few have focused on gas dispersion properties.

Pugh et al. (1997) performed batch graphite flotation tests in the presence of a series of inorganic salts. They demonstrated that recovery increased with the addition of certain inorganic salts (Figure 1). The salts divided into three groups (A, B and C). Group A consisted of divalent and trivalent salts (such as $CaCl_2$, $MgCl_2$ and $MgSO_4$) which showed high flotation response. Group B consisted of monovalent salts (such as KCl and NaCl) which showed medium flotation response. Group C consisted of strong acids and perchlorate salts which showed low flotation response up to 0.3 M concentration.

Pugh et al. (1997) discussed the findings by considering, for example, electrostatic interactions and gas solubility in electrolyte solutions. They did note that increased recovery appeared correlated with decreased bubble size in the froth, as did Kurniawan et al. (2011) in the case of coal flotation. No quantitative examination of bubble size in the pulp zone (i.e., gas dispersion) was undertaken.

The trend in Figure 1 was reminiscent of the trend in gas holdup with inorganic salt concentration reported by Quinn et al. (2007) which stimulated the present enquiry. Gas holdup reflects bubble size, increasing as bubble size decreases. It is thus a convenient surrogate for bubble size being easier to measure, at least in two-phase (water-air) systems. A case can be made for gas holdup being related to flotation kinetics. Gorain et al. (1997; 1998) have proposed flotation kinetic models which correlate the rate constant with bubble surface area flux,

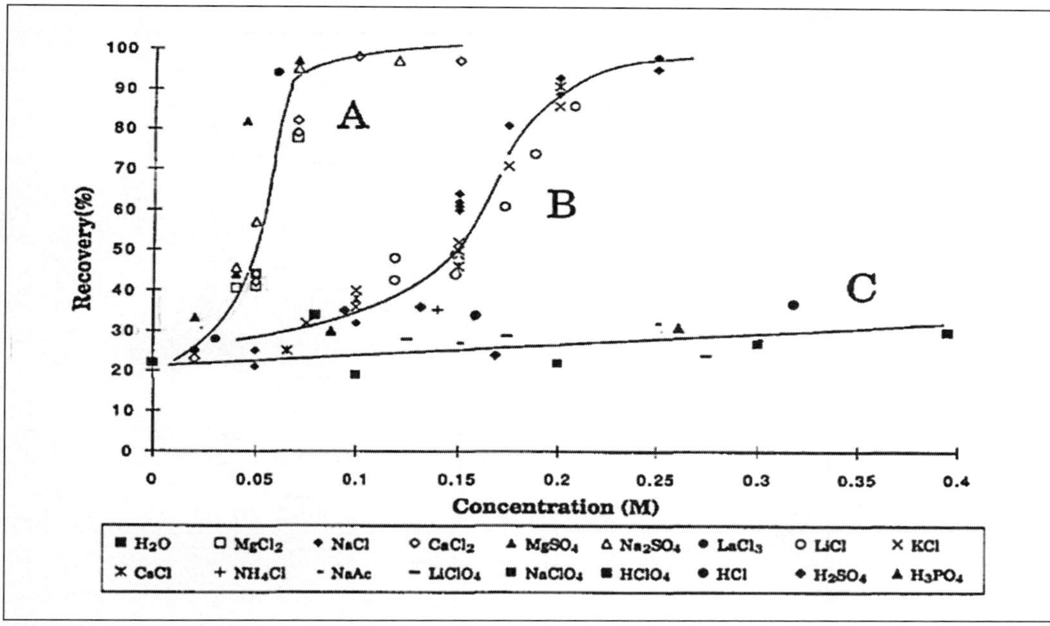

Figure 1. Recovery as a function of inorganic salt concentration (reprinted from *International Journal of Mineral Processing*, 51, Pugh, R.J., Wesissenborn, P., and Paulson, O., Flotation in inorganic electrolytes; the relationship between recover of hydrophobic particles, surface tension, bubble coalescence and gas solubility. 125–138, 1997, with permission from Elsevier)

S_b (Finch and Dobby, 1990). Finch et al. (2000) showed that bubble surface area flux and gas holdup (ε_g) are roughly proportional according to Equation 1 (ε_g in %, S_b in s^{-1}):

$$\varepsilon_g \approx 5.5 S_b \qquad (1)$$

We would, therefore, expect flotation recovery (at a fixed time) to be related to gas holdup. Based on this principle, the purpose of this paper is to determine gas holdup in inorganic salt solutions and correlate with the graphite flotation results of Pugh et al. (1997).

EXPERIMENTAL

Two-phase (water-air) gas holdup tests were undertaken in a bubble column with the general setup shown in Figure 2. The column had an internal diameter of 7.62 cm and a height of 398 cm. Air was introduced to the base of the column through a cylindrical porous sparger (vertical orientation). Gas holdup was measured using a Bailey differential pressure transducer (method detailed in Finch and Dobby (1990)). The differential pressure was measured by tapping between heights 193 cm and 283 cm from the base of the column. Air flow rate was controlled using a 0–5 L/min MKS mass flow meter. Gas holdup was measured at a gas superficial velocity, J_g, of ca. 0.95 cm/s (equivalent to 3 L/min air flow rate). Gas velocity was corrected for temperature and pressure to the mid-point of the gas holdup measurement.

Before and after each experiment, the tank and column were thoroughly rinsed with tap water to limit contamination. The column was filled to a height of 298 cm. The salts tested were KCl, NaCl, Na_2SO_4, $MgCl_2$, $CaCl_2$ and $MgSO_4$ (supplied by Fisher—A.C.S. grade) to

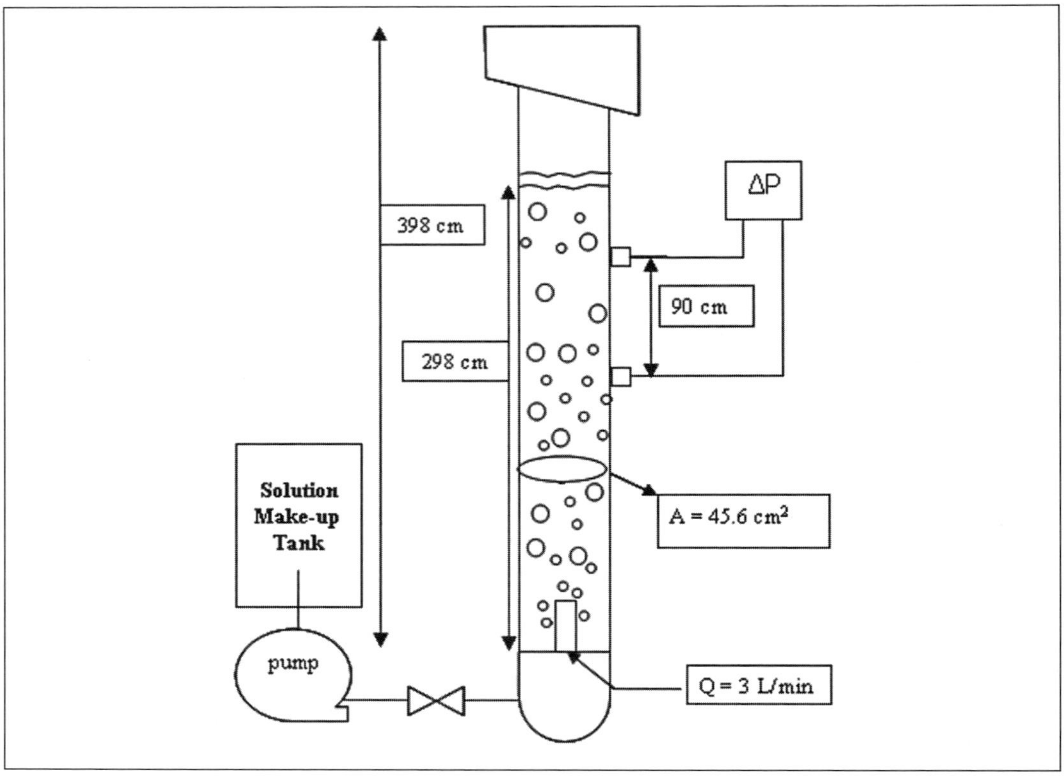

Figure 2. Laboratory bubble column

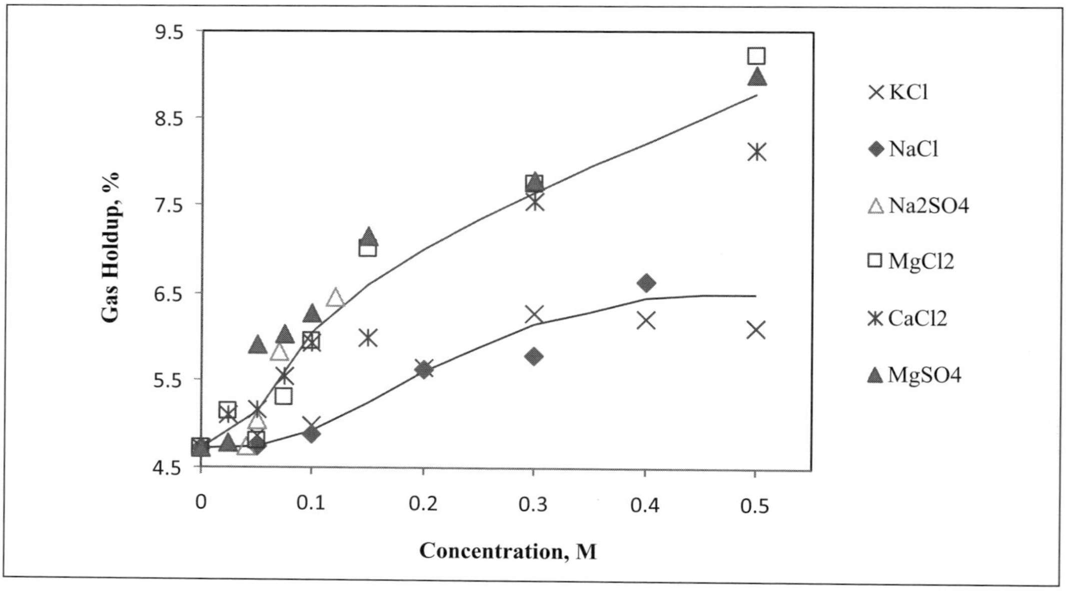

Figure 3. Gas holdup as a function of inorganic salt concentration

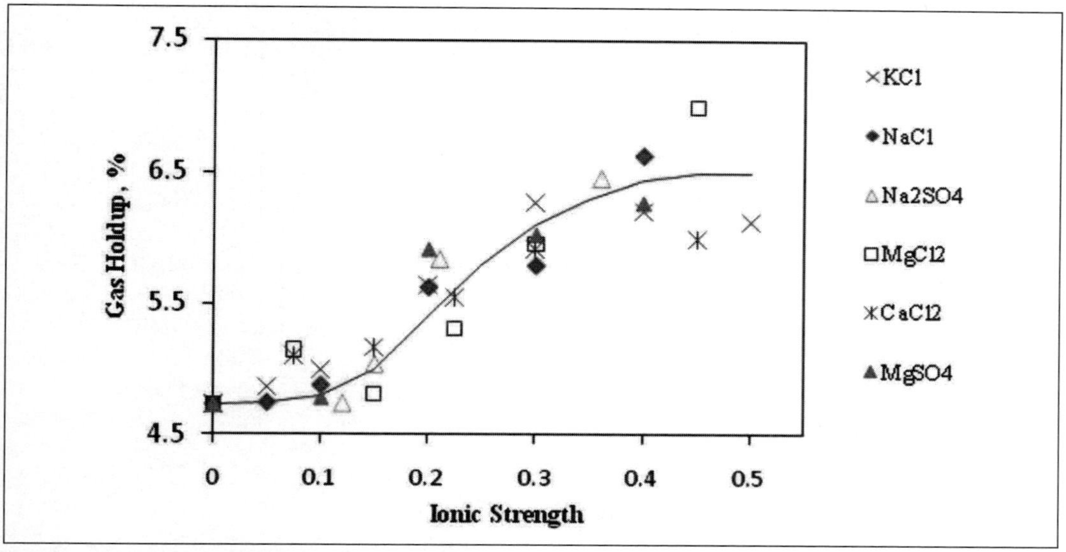

Figure 4. Gas holdup as a function of ionic strength

overlap with those used by Pugh et al. (1997). They are all coalescence inhibiting salts (Craig et al., 1993).

RESULTS AND DISCUSSION

Figure 3 shows gas holdup increased with concentration of salt, indicating a decrease in bubble size. The salts divided into two groups based on the specific salt cation-anion pair: the two 1-1 salts, KCl and NaCl, showed a moderate increase in gas holdup over the concentration range tested; and the 1-2, 2-1 and 2-2 salts had a more profound effect, increasing gas holdup at lower molar concentrations. The results are similar to those reported previously (Quinn et al. 2007).

Due to the noted valence effect, several authors have shown a correlation between gas dispersion parameters and solution ionic strength (Zieminski and Whittemore 1971; Keitel and Onken 1982; Quinn et al. 2007). Ionic strength (μ) is defined as

$$\mu = \frac{1}{2}\sum C_i Z_i^2 \qquad (2)$$

where C_i is the molar concentration of the ith species, Z_i the charge of the ith species and the summation is over all species in solution. Figure 4 shows gas holdup as a function of ionic strength. Applying the same reasoning, Figure 5 re-plots the graphite recovery data in Figure 2 for the six salts tested in the current study as a function of ionic strength. The trends in Figures 4 and 5 have similar shape and in both cases, the data appear to show a relationship independent of salt type.

Since not all salt concentrations were in common between the two studies, the gas holdup data were fitted to a function to interpolate: examples for KCl and $CaCl_2$ are shown in Figure 6.

The recovery data were then plotted against gas holdup: Figure 7 shows the recovery-gas holdup correlation for KCl and $CaCl_2$; and Figure 8 shows the relationship for all (6) salts tested (interpolated data were used to estimate gas holdup for the concentrations tested by Pugh et al. (1997) but not tested in the current gas holdup work).

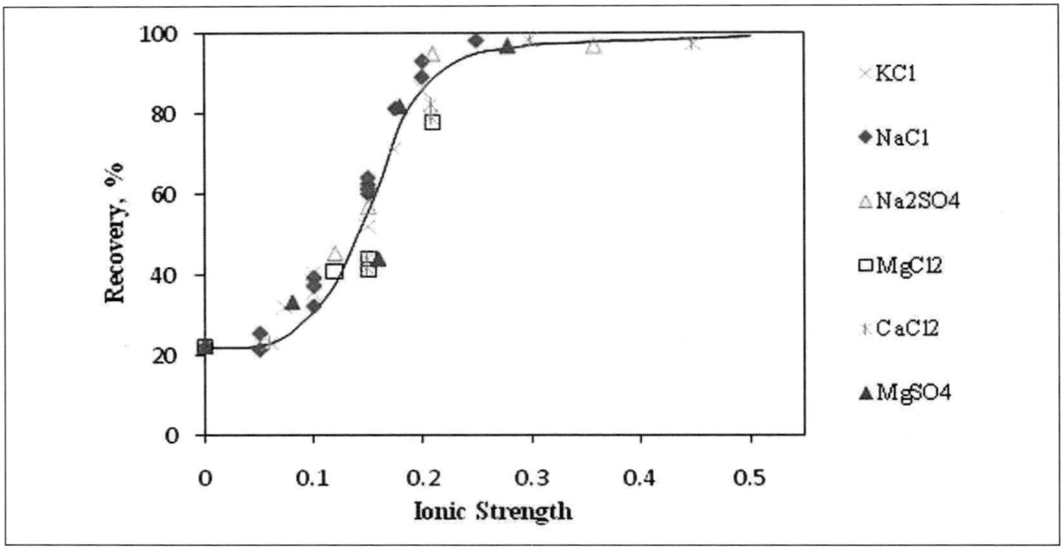

Figure 5. Graphite recovery (data from Pugh et al. 1997, Figure 2) as a function of ionic strength

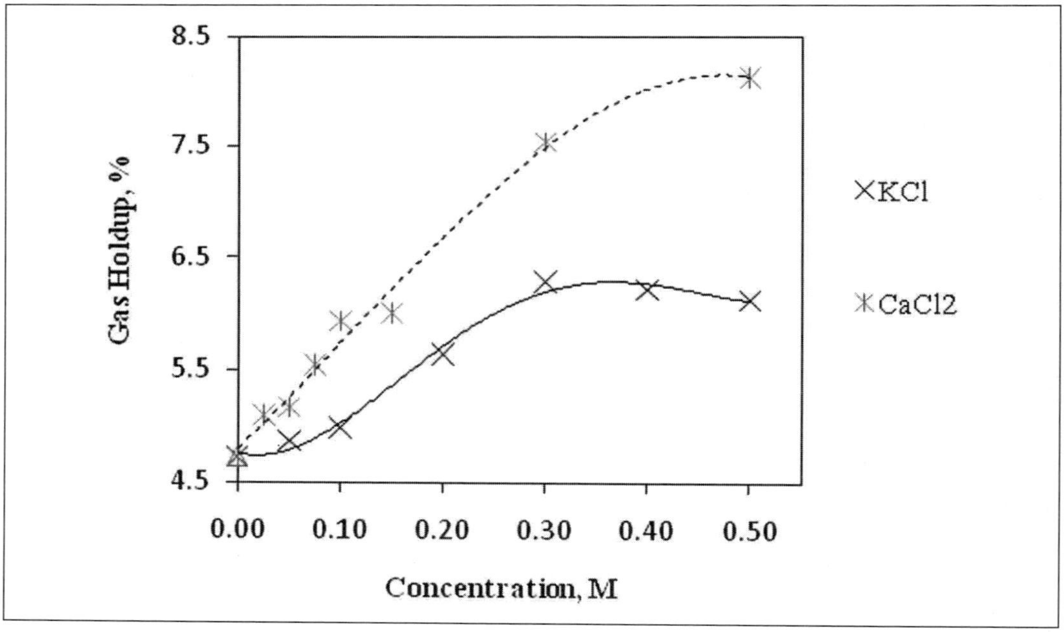

Figure 6. Model fit of gas holdup as a function of salt concentration for KCl (solid line) and $CaCl_2$ (dashed)

The individual salts support a linear trend (Figure 7) and although there is scatter when all data are considered (Figure 8), it is evident that salts that increased gas holdup resulted in higher flotation recovery. This implies that a significant component of the recovery is the impact of the salts on gas dispersion, i.e., bubble size and surface area flux as reflected in the gas holdup in the pulp zone. The observations relating to bubble size in the froth (Pugh et al.

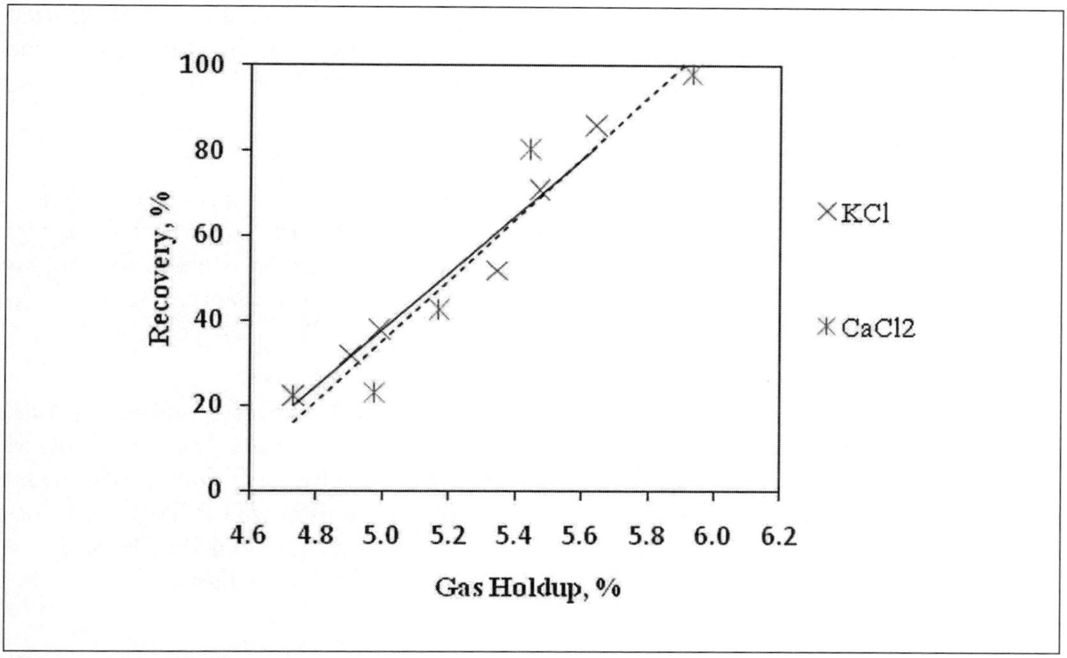

Figure 7. Graphite recovery as a function of gas holdup for KCl (solid trend line) and $CaCl_2$ (dashed trend line)

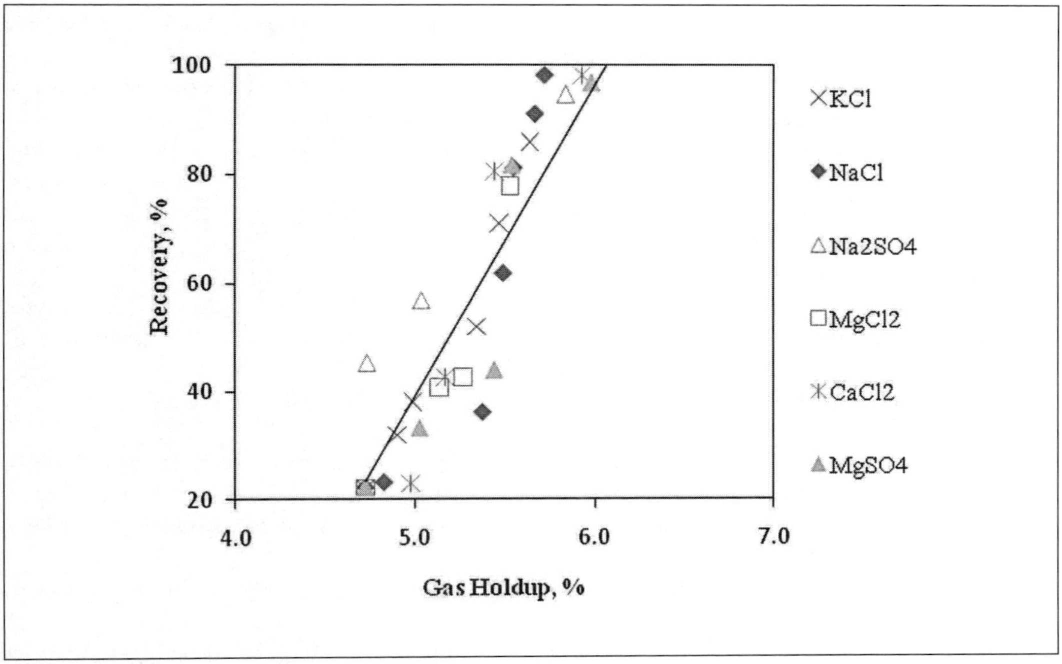

Figure 8. Recovery as a function of gas holdup for all salts tested

1997; Quinn et al. 2007; Kurniawan et al. 2011) are likely an extension of what has occurred in the pulp zone. Other effects, such as electrostatic interactions, could be explored once the impact of gas dispersion is accounted for.

CONCLUSIONS

Gas holdup as a function of concentration was measured for six inorganic salts. A dependence on ionic strength was demonstrated. The gas holdup results correlated with the graphite flotation data taken from Pugh et al. (1997). In the presence of coalescence inhibiting salts, gas dispersion appears to play a significant role in dictating flotation performance.

ACKNOWLEDGMENTS

Funding for this work was through the Chair in Mineral Processing sponsored by Vale, Teck, Xstrata Process Support, Barrick Gold, Shell Canada, SGS Lakefield Research, COREM and Flottec under the NSERC (Natural Sciences and Engineering Research Council of Canada) CRD (Collaborative Research and Development) program. We also thank R. Pugh for bringing his paper to our attention. S. Alexander is an exchange student from the University of Newcastle, Australia, and we thank Professor G. Jameson for this arrangement.

REFERENCES

Craig, V.S.J., Ninham, B.W., and Pashley, R.M. 1993. The effect of electrolytes on bubble coalescence in water. *Journal of Physical Chemistry*, 97:10192–10197.

Craig, V.S.J. 2004. Bubble coalescence and specific-ion effects. *Current Opinion in Colloid and Interface Science*, 9(1-2):178–184.

Finch, J.A., Xiao, J., Hardie, C., and Gomez, C.O. 2000. Gas dispersion properties: Bubble surface area flux and gas holdup. *Minerals Engineering*, 13(4):365–372.

Foulk, C.W., and Miller, J.N. 1931. Experimental evidence in support of the balanced-layer theory of liquid film formation. *Industrial and Engineering Chemistry*, 23(11):1283–1288.

George, C. 1996. The Mt. Keith operation. In: Grimsey, E.J., and Neuss, I. (eds.). *Australasian Inst. Min. Metall.*, Melbourne, 9–23.

Gorain, B.K., Franzidis, J.P., and Manlapig, E.V. 1997. Studies on impeller type, impeller speed and air flow rate in an industrial scale flotation cell. Part 4: Effect of bubble surface area flux on flotation performance. *Minerals Engineering*, 10(4):367–379.

Gorain, B.K., Napier-Munn, T.J., Franzidis, J.P., and Manlapig, E.V. 1998. Studies on impeller type, impeller speed and air flow rate in an industrial scale flotation cell. Part 5: Validation of k-S_b relationship and effect of froth depth. *Minerals Engineering*, 11(7):615–626.

Haig-Smillie, L.D. 1974. Sea water flotation. *Proceedings Canadian Mineral Processors Conference*, 263–281.

Hofmeier, U., Yaminsky, V.V., and Christensen, H.K. 1995. Observations of solute effects on bubble formation. *Journal of Colloid and Interface Science*, 174:199–210.

Keitel, G., and Onken, U. 1982. Inhibition of bubble coalescence by solutes in air/water dispersions. *Chemical Engineering Science*, 37(11):1635–1638.

Klassen, V.I., and Vlasova, N.S. 1967. The effect of reagents in coal flotation. *Journal of Mining Science*, 3(5):504–510.

Kurniawan, A.U., Ozdemir, O., Nguyen, A.V., Ofori, P., and Firth, B. 2011. Flotation of coal particles in $MgCl_2$, NaCl, and $NaClO_3$ solutions in the absence and presence of Dowfroth 250. *International Journal of Mineral Processing*, 98(3-4):137–144.

Laskowski, J. 1966. The flotation of naturally hydrophobic minerals in solution with a raised concentration of inorganic salts (In Polish). Technical University, Gliwice.

Laskowski, J.S., Cho, Y.S., and Ding, K. 2003. Effect of frothers on bubble size and foam stability in potash ore flotation systems. *Canadian Journal of Chemical Engineering,* 81:63–69.

Lessard, R.D., and Zieminski, S.A. 1971. Bubble coalescence and gas transfer in aqueous electrolytic solutions. *Industrial and Engineering Chemistry, Fundamentals*, 10:260–289.

Marrucci, G., and Nicodemo, L. 1967. Coalescence of gas bubbles in aqueous solutions of inorganic electrolytes. *Chemical Engineering Science,* 22:1257–1265.

Nesset, J.E., Finch, J.A., and Gomez, C.O. 2007. Operating variables affecting the bubble size in forced-air mechanical flotation machines. *Ninth Mill Operators' Conference*. Fremantle, WA.

Paulson, O., and R.J. Pugh. 1996. Flotation of inherently floatable particles in aqueous solutions of inorganic electrolytes. *Langmuir,* 12:4808–4813.

Peng, Y., and Seaman D. 2011. The flotation of slime—Fine fractions of Mt. Keith pentlandite ore in de-ionised and saline water. *Minerals Engineering,* 24(5):479–481.

Pugh, R.J., Wesissenborn, P., and Paulson, O. 1997. Flotation in inorganic electrolytes; the relationship between recover of hydrophobic particles, surface tension, bubble coalescence and gas solubility. *International Journal of Mineral Processing,* 51:125–138.

Quinn, J.J., Kracht, W., Gomez, C.O., Gagnon, C., and Finch, J.A. 2007. Comparing the effect of salts and frother (MIBC) on gas dispersion and froth properties. *Minerals Engineering,* 20:1296–1302.

Yoon, R.H., and Sabey, J.B. 1989. Coal flotation in inorganic salt solutions. Surfactant Science Series: Interfacial Phenom. *Coal Tech.* 32:87–114.

Zahradnik, J., Fialová, M., and Linek, V. 1999. The effect of surface additives on bubble coalescence in aqueous media. *Chemical Engineering Science,* 54:4757–4766.

Zieminski, S.A., and Whittemore, R.C. 1971. Behavior of gas bubbles in aqueous electrolyte solutions. *Chemical Engineering Science,* 26:509–520.

Foaming Properties of Flotation Frothers at High Electrolyte Concentrations

S. Castro
Department of Metallurgical Engineering, University of Concepción, Chile

P. Toledo
Department of Chemical Engineering, University of Concepción, Chile

J.S. Laskowski
Department of Mining Engineering, University of British Columbia, Canada

ABSTRACT

The effect of ionic strength (NaCl solutions, and sea water) on the foaming properties of the flotation frothers (MIBC and DF-250) have been investigated. It was found that the foamability of these frothers is higher in electrolyte solutions and sea water than in distilled water. These results are in a good agreement with bubble size measurements, which demonstrate that bubble coalescence is inhibited in electrolyte solutions and in sea water. The effect of MIBC and DF-250 on surface tension of aqueous solutions varying in ionic strength is found to be very different. The presence of NaCl in aqueous solutions increases surface tension of water over a broad MIBC concentrations ranges of up to 120 ppm. In the case DF-250 this is observed only in the frother concentration range of up to 1.2 ppm; at higher DF-250 concentrations the surface tension of the aqueous solutions is reduced below the values measured in the absence of NaCl. This by itself indicates much higher surface activity of DF-250 in comparison with that of MIBC. Foamability measurements for the two tested frothers carried out at varying NaCl concentrations confirm these conclusions. Increasing NaCl concentration was found to increase dynamic foamability index (DFI), this increase is much more pronounced for DF-250 than for MIBC.

INTRODUCTION

Frothers play a very important role in flotation process. Frother's foaming/frothing properties are affected by the quality of process water. In many flotation plants processing water is a concentrated electrolyte solution (Quinn et al., 2007) and the effect of inorganic electrolytes on the foaming properties of frothers became an important research topic.

The size of bubbles and foam stability are determined by bubble coalescence (Cho and Laskowski, 2002a,b; Laskowski, 2003). Flotation frothers applied in concentrated electrolyte solutions may exhibit quite different properties. In concentrated electrolyte solutions (e.g., sea water) the bubble coalescence is a particularly complex phenomenon since both surface active compound (frother) and surface inactive compound (inorganic salt) are able to stabilize bubbles against coalescence (Castro et al., 2010).

As shown by Traube (1926), the surface-active substances are displaced from their aqueous solutions by surface-inactive substances, and this additionally decreases surface tension of water. In the recent paper (Castro et al., 2011) we introduced the term *surface tension switch point* (s.t.s.p.). As Figure 1 explains, since the slope of the curve when surface tension is plotted

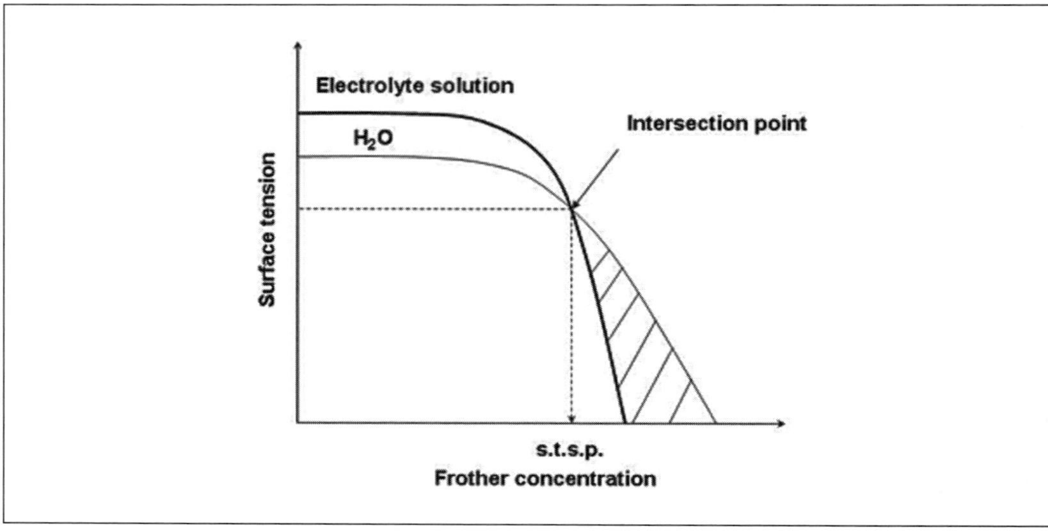

Figure 1. Surface tension switch point concept

versus frother concentration changes with electrolyte concentration, the two curves shown in Figure 1 intersect at a given frother concentration. This concentration is the *surface tension switch point*. The change of the slope of the curve surface tension = f(c) indicates that a relatively weakly surface active compound changes into a stronger surface active compound at increased electrolyte concentration.

The results reported in this paper constitute a part of the larger project aimed at utilization of sea water in flotation of Cu-Mo sulfide ores. The aim of this work is to study the effect of inorganic electrolyte concentration (NaCl) on the foaming properties of the flotation frothers (MIBC and DF-250). The dynamic foamabilty tests at different electrolyte concentrations, carried out to delineate its foaming properties, have been run in parallel with the surface tension measurements, and the examination of bubble coalescence characterized from the bubble size data.

EXPERIMENTAL

Materials

Methyl isobutyl carbinol (MIBC) was provided by Cytec-Chile; DF-250, DF-400 and DF-1012 polyglycol frothers were from Moly-Cop Chile S.A. These were technical commercial grade products. All inorganic salts employed in the project were of reagent grade (Merck).

A local sample of sea water from Concepcion city (*Bellavista-Tomé*) with salinity of 33.5‰ was employed.

Methods

KSV Sigma 700 tensiometer, with a Pt Du Noüy ring, was employed in the surface tension measurements. Bi-distilled water was used for preparation of solutions. The Wilhelmy plate method was used in several cases to cross-check the results. Low frother concentration solutions were prepared by successive dilution of a 5 ppm frother solution. All the measurements were carried out at ambient temperature and resulting pH.

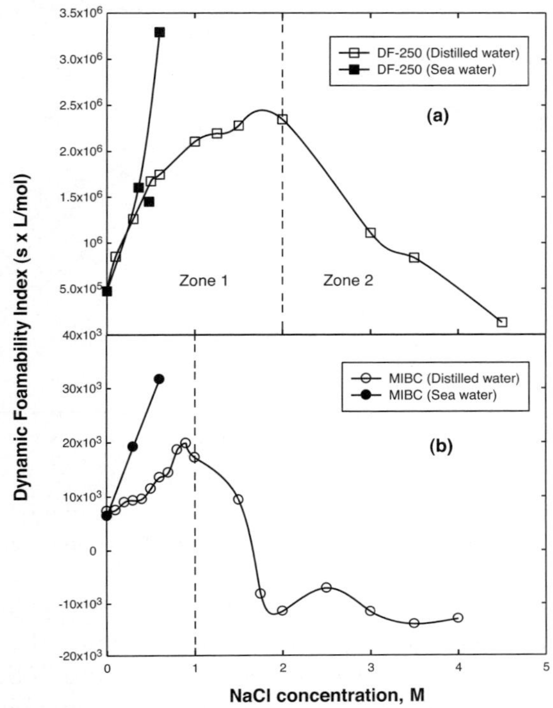

Figure 2. Effect of NaCl concentration on the dynamic foamability index of the frothers DF-250 and MIBC (Castro et al., 2010); in some tests, sea water was used

The bubble size analyser developed at the Cape Town University (UCT) was employed. A detailed description of the instrument has been published by Randall et al. (1989).

The size of bubbles was measured in a 4.5 cm inner diameter column (the same column was employed in the dynamic foamability measurements in which nitrogen gas was bubbled through the sinter glass frit either in water or in electrolyte solutions. The gas flow rate was 275 cm^3/min.

Determination of the Dynamic Foamability Index (DFI) requires measurement of gas retention time in a column (in which gas is bubbled through sintered disc) as a function of gas flow rate and frother concentration (Malysa et al., 1981; Czarnecki et al., 1982). What is measured is the total height of the solution plus the foam phase. These tests were carried out in a glass column of 101 cm height and 4.5 cm inner diameter, with the pore size range being from 40 to 100 µm. 500 mL of solution was used in each experiment (Castro et al., 2010).

RESULTS

Dynamic Foamability Index Measurements

To characterize the effect of electrolyte concentration on foamability of a given frother the DFI was determined in NaCl solutions and sea water. These measurements are based on the bubble retention time (rt), which is defined as the average time necessary for an average bubble to pass trough the system consisting of solution and foam (Czarnecki et al., 1982; Wantke et al., 1994). The results of the Dynamic Foamability Index measurements are given in Figure 2.

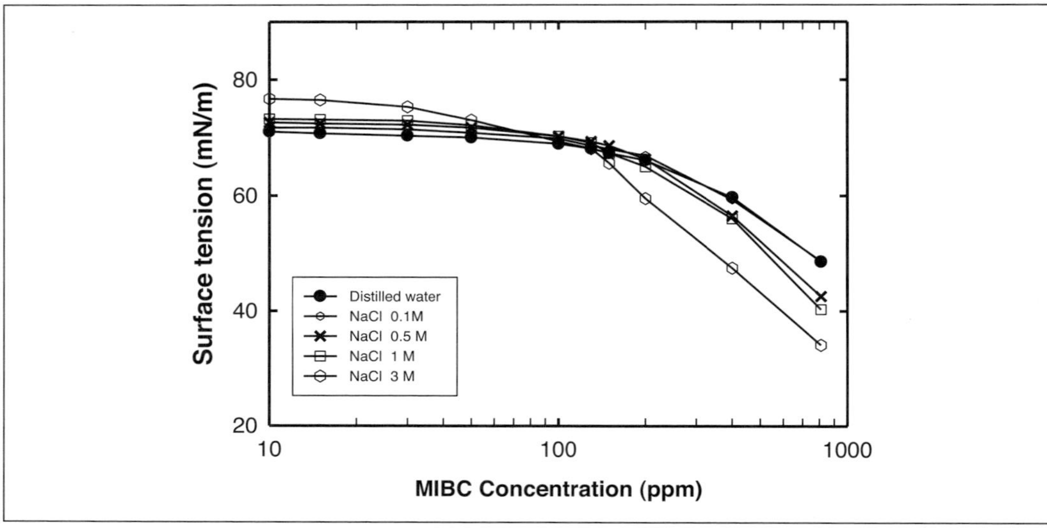

Figure 3. Effect of MIBC concentration on surface tension of aqueous solutions of NaCl

As Figure 2 demonstrates, the foamability of DF-250 is higher from the foamability of MIBC solutions, both in distilled water and sea water. The figure indicates that NaCl concentration has a pronounced effect on the DFI values for both tested frothers. In both cases, the foamability increases up to a maximum point, which is situated around 2M NaCl for DF-250 and 1M NaCl for MIBC. In Zone 1, which is on the left side from that maximum, the DFI values increase quite rapidly with NaCl concentration. In Zone 2, which is on the right side of the maximum, a reverse effect is observed. Since the salinity of sea water is below 0.6M NaCl, it can be concluded that the dynamic foamability for the two common flotation frothers is much higher in sea water than in distilled water.

Surface Tension Measurements

Figure 3 shows the influence of NaCl concentration on surface tension of MIBC aqueous solutions. As this figure indicates for MIBC the surface tension switch point (s.t.s.p.) is situated around 120 ppm. This point for DF-250 is situated around concentration of 1.2 ppm (Figure 4).

As Figure 5 demonstrates, for all tested polyglycol frothers the s.t.s.p's are situated below the concentration of 2 ppm.

Bubble Size Measurements

The bubble size measurements are shown in Figure 6. In accordance with literature, in distilled water the critical coalescence concentration (CCC) values for DF-250 are much smaller (0.042 mmol/L) than for MIBC (0.089 mmol/L). If CCC is determined in the same way for NaCl its value is around 0.78 M. As this figure indicates, the bubbles generated in NaCl solutions at $c > CCC$ are smaller from those generated in either MIBC or DF-250 solutions.

The bubble size measurements carried out in NaCl solutions as a function of frother concentration reveal entirely different trends from those tests which were carried out in NaCl solutions (Figures 7 and 8).

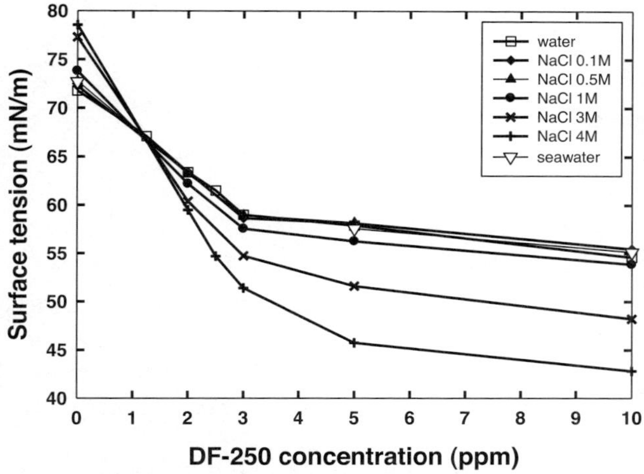

Figure 4. Effect of DF-250 concentration on surface tension of aqueous solutions of NaCl

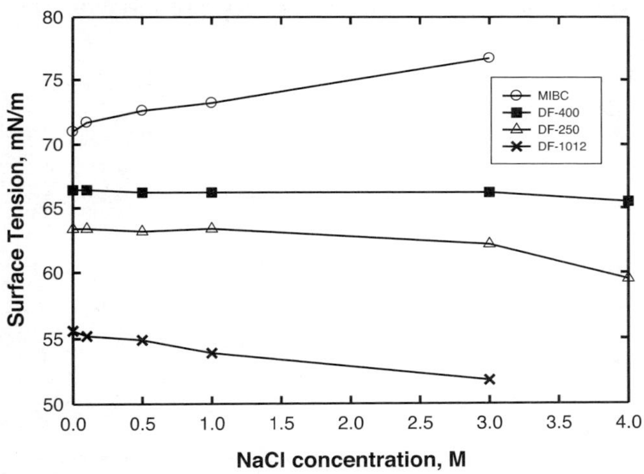

Figure 5. Effect of NaCl concentration on surface tension for the solutions of polyglycol frothers (DF-250, DF-400, DF-1012) and for MIBC. The curves shown for the polyglycols are for 2 ppm solutions; for MIBC it is for 10 ppm solution.

The results of bubble size measurements in sea water are given in Figure 9. These results indicate that the size of bubbles generated in sea water in the presence of MIBC is finer from the bubbles generated when polyglycols are utilized.

DISCUSSION

On the basis of DFI measurements, it can be concluded that the foamability of DF-250 and MIBC is higher in electrolyte solutions and sea water than in distilled water. These results are in a good agreement with direct foam layer thickness determined in a modified laboratory flotation cell (Castro et al., 2012). These results are also supported by bubble size

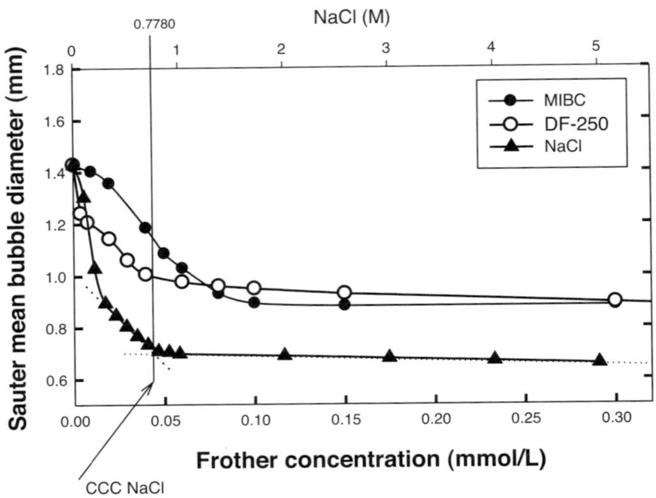

Figure 6. Bubble size as a function of frother concentration for MIBC and DF-250 (in distilled water); and as a function of NaCl concentration in a separated solution

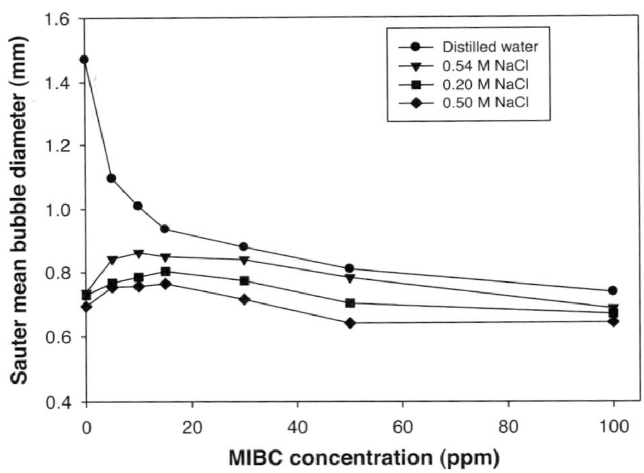

Figure 7. Bubble size as a function of MIBC concentration in distilled water and in NaCl solutions

measurements, which demonstrate that bubble coalescence is inhibited in electrolyte solutions and in sea water.

The surface tension measurements in distilled water indicate that polyglycol frothers are more surface active than MIBC. The term "surface activity" indicates the preference of such compounds to accumulate at liquid/gas interface. Hence, this preference is much larger for DF-250 than for MIBC. The surface tension measurements in NaCl solutions reveal that at as low concentration of DF-250 as 1.2 ppm (s.t.s.p.), the presence of the frother dominates over the effect of NaCl. For MIBC this is observed only at MIBC concentrations that are 100 times higher (120 ppm). Over the concentration range from 0 to 120 ppm (MIBC), the surface tension curve is dominated by the surface-inactive compound (NaCl) and only at the

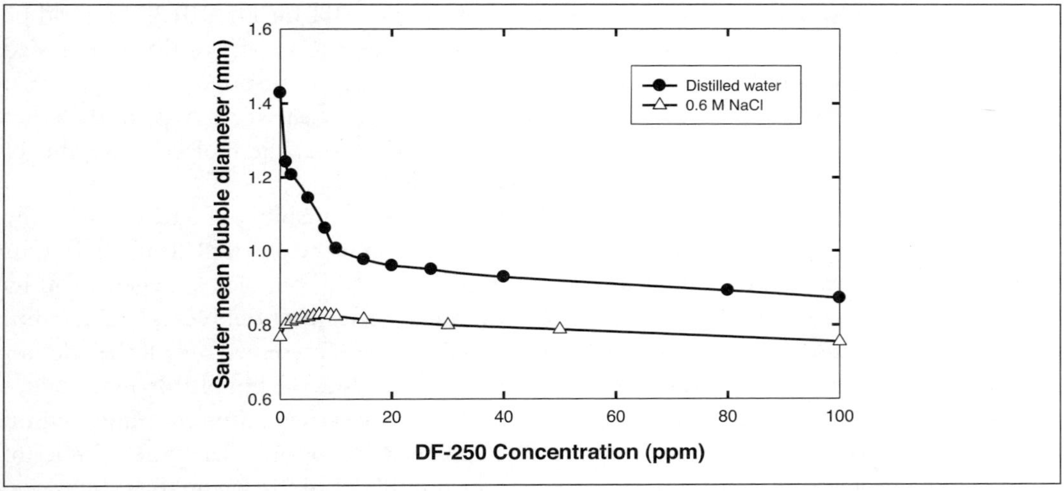

Figure 8. Bubble size as a function of DF-250 concentration in distilled water and in NaCl solutions

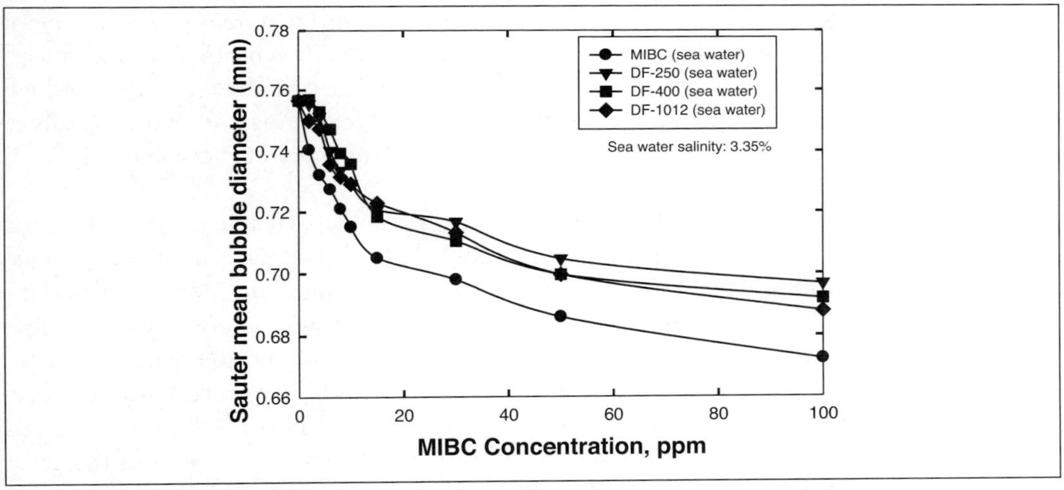

Figure 9. Bubble size as a function of concentration for the tested frothers in sea water

MIBC concentration higher than the s.t.s.p. the plot starts resembling the situation in which a surface-active compound is present in the system.

The results of surface tension measurements correlate rather well with the foamability tests (Figure 2). The foamability of DF-250 in the environment of sea water is much better than that of MIBC. The surface tension measurements indicate that the accumulation of the frother at the interface increases with increasing electrolyte concentration. This initially improves foaming. The adsorbed frother bonds some water molecules by hydrogen bonding but, apparently, the amount of such water in a bubble "hydration shell" is reduced at higher electrolyte concentrations. This seems to be explaining the observed reduced foaming when electrolyte concentration is very high.

In this paper the samples of commercial frothers were utilized. Nominal formulas of the tested frothers are as follows: $CH_3CH(CH_3)CH_2CH(OH)CH_3$ for MIBC, $CH_3(OC_3H_6)_4OH$

for DF-250, $H(PO)_{6.5}OH$ for DF-400 and $CH_3(OC_3H_6)_{6..3}OH$ for DF-1012. As could be expected, the surface activity of DF-1012 is much higher than the activity of the other tested frothers. The presence of H in DF-400 instead of CH_3 group (as in the other two polyglycols) makes the surface activity of this compound much lower from the surface activity of the other two tested polyglycols (Figure 5). It is interesting to point out that the molecular weights of DF-1012 and DF-400 frothers are practically identical (400).

In spite of much higher foamability of DF-250 from the foamability of MIBC, especially in sea water, finer bubbles are produced in sea water in the presence of MIBC (Figure 9). Our bubble size measurements were carried out in a column in which bubbles were generated by passing nitrogen under pressure through a glass frit. Since a stream of bubbles is generated in such a device the bubbles collide with each other and their size is preserved only if they do not coalesce. Only at frother concentrations exceeding the CCC values the bubbles do not coalesce and assume uniform size (Figure 6). Our foamability tests were carried out in the same column which was used for bubble size measurements. The foam starts to form when the number of bubbles arriving at the liquid/gas interface exceeds the number of rupturing bubbles. However, this coalescence process now takes place in the foam and the properties and stability of the liquid films separating bubbles in the foam are very different from the "films" separating colliding bubbles in aqueous solution. The experimental results indicate that while the bubble coalescence characterized by bubble size measurements were not very sensitive to a type of frother, the results of the foamability tests turned out to be very different. In comparison with MIBC, the foamability is obviously better when DF-250 frother is utilized, especially at high salt concentrations. This confirms importance of dynamic effects in foam formation (Malysa, 1992; Malysa and Lunknheimer, 2008) and possibly different types of the stabilizing forces which dominate under studied conditions.

Surface elasticity forces are induced and operate only under dynamic conditions. It is a "self-healing" capacity of the film against external disturbances. The elasticity effects are possible only if a layer with different properties is formed at the interface, the layer in which there is a gradient of concentration. In the case of surface-active agents this is very easy to visualize: such molecules accumulate at the liquid/gas interface (adsorption) and the gradient of concentrations at the interface is clearly established. The movement or any other dynamic disturbance that affects this gradient will immediately initiate the action of the "self-healing" forces. These are Marangoni dynamic elasticity effects. In the case of aqueous solutions of inorganic electrolytes the concentration of ions is lower in the surface layer than in the bulk (increasing surface tension), and the concentration gradient within this surface layer is also established. The elasticity effects should then also be possible. However, if elasticity forces are responsible for stabilization of a dynamic system in which bubbles collide with each other, the stability of the foam is rather determined by other phenomena.

The bubble size measurements plotted versus frother concentration provide the value of CCC which is obtained graphically from such a plot as the point of intersection of the linear segments of the experimental curves. The CCC values, determined in that way from Figure 6, are 0.042 mmol/L for DF-250 (11.1 ppm) and 0.088 mmol/L for MIBC (9 ppm). This value for NaCl is 0.78 M. Quinn et al. (2007) showed that gas dispersion in 0.4 M NaCl and 10 ppm MIBC solutions are quite similar.

The two cases: frother in distilled water that is the system with a surface-active compound, and electrolyte solution that is the system with surface-inactive compound, are easy to visualize. The system in which both these compounds are present is much more complicated. As Figures 7 and 8 demonstrate, while frother is needed to reduce bubble size in distilled water, in electrolyte solutions the bubbles are stabilized against coalescence even in the absence of the frother. This effect has been well studied (Marrucci and Nicodemo, 1967; Lessard and Zieminski, 1971; Zieminski and Whittemore, 1971; Craig et al., 1993; Laskowski et al., 2003). The curves in the region where the frother concentration is lower than the CCC show the effect of frother on bubble coalescence. What is interesting here is that the bubble size in the region where the frother concentration exceeds the CCC may still be finer in the mixed systems with both frother and electrolyte present (Figures 7 and 8). As these figures demonstrate, introduction of a frother to the electrolyte solution initially leads to some disturbances and bubble size initially seems to slightly increase when the forther is added. However, with further increase in frother concentration even finer bubbles are produced. This size of the bubbles is produced by the sparging mechanism and the results indicate the effect of frother and electrolyte on bubble breaking-up (Grau and Laskowski, 2006). Kracht and Finch (2009) discussed such effects further.

CONCLUSIONS

The properties of foams in NaCl solutions generated in the presence of common flotation frothers under dynamic conditions may provide valuable information for copper sulfide flotation in sea water. The dynamic foamability of MIBC and DF-250 is much higher in NaCl solutions and sea water than in distilled water.

The surface tension of a solution of surface active agents and an inorganic electrolyte is characterize by the *surface tension switch point (s.t.s.p.)*, which indicates the frother concentration at which the forther changes into a stronger surface active compound in electrolyte solutions, compared with its behaviour in distilled water. This parameter is much higher for MIBC (120 ppm) than DF-250 (1.2 ppm).

Bubble size measurements in NaCl solutions and sea water indicate that electrolytes are able to reduce bubble size by itself, without any surface active agent. In distilled water, the CCC values are slightly lower for DF-250 than MIBC (11 ppm for DF-250 and 9 ppm for MIBC), but at the concentration higher than CCC the bubble size reaches similar values for both frothers. However, when bubble size is measured in a mixed solution of NaCl and a frother, the resulting bubble size is finer than that of the same frother in distilled water.

ACKNOWLEDGMENTS

Funding for this project was provided by the CORFO-INNOVA CHILE (Project 08CM01-18), with industrial sponsorship from: Antofagasta Minerals, (Minera *Esperanza*); BHP Billiton (*Minera Escondida* Ltd.); Anglo American Chile (*Mantos Blancos*); and Teck (*Carmen de Andacollo*), through AMIRA-International operating Research Grant P968.

REFERENCES

Castro, S., Venegas, I., Landero, A., and Laskowski, J.S. 2010. Frothing in seawater flotation systems. *Proc. 25th Int. Mineral Processing Congress*, Brisbane, pp. 4039–4047.

Castro, S., Miranda, C., Toledo, P., and Laskowski, J.S. 2011. Effect of frothers on bubble coalescence and foaming in electrolyte solutions and seawater. (Forthcoming).

Castro, S., Ramos, O., Cancino, J.P., and Laskowski, J.S. 2012. Frothing in the flotation of copper sulfide ores in sea water (*This symposium*).

Craig, V.S.J., Ninham, B.W., and Pashley, R.M. 1993. The effect of electrolytes on bubble coalescence in water. *J. Phys. Chem.*, 97(39):10192–10197.

Cho, Y.S., and Laskowski, J.S. 2002a. Effect of flotation frothers on bubble size and foam stability. *Int. J. Miner. Process.*, (64):69–80.

Cho, Y.S., and Laskowski, J.S. 2002b. Bubble coalescence and its effect on dynamic foam stability. *Can. J. Chem. Eng.*, (80):299–305.

Grau, R.A., and Laskowski, J.S. 2006. Role of frothers in bubble generation and coalescence in a mechanical flotation cell. *Can. J. Chem. Eng.*, 84:170–182.

Czarnecki, J., Malysa, K., and Pomianowski, J. 1982. Dynamic frothability index. *J. Coll. Interface Sci.*, 86(2):570–572.

Kracht, W., and Finch, J.A. 2009. Bubble break-up and the role of frother and salt. *Int. J. Miner. Process.*, 92:153–161.

Laskowski, J.S. 2003. Fundamental properties of flotation frothers. *Proc. 22nd Int. Mineral Processing Congress*, Cape Town, 2:788–797.

Laskowski, J.S., Cho, Y.S., and Ding, K. 2003. Effect of frothers on bubble size and foam stability in potash ore flotation systems, *Can. J. Chem. Eng.*, 81:63–69.

Lessard, R.R., and Zieminski, S.A. 1971. Bubble coalescence and gas transfer in aqueous electrolytic solutions. *Ind. Eng. Chem. Fundam.*, 10(2):260–269.

Małysa, K., Cohen, R., Exerowa, D., and Pomianowski, A. 1981. Steady-state foaming and the properties of thin liquid films from aqueous alcohol solutions. *J. Coll. Interface Sci.*, 80:1–6.

Malysa, K. 1992. Wet foams: Formation, properties and mechanism of stability. *Adv. Coll. Interface Sci.*, 40:37–83.

Malysa, K., and Lunkenheimer, K. 2008. Foams under dynamic conditions. *Curr. Opin. Coll. Interface Sci.*, 13:150–162.

Marrucci, G., and Nicodemo, L. 1967. Coalescence of gas bubbles in aqueous solutions of inorganic electrolytes. *Chem. Eng. Sci.*, 22:1257–1265.

Quinn, J.J., Kracht, W., Gomez, C.O., Gagnon, C., and Finch, J.A. 2007. Comparing the effect of salts and frother (MIBC) on gas dispersion and froth properties. *Miner. Eng.*, 20:1296–1302.

Randall, E.W., Goodall, C.M., Fairlamb, P.M., Dold, P.L., and O'Connor, C.T. 1989. A method for measuring the sizes of bubbles in two- and three-phase systems. *J. Phys. E: Sci. Instrum.*, (22): 827–833.

Traube, J. 1926. Attraction intensity or attraction pressure. *Colloid Chemistry. Edited by J. Alexander. Chemical Catalog Co.*, pp. 640–646.

Wantke, K., Malysa, K., and Lunkenheimer, K. 1994. A relation between dynamic foam stability and surface elasticity. *Coll. Surf. A*, 82:183–191.

Zieminski, S.A., and Whittemore, R.C. 1971. Behavior of gas bubbles in aqueous electrolyte solutions. *Chem. Eng. Sci.*, 26:509–520.

Role of Saline Water in the Selective Flotation of Fine Particles

Yongjun Peng and Shengli Zhao
School of Chemical Engineering, University of Queensland, St. Lucia, Brisbane, Australia

Dee Bradshaw
Julius Kruttschnitt Mineral Research Centre, University of Queensland, Indooroopilly, Brisbane, Australia

ABSTRACT

The beneficial effect of electrolytes on mineral flotation has been observed and attributed to the improved bubble-particle attachment in a number of studies. However, the behavior of gangue minerals in electrolyte solutions has not been studied. In fine particle flotation, gangue minerals play an important role in determining both the recovery and grade of valuable minerals as results of slime coating and high mechanic entrainment. In this study, the selective flotation of fine particles from a nickel ore deposit and a secondary copper ore deposit is studied in both fresh water, and bore water having high ionic strength. Flotation results from the two ore systems show the same pattern. In fresh water, slime coats on valuable mineral particles resulting in the low flotation recovery of the valuable minerals. At the same time the mechanic gangue entrainment is high due to the fine particle size. However, in bore water, slime coating is mitigated and the entrainment of gangue minerals is reduced simultaneously, resulting in the improved flotation separation. Mechanisms responsible for the improvement in electrolyte solutions are investigated as well. It seems that the screening of electrical double layer forces is the main contributor.

INTRODUCTION

Saline water is an important part of the Australian mining industry. In Western Australia, bore water with high ionic strength has to be used on mine sites for production, site rehabilitation, and downstream processing because fresh water is not available locally. In Queensland, most mine sites have adopted water re-use as a means for making freshwater savings. However, water re-use results in increased salinity in site water stores, which is driven largely by evaporation and ongoing salt inputs from spoil, minerals and groundwater. Flotation relies on a large amount of water and therefore the impact of saline water on flotation performance has gained more and more attention. Despite a number of studies, the role saline water plays in flotation is still not clear. In the past years, there has been debate over whether fresh water should be introduced to flotation plants since fresh water is normally used and almost all flotation chemicals are designed for use in fresh water.

The beneficial effect of saline water on coal flotation has been observed and a number of studies have been conducted to investigate the mechanisms. In general, the improved coal flotation in saline water compared to fresh water has been attributed to the increased bubble-particle attachment. Yoon (1982) and Paulson and Pugh (1996) proposed that reduced bubble sizes and increased population in electrolyte solutions increased the encounter efficiency of bubble-particle attachment. Fuerstenau et al. (1983) and Yoon and Sabey (1989) attributed the increased bubble-particle attachment to the reduction of zeta potential of both bubbles and particles resulting from the compression of electrical double-layer in the presence of electrolytes.

Another mechanism proposed is that the inorganic electrolytes destabilize the hydrated layers surrounding coal particles and reduce their surface hydration therefore enhancing the bubble-particle attachment (Klassen and Mokrousov, 1963). Nanobubbles or nanopancakes were also observed to form on coal surfaces in electrolyte solutions facilitating bubble–particle attachment (Mishchuk, 2005; Zhang and Ducker, 2007).

Studies in coal flotation with saline water have not taken into account the behavior of gangue minerals. In fine particle flotation gangue minerals play an important role. It has been established that fine gangue minerals may coat on the surfaces of valuable minerals preventing the adsorption of collectors and then depressing the flotation. The electrostatic interaction between gangue and valuable minerals has attributed to slime coating. The typical example is the flotation of pentlandite in the presence of serpentine in fresh water. In neutral and weakly alkaline solutions, pentlandite is negatively charged, but serpentine is positively charged, resulting in an electrostatic interaction between them (Edwards et al., 1980; Bremmell et al., 2005). The coating of serpentine minerals on pentlandite surfaces explains the poor flotation behavior of pentlandite (Edwards et al., 1980; Bremmell et al., 2005). The depressed flotation of coal by clay minerals, and sphalerite by fine silica minerals has also been contributed to slime coating (Arnold and Aplan, 1986; Duarte and Grano, 2007). In fact, both sphalerite and silica may be negatively charged in neutral and weakly alkaline solutions, but the dissolved zinc ions, added copper ions and collector molecules may increase particle interactions between silica and sphalerite (DiFeo, and Finch, 2002; Duarte and Grano, 2007).

High gangue entrainment is another important issue in fine particle flotation resulting in the low quality of flotation concentrate. The collection of gangue minerals into the flotation concentrate is through the water layers in the laminar borders between gas bubbles. The smaller the particles, the higher the entrainment (Trahar and Warren, 1976; Warren, 1985; Ross and Van Deventer, 1988). Liu et al. (2006) and Gong et al. (2010) demonstrated that the enlargement of the size of fine gangue particles significantly reduced their entrainment in flotation.

In this study, the effect of saline water on the behavior of gangue minerals in fine particle flotation in terms of slime coating and entrainment was investigated by using two types of ores, a nickel ore and a secondary copper ore.

EXPERIMENTAL

Materials and Reagents

The nickel ore or copper ore sample was crushed to a size of –2.36 mm before grinding and flotation. The nickel grade of the nickel ore is about 0.6%. X-Ray Diffraction (XRD) analysis indicates that the nickel ore contains about 2% pentlandite and 98% serpentine, predominantly by lizardite. The copper grade of the copper ore is about 0.8%. As indicated by the XRD analysis, the copper ore contains 0.4% chalcopyrite, 0.7% chalcocite, 2.8% pyrite, 1.9% goethite, 66.1% quartz, 22.8% dolomite and 5.2% bentonite. About 85% Cu originates from chalcocite and 15% Cu from chalcopyrite. The Cu from chalcocite is cyanide soluble, while the Cu from chalcopyrite is cyanide insoluble (Sheffel, 2002). Pentlandite, lizardite, chalcopyrite, chalcocite, quartz and bentonite single minerals were obtained from Ward's Natural Science Establishment (US). All of them have 98% purity analyzed by XRD.

The chemical analysis of the bore water is shown in Table 1. The conductivity and TDS (total dissolved solids) are about 12.8 mS/cm and 77.1 g/L, respectively. Sodium ethyl xanthate

Table 1. Composition of the bore water (mg/L)

Na	K	Ca	Mg	SiO_2	Cd	Cr	Cu	Ni	Cl	SO_4
20000	940	400	5100	4.9	0.002	0.001	<0.005	0.033	32000	23000

and H407 (polypropylene glycol ether blend), industrial grade, were used as collector and frother, respectively in the nickel flotation. Potassium amyl xanthate (PAX) and IF56, industrial grade, were used as collector and frother, respectively in the copper flotation.

Grinding and Flotation

A 1-kg crushed nickel ore sample was ground in a laboratory stainless steel rod mill at 40% solids to obtain 80% particles passing 125 μm to match the nickel distribution produced in the plant. The mill discharge was then transferred and run through a 1" Mozley cyclone. The overflow stream with about 400 g solids was filtered and then floated. The P80 and P50 of the stream are 25 μm, and 8 μm, respectively, measured by Malvern Mastersizer. The nickel grade of the cyclone overflow stream was about 0.43%. Flotation was carried out in a 2.5 L Agitair flotation cell using an agitation speed of 750 rpm. The solid density in the flotation cell was 15%. During the conditioning, collector (150 g/t) and frother (30 g/t) were added. Flotation was performed for 32 min and 6 concentrates were collected after cumulative times of 1, 2, 4, 8, 16 and 32 min. Both de-ionized and bore water were tested in this study. The chosen water type was utilized in all stages of grinding, cycloning and flotation. The pH during grinding and flotation was constant, about 8.8 due to the buffering effect of the ore.

A similar grinding and flotation procedure applied to the copper ore sample. A 1-kg ore sample was ground at 40% solids until 80 wt.% of the particles present were less than 38 μm in diameter. The P50 of the mill discharge was 12 μm measured by Malvern Mastersizer. The mill discharge was transferred to the flotation cell and conditioned with 100 g/t collector and 20 g/t frother. Lime was used to control pH 9.0 at the end of grinding and during flotation. In flotation, four concentrates were collected after cumulative times of 0.5, 2.0, 4.0 and 8.0 min.

Zeta Potential Measurement

Particle electrophoretic mobility was measured with a Rank Brothers Microelectrophoresis Mark II apparatus. Single mineral particles were ground in a ceramic mill, screened to obtain –38 μm and then conditioned in a 1.25 dm^3 reaction vessel with de-ionized water containing 10^{-3} M KCl solution or mixed with bore water at pH 9.0 for 30 minutes. The pH was maintained by the addition of sodium hydroxide solutions. 10 mobility measurements at each of the two stationary planes were performed at each pH. The average mobility was converted to zeta potentials using the Smoluchowski equation.

Settling Test

The flotation feed was conditioned in the flotation cell and then transferred to a 2.5 L graduated cylinder. De-ionised or bore water was added to bring the water level to 2.5 L. The cylinder was then stoppered and inverted 10 times to further mix the slurry, and then set still in the upright position. The bed volume was monitored as a function of time. The settling rate is an indication of pulp stability or the extent of mineral aggregation. The quicker the bed settles, the lower the pulp stability and the stronger the mineral aggregation.

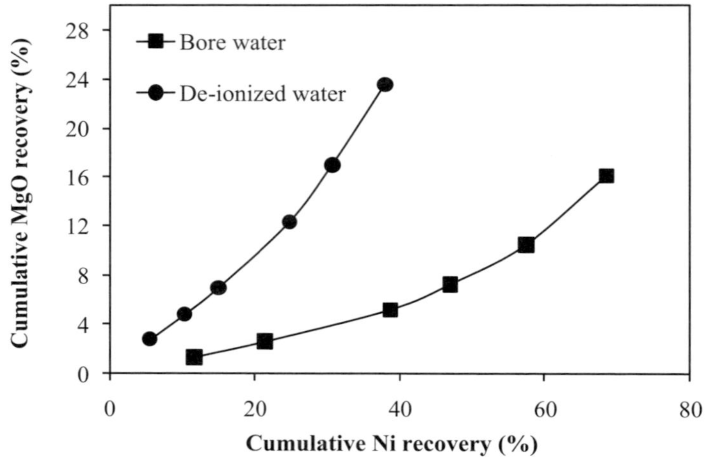

Figure 1. Ni recovery as a function of MgO recovery from the nickel ore flotation in de-ionized water and bore water

RESULTS

Flotation

Figure 1 shows the effect of water types on Ni and MgO recovery in the flotation of the nickel ore. In bore water, Ni recovery was very low, about 38% in the completion of 32 min of flotation. This is consistent with other researchers demonstrating poor pentlandite flotation in the presence of serpentine minerals (Edwards et al., 1980; Bremmell et al., 2005). The electrostatic interaction between pentlandite and serpentine minerals results in slime coating (Edwards et al., 1980; Bremmell et al., 2005). Meanwhile, MgO recovery was high in de-ionized water, reaching 24% in the completion of 32 min of flotation. This is also in good agreement with high gangue entrainment in fine particle flotation observed by other researchers (Trahar and Warren, 1976; Warren, 1985; Ross and Van Deventer, 1988). However, in bore water, Ni recovery was increased significantly, while MgO recovery was decreased significantly. In the end of flotation, Ni and MgO recoveries were 70% and 16%, respectively. It seems that the adverse effect of serpentine on pentlandite flotation was mitigated in saline water, while the entrainment of fine gangue minerals was reduced.

Figure 2 shows MgO and Ni recoveries as a function of water recovery in de-ionized and bore water. In both de-ionized and bore water, a linear relationship existed between MgO recovery and water recovery, indicating that the collection of serpentine gangue minerals to the concentrates was through water entrainment (Trahar and Warren, 1976; Warren, 1985; Ross and Van Deventer, 1988). The relationship between Ni recovery and water recovery deviated from the MgO-water recovery line slightly in de-ionized water but remarkably in bore water. A small difference occurred between Ni and MgO recoveries versus water recovery in de-ionized water indicating a low level of true flotation of pentlandite. However, in bore water, the true flotation of pentlandite was the predominant mechanism since the difference between Ni and MgO recoveries was significant. The different levels of true flotation in de-ionized and bore water may also be examined by nickel enrichment. In de-ionized water, nickel grade was upgraded from 0.43% in the flotation feed (the cyclone overflow fraction) to 0.68% in the

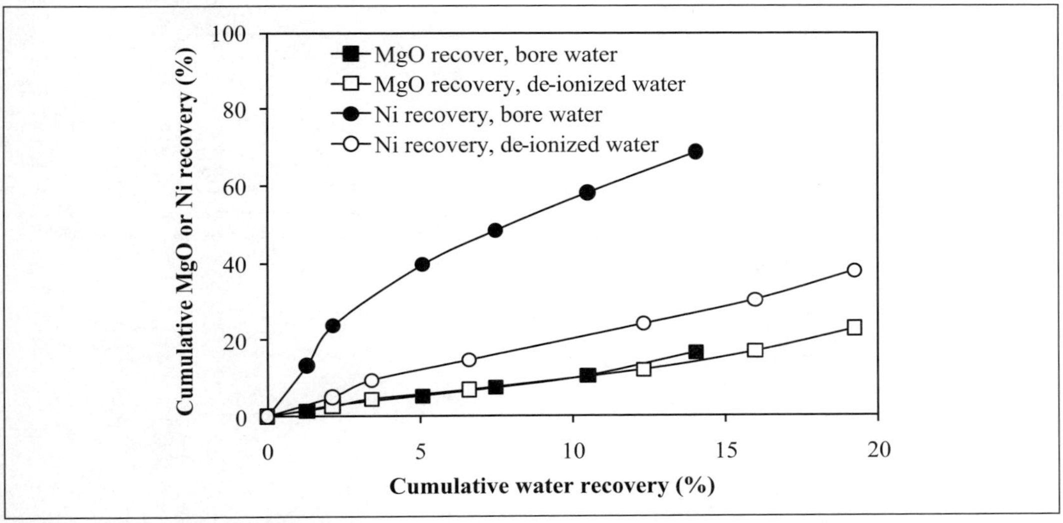

Figure 2. MgO or Ni recovery as a function of water recovery from the nickel ore flotation in de-ionized water and bore water

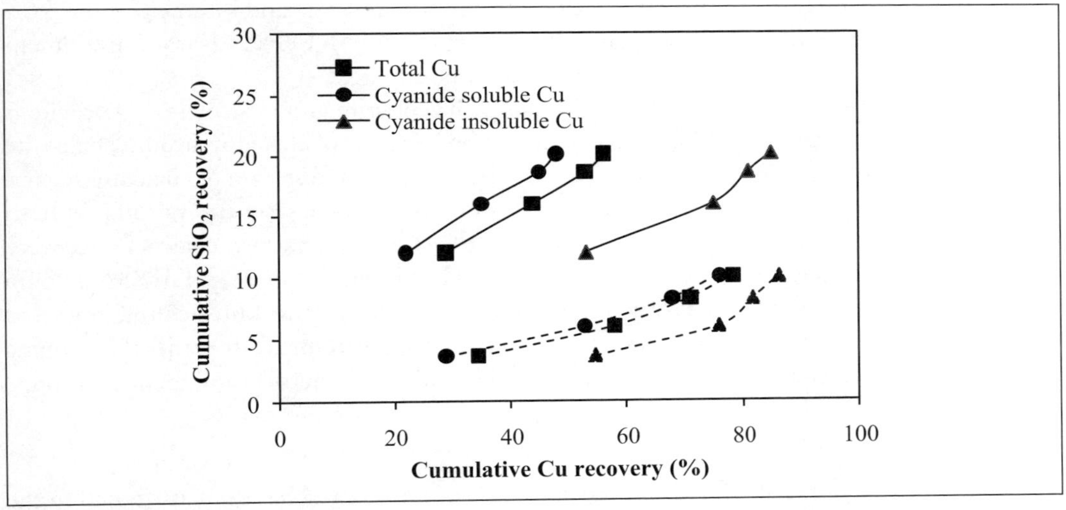

Figure 3. Cu recovery as a function of SiO_2 recovery from the Cu ore flotation in de-ionized water (solid lines) and bore water (dashed lines)

concentrate, while in bore water, nickel grade was upgraded to 1.20%. Apparently, electrolytes in the bore water increased the true flotation of pentlandite.

A similar trend was observed in the flotation of the copper ore. As shown in Figure 3, the total Cu recovery was increased from 56% in de-ionized water to 78% in bore water in the completion of 8 min of flotation, while SiO_2 (from quartz and clay minerals) recovery was decreased from 20% to 10%. The response of both valuable and gangue minerals in nickel and copper ores to the water type is exactly the same. It seems that the gangue minerals in the copper ore may coat on copper mineral surfaces resulting in the low Cu recovery in de-ionized water. This coating may be mitigated in bore water resulting in the improved copper flotation.

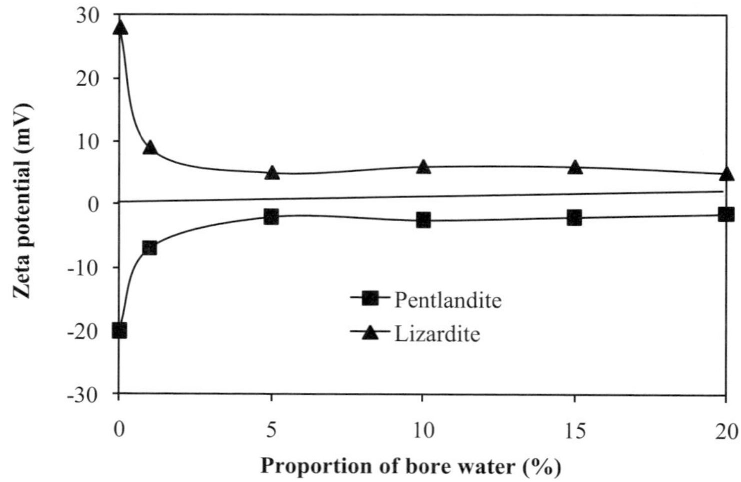

Figure 4. Zeta potential of pentlandite and lizardite as a function of the proportion of bore water at pH 9.0

In fact, the deleterious effect of clay minerals on the flotation of coal and bitumen by the coating has been observed (Arnold and Aplan 1986; Liu et al., 2005). Obviously, the entrainment of gangue minerals in the copper ores were halved in bore water.

The copper ore contains two copper minerals, chalcopyrite and chalcocite. Chalcocite is cyanide soluble, but chalcopyrite is not. Therefore, the behavior of chalcopyrite and chalcocite in flotation may be examined separately. Figure 3 shows that the response of chalcopyrite and chalcocite to the water type was completely different. Chalcopyrite recovery was high in both de-ionized and bore water and the water type had little effect. In contrast, chalcocite recovery was low in de-ionized water, about 48% in the end of flotation, but increased to 78% in bore water. The increased total Cu recovery in bore water was mainly contributed to the improved chalcocite flotation since chalcocite was the dominant copper mineral. It seems that gangue minerals have stronger affinity to chalcocite in de-ionized water and deteriorate its flotation.

Zeta Potential Measurements

The improved flotation of hydrophobic minerals in saline water has been attributed to the inhibition of bubble coalescence (Craig et al., 1993; Weissenborn and Pugh, 1995; Miklavcic, 1996) and the increase of bubble-particle attachment (Yoon, 1982; Paulson and Pugh, 1996). While the inhibition of bubble coalescence and the increase of bubble-particle attachment may improve pentlandite flotation in bore water in this study, the direct factor should be the change in the electrostatic interaction between serpentine and pentlandite since the electrostatic interaction plays an important role in pentlandite flotation (Edwards et al., 1980; Bremmell et al., 2004). However, all the flotation tests and electrophoresis measurements in previous studies were conducted in fresh or de-ionized water with low ionic strength. In the current study, zeta potentials of single minerals were measured in both de-ionized and bore water.

Figure 4 shows the zeta potential of pentlandite and lizardite, the main minerals in the nickel ore, as a function of the proportion of bore water at pH 9.0. In de-ionized water corresponding to zero proportion of bore water, pentlandite was strongly negatively charged with

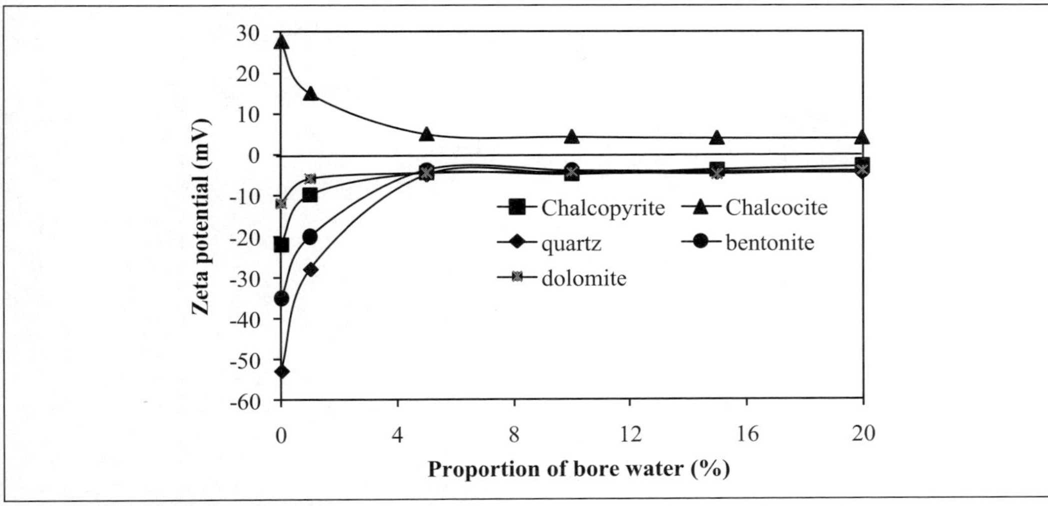

Figure 5. Zeta potential of chalcopyrite, chalcocite, quartz, bentonite, and dolomite as a function of the proportion of bore water at pH 9.0

a zeta potential about –20 mV, while lizardite was strongly positively charged with a zeta potential about 28 mV. This is consistent with the observations by Edwards et al. (1980) and Bremmell et al. (2004). With the addition of bore water, the magnitude of the zeta potential was reduced significantly. With 5% saline water, the zeta potentials reached a minimum. Clearly, the electrostatic interaction between pentlandite and lizardite was minimized in bore water. This may explain the improved pentlandite flotation from the nickel ore in bore water through the mitigation of slime coating.

Figure 5 shows the zeta potential of chalcopyrite, chalcocite, quartz, bentonite and dolomite, the main minerals in the copper ore, as a function of the proportion of bore water. The three gangue minerals were negatively charged. In particular, the zeta potential of quartz and bentonite has a great magnitude. This has been observed by many other researchers (Chen and Tao, 2004; Liu, et al. 2005; Herrera-Urbina and Fuerstenau, 1995). However, in de-ionized water, chalcopyrite and chalcocite displayed different electrical properties. Chalcopyrite was negatively charged whilst chalcocite was positively charged although the two samples were prepared exactly the same in air. Fullston et al. (1999) studied the oxidation of copper sulfide minerals by measuring their zeta potentials. Without oxidation, both chalcopyrite and chalcocite were negatively charged at a large pH range. Chalcopyrite was inert to the oxidizing condition. Oxygen purging only slightly moved the zeta potential curve to a more positive direction. In contrast, chalcocite was sensitive to the oxidizing condition. Oxygen purging reversed the sign of the zeta potential of chalcocite with a positive value at pH 9.0 (Fullston et al. 1999). Lascelles and Finch (2002) also demonstrated that chalcocite was more easily oxidized than chalcopyrite producing 50 times more copper ions. It seems that the surface oxidation of copper minerals governs their electrical properties. At the normal flotation condition, the zeta potential on the chalcopyrite surface may remain negative and repulsive from the slime gangue minerals corresponding to the good flotation behavior. Unlike chalcopyrite, chalcocite may become positively charged at weakly alkaline solutions due to surface oxidation and attractive to slime gangue minerals, resulting in the decreased flotation.

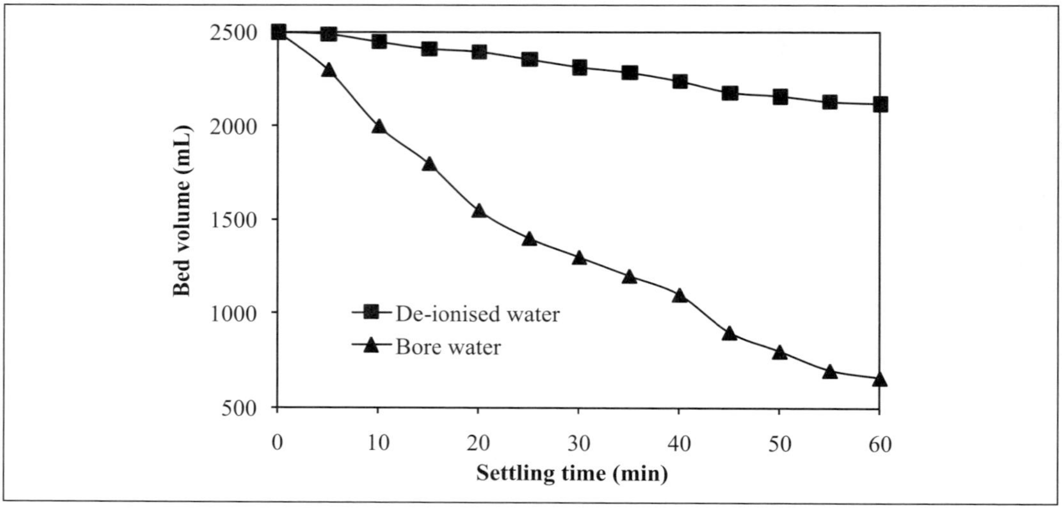

Figure 6. Bed volume of the flotation feed of the nickel ore as a function of settling time in de-ionized and bore water

Settling Tests

In the flotation of the nickel or copper ore with bore water, not only nickel or copper flotation was improved but also the gangue entrainment was reduced. The improved nickel and copper flotation in bore water may be attributed to the mitigation of slime coating. However, the mitigation cannot explain the reduced gangue entrainment. Liu et al. (2006) correlated fine gangue entrainment with the aggregation measured by settling tests. The quicker the settling of gangue minerals in the presence of high molecular weight polymers, the less the entrainment. This was attributed to the enlarged particle size of gangue minerals. In this study, the settling rate of the nickel and copper ores were measured.

Figure 6 shows the settling of the cyclone overflow fraction of the nickel ore in de-ionized and bore water. Clearly, the settling of the mineral suspension in de-ionized water was slow, whilst the solids settled quite fast in bore water. For example, the solids settled 1840 mL in bore water, but only 380 mL in de-ionized water. The same trend was observed in the settling of the copper ore as indicated in Figure 7. The copper ore settled much quicker in bore water than de-ionized water. In fact, the phenomenon that electrolytes promote solid aggregation and subsequent fast settling has been observed by other researchers (Rattanakawin and Hogg, 2001). It has been proposed that van der Waals attraction forces are predominant in electrolyte solutions and cause the aggregation (Rattanakawin and Hogg, 2001). The aggregated fine gangue minerals in the flotation of the nickel or copper ore in bore water may be responsible for the reduced gangue entrainment.

DISCUSSION

In this study, two difficult ores, nickel and copper ores, are examined in de-ionized and bore water. The water type had a significant effect on the selective flotation. In de-ionized water a low level of true flotation occurred. Ni or copper recovery and the selectivity against gangue minerals were low as well. In bore water, the true flotation became predominant with reduced gangue entrainment. As a result, the selective flotation was improved remarkably in

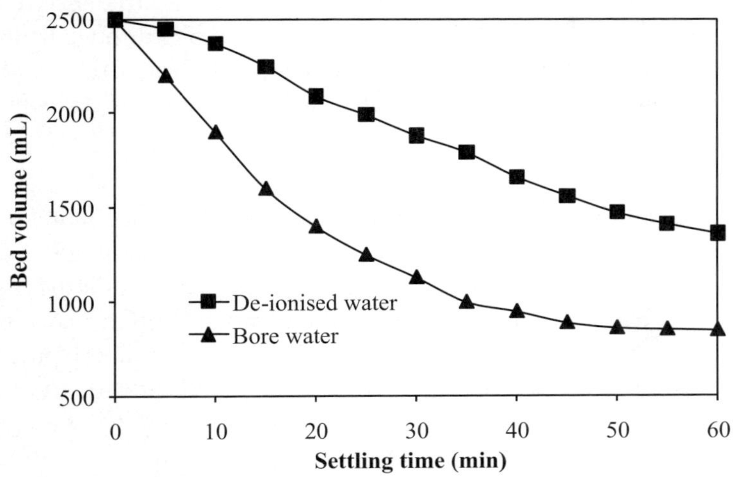

Figure 7. Bed volume of the flotation feed of the copper ore as a function of settling time in de-ionized and bore water

bore water. The different flotation behavior of the nickel and copper ores in de-ionized and bore water may be dictated by the inter-particle interactions.

Based on the DLVO theory, when two particles are brought into contact they are subjected to van der Waals and electrostatic forces (Liang et al., 2007). Van der Waals forces are due to the interaction between two dipoles that are either permanent or induced. They are weak, attractive and insensitive to variations in electrolyte strength and pH. Electrostatic forces depend on the charges present on the particles. They can be attractive or repulsive. In this study, in the nickel ore, serpentine and pentlandite are oppositely charged and an electrostatic attraction occurs between them. For the same reason, an electrostatic attraction occurs between chalcocite and other gangue minerals in the copper ore. Between two particles with the same sign of charges, the electrical double layer repulsive forces occur. Since the gangue minerals are predominant with more than 90% content in both the two ore systems, the strong electrical double layer repulsive forces between the gangue particles may play an important role in the high stability of suspensions in de-ionized water. The high stability of suspensions may result in the high entrainment of fine gangue minerals in flotation. Meanwhile, the coating of fine gangue minerals on the pentlandite or chalcocite surface through the electrostatic attraction may result in the low nickel or copper flotation.

A different scenario occurs in bore water. The presence of charged species causes the screening of surface charges on particles and the electrostatic forces between particles may be minimized. As a result of the minimization of the electrostatic forces (both the electrical double layer repulsion between similar particles and the electrostatic attraction between dissimilar particles), the coating of fine gangue minerals on nickel or copper minerals due to the electrostatic attraction may be mitigated, while the attractive van der Waals forces may promote mineral aggregation. The former explains the improved nickel and copper flotation in bore water, while the latter explains the reduced gangue entrainment possibly through the enlarged particle size.

In bore water, it is also possible that the attractive van der Waals forces promote the aggregation between fine valuable minerals, and between fine valuable minerals and gangue minerals. The aggregation between fine valuable minerals may improve the flotation due to the increased bubble-particle collision efficiency (Ralston and Fornasiero, 2010). The aggregation between fine valuable minerals and gangue minerals may have little effect on the flotation because the interaction between valuable minerals and air bubbles through hydrophobic forces is stronger than van der Waals forces. The detailed forces between particles and air bubbles in electrolyte solutions and their effect on mineral flotation are under investigation.

CONCLUSIONS

The current study indicates the importance of the mineral interaction in fine particle flotation. In de-ionized water, the coating of fine gangue minerals on the surface of valuable minerals may occur as a result of the electrostatic attraction and therefore limit the true flotation. As normal, high gangue entrainment is expected in the flotation. In bore water, the flotation of valuable minerals may be improved and gangue entrainment reduced both probably due to the screening of the electrostatic interaction between particles by electrolytes in the solution. The screening of the electrostatic interaction may mitigate the slime coating of gangue minerals on valuable minerals, while promoting the aggregation of gangue minerals.

REFERENCES

Arnold, B.J., and Aplan, F.F. 1986. The effect of clay slimes on coal flotation, Part ii: The role of water quality. *Int. J. Miner. Process.* 17: 243–260.

Bremmell, K.E., Fornasiero, D., and Ralston, J. 2005. Pentlandite–lizardite interactions and implications for their separation by flotation. *Colloids Surf. A* 252, 207–212.

Chen, G., and Tao, D. 2004. Effect of solution chemistry on flotability of magnesite and dolomite. *Int. J. Miner. Process.* 74: 343–357.

Craig, V.S.J., Ninham, B.W., and Pashley, R.M. 1993. The effect of electrolytes on bubble coalescence in water. *J. Phys. Chem.* 97: 10192–10197.

DiFeo, A.J., and Finch, A. 2002. Sphalerite/silica interactions: Model predictions. *Int. J. Miner. Process.* 64: 219–227.

Duarte, A.C.P., and Grano, S.R. 2007. Mechanism for the recovery of silicate gangue minerals in the flotation of ultrafine sphalerite. *Miner. Eng.* 20: 766–775.

Edwards, G.R., Kipkie, W.B., and Agar, G.E. 1980. The effect of slime coatings of the serpentine minerals, chrysotile and lizardite on pentlandite flotation. *Int. J. Miner. Process.* 7, 33–42.

Fuerstenau, D.W., Rosenbaum, J.M., and Laskowski, J. 1983. Effect of surface functional groups on the flotation of coal. *Colloids Surf.* 8: 153–173.

Fullston, D., Fornasiero, D., and Ralston, J. 1999. Zeta potential study of the oxidation of copper sulfide minerals. *Colloids Surf. A* 146: 113–121.

Gong, J., Peng, Y., Yeung, A., and Liu, Q. 2010. Reducing quartz gangue entrainment in sulphide ore flotation by high molecular weight polyethylene oxide. *Int. J. Miner. Process.* 97: 44–51.

Herrera-Urbina, R., and Fuerstenau, D.W. 1995. The effect of Pb(II) species, pH and dissolved carbonate on the zeta potential at the quartz/aqueous solution interface. *Colloids Surf. A* 98: 25–33.

Klassen, V.I., and Mokrousov, V.A. 1963. *An Introduction to the Theory of Flotation.* London: Butterworths.

Lascelles, D., and Finch J.A. 2002. Quantifying accidental activation, Part I. Cu ion production. *Miner. Eng.* 15: 567–571.

Liang, Y., Hilal, N., Langston, P., and Starov, V. 2007. Interaction forces between colloidal particles in liquid: theory and experiment. *Adv. Colloid Interface Sci.* 134: 151–166.

Liu, J., Xu, Z., and Masliyah, J. 2005. Interaction forces in bitumen extraction from oil sands. *J. Colloid Interface Sci.* 287: 507–520.

Liu, Q., Wannas, D., and Peng, Y. 2006. Exploiting the dual functions of polymer depressants in fine particle flotation. *Int. J. Miner. Process.* 80: 244–254.

Miklavcic, S.J. 1996. Deformation of fluid interfaces under double-layer forces stabilizes bubble dispersions. *Phys. Rev. E.* 54: 6551–6556.

Mishchuk, N. 2005. The role of hydrophobicity and dissolved gases in non-equilibrium surface phenomena. *Colloids Surf. A* 267: 139–152.

Paulson, O., and Pugh, R.J. 1996. Flotation of inherently hydrophobic particles in aqueous solutions of inorganic electrolytes. *Langmuir* 12: 4808–4813.

Ralston, R., and Fornasiero, D. 2010. The limits of fine particle flotation. *Miner. Engi.* 23: 420–437.

Ross, V.E., and Van Deventer, J.S.J. 1988. Column flotation. In *Proceedings of an International Symposium.* Edited by K. Sastry. Littleton, CO: SME.

Scheffel, R.E. 2002. Copper heap leach design and practice. In *Mineral Processing Plant Design, Practice, and Control.* Edited by A.L. Mular, D.N. Halbe, and D.J. Barratt. Littleton, CO: SME.

Trahar, W.J., and Warren, L.J. 1976. The floatability of very fine particles—A review. *Int. J. Miner. Process.* 3: 103–131.

Warren, L.J. 1985. Determination of the contributions of true flotation and entrainment in batch flotation tests. *Int. J. Miner. Process.* 14: 33–44.

Weissenborn, P.K., and Pugh, R.J. 1995. Surface tension and bubble coalescence phenomena of aqueous solutions of electrolytes. *Langmuir* 11, 1422–1426.

Yoon, R.H. 1982. Flotation of coal using micro-bubbles and inorganic salts. *Min. Congr. J.* 68: 76–80.

Yoon, R.H., and Sabey, J.B. 1989. Coal flotation in inorganic salt solutions. In *Interfacial Phenomena in Coal Technology.* Edited by G.D. Botsaris and Y.M. Glazman. New York: Marcel Dekker.

Zhang, X.H., and Ducker, W. 2007. Formation of interfacial nanodroplets through changes in solvent quality. *Langmuir* 23: 12478–12480.

Induction Time Measurements for Air Bubbles on Chalcopyrite, Bornite, and Gold in Seawater

Jaroslaw Drelich
Dept. of Materials Science and Engineering, Michigan Technological University, Houghton, MI, USA

Jan D. Miller
Dept. of Metallurgical Engineering, University of Utah, Salt Lake City, UT, USA

ABSTRACT

The kinetics of air bubble attachment to chalcopyrite, bornite and gold surfaces were recorded with a high-speed video system in 1 mM NaCl solution and simulated seawater of neutral and alkaline pH to measure the induction time. Collectors at a concentration of 20 mg/L included PAX (potassium amyl xanthate), S-701 (dialkylsulfide), Aero 6098 (thionocarbamate) and Aero 3477 (dithiophosphate). The experimental results show that the flotation chemistry variables have a strong effect on the kinetics of bubble attachment. The measurements indicated a positive effect of seawater on the kinetics of air bubble attachment to chalcopyrite and gold, particularly in the presence of PAX collector. The use of seawater with S-701 collector does not guarantee an improvement in the kinetics of bubble attachment to chalcopyrite, bornite and gold.

INTRODUCTION

Continuously rising human population and expanding industrial and agricultural activities inflict significant stresses on already limited fresh water resources. Availability and more efficient water use have become priorities for the mining and mineral processing industries in many regions of the world. Remote areas of potential mining activities, including southern Peru and northern Chile, often have limited water resources. Therefore, existing operations and new projects invest into technologies with reduced water use and sustainable options such as brine solutions and seawater.

About 97% of the water on the earth is salty water, and use of this water in mining and mineral processing operations must be considered when possible. Seawater has been used in ore beneficiation since the 1920s (Castro et al. 2010). Examples of mining operations where salty water is utilized are listed in Table 1. Additionally, exploration and recovery of economic mineral resources from the marine environment is expected to increase in years to come. About 95% of the oceans remain unexplored and the potential for the discovery of valuable minerals is high (Cruickshank 2011). Marine mining inevitably will rely on either raw or processed seawater.

Fundamental research on the processing of minerals in seawater is not well developed but it has gained interest in response to a growing mining industry demand. Castro et al. (2010) briefly reviewed the past literature on flotation studies with seawater at the XXV International Mineral Processing Congress. Several related papers summarizing current and past research studies on flotation systems with seawater are also included in these proceedings.

Table 1. Mining operations with the use of sea water or saline water (based partially on Philippe et al. 2011)

Project	Company	Country	Water Source
Batu Hijau	Newmont	Indonesia	Sea and fresh water
Las Luces	Minera Las Cenizas	Chile	Sea water
Michilla	Antofagasta	Chile	Sea water
KCGM	Barrick/Newmont	Australia	Saline
Mt Keith	BHP Billiton	Australia	Saline
Raglan	Xstrata	Canada	Saline
Texada	closed	Canada	Sea water
Tocopilla	closed	Chile	Sea water
Esperanza—under development	Antofagasta	Chile	Sea water

The flotation process is controlled by the thermodynamics, kinetics and hydrodynamics of the system (Laskowski 1974; 1986; Schulze 1984). Fundamental aspects of the flotation event include: (i) creation of a hydrophobic surface for specified particles (thermodynamic condition); (ii) provision of sufficient time for bubble attachment (kinetic condition); and (iii) stabilization of bubble/particle aggregates under the prevailing flow (hydrodynamic condition). All these criteria are important and have received extensive study. Thermodynamics and kinetics of particle-bubble attachment can be controlled by changing the solution chemistry. In this paper, we only report the kinetics of bubble attachment through induction time measurements. Such measurements were proven in our laboratory to be useful in the selection of appropriate solution chemistry conditions for the flotation of sulfide and gold minerals in seawater. Importantly, studies on the effect of seawater on the kinetics of bubble attachment at a mineral surface have not been reported previously. The results presented summarize part of our 1995/1996 project on surface chemistry of chalcopyrite, bornite and gold in seawater and have not been previously published.

EXPERIMENTAL

Reagents

Collectors used in this study included: potassium amyl xanthate (PAX, Tessenderlo Kerley Inc.), dithiophosphate (Aero 3477, Cytec), thionocarbamate (Aero 6098, Cytec), and dialkylsulfide (S-701). Fresh solutions of these collectors were prepared in deionized water (200 mg/L) before experiments. In all experiments the concentration of collector was adjusted to 20 mg/L, which is approximately equivalent to 40 gpt in a flotation circuit (assuming 33 wt% solids). S-701 collector does not dissolve completely at a concentration of 20 mg/L and a certain volume of this collector remained in an emulsified state during experiments.

A mixture of frothers was prepared using 40 vol% of Dow 200, 40 vol% of Dow 250, and 20 vol% of Dow 1012. A solution of the blended frother was prepared in deionized water (100 mg/L) before experiments. The frother concentration for experiments was adjusted to 5 mg/L or the equivalent of about 10 gpt in a flotation circuit (assuming 33 wt% solids).

Table 2. Coralite scientific-grade marine salt chemical composition

- Concentrations are expressed in mg/L (ppm)
- Approximate specific gravity 1. 023 at 25°C (77°F)
- One hour after dissolving in deionized water, pH = 8.3

Ingredients:

• Chloride—19,000	• Magnesium—0.10
• Sodium—11,000	• Iron—0.08
• Sulfate—2,500	• Molybdate—0.068
• Magnesium—1,350	• Zinc—0.035
• Calcium—420	• Copper—0.030
• Potassium—410	• Selenium—0.004
• Bromide—60	• Lead—<0.0004
• Strontium—9.0	• Aluminium—<0.0004
• Silicon—1.2	• Nickel—<0.0002
• Lithium—0.2	• Cobalt—<0.0002
• Iodide—0.12	• Cadmium—<0.0002

Other chemicals used in this study included analytical grade sodium chloride, potassium chloride, and lime. The pH of the aqueous phase was adjusted with lime and hydrochloric acid (technical grade). Mili-Q water (resistivity +18 MΩ) was used in preparation of all solutions.

Simulated seawater was prepared by dissolving 36 g of commercial sea salt formula, Coral Sea (Kordon) in 1 L of deionized water. The chemical composition of this formula as specified by the manufacturer is shown in Table 2.

Mineral Samples

Natural specimens of chalcopyrite (Durango, Mexico) and bornite (Montana, USA) were purchased from Ward's Natural Science Establishment, Inc. The specimens were cut into the approximate dimensions of 36×20×10 mm using a Buehler Isomet low-speed diamond saw. Once cut, the samples were polished with a series of polishing papers (grit: #200, #300, #400, #800) and with alumina powder (0.05 micron). Polished surfaces were cleaned from residual fine alumina powder ultrasonically in the Bransonic 211 bath and washed with deionized water. Samples were repolished and cleaned before each experiment.

One-hundred-nanometer-thick gold films were prepared by electron-beam evaporation of gold (Materials Research Co.; 99.999%) onto silicon test wafers. Before each experiment the gold samples were washed with deionized water and cleaned with argon plasma in a Tegal Plasmod chemical reactor for 40 min. Using Teflon-coated tweezers, the samples were immersed and stored in deionized water before experiments in order to avoid/reduce contamination with air-born hydrocarbons and particulates.

Induction Time Measurements

The kinetics of air bubble attachment to the mineral surfaces were examined using the experimental set-up shown in Figure 1 (Drelich et al. 1997). A high-speed camera of 1000 frames per second and 10 μs exposure time (Kodak EktaPro 1000 High-Speed Video System) was coupled with a photographic lens. The sample was placed on two stable supports with flat surfaces in a rectangular glass chamber, and the chamber was filled with the desired solution. The mineral sample was equilibrated for 10 to 15 min in the solution before

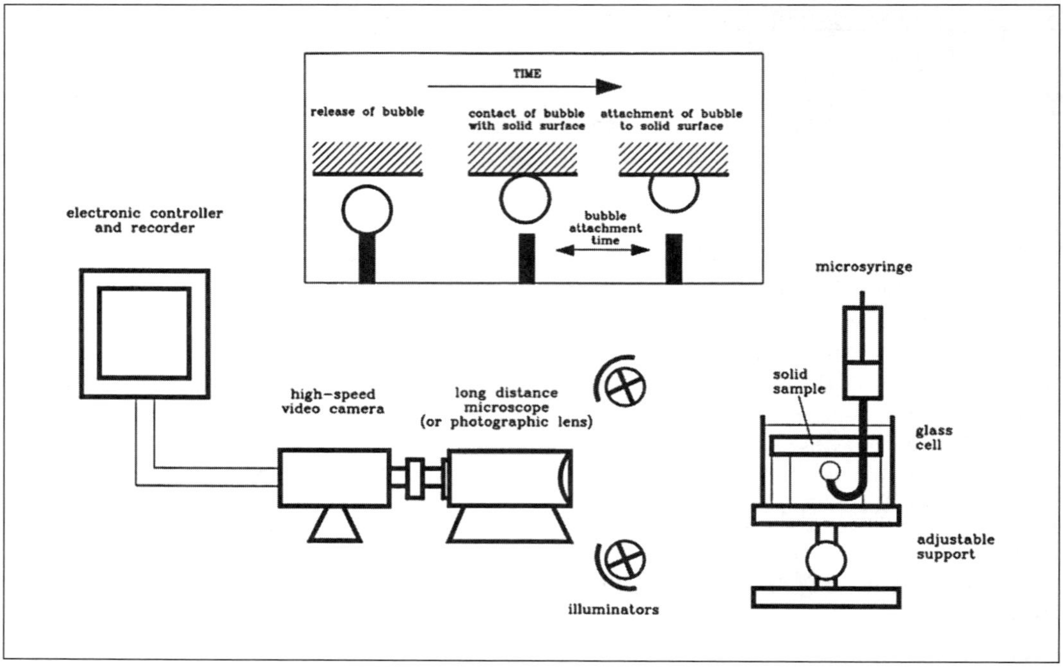

Figure 1. Experimental high-speed video set-up used for recording bubble attachment to mineral samples

the gas bubble was injected. A small air bubble (~1.5 mm diameter) was made at the tip of a U-shaped needle using a microsyringe (at a distance less than 1 mm from the sample surface) and was then released from the needle tip. Released bubbles rested at the mineral surface. The entire event starting with the release of the bubble from the needle, rest of the bubble and its attachment were recorded with the video camera. The bubble attachment process was viewed in slow motion and the bubble induction time was determined with an accuracy of 1 to 5 ms, depending on the speed of recording.

All bubble induction time measurements were conducted at room temperature (21–22°C). At least 40 measurements were performed for each experimental condition in order to establish the distribution of induction times.

Contact Angle Measurements

The rectangular glass chamber with a mineral sample and experimental solution which was used in bubble attachment experiments was placed on the table of a Rame-Hart goniometer and the receding contact angle was measured using the captive-bubble technique (Drelich et al. 1996). An air bubble with a diameter of about 1.5 mm was made at the tip of a U-shaped needle using a microsyringe. After attachment, the three-phase contact line of the air bubble was made to advance by adding a small volume of air, and the receding contact angle (as measured through the aqueous phase) was measured by a goniometer within 30–45 seconds on both sides of the bubble. The contact angles were measured for a bubble base diameter of approximately 3–4 mm. The needle remained in contact with the bubble during all measurements. Care was taken to insure that the needle did not distort the bubble shape. The contact angle measurements were carried out for at least five bubbles (at both sides) at room

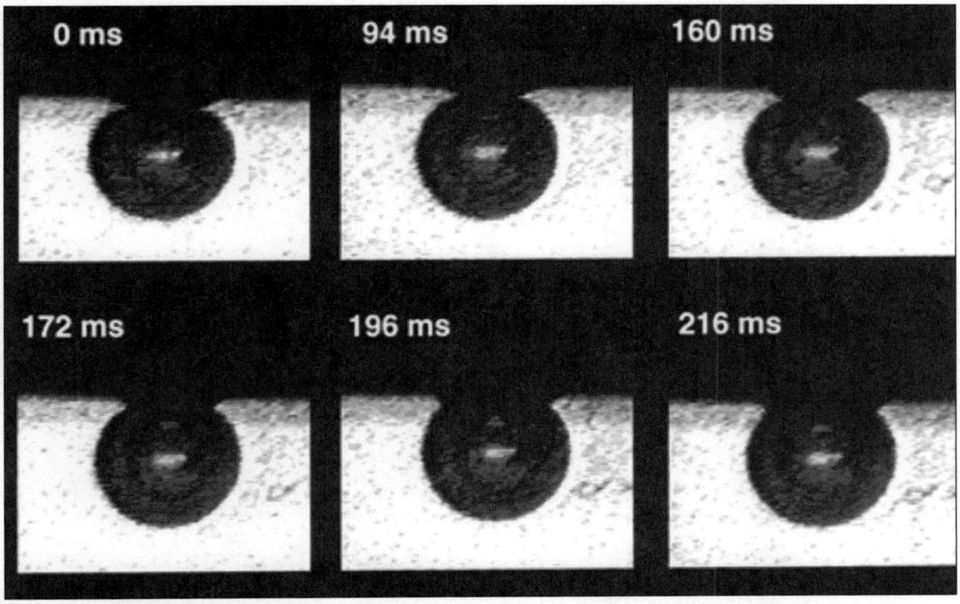

Figure 2. Air bubble attachment at the chalcopyrite surface in deionized water as captured with a high-speed video system. Upper row of frames presents the bubble induction time sequence, from the moment that the bubble rests at the mineral surface to the moment when attachment takes place. Lower row of frames presents the kinetics of bubble base expansion.

temperature (21–22°C). The ranges of measured receding (water) contact angle values, from smallest to largest ones, are reported in this communication.

RESULTS AND DISCUSSION

Figure 2 presents an example of the air bubble attachment event at the chalcopyrite surface in deionized water. This entire sequence of frames provides the snapshot of three fundamental stages of flotation: (1) water film thinning, (2) water film rupture to the nucleation of a hole and formation of the three-phase contact line, and (3) expansion of the three-phase contact line to a stable configuration. These three stages of bubble attachment are sometimes combined to represent the *induction time* (Gu et al. 2003). However, in this contribution we use the definition of the induction time stated previously (Ye et al. 1989): *the time from the moment of the bubble stationary rest at the mineral surface (first contact) to the moment when the aqueous film, which separates the gas bubble from the mineral surface, ruptures.*

For example, the sequence of three high-speed video frames in the upper row represents the induction time at the chalcopyrite surface which is reported in this contribution. The lower row of high-speed video frames in Figure 2 shows the kinetics of the displacement of the aqueous film by the air bubble from the moment the aqueous film breaks to the final stable state. The rate of movement of the three-phase contact line after rupture of the aqueous film is not discussed in this communication.

From 40 to 50 events of bubble attachment at the mineral surface were recorded for each system. The experimental results were plotted as cumulative distributions such as shown in Figure 3. The numbers in parentheses in Figure 3 represent the range of measured receding contact angles. The receding contact angles were preferred for flotation systems because these contact angles are supposed to correlate better with flotation response than other measures of

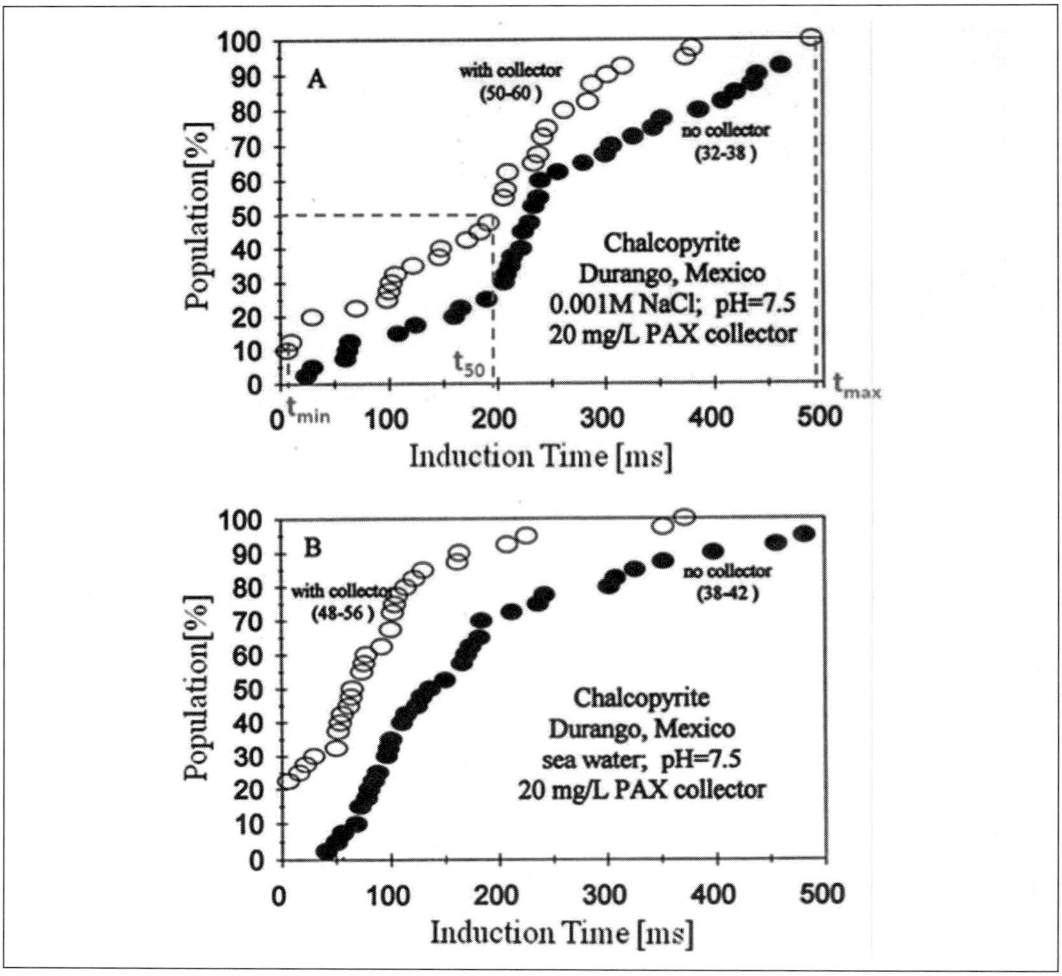

Figure 3. Effect of PAX collector on induction time at a chalcopyrite surface for 1 mM NaCl solution (a) and simulated seawater (b) at pH 7.5. The range of measured receding contact angles is reported in parentheses.

contact angles (Laskowski 1994). This data allowed us to track changes in wetting characteristics of mineral specimens in solutions of different chemistry.

Any point on the cumulative distribution curves represents the percentage of the total number of observations which have a bubble induction time less that the indicated time. For example, for the chalcopyrite in 1 mM NaCl with 20 mg/L PAX collector (Figure 3A; open circles), the minimum (t_{min}) and maximum (t_{max}) induction times are 5 and 491 ms, respectively. 50% of the total number of measurements (50% population) had induction times corresponding to $t_{50} \cong 197$ ms, as compared to $t_{50} \cong 233$ ms for chalcopyrite without collector. The t_{50} value corresponds to the time for which half of the induction time measurements are less than the t_{50} value and half of the measurements are greater than the t_{50} value. Minimum and maximum times refer to the shortest and longest, respectively, bubble induction times recorded in a particular experiment. In the further part of this communication, t_{min}, t_{max}, and t_{50} values are reported instead of plots with cumulative distribution curves.

Table 3. A comparison of average induction times for flat specimens of chalcopyrite and a particle (–212+150 µm) bed (Drelich et al. 1997)

		Induction Time (ms)	
		At Polished Specimen	At Particle Bed
Solution Chemistry	pH	t_{50}	t_{50P}
Deionized water	5.6	320	90
	2.4	110	<1
0.01 M NaCl	5.6	240	40
1M NaCl	5.6	225	25
20 mg/L PAX	7.5	180	<1
	9.5	180	<1
20 mg/L S-701	7.5	190	<1
	9.5	370	55

Electronic induction timers are more commonly used in examination of flotation systems (Ye et al. 1989; Gu et al. 2004; Wang et al. 2005; Albijanic et al. 2010; Min 2010) than high-speed video systems such as used in this study due to simplicity of instrumentation and measurements, and involvement of particles instead of flat larger specimens. An important disadvantage of the induction time measurements at a particle bed is, however, that for many systems the induction time is less than the detection limit for the instrument. This problem was illustrated in our previous contribution (Drelich et al. 1997). We compared the t_{50} values measured for a polished specimen with the induction times (t_{50P}) measured for –212+150 µm particles of the same chalcopyrite in different solutions. As shown in Table 3, in several solutions the induction time t_{50P} was too short, less than 1 ms, to be measured precisely. In contrast, the induction times measured on flat specimens are one to two orders of magnitude greater. This is the result of the fact that gas bubbles need to displace the liquid film from the much larger area of the polished surface than from the particle(s) surface.

As indicated by the data in Table 3, both t_{50} and t_{i50} values follow a similar pattern of changes in response to different solution chemistry. It should be recognized, however, that selection of a representative mineral specimen and its preparation for bubble attachment time measurements can have a significant impact on the results. Too large and/or too many inclusions (inhomogenities) in the area of bubble attachment can significantly shift the t_{50} value. It is often necessary to use the same quality specimen to test the effects of different solution chemistry in order to optimize the selection of flotation reagents for short bubble attachment/induction times.

Chalcopyrite

In the absence of collector chalcopyrite exhibits modest hydrophobicity as revealed by a receding contact angle of 30–40 degrees as presented in Table 4. The effect of seawater on bubble induction time at the chalcopyrite surface for different flotation chemistry conditions was examined. The experimental results are summarized in Tables 4 and 5.

Four collectors, PAX, S-701, 6098, and 3477, were tested with chalcopyrite (Table 4). As expected, the PAX collector improved the hydrophobicity of the chalcopyrite surface and bubble induction time was reduced from t_{50} =233 to 197 ms in 1 mM NaCl solution at pH = 7.5. Similar positive effects were observed for solutions with two other collectors, S-701

Table 4. Effect of seawater on induction time and receding contact angle measured at a chalcopyrite surface in solutions of different collectors and frother at pH = 7.5

Mineral	pH	Solution	Collector (20 mg/L)	θ_{rec} (deg)	Induction Time (ms) t_{min}	t_{max}	t_{50}
Chalcopyrite	7.5	1 mM NaCl $I=1\times10^{-3}$	None	32–38	23	>500	233
			PAX	50–60	5	491	197
			S-701	35–41	22	>500	187
			6098	42–48	16	>500	176
			3477	<20	54	>500	347
			Frother*	38–42	32	>500	308
		Seawater I=0	None	38–42	38	>500	136
			PAX	48–56	5	373	63
			S-701	42–46	28	450	149
			6098	44–50	5	>500	84
			3477	<15	>1000	>>1000	>1000
			Frother*	44–50	128	>800	297

*Without collector.

Table 5. Comparison of induction time and receding contact angle values measured for air bubbles at a chalcopyrite surface in 20-mg/L solutions of PAX and S-701 collectors

Mineral	Collector (20 mg/L)	Solution	pH	θ_{rec} (deg)	Induction Time (ms) t_{min}	t_{max}	t_{50}
Chalcopyrite	PAX	1 mM NaCl	7.5	50–60	5	491	197
			9.5	50–60	5	500	190
		Seawater	7.5	48–56	5	373	63
			9.5	64–68	5	365	58
	S-701	1 mM NaCl	7.5	35–41	45	735	173
			9.5	37–45	10	>2000	372
		Seawater	7.5	42–46	30	460	147
			9.5	30–33	85	>2000	537

and 6098. In contrast, a negative effect was observed for the 3477 collector and this collector reduced the hydrophobicity of chalcopyrite and increased the t_{50} value. Also frother stabilized the water film at the chalcopyrite surface and prolonged the time needed for the attachment of air bubbles in 1 mM NaCl solutions at pH = 7.5.

The series of experiments with simulated seawater (both with and without collector) showed a positive effect of high ionic strength on the destabilization of the water film at the chalcopyrite surface as can be concluded from the t_{50} values in Tables 4 and 5. At pH = 7.5, the induction times were reduced significantly for air bubbles attaching to the chalcopyrite surface without collector, and with 20 mg/L of PAX and 6098. A positive, but smaller effect, was also observed for the solution with 20 mg/L of S-701. The negative activity of the 3477 collector

Table 6. Effect of seawater on induction time and receding contact angle measured at a bornite surface in 20 mg/L solutions of PAX and S-701 collectors at pH = 7.5 and 9.5

Mineral	pH	Solution	Collector (20 mg/L)	θ_{rec} (deg)	Induction Time (ms) t_{min}	t_{max}	t_{50}
Bornite	7.5	1 mM NaCl	None	50–60	10	273	105
			PAX	80–90	9	78	<9
			S-701	42–50	10	250	127
		Seawater	None	45–55	10	82	36
			PAX	75–85	9	121	17
			S-701	40–50	10	480	66
	9.5	1 mM NaCl	None	45–55	10	173	81
			PAX	75–80	10	250	42
			S-701	44–48	10	350	127
		Seawater	None	42–52	10	92	<10
			PAX	70–80	9	300	123
			S-701	<20	10	>2000	>2000

was further enhanced in seawater. Seawater had negligible, if any, effect on the stability of water films at the chalcopyrite surface stabilized by the Dow frother.

The effect of seawater at pH = 9.5 was only tested for two collectors, PAX and S-701 (Table 5). It was observed that seawater had a negligible effect on the performance of the PAX collector. However, the activity of the S-701 collector was reduced in seawater at pH = 9.5. As noted in the experimental section, the S-701 collector could not be dissolved completely in the aqueous phase at a concentration of 20 mg/L and thus remained partially in an emulsified state. Seawater could affect the stability of the emulsified collector.

Bornite

Bornite exhibits even greater natural hydrophobicity than chalcopyrite as revealed by the receding contact angles in Table 6. The receding contact angle for chalcopyrite in the absence of collector was 30–40 degrees whereas the receding contact angle for bornite in the absence of collector was 45–60 degrees. The PAX collector significantly improved the hydrophobicity of the bornite surface and the receding contact angle increased to 80–90 degrees. On the other hand, S-701 reduced the receding contact angle to 42–50 degrees.

The pH had no detectable effect on the natural hydrophobicity of bornite and it had only a small impact on the bubble induction time (Table 6). The induction time appeared to be reduced at pH = 9.5 as compared to pH = 7.5 in the absence of a collector. The system showed a different response when PAX collector was added to the system. An increase in the alkalinity of the solution had a negative effect on the kinetics of bubble attachment to the bornite surface. Also, the value of the bubble induction time for systems with S-701 collector was significantly increased when the pH of the solution was increased from pH = 7.5 to pH = 9.5, consistent with observations for chalcopyrite. These results indicate that for some collectors the correlation between pH of the solution and bubble induction time is not simple and the alkalinity might affect both the stability of the water film and the dissociation/activity of collector.

The series of experiments with simulated seawater showed a positive effect of high ionic strength on the destabilization of water films at the natural bornite surface. On the other hand, seawater showed rather a negative effect on the induction time when PAX and S-701 were used in the bubble induction time experiments. It appears that activity of collectors in seawater, particularly at pH = 9.5, was reduced. This is consistent with observations during experiments with chalcopyrite and S-701 collector but in disagreement with experiments involving chalcopyrite and PAX collector. It should be recognized however, that t_{50} values for bornite are typically much smaller than for chalcopyrite, suggesting that the seawater effect might be overall negligible in the flotation of bornite.

Gold

Gold is a naturally hydrophilic mineral if its surface is protected from air-born contaminants (Smith 1980). Freshly sputtered gold used in this study also stabilized aqueous films, and air bubbles could not attach to the gold surface (Table 7). A hydrophobic state was created at the gold surface by PAX, S-701, 6098, and 3477 collectors in alkaline solutions. Virtually only the PAX collector was capable of making the gold hydrophobic at pH = 7.5 when used at 20 mg/L concentration.

As shown in Table 7, for 1 mM NaCl at both pH = 7.5 and 9.5 the induction time between the air bubble and the gold surface strongly depended on the collector type. The ability of the collector to reduce the bubble attachment time was in the following order:

at pH = 7.5	PAX >> S-701 >> 6098 ~ 3477
at pH = 9.5	S-701 > PAX >> 6098 ~ 3477

An increase in pH value from 7.5 to 9.5 reduced the induction time for systems involving collectors S-701, 6098 and 3477, but slightly prolonged the induction time for the PAX collector.

Using seawater instead of 1 mM NaCl solution improved the destabilization of water films at the gold surface when PAX collector was used (Table 7). On the other hand, the activity of the S-701 collector changed drastically in seawater. Although the S-701 collector improved the hydrophobicity of the gold surface, the induction time increased significantly when compared to the system with 1 mM NaCl solution. A similar effect of seawater on the activity of the S-701 collector was observed for chalcopyrite.

SUMMARY AND CONCLUSIONS

A high-speed video system was used to record bubble attachment events to mineral surfaces. The bubble induction time is 1–2 orders of magnitude larger than induction time measured for particles using an electronic induction timer. The high-speed video observations of bubble attachment to polished surfaces appeared to be especially useful for such systems where bubble attachment of particles is below the limit and resolution of the electronic instrument (<1 ms for the MCT-100 induction timer).

The kinetics of air bubble attachment with a time resolution of 1–5 ms at chalcopyrite, bornite and gold surfaces in aqueous solutions of varying pH (pH = 7.5 vs. 9.5) and ionic strength (1 mM NaCl vs. seawater) were studied for completely dissolved collectors (PAX, 6098, 3477) and for the partially dissolved collector (S-701) at a fixed concentration

Table 7. Effect of seawater on induction time and receding contact angle measured at a gold surface in solutions of different collectors at pH = 7.5 and 9.5

Specimen	Solution	pH	Collector (20 mg/L)	θ_{rec} (deg)	Induction Time (ms) t_{min}	t_{max}	t_{50}
Gold	1 mM NaCl	7.5	No	<15	>2000	>2000	>2000
			PAX	52–58	10	680	157
			S-701	11–15	100	>2000	920
			6098	<10	>2000	>2000	>2000
			3477	<10	>2000	>2000	>2000
		9.5	No	<15	>2000	>2000	>2000
			PAX	78–82	5	910	248
			S-701	55–60	5	433	90
			6098	40–45	45	>2000	980
			3477	68–72	87	>2000	1270
	Seawater	7.5	No	<15	>2000	>2000	>2000
			PAX	45–50	41	540	135
		9.5	No	<15	>2000	>2000	>2000
			PAX	75–80	6	800	345
			S-701	68–72	100	2000	887

of 20 mg/L. The systems examined were also characterized by the receding contact angle as measured with the captive-bubble contact-angle measurement technique using a goniometer.

The solution chemistry used affected both the wetting characteristics of minerals and the kinetics of bubble attachment to these mineral surfaces. No direct correlation was found between the bubble induction time and the value of the receding contact angle. The results, summarized in Table 8, are as follows.

Chalcopyrite

1. Ionic strength had a significant effect on both bubble induction time and natural wettability of chalcopyrite. It was found that the induction time decreased and the receding contact angle increased for seawater. Simulated seawater had a positive effect on the attachment of air bubbles to the chalcopyrite surface as compared to the 1 mM NaCl solution at pH values pH = 7.5 and 9.5, for most of the systems studied.
2. Natural hydrophobicity of chalcopyrite is only slightly sensitive to the pH of the aqueous phase. A significant reduction in the induction time was observed when the S-701 collector was used and the pH value was reduced from pH = 9.5 to 7.5. On the other hand, no effect of pH on bubble induction time was observed when PAX solutions were used.
3. The PAX collector significantly enhanced the hydrophobicity of chalcopyrite at both pH = 7.5 and 9.5. Also S-701 and 6098 collectors improved the hydrophobicity of the chalcopyrite surface at pH = 7.5. On the other hand, the 3477 collector reduced the hydrophobicity of the chalcopyrite surface at pH = 7.5.

Table 8. Summary of flotation chemistry findings

Mineral	Variable	Effect
Chalcopyrite	Collector type	PAX > S-701 ~ 6098 >> 3477
	pH = 7.5 vs pH = 9.5	No collector: negative effect of higher pH with collector: PAX—no effect; S-701—negative effect
	Seawater effect	No collector: positive effect with collector: PAX—positive effect; S-701—negative effect
Bornite	Collector type	PAX > S-701
	pH = 7.5 vs pH = 9.5	No collector: With collector: PAX & S-701—negative effect
	Seawater effect	No collector: positive effect With collector: PAX & S-701—negative effect
Gold	Collector type	PAX > S-701 >> 6098 ~ 3477
	pH = 7.5 vs pH = 9.5	No collector: no effect With collector: PAX—negative effect; S-701—positive effect
	Seawater effect	No collector: no effect With collector: PAX—positive effect; S-701—negative effect

Bornite

1. Solution chemistry significantly affected the kinetics of bubble attachment to a bornite surface. A positive effect of using simulated seawater on the kinetics of bubble attachment to a natural surface was observed.
2. The use of seawater with PAX and S-701 collectors does not guarantee an improvement in the flotation kinetics of bornite. The selection of an appropriate collector for the effective flotation of bornite might be facilitated based on the results from induction time measurements.
3. The PAX collector seems to be a much more promising collector for the flotation of bornite than the S-701 collector.

Gold

1. Gold is naturally hydrophilic and thus can only be floated in the presence of collectors. Among four collectors tested PAX and S-701 reduced significantly the induction time between an air bubble and the gold surface in aqueous solutions at pH = 7.5 and 9.5. The PAX collector was more effective than the S-701 collector at neutral pH (pH = 7.5) and for the 1 mM NaCl solution, whereas the S-701 collector showed a greater impact on the induction time in lime solutions (pH = 9.5). This, however, changed when simulated seawater replaced the 1 mM NaCl solution. Collectors 6098 and 3477 appeared to be ineffective in destabilization of water films at the gold surface.
2. The seawater had a small effect on the stability of water films at the gold surface in the presence of the PAX collector. The induction time was reduced for the system with seawater when compared to the system with 1 mM NaCl. A reverse tendency was observed when the S-701 collector was used and the induction time increased in seawater as compared to the 1 mM NaCl solution.

3. The pH of the aqueous phase had a significant effect on the stability of water films at the gold surface in the presence of PAX and S-701 collectors. The water film was more stable in the alkaline solution (pH = 9.5) of the PAX collector than in the neutral solution (pH = 7.5). On the other hand, the stability of the water film at the gold surface was enhanced in the neutral solution of the S-701 collector as compared to the case in an alkaline solution.

ACKNOWLEDGMENT

The authors gratefully acknowledge the financial support which was received from Newmont Mining Company for this research.

REFERENCES

Albijanic, B., Ozdemir, O., Nguyen, A.V., and Bradshaw, D. 2010. A review of induction and attachment times of wetting thin films between air bubbles and particles and its relevance in the separation of particles by flotation. *Adv. Colloid Interface Sci.* 159: 1–21.

Castro, S., Venegas, I., Landero, A., and Laskowski, J. 2010. Frothing in seawater flotation systems. In *Proceedings of the XXV International Mineral Processing Congress*, Brisbane, Qld., Australia, September 6–10, 2010: 4039–4047.

Cruickshank, M. 2011. Marine mining. An area of critical national need. *Min. Eng.* 63(5): 89–93.

Drelich, J., Miller, J.D., and Good, R.J. 1996. The effect of drop (bubble) size on advancing and receding contact angles for heterogeneous and rough solid surfaces as observed with sessile-drop and captive-bubble techniques. *J. Colloid Interface Sci.* 179: 37–50.

Drelich, J., Miller, J.D., Li, J.-S., and Wan, R.-Y. 1997. Bubble attachment time measurements at a chalcopyrite surface using a high-speed video system. In *Proceedings of the XX International Mineral Processing Congress*, September 21–26, 1997, Aachen, Germany. Vol. 3: Flotation and other physical-chemical processes. GMDB Gesellschaft fur Bergbau, Metallurgie, Rohstoff- und Umwelttechnik, Clausthal-Zellerfeld, Germany: 53–64.

Gu, G., Xu, Z., Nandakumar, K., and Masliyah, J. 2003. Effects of physical environment on induction time of air-bitumen attachment. *Int. J. Miner. Process.* 69: 235–250.

Gu, G.X., Sanders, R.S., Nandakumar, K., Xu, Z., and Masliyah, J.H. 2004. A novel experimental technique to study single bubble-bitumen attachment in flotation. *Int. J. Miner. Process.* 74(1-4): 15–29.

Laskowski, J.S. 1974. Particle-bubble attachment in flotation. *Miner. Sci. Eng.* 6(4): 223–235.

Laskowski, J.S. 1986. The relationship between floatability and hydrophobicity. In *Advances in Mineral Processing*. Edited by P. Somasundaran. Littleton, CO: SME.

Laskowski, J.S. 1994. Coal surface chemistry and its role in fine coal beneficiation and utilization. *Coal Prep.* 14: 115–131.

Min, M.A. 2010. Measuring the floatability of sulphide minerals and ores: The captive bubble attachment times of galena, sphalerite and Cannington lead-zinc ore particles floating at different rates. Ph.D. thesis, University of Queensland, Brisbane, Australia.

Philippe, R., Dixon, R., and Dal Pozzo, S. 2011. Seawater supply options for the mining industry. 2nd International Seminar on Geology for the Mining Industry.

Schulze, H.J. 1984. *Physico-Chemical Elementary Processes in Flotation*. Amsterdam: Elsevier.

Smith, T. 1980. The hydrophilic nature of a clean gold surface. *J. Colloid Interface Sci.* 75(1): 51–55.

Wang, W.X., Zhou, Z., Nandakumar, K., Masliyah, J.H., and Xu, Z. 2005. An induction time model for the attachment of an air bubble to a hydrophobic sphere in aqueous solutions. *Int. J. Miner. Process.* 75(1-2): 69–82.

Ye, Y., Khandrika, S.M., and Miller, J.D. 1989. Induction-time measurements at a particle bed. *Int. J. Miner. Process.* 25: 221–240.

Removal of Ions from Water with Electrosorption Technology

Jiann-Yang Hwang
Michigan Technological University, Houghton, MI, USA

Xiaowei Sun
EST Water and Technologies Co., Ltd., China

ABSTRACT

Electrosorption is defined as potential-induced adsorption of ions onto the surface of charged electrode. When an electrical potential was applied to electrode, charged ions migrated to the electrode and are held in the electric double layer. Once external field is removed, the ions are quickly released back to bulk solution. Compared to conventional desalination technologies, electrosorption is an energy-efficient desalination process due to it operates at lower electrode potential (about 1–1.5 V) at which no electrolysis reactions occur. In addition, this process is environmentally attractive because it requires no chemicals for regeneration. Activated carbon is an effective electrode material for electrosorption. The material is inert, conductive, inexpensive, abundant and has high surface area to provide sufficient adsorption sites. Applications of electrosorption on ions removal for various kinds of water streams are introduced in this study.

INTRODUCTION

Water shortage is widely experienced in the whole world. It is particularly severe in China, which could even cause security problem in the world [1]. The water treatment market is enormous in China. This has naturally brought fierce competition for all major water companies all over the world. It is also the place where new technology has to be developed to meet all market challenges.

China has been experiencing high economic growth rate. However, water shortage is getting worse with the industrialization and the expansion of cities. China has over 600 cities (population above half million per city) in shortage of water. The amount in shortage is about 40 billion metric tons in 2006. In the rural area, more than 60 million farmers are drinking water with fluorine contents exceeding the limits and about 37 million farmers are drinking brackish water. The water needs to be treated for their drinking exceeds 15 billion metric tons per year [2, 3].

The World Bank has estimated that the population of China will reach 1.6 billion in 2030. The water demand will be 3 times of that in 1980. The water pollution problems in China are getting worse recently. Water shortage and water pollution have been identified as a serious threat to the public safety by the government. Many policies are made. These policies will promote the development of the water business continuously.

According to the governmental statistics, the water treatment market in China grows from $300 billion RMB in 2007 to $2,000 billion RMB in 2010. The annual growth rate is estimated at 15% for the next 30 years. This enormous market has brought competition from major companies all over the world.

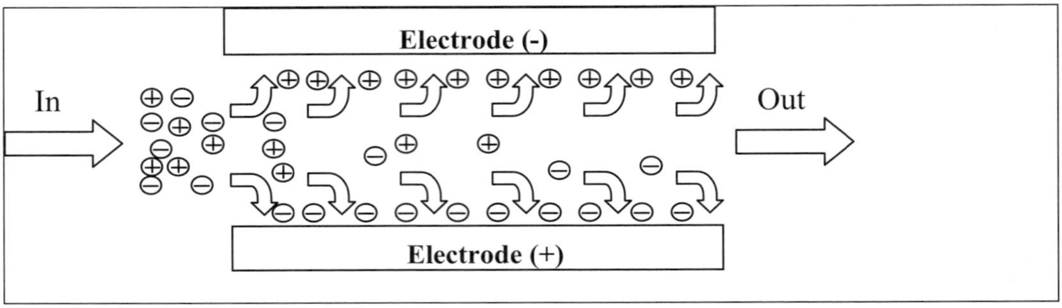

Figure 1. Capacitive deionization

Technologies are the fundamental for all the competitions. There have been many water treatment technologies developed. Depending on the applications, there are different advantages and disadvantages for different technologies. In general, there are various sedimentation, flotation and filtering technologies for the removal of suspended solids and oil droplets. Organics are frequently digested through various biological systems or removed with adsorbents. The removal of ions (desalination) is only considered when necessary due to its high cost.

The common desalination technologies include distillation, ion exchange, electrodialysis, and reverse osmosis [4–11]. High purity water can be obtained when these technologies are applied. However, there are in many cases we do not need to remove all the salts to generate the high purity water. For different applications, such as the drinking water, tap water, cooling water, irrigation water, paper mill water, steel mill water, petrochemical plant water, and electronic plant water, etc., different purities are required. There is definitely a need of a technology that is capable to adjust the amount of salt removed at will.

ELECTROSORB TECHNOLOGY

The electrosorb technology (EST) is developed based on the capacitive deionization (CDI) technology. The CDI technology was first patented by Becker in 1957 [12]. The fundamental theory is based on the electric double layer phenomenon in water. On a charged surface, counter ions are adsorbed preferably in the Helmholtz layer and the Diffuse layer on the surface. Therefore, a pair of electrodes will be able to adsorb both the cations and anions, as shown in Figure 1. When the charge is removed, cations and anions will be released from the electrodes. This mechanism allows one to design and engineering the apparatus capable to separate ions from water. Since the adsorption of ions is related to the voltages applied, the gap between the electrodes, the surface area of electrodes, the flow velocity, the retention time, etc., there are plenty parameters to adjust the amount of ions adsorbed during the process. Hence this technology has the potential to meet the needs on adjusting effluent salt contents at will.

There have been many developments on the CDI technology [13–21]. A good review can be found in a recent article by Marc Anderson [15]. Most of the developments are related to the electrode materials, which include graphite, activated carbon, carbon aerogels, carbon nanotubes, carbon cloth, etc.

Most developments of the CDI technology ended in the laboratory, with only a few making the effort to commercialize. Among those commercial developments, there is only one reached the full commercial stage with mass production capability, while the rests are still in the pilot to small capacity unit stages.

Figure 2. EST manufactured electrode modules

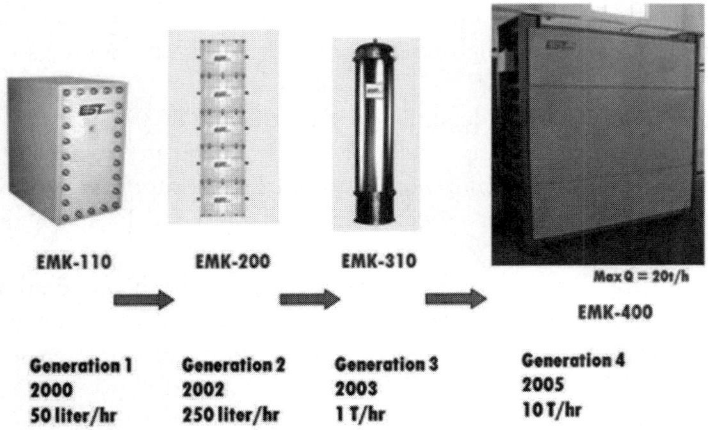

Figure 3. Advancement of EST treatment module

EST Water and Technologies Co., Ltd., is the company that so far has advanced the CDI technology to the arena that can compete commercially with all other technologies. The company is headquartered in Changzhou, China, where its manufacturing facility (Figure 2) is also located. Its marketing division and R&D Center are located in Beijing.

The company was established in 2000. Its R&D has made tremendous effort to refine the technology on both the electrodes and the systems to scale up the capability. As shown in Figure 3, the company took a module manufacturing approach. Its first generation product is only capable to treat 50 liter per hour. The capacity per module has been increased to 250 liter per hour in 2002, 1 ton per hour (1 cubic meter per hour or 6,341 gallons per day) in 2003, and 10 ton per hour in 2005. With the capability to mass produce the 10 ton per hour module, the company is able to compete with other technologies for large scale industrial water market by using multiple modules.

Activated carbon is utilized to make the electrodes. Activated carbon has high surface area, excellent permeability, good conductivity, and is inert to the reaction. It is abundant and is not too expensive. These characteristics made it an ideal electrode material.

Each module contains multiple pairs of electrodes. The gap between the electrodes is about 1 to 2 millimeters so that ions do not have to travel far distance. Figure 4 shows the schematics of the EST water treatment system. Mechanical filters are placed before the EST modules to prevent the entrance of large particles which may cause the blocking of the electrodes. Water is pumped into the EST modules from the bottom of the module. Conductivity is measured continuously for both the influent and the effluent. Voltage, flow velocity and retention time are the parameters utilized to ensure the effluent meeting the water salinity specifications. When the conductivity exceeds the specifications, the valve to the clean effluent is closed. Voltage is then turned off to release ions from the electrodes. A small amount of water is fed into the module to flush the ions out, which is collected as the concentrate. When disinfection is required, UV, ozone, or chlorination can be installed after the desalination.

APPLICATIONS

The EST system has been widely applied in a variety of industries in China. Its robustness and cost effectiveness have been demonstrated. Examples include applications in municipals, groundwater, petrochemicals, steel mills, thermoelectric powers, coal chemicals, paper mills, fertilizers, and high fluorine and high arsenic brackish water, etc. Some of the examples are presented here.

Figure 5 shows the picture of a desalination plant treating 80,000 tons per day (21 MGD) wastewater from the municipal sewage and coal chemical plant in Taiyuan, China. Taiyuan is the coal capital in China with 3 million populations. Its water supply is in extremely shortage. The Taiyuan Chemicals, a coal chemical company, has to look for alternative resources to meet its water demand.

The plan to use alternative water is shown in Figure 6. With cooperation from the city government, the company is able to obtain the city sewage water. Together with the wastewater from the coal company, the water is treated with conventional primary and secondary process and then feed to the EST modules.

Many technologies have been evaluated for this application. After a series of round robin pilot testing, the EST technology is determined as the only one that meets the challenges from both the technical and economic points. The organics from both the sewage and chemical plants are too hazardous to the membrane technologies, which causes the quick fouling of the membranes of reverse osmosis and other membrane based approaches.

The plant was built in 2007 and has been operating smoothly. Technical results of the operation are shown in Table 1. It is designed to remove 75% salts at 75% water yields. This has been easily achieved. The conductivity is reduced from 1174 μS/cm to 197 μS/cm, equivalent to 83% salt removal. The COD is reduced from 20 mg/L to 8 mg mg/L. Color is reduced from 26 degree to 9 degree. Elements such as chloride, sulfate, nitrogen, calcium, iron etc., are all removed at satisfactory level. This plant is planned to be expanded later to 8 times bigger to treat the whole city's wastewater.

In another example, the wastewater from the cold rolling mill of Bao Steel in Shanghai, China was recycled. The water contains oils and has been treated with catalytic oxidation and Membrane Bioreactor (MBR). The effluent of MBR is rich in salt, with conductivity around 3200 μS/cm. This water needs to be desalinated before it can be reutilized. The specification is to reduce the conductivity to less than 1500 μS/cm at a yield of 75%.

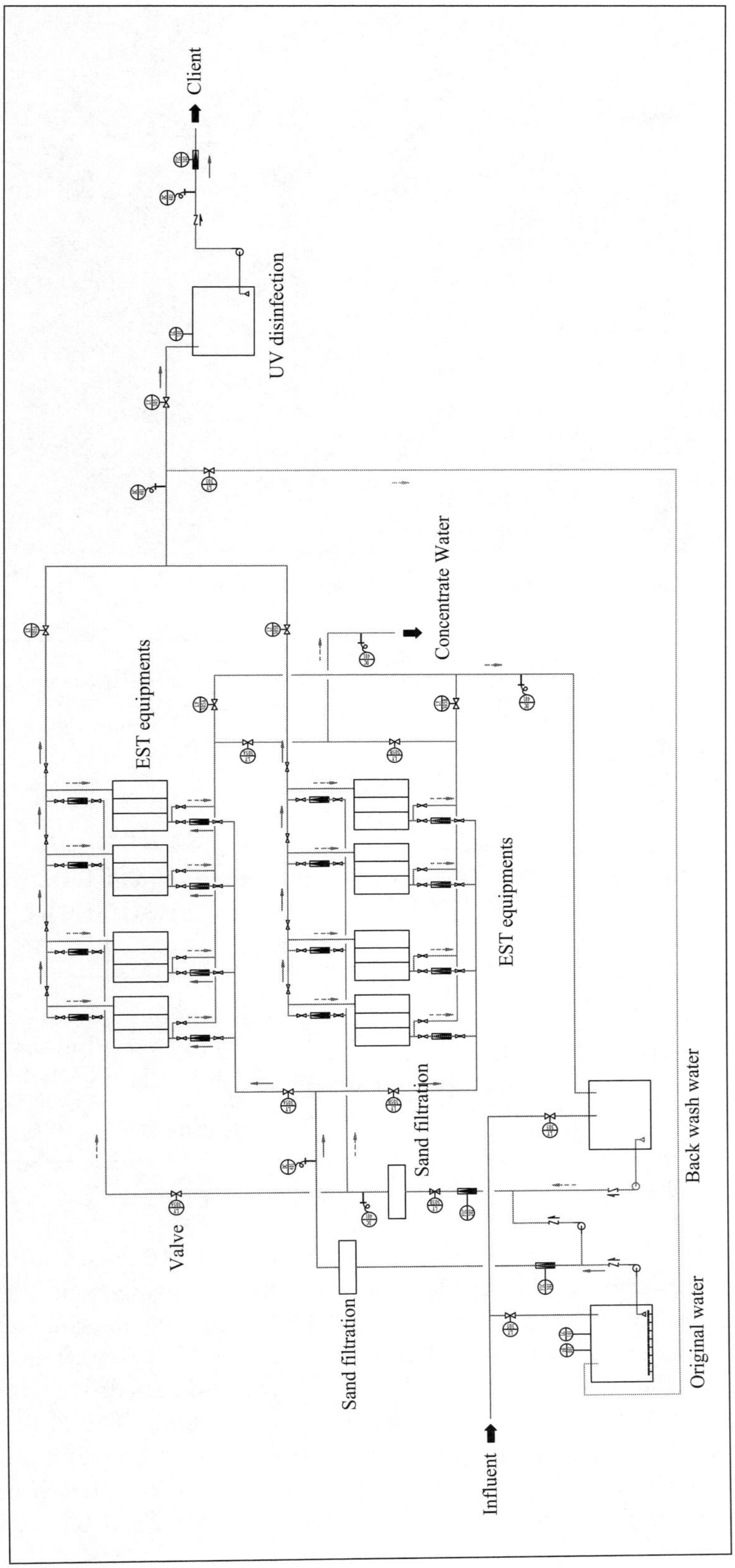

Figure 4. EST desalination process

Figure 5. An EST plant at 80,000-TPD (21-MGD) capacity

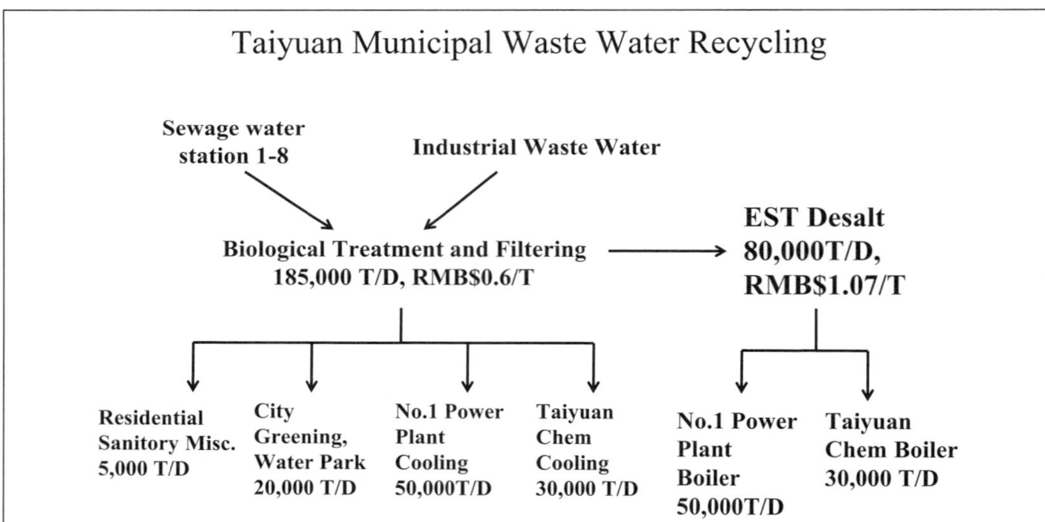

Figure 6. Treatment plan to recycle municipal wastewater

Figure 7 shows the EST plant operation data of the water for 16 days. During the operation, the water comes into the EST module has conductivity ranging from 2980 μS/cm to 3490 μS/cm. The average is 3213 μS/cm. After the EST treatment, the conductivity of the water is reduced to about 1100 μS/cm, with a range of 910 to 1254 μS/cm. This is well below the 1500 μS/cm specification. The 75% yield requirement is also exceeded.

Arsenic is an element that has troubled many water operations. This element has been linked to cancer of the bladder, lungs, skin, kidney, nasal passages, liver, and prostate. Non-cancer effects can include thickening and discoloration of the skin, stomach pain, nausea, vomiting; diarrhea; numbness in hands and feet; partial paralysis; and blindness. EPA has

Table 1. Taiyuan municipal wastewater treatment results

No.	Item	Unit	Influent	Effluent	Removal (%)
1	Conductivity	μS/cm	1174	197	83.2
2	COD_{Cr}	mg/L	20	8	60.0
3	Color	degree	26	9	65.4
4	Turbidity	NTU	6	1	81.4
5	Nitrogen	mg/L	10.1	3.7	63.8
6	Hardness	mg/L($CaCO_3$)	336.7	44.9	86.7
7	Alkalinity	mg/L($CaCO_3$)	248.4	77.3	68.9
8	Chloride	mg/L	132.2	7.1	94.6
9	Sulfate	mg/L	144.5	36.4	74.8
10	Total iron	mg/L	0.130	0.053	58.9
11	TDS	mg/L	882	224	76.3

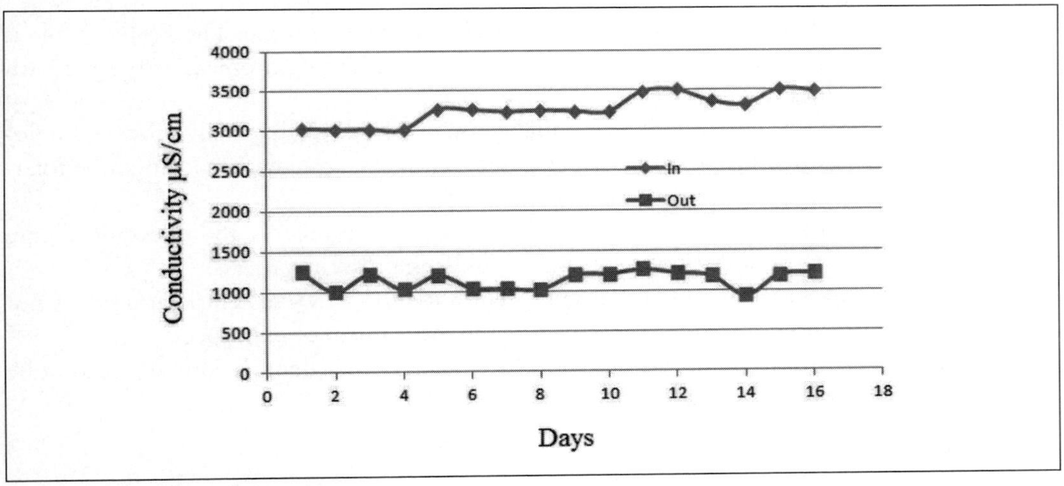

Figure 7. Desalination results at Bao Steel cold rolling mill

set the arsenic standard for drinking water at 0.010 parts per million (10 parts per billion) to protect consumers.

EST system has been installed for arsenic removal from the spring water by a bottling company. Except the arsenic, this water is rich in trace elements which is considered beneficial to the client's health. The conductivity of the water is about 450 μS/cm. The goal of this company is to reduce the arsenic to below the 10 ppb level.

Table 2 shows the operation data for this plant. The goal of arsenic reduction to below 10 ppb is well achieved. In addition, the conductivity is reduced to a level of about 90 μS/cm. This level gives the best taste of water. The system is still in good condition after 8 years operation.

CONCLUSIONS

EST has established its position in the water and wastewater treatment businesses. It has demonstrated the following technical advantages:

Table 2. Plant data on arsenic removal

	Arsenic (mg/L)		Conductivity (μS/cm)		pH	
No.	Influent	Effluent	Influent	Effluent	Influent	Effluent
1	0.06	<0.01	450	90	7.2	6.5
2	0.06	<0.01	446	90	7.4	6.5
3	0.06	<0.01	447	92	7.5	6.5
4	0.07	<0.01	447	93	7.2	6.5
5	0.07	<0.01	446	95	7.2	6.5

1. It consumes less energy than other desalination technologies. The low pressure drop nature of the process requires only low pumping energy. The flow path design is simple and is independent of the other structure factors. It does not require thermal energy.
2. The technology offers flexible and adjustable salt removal. The adjustment can be achieved by adjusting operation parameters such as voltage applied, flow velocity, and retention time, etc.
3. The EST technology is quite robust. It needs low maintenance. The EST module is very sturdy and can last for more than 5 years. The fouling problem associated with the membranes is not involved.
4. Little restrictions on influent water quality are involved. Unlike many other technologies, water containing COD, oils, chlorine, hardness, solvents, etc., can all be tolerated to some extents in the EST system.
5. No chemical additions are necessary. This is a physical process and does not require chemicals such as chelating agents or salt exchangers in the process.
6. It works in a broad temperature range. The variation of water temperature does not change the function of the equipment or the mechanism of the process.
7. The yield of water is typically above 75%. This is higher than the others and is quite important for areas where water is in shortage.

REFERENCES

[1] Brown, Lester R.. and Brian Halweil. 1998. China's Water Shortage Could Shake World Food Security. *World Watch*, (July):10–21.

[2] Steven Mufson. As Economy Booms, China Faces Major Water Shortage. *The Washington Post*, March 16, 2010.

[3] National Geographic. 2009. Bitter Waters (January).

[4] Derickson, Russell, Fred Bergsrud, and Bruce Seelig. 1992. Treatment Systems for Household Water Supplies: Distillation. www.ag.ndsu.edu/pubs/h2oqual/watsys/ae1032w.htm

[5] Remco Engineering. Ion Exchange. www.remco.com/ix.htm.

[6] Derickson, Russell, Bruce Seelig, and Fred Bergsrud. Reverse Osmosis. www.ag.ndsu.edu/pubs/h2oqual/watsys/ae1047w.htm.

[7] Buros, O.K. 1987. An Introduction to Desalination. In *Non-Conventional Water Resources Use in Developing Countries*. Natural Resources/Water Series No. 22. New York, United Nations, pp. 37–53.

[8] California Coastal Commission. Seawater Desalination in California. www.coastal.ca.gov/desalrpt/dchap1.html.

[9] GE Water and Process Technologies. Reverse Osmosis Equipment (Spiral Membranes). www.gewater.com/products/equipment/spiral_membrane/index.jsp.

[10] Eisenberg, Talbert N., and E. Joe Middlebrooks. 1992. A Survey of Problems with Reverse Osmosis Water Treatment. *Journal AWWA*, 76(8):44.

[11] Birkett, J.D. 1987. Factors Influencing the Economics of Desalination. In *Non-Conventional Water Resources Use in Developing Countries*. Natural Resources/Water Series No. 22. New York, United Nations, pp. 89–102.

[12] Becker, H.I. 1957. U.S. Patent 2800616.

[13] Johnson, A.M., and John Newman. 1971. Desalting by Means of Porous Carbon Electrodes. *J. Electrochem. Soc.: Electrochem. Technol.* 118(3):510–517.

[14] Williams, J.M. 1975. U.S. Patent 3,859,195.

[15] Anderson, Marc A., Ana L. Cuderob, and Jesus Palmab. 2010. Capacitive Deionization as an Electrochemical Means of Saving Energy and Delivering Clean Water. Comparison to Present Desalination Practices: Will it compete? *Electrochim. Acta*, 55:3845–3856.

[16] Daoduo, Q., Z. Linda, and H. Eric. 2007. *Res. J. Chem. Environ.* 11:92.

[17] Farmer, J., D. Fix, G. Mack, R. Pekala, and J. Poco. 1996. *J. Appl. Electrochem.* 26:1007.

[18] Farmer, J.C., D.V. Fix, G.V. Mack, R.W. Pekala, and J.F. Poco. 1996. *J. Electrochem. Soc.* 143:159.

[19] Farmer, J.C., S.M. Bahowick, J.E. Harrar, D.V. Fix, R.E. Martinelli, A.K. Vu, and K.L. Carroll. 1997. *Energy Fuels* 11:337.

[20] Pekala, R.W., J.C. Farmer, C.T. Alviso, T.D. Tran, S.T. Mayer, J.M. Miller, and B. Dunn. 1998. *J. Non-Cryst. Solids* 225:74.

[21] Welgemoed, T.J., and C.F. Schutte. 2005. *Desalination* 183:327.

Research on Primary Water Hardness of Coal Slurry

Jiong-tian Liu and Ming-qing Zhang
China University of Mining and Technology, Xuzhou, Jiangsu, China

ABSTRACT

We analyzed the composition, size of fine coal and ion concentration in coal slurry from 18 coal dressing plants. The results show that the water hardness of coal slurry can either change (increase or decrease) or remain stable during coal washing. This finding violates the common understanding that water hardness can only increase with coal slurry recycling. In this study, we analyzed the relations between water hardness and the settling performance of coal slurry, and calculated the interaction energies between coal particles in coal slurry suspensions under the different conditions of water hardness. The results show that water hardness is the determining factor in slurry settling performance; the higher the water hardness, the better the settling performance. On the basis of this research, we propose the new concept of primary water hardness and means of classification. Based on primary water hardness versus settling performance, we developed a mathematical model of calculating primary water hardness of coal slurry in coal preparation.

INTRODUCTION

Water is the working liquid medium in coal preparation plants, and water chemistry has a strong effect on coal classification. The settling performance of coal not only determines the natural clarification of coal slurry water, but also directly affects other production steps. Prior research showed that in the case of stable coal slurries, the use of circulated water with mud as the washing medium deteriorates the quality of product by 1–2 grades and production decreases 3–5%. In the separation stage, circulating water with mud adversely affects production quality and recovery, and coal flotation often cannot be conducted. High coal slurry stability is a cause of production interruptions in coal preparation plants. To maintain production, extra water is used to dilute the high-density coal slurry, which makes the load rate and operation rate only about 70% (Zhang et al. 2009). Therefore, good settling performance of coal slurry water is necessary to improve coal quality and increase the profits of coal preparation operations.

In fact, there are big differences in the settling performance of various coal slurries. The easily settling coal slurries can clarify water quickly without adding any reagents, but poorly settling coal slurries can be stable for many weeks or months if reagent is not added. At present, the main index used to describe the natural settling performance of coal slurry water is settling velocity. The settling velocity refers to the speed of settling, but it is not explicitly a criterion for settling performance. The settling velocity of coal slurry water often has great uncertainty because of various factors.

In this paper, we propose a new characterization index for natural settling performance of coal slurry in terms of water hardness. The influential main factors for primary hardness of coal slurry water are analyzed in terms of coal mineral composition and its chemical reaction in aqueous solution. The classification method based on primary hardness of settling performance is proposed and a computational model of primary hardness theory is introduced.

Using our approach, it is possible to explain settling performance of coal slurry water and provide some guidelines for technological design of coal slurry water treatment, equipment selection, and design of other related processes.

Emerging Background of Primary Hardness

Coal Mineral Composition and Settling Performance of Coal Slurry Water

Beside organic maceral, coal is made of clay, soluble minerals, quartz and other minerals. Organic maceral is relatively stable, but it easily breaks to smaller particles. Due to a strong hydrophobic attracting energy, coal particles have a tendency to agglomerate in water. Soluble minerals, on the other hand, can dissolve in water. Quartz and other inorganic minerals remain stable in water; their high densities make it easy for them to settle in water. Clay particles expand and disperse with electric charges on their surfaces. These clays remain suspended in water and their concentration increases in recycled water, thus becoming the main mineralogical component of coal slurry water (Zhang et al. 2008). For example, Sabah and Cengize (2004) found that suspended minerals in coal slurry water from the Tuncbiled coal preparation plant contained 50 wt% clay minerals. Further, during studies on settling performance of coal slurry water in Wanbo Mining Corporation, de Drester and Boger (1995) discovered that the presence of montmorillonite in the coal slurry water notably decreases both cohesion and settling performance. Moreover, the study by Arnold and Aplan (1986) showed that clay minerals account for more than 80% of suspended particles in the recycled coal slurry water. In general, when clay content in coal slurry is too high, slurry water appears either yellow or white and coal slurries have difficulty settling.

Chemical Features and Settling Performance of Coal Slurry Water Solution

Suspended particles in coal slurries carry electric charges on their surfaces. Positive ions are needed to neutralize surface charges and accelerate settling of particles and improve the overall performance. The main physical and chemical processes that occur in water include:

1. Dissolution of soluble and semi-soluble minerals including carbonates (mainly gypsum and calcite), chlorides, and sulphates in water. Dissolved ions increase the ionic strength of water.
2. Oxidation-reduction reaction. The reaction is between elemental sulphur and pyrite. It forms a weak acid solution that is beneficial in dissolving other inorganic minerals, contributing to a further increase in the ionic strength of water.
3. Ion exchange and adsorption/absorption reactions in clays.

After the above reactions (1 through 3), the main positive ions in coal slurry include Al^{3+}, Fe^{3+}, Ca^{2+}, Mg^{2+}, K^+ and Na^+. The Al^{3+} and Fe^{3+} can hydrolyze in the pH range from 6.5 to 8.5; therefore, their concentrations are very low and there is typically no significant effect of Al^{3+} and Fe^{3+} on the settling performance of coal slurry water. The concentrations of K^+ and Na^+ in coal slurry water are much higher. Some researchers show that the presence of K^+ is beneficial to dissolution processes and the dispersion of clay particles (Liu et al. 2010). So in some coal slurry water, the K^+ and Na^+ concentrations are higher, but slurry settling performance is poor. Both Ca^{2+} and Mg^{2+} adsorb on clays and have outstanding effects on the settling performance of coal slurry water. When the concentrations of Ca^{2+} and Mg^{2+} are higher, the coal slurry water settles much more quickly (Zhang et al. 2008).

We performed a systematic study on the effect of solution chemistry and the natural settling performance of coal slurry for hundreds of coal preparation plants in China. "Easy" natural setting performance expresses that it can naturally clarify without adding any reagent. "Middle" settling performance shows that the time of natural clarification is longer and some flocculants need to be added to ensure steady operation. "Difficult" settling means that the slurry cannot be clarified without the introduction of high flocculent dosages. Results for 18 coal preparation plants are shown in Table 2. The data in Table 2 basically verify the above classification.

It can be seen from Table 1 that the water hardness is the determining factor for natural settling performance of coal slurry water.

In the design and selection of classification equipment, the solid content and particle size distribution are the most important factors. In the design of thickener, the area of thickener F is calculated from the following formula (Hao and Li 1993):

$$F = Gf \tag{1}$$

where G is the dry slurry content in the feed and f is the area of thickener needed in unit productivity of dry slurry. f is determined by particle size of separation and density of coal slurry water. For the same particle size of separation and density of coal slurry water, f is the same.

The settling performance of coal slurry water not only has the relation with particle size of separation and density, but also water hardness and the particle content in water. When the water hardness is high and clay mineral content is low, f is small. For the opposite case, f is big. It is obvious that the factors considered in the above analysis are incomplete. Therefore, it is necessary to accurately predict the influences of natural settling performance of coal slurry water on the design and selection of clarifying equipment, the choice of classification reagent, and systematic management.

Concept of Primary Hardness

The primary hardness of coal slurry water is a new concept, which analyzes the correlation between water hardness and settling performance of a coal slurry. The hardness is determined by the various physical and chemical processes that occur in the coal slurry, as discussed earlier. The specific definition is: the hardness of recycled coal slurry water, without adding flocculants, which is formed by a series of dissolution events and adsorption/absorption of organic matter and inorganic minerals in the slurry.

Determining Factors of Primary Hardness

The primary hardness is determined by the mineral composition of coal and the condition of filling water. Under the conditions that the pH of the filling water and its ionic composition remain stable, it represents inherent characteristics of coal. Therefore, according to our concept, the determining factor of primary hardness is, indirectly, the mineralogical composition of coal.

The main minerals of coal include various soluble/semi-soluble minerals and clays. The dissolution of carbonate (such as gypsum and calcite), chloride and sulphate make the hardness increase. The oxidation-reduction reactions between elemental sulphur and pyrite form a weakly acidic environment, prompting minerals dissolution further and adding more ions to recycled water. The ion exchange and adsorption of K^+, Na^+ and Ca^{2+} and Mg^{2+} occur in clay minerals, which make the water hardness decrease. The balance of these three reactions finally determines the primary hardness of coal slurry water. Thus, the final primary hardness of coal

Table 1. Filling and circulating water quality and settling performance in some coal preparation plants

Plant	Water	Na^+/K^+ (mg/L)	Ca^{2+} (mg/L)	Mg^{2+} (mg/L)	Al^{3+} (mg/L)	Fe^{3+} (mg/L)	Water Hardness (mg $CaCO_3$/L)	Settling Performance
Dutun	Circulating water	406.0	454.9	128.9	0.6	2.7	1664.9	Easy
	Filling water	249.6	520.2	147.0	0.7	2.8	1912.5	
Zazhuang	Circulating water	268.4	594.8	320.6	—	—	1398.9	Easy
	Filling water	76.3	567.6	223.9	—	—	1056.1	
Konzhuang	Circulating water	562.2	143.0	70.7	1.0	2.6	647.6	Middle
	Filling water	324.9	178.7	82.6	0.3	2.6	786.1	
Chaili	Circulating water	337.3	594.8	201.3	—	—	901.4	Easy
	Filling water	355.8	73.8	17.7	—	—	128.3	
Taixi	Circulating water	124.9	245.7	85.5	—	—	836.9	Middle
	Filling water	90.7	24.8	7.1	—	—	92.0	
Linhuan	Circulating water	424.2	13.6	9.9	0.6	2.8	74.7	Difficult
	Filling water	216.7	69.5	53.6	1.3	—	393.9	
Jianzhuang	Circulating water	336.4	118.5	17.6	—	—	201.0	Middle
	Filling water	137.3	229.2	33.0	—	—	353.5	
Xiaqiao	Circulating water	21.7	246.4	76.9	0.1	0.5	936.4	Easy
	Filling water	17.8	187.6	50.9	0.1	2.2	681.3	
Jiahe	Circulating water	328.7	36.9	4.2	1.3	2.5	109.6	Difficult
	Filling water	233.2	21.4	31.1	1.8	2.3	181.3	
Shitai	Circulating water	311.5	21.0	9.3	0.1	3.5	91.1	Difficult
	Filling water	50.0	107.6	31.3	0.4	—	397.3	
Dawukou	Circulating water	99.4	39.1	16.1	—	—	489.7	Middle
	Filling water	64.2	48.9	12.1	—	—	172.5	
Pingba	Circulating water	186.2	47.8	22.1	1.3	2.3	210.9	Middle
	Filling water	118.0	58.3	17.4	1.1	2.4	217.3	
Bayi	Circulating water	161.6	75.8	8.3	—	—	111.6	Difficult
	Filling water	96.6	73.8	11.8	—	—	116.2	

— shows that the content is low and cannot be tested.

slurry water is the result of clay mineral content and the presence of calcium-magnesium minerals in coal.

The types of minerals and content of coal maceral are closely related with the conditions of coal formation. Therefore, the coal can be classified as follows (Table 2).

Settling Performance Classification of Coal Slurry Water Based on Primary Hardness

Coupling the practical production experience with our proposed model, we put forward the method of natural settling performance classification of coal slurry water based on primary hardness, which is shown in Table 3. The natural settling performance of coal slurry water can be divided into three categories, including easy settling, middle settling and difficult settling. The easy settling slurries represent suspended particles that are in cohesive state; coal slurry

Table 2. Classification of coals

Classification	Minerals in Coal	Formation Conditions
Easy settling coal	A lot of carbonate, sulfide ore and sulphate	Marine sedimentary coal
Middle settling coal	The clay minerals are less. A certain amount of carbonate, sulfide ore and sulphate	Various sedimentation of coal
Difficult settling coal	A lot of clay minerals	Continental sedimentation of coal

Table 3. Settling performance classification of coal slurry water

Classification	Primary Hardness (mg $CaCO_3$/L)
Easy Settlement	>900
Middle Settlement	200~900
Difficult Settlement	<200

water can clarify naturally without adding any flocculants. The difficult settling slurries represents those with suspended particles remaining in the disperse state, and more particles accumulate easy in water. This slurry circulates well but its density constantly increases. Suspended particles are resistant to full cohesive even after addition of flocculants. Adding too much flocculent typically enhances the stability of the slurry and the settling reagent regime of coal slurry water mainly consists of flocculants. It can make the coal slurry water clarify, but the cost of the reagent is high. Finally, the middle settling coal slurry water represents a suspension in between the above two types. The reagent regime and dosage are also between the above two types.

A Computing Model on Primary Hardness of Coal Slurry Water

As discussed earlier, various ions in water can be balanced by various microeffects. The change of hardness has three variation tendencies shown in Figure 1, including rising type, reducing type and basic stable type. No matter which tendency, the water hardness tends to plateau.

Taking the difficult to settle coal slurry water as an example, its changing tendency in hardness is the reducing type. And during this process, the adsorption contents of Ca^{2+} and Mg^{2+} are more than dissolution capability. Therefore, the process of dissolution can be neglected in the model.

The basic equation of absorption dynamics is as follows:

$$q_t = \frac{q_e^2 \cdot t}{q_e \cdot t + K} \tag{2}$$

where q_t is the density variation of Ca^{2+} and Mg^{2+} at t time during recirculation the coal slurry water, mmol/L; t is the time of absorption, h; K is the speed constant of absorption.

Then the equation of absorption dynamics at many recirculation cycles is as follows:

$$q_{tn} = \frac{q_{en}^2 \cdot t}{q_{en} \cdot t + K_n} \tag{3}$$

where q_t is the density variation of Ca^{2+} and Mg^{2+} at t time after n cycles of coal slurry water recycling, mmol/L; K_n is the speed constant of absorption after n cycles.

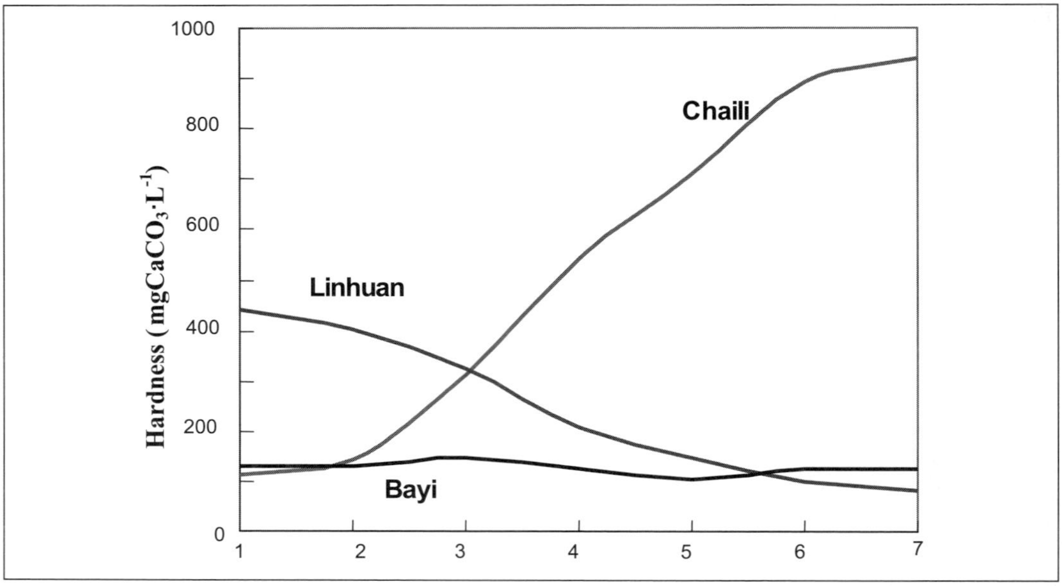

Figure 1. Coal slurry hardness along flow path for three plants

Assuming that the time of each circulation is t_0, the density of Ca^{2+} and Mg^{2+} at any moment in each cycle is obtained.

The density of Ca^{2+} and Mg^{2+} after the 1st cycle:

$$C_1 = C_0 - \frac{q_{e1}^{\ 2} \cdot t_0}{q_{e1} \cdot t_0 + K_1} \tag{4}$$

where C_0 is the density of Ca^{2+} and Mg^{2+} of filling water, mmol/L; t_0 is the time of accomplishing a cycle, h is density of Ca^{2+} and Mg^{2+} after n cycles.

$$C_n = C_{n-1} - \frac{q_{en}^{\ 2} \cdot t_0}{q_{en} \cdot t_0 + K_n}$$

$$= C_0 - \frac{q_{e1}^{\ 2} \cdot t_0}{q_{e1} \cdot t_0 + K_1} - \frac{q_{e2}^{\ 2} \cdot t_0}{q_{e2} \cdot t_0 + K_2} - \cdots - \frac{q_{en}^{\ 2} \cdot t_0}{q_{en} \cdot t_0 + K_n} \tag{5}$$

When n approaches infinity, the C_n is primary hardness.

Besides the above methods of computing primary hardness, it also can be obtained by considering the mineralogical composition of raw coal, which is neglected in this analysis.

CONCLUSIONS

The high clay mineral content is the basic reason that coal slurry water is difficult to settle and the water hardness is the determining factor for natural settling performance of coal slurry water. When the water hardness is high, the settling performance of coal slurry water is good. When the water hardness is low, the rate of particle settling is low and settling performance is poor. The primary hardness is determined by various physical and chemical processes occurring in slurries that affect solution chemistry. The natural settling performance of coal slurry water can be divided into three categories, including easy settling, middle settling and difficult settling slurries. The primary hardness of coal slurry water may be computed by the theoretical model presented in this paper.

REFERENCES

Arnold, B.J., and Aplan, F. 1986. The effect of clay slimes on coal flotation: Part A. The nature of the clay. *International Journal of Mineral Processing* 17(3):201–205.

de Dretser, R.G., and Boger, D.V. 1995. Improvement of coal tailings disposal via understanding of clay chemistry. The AusiMM Annual Conference. Newcastle, 23–26.

Hao, F.-Y., and Li, W.-L. 1993. *Coal Preparation Handbook*. Haerbin: Haerbin Industry Publishing, 34–35.

Liu, J.-T., Zhang, M.-Q., Zeng, Y., et al. 2010. Effects of different type clays on the dispersion of fine particles in coal slurry. *Journal of China University of Mining & Technology* 39(1):60–65.

Sabah, E., and Cengize, I. 2004. An evaluation procedure for flocculation of coal preparation plant tailings. *Water Research* 38:1542–1549.

Zhang, M.-Q., Liu J.-T., Han, B.-P., et al. 2008. Dissipation Structure Characteristics of Coal Processing System. *Journal of China University of Mining & Technology* 30(6):735–739.

Zhang, M-Q., Liu, J-T., He, W., et al. 2009. Fractal Characteristics of Flocs in Coal Slurry Treated Using PAM. *Research of Environmental Sciences* 29(8):956–960.

Zhang, M.-Q., Liu, J.-T., and Wang, Y.-T. 2008. Effects of water hardness on the dispersion of fine coal and kaolinite in coal slurry. *Journal of China University of Mining & Technology* 33(9):1068–1062.

Membrane Operations in Water Treatment and Reuse

Enrico Drioli
University of Calabria, Arcavacata di Rende (CS), Italy
WCU Energy Engineering Department, Seoul, South Korea

Francesca Macedonio
University of Calabria, Arcavacata di Rende (CS), Italy

ABSTRACT

In the last decades membrane operations have gained great success in water treatment and reclamation because their efficiency has been proven from a technical, economical and environmental point of view. Due to their unique advantages like high recovery factor, decreased footprint, low energy requirement and relatively high selectivity, membrane processes are being increasingly used in water treatment and their costs have decreased rapidly.

This manuscript gives an overview of the membrane operations used in water treatment, some details regarding their theoretical background and various examples of applications. Then, the advantages in the use of integrated membrane systems and the future development of membrane technology are presented.

INTRODUCTION

Nowadays membrane processes are essential operations for a wide range of applications, including energy generation, tissue repair, pharmaceutical production, food packaging, and the separations needed for the manufacture of chemicals, electronics and a range of other products (Rios et al. 2007). However, water treatment is the field that epitomizes the success of membrane technology, both for overcoming water scarcity and for preventing water pollution.

The modular design of membrane operations, the operational simplicity and flexibility, the compactness and small carbon footprint of these plants (Drioli and Curcio 2007; Peters 2010), the short construction time, the relatively high selectivity and permeability for the transport of specific components, the good stability under a wide spectrum of operating conditions justify the success of membrane science and membrane engineering. The low energy requirement as well as the economic and long term operational reliability are further advantageous aspects. These features result from the intrinsic properties of the membranes, from their combination with an appropriate module configuration and plant design, and from the work of researchers able to face the basic problems related to the understanding the final morphology of dense and porous membranes, their transport mechanisms, and to develop new membrane operations for molecular separation, mass and energy transfer between different phases.

Membrane operations have been able to show their potentialities in the rationalization of water treatment systems and the use of membrane systems has increased significantly: in 1997 the membrane sales were only US$ 900 million (Peter-Varbantes et al. 2009). Between 2006 and 2010, the global market in membrane systems for water and wastewater applications grew from US$ 6.7 billion to US$ 10 billion (Peter-Varbantes et al. 2009). 60% of the total worldwide desalination installed capacity is based on reverse osmosis (RO) technology. Conventional wisdom in the desalination industry says that thermal technology is in terminal decline because of the higher energy consumption and lower recovery factor with respect to membrane based operations. Moreover, water price with RO is 23% cheaper than thermal processes (IDA Desalination Yearbook 2010–2011).

In the next sections, first an overview of the existing membrane-based water treatment processes will be given. Then, the future development of membrane technologies will be illustrated.

MEMBRANE PROCESSES FOR WATER TREATMENT

In many regions of the world, due to large population, tourist infrastructure and industrial development increase, it is no longer possible to satisfy the growing water demand by conventional methods of water supply and processing. Therefore, keeping in mind that water has a value and not only a price, in addition to a more conscientious use of natural resources, an increased utilization of advanced separation techniques such as membrane operations was witnessing.

Membranes are used to produce potable water from the sea, to clean industrial effluents, and to recover and reuse municipal wastewaters. The production of potable water from saline and polluted waters is essential for increasing the amount of available good quality water. The purification and reuse of wastewaters is crucial for rising the exploitation of potable water and reducing its consumption. The treatment and recycling of process waters (e.g., the purification of the effluent from sewage treatment plants) is indispensable for preventing further contamination of water resources.

In general, membrane processes are characterized by the use of a semi-permeable thin barrier (i.e., the membrane) that controls the exchange between two phases on the basis of the applied driving forces, the fluid properties and through the intrinsic characteristics of the membrane material itself. The driving force can be a difference in pressure, concentration, temperature or electric potential. Most membrane processes are pressure-driven (microfiltration, ultrafiltration, nanofiltration, reverse osmosis and membrane bioreactor), where the driving force is an hydrostatic pressure difference across the membrane. In water treatment, however, electrically driven (e.g., electrodialysis) and thermally driven processes (e.g., membrane distillation) are also used.

In a membrane process the water to be treated is separated into a stream of filtrate (or permeate) and one of retentate (or concentrate or brine, containing the components of the feed water rejected by the membrane). The quality of the produced filtrate depends on the characteristics of the membrane technologies in relation to the water-borne contaminants:

- Microfiltration (MF) is used, e.g., for suspended solids and large bacteria removal from waters. MF membranes are usually symmetric microporous structures, with pore size in the range of 10–0.05 μm (Mulder 1996). The membrane thickness can extend from 10 to more than 150 μm. Separation is accomplished by MF membranes via mechanical sieving and particles are separated solely according to their dimensions with respect to those of membrane pores. The hydrostatic pressure difference used as driving force is low (usually less than 2 bar).
- The membranes used in ultrafiltration (UF) have an asymmetric structure with pore size (1–50 nm) small enough to ensure high removal of all kinds of microbiological hazards such as Cryptosporidia, Giardia and total bacterial counts (Hagen 1998); suspended solids, large bacteria, dissolved macromolecules, colloids and smaller bacteria are retained; turbidity and suspended solids are completely removed. Substantial virus removal can be also attained with UF membranes since the size of viruses is in the range of 30–300 nm (Peters 2010). The operating pressure is up to 10 bar.

- Nanofiltration (NF) and reverse osmosis (RO) can be used to remove low molecular weight solutes (such as inorganic contaminants, inorganic salts or small organic molecules) from waters. Both processes are considered as one process since the basic principles are the same. RO is operated with tighter type of membrane than NF. Most NF membranes are effective for color, sugar and dye removal or for removing hardness, sulfate, bivalent ions (typical retention > 90%). RO membranes are required for monovalent ions removal (e.g., desalination of seawater or brackish water is currently performed with RO membranes), and for the separation of dissolved salts and ions with a molecular weight of less than 200 g/mol (Peters 2010). Compared to MF and UF, in NF and RO the organic and inorganic molecules are separated from a feed solution by a solution diffusion process. The operating pressure is up to 25 bar for NF and up to 80 bar for RO.
- Membrane bioreactor (MBR) is a separation process combining membrane filtration with biological treatment which is finding increasing applications for industrial and municipal water treatment.

  Conventional MBR uses low-pressure membrane filtration, either MF or UF, to retain the mixed liquor of the bioreactor, and delivers particle-free treated effluent (Lay et al. 2010). Because the membrane is an absolute barrier for bacteria and in the case of UF also for viruses, the MBR process provides a considerable level of physical disinfection. The resulting high quality and disinfected effluent implies that MBR processes can be especially suitable for reuse and recycling of wastewater.

  Advantages in the use of MBR are as follows: (i) the technology permits bioreactor operation with considerably higher mixed liquor suspended solids (MLSS) concentration than conventional activated sludge (CAS) systems, which are limited by sludge settling; (ii) compactness (up to 5 times more compact than a CAS plant).

  With respect to costs, MBR is considered a high tech process, with higher initial investment costs than conventional wastewater treatment. Moreover, the energy demand to cope with membrane fouling is the main contribution to the overall operating costs (Kraume and Drews 2010).

  Membrane bioreactors are by now almost exclusively used in wastewater treatment. However, the great potential of MBRs to produce high quality effluent could also be of great interest in the removal of a variety of anthropogenic organic pollutants and fouling agents that are increasingly present in sea/brackish-water.
- In electrodialysis (ED) charged membranes are used to remove ions from aqueous solution. A number of cation- and anion-exchange membranes are placed in an alternating pattern between a cathode and an anode. When a direct current is applied, the positively charged ions migrate to the cathode and the negatively charged ions to the anode. Therefore the cation-/anion-exchange membranes are ion-selective membranes which control the movement of ions. Thus, the concentration of ionic species is reduced in the so-called *diluted* compartments and increased in the *concentrated* compartments.

  In commercial applications several hundreds of cell pairs are assembled in a stack and in this way the applied driving force is very effective.

  ED has been in commercial use for desalination of brackish water for the past three decades, particularly for small- and medium-scale processes (AlMadani 2003; Charcosset 2009).

Table 1. Membrane distillation applications

Desalination and pure water production from seawater and/or brackish water
Nuclear industry (concentration of radioactive solutions and wastewater treatments; pure water production)
Textile industry (removal of dyes and wastewater treatment)
Industrial and municipal used waters (removal of small size and persistent contaminants)
Chemical industry (concentration of acids, removal of VOCs from water, separation of azeotropic aqueous mixtures such as alcohol/water mixtures and crystallization)
Pharmaceutical and biomedical industries (removal of water from blood and protein solutions, wastewater treatment)
Food industry (concentration of juices and milk processing) and in areas where high temperature applications lead to degradation of process fluids

ED process is non-economical for waters with high salts concentrations (Van der Bruggen and Vandecasteele 2002; Fritzmann 2007), but is competitive for brackish waters with up to 3000 ppm salt.

- Membrane distillation (MD) is a thermally driven membrane process which operates on the principle of vapour-liquid equilibrium. In this process, two solutions at different partial pressure are separated by a hydrophobic porous membrane and the solutions must not wet the membrane. When a vapour pressure difference exists across the membrane, the vapour molecules are transported through the membrane pores, from the high vapour pressure side to the low vapour pressure side. Only volatile components are transferred through the membrane, therefore 100% (theoretical) of non-volatile components are rejected and a high quality permeate can be obtained.

  MD is a membrane process in which the membrane does not distinguish between solution components on a chemical basis, does not act as a sieve and does not react electrochemically with the solution. Membrane in MD acts merely as a support for the vapour-liquid interface and the selectivity is determined by the vapour-liquid equilibrium involved.

  MD applications are essentially determined by the wettability of the membrane. To avoid wetting, the surface energy of the polymer must be low (polypropylene, polytetrafluoroethylene and poly(vinylidene fluoride) are usually used), the maximum pore size small (pore size in the range of 0.2–1.0 μm is desiderable), the surface tension of the liquid high (e.g., water). The latter implies that mainly aqueous solutions containing inorganic solutes can be treated. MD can be used for (Table 1): (i) water desalination, (ii) the treatment of water for the semiconductor industry or power plants, (iii) the treatment of wastewaters, (iv) the removal of volatile bioproducts.

POTENTIAL FOR INTEGRATION

Water treatment processes were revolutionized by the introduction of membrane operations. In particular seawater desalination was transformed by RO technology. In comparison to conventional water treatment, the main advantages of membrane processes are that, in principle, water can be treated in one stage without chemicals or utilities, the footprint is relatively small, the membrane system can be built in a modular form which enables easy adaptation of process scale. However, the main limitation of membrane systems is fouling. Components present in the feed water can be deposited and/or absorbed on the membrane surface. This inevitable phenomenon, called membrane fouling, makes necessary to perform proper pre-treatment of the feed and periodic membrane cleanings in order to maintain the economic feasibility of membrane operations.

Also the handling of the produced retentate is not an easy task. The solution chosen for the brine rejection depends on several parameters, such as chemical composition, flow and dilution- rates of the concentrate, etc.

In the case of seawater desalination, the brine can be discharged directly in the natural environment (e.g., for desalination plants located near coastal areas the concentrate stream is discharged to the sea). In this case particular attention should be given to the possibility of dangerous ecosystem modifications. When the concentrate cannot be directly discharged (such as in some brackish water desalination plants), it is frequently discharged into solar evaporation ponds.

While well known technical solutions are usually available for the design and the manufacture of a membrane based unit, the pretreatment of the water to be processed as well as the handling of the retentate (here referred as post-treatment) represent crucial aspects of each water treatment process, those determining the success or the failure of a plant. Pre-treatment and post-treatment have to be adapted to the specific conditions at the construction site of a plant. These can differ over a wide range, depending also on the raw water characteristics. They include the systems for dosage and the handling of chemical agents for pretreatment and for the cleaning of membranes.

In addition, the wastewater generated in the pretreatment and the wastewater generated during membrane cleaning are polluted effluents whose final destination must be controlled. It is advisable to treat them separately and not to mix them for discharge (as it has been usual practice in the past). On one hand, the separate handling allows for recycling possibilities for certain partial streams and, on the other hand, an environmentally sustainable operation that avoids the contamination of the receiving water body is achieved (Peters 2010).

A significant possibility for improving further current water treatment systems is offered by the mutual compatibility of different membrane operations for integration. Integrating diverse membrane operations means coupling various membrane processes for (i) overcoming the limits of the single units and (ii) using their synergic effects in terms of better performance of the overall system. Integrated membrane systems offer new opportunities in the design, rationalization and optimization of industrial processes.

Seawater desalination is probably the clearest example of what can be achieved through integrated membrane systems: cheaper, better quality and more abundant water, with less brine production. Figure 1 shows a schematic integrated membrane based desalination system.

As earlier illustrated, RO pre-treatment and post-treatment steps are required to condition water before and after the membrane process.

The implementation of MF and/or UF technologies for the RO pre-treatment in place of the conventional pre-treatment is now expanding worldwide. Membrane pre-treatment can provide (i) high levels of contaminants removal (including particulates, colloids and pathogens) with lower chemicals addition and, therefore, with a lower environmental impact, (ii) reduced cost and foot-print, (iii) better capability to handle wide fluctuations in raw water quality, (iv) operation with a high and stable permeate flux during long term operation, (v) low energy consumption.

Moreover, it can be necessary to add a coagulation and settling/flotation for the treatment of very bad water quality. This coagulation and settling is considered "the pre-treatment

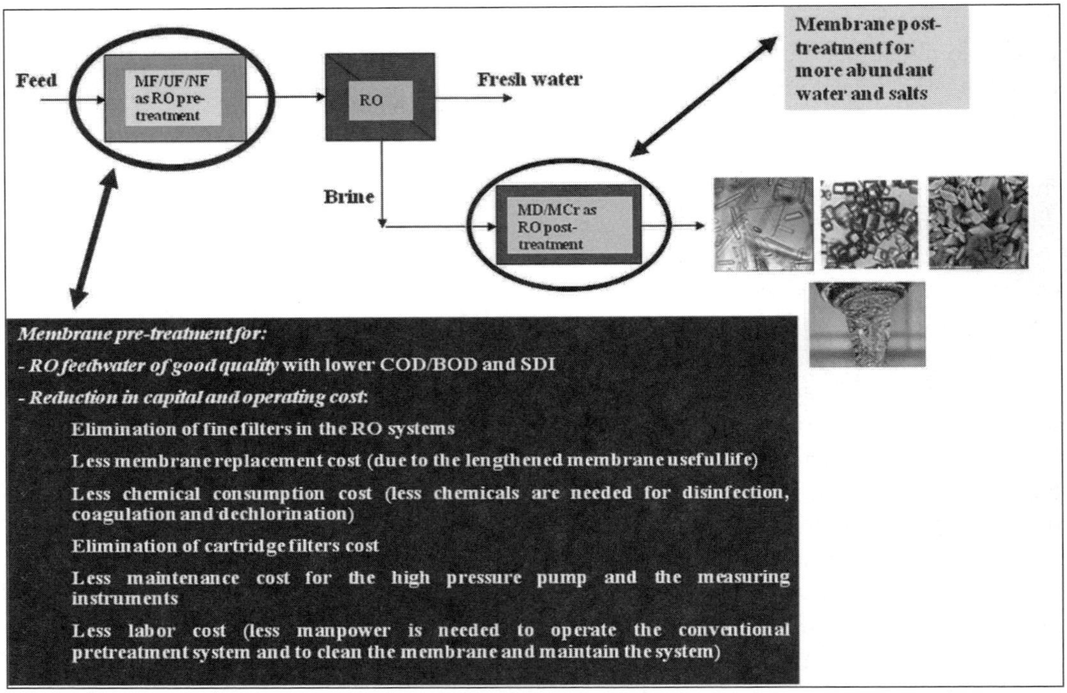

Figure 1. Scheme of an integrated membrane desalination system

of the pre-treatment" to face bad seawater quality, high turbidity, algae counts (red tide and blooms), hydrocarbon pollutions. It must be noted that the association pre-clarification plus UF/MF is more expensive than conventional pre-treatment. UF/MF and pre-coagulation is justified when associated with better control of membrane fouling and biofouling issue, reduction of RO cost, higher RO flux and/or higher recovery. The latter typically ranges between 40–50% because of the limitations imposed by scaling problems. Appropriate antiscalant dosages coupled to NF remove the large part of multivalent ions present in the feed water, thus decreasing the potential for salts precipitation even if the RO unit operates at higher recovery factors (Drioli et al. 1999).

To accomplish the ambitious objective of reaching recovery factor higher than 85–90%, conventional membrane operations (such as MF, UF, NF and RO) need to be combined with other innovative membrane processes (such as MD). With respect to pressure-driven membrane processes, MD does not suffer limitations arising from concentration polarization and can therefore be employed whenever high permeate recovery factors are requested. Actual experimental data and design studies in literature suggest that, if MD is operated on the RO retentate, the total amount of desalinated water represents almost 88% of the feed water (Drioli et al. 2006; Macedonio et al. 2007). Additionally, the use of a Membrane Crystallization (MCr, an interesting extension of MD concentration up to supersaturation) allows to exploit brine added value, not only increasing plant recovery factor, but also extracting the salts naturally present in the brine streams of the desalination plants thus decreasing brine disposal problem and its negative environmental impact (Macedonio et al. 2009; Macedonio and Drioli 2010).

The overall water recovery factor of an integrated desalination system constituted by MF-NF-RO and MCr working on both NF and RO brine can increase up to about 93%.

The drawback of the processes with MD/MCr units is their higher operating temperature with respect to RO. On the other hand, the MD/MCr required operating temperatures (typically below 70°C) are much lower than those of conventional thermal desalination processes, therefore low-grade waste and/or alternative energy sources can be coupled with MD/MCr for a cost and energy efficient water treatment system.

The benefits of MD, however, are not limited to the treatment of brine solutions of desalination plants. More recently membrane distillation has been also used in membrane bioreactor configuration (MDBR) for the treatment of industrial and municipal used waters (Fane et al. 2005; Phattaranawik et al. 2008). The development of MDBR system was due to the fact that, in a conventional MBR, the molecular weight cut-off of the utilized MF/UF membranes delivers a portion of the organic species of the feed. Lay at al. (Lay et al. 2010) reports that the effect of this is that recalcitrant organics may not be well degraded, and the direct reuse potential of the permeate may be limited. For overcoming this, the development of a number of innovative high retention membrane bioreactors (HRMBRs) have been developed such as membrane distillation MBR, where the MD membrane is used in place of MF/UF. Other examples are nanofiltration MBR (NFMBR) (Choi et al. 2002; Choi et al. 2007] and the osmotic MBR (OMBR) (Cornelissen et al. 2008; Oo et al. 2008]. The HRMBR systems are able, in principle, to retain effectively small size and persistent contaminants, which facilitates their biodegradation in the bioreactor, thereby producing higher quality product water (Lay et al. 2010). Moreover HRMBR systems potentially possess comparative economical advantage in removing pollutants from the used water more effectively. However, as these high rejection systems retain dissolved solids, they need to be operated under elevated salt condition (Lay et al. 2010).

FUTURE DEVELOPMENT OF MEMBRANE TECHNOLOGY

At present, important developments are taking place in industrial membrane applications focused on the integration of different membrane processes. Integrated systems offer new opportunities for reaching better quality products, more compact production plants, more sustainable and efficient processes with reduced energy consumption.

The future development of membrane technology are influenced by factors such as:

1. The development of membranes particularly adapt to specific applications, able to achieve the most increasingly stringent degree of purification of the raw feed waters as expected from the customer or imposed by law;
2. The handling and control of concentrate discharge;
3. The development of water treatment systems coupled with renewable energy sources for a significant reduction in energy consumption;
4. The realization of advanced integrated water management systems, at closed-circuit, that follow all the lines, from the water intake to water distribution and reuse (Macedonio et al. *accepted*), and based on *graduated quality requirements* (Peters 2010; Macedonio et al. *accepted*), i.e., on supplying water of diverse quality to the final users depending from their requirements (drinking, washing, agriculture, irrigation and industrial use);
5. The necessity to further reduce water treatment costs.

Among the technological developments, carbon nanotubes, fullerene, aquaporin channels, new protein-based membranes and graphene membranes are emerging in the recent years as innovative water technologies, as developed membranes with superior permeability, durability and selectivity for water purification (Macedonio et al. *accepted*).

CONCLUSIONS

Membrane Engineering is playing a dominant role in water treatment and reuse. This is mainly due to the operational simplicity and flexibility of membrane systems, their modular design, compactness, long term operation, high separation capacity and energy efficiency. Moreover, the experience gained in the last decades (in particular during the operation of reverse osmosis desalination systems), the improvements in material selection for membranes manufacturing as well as the mutual compatibility of different membrane operations for integration justify the success of membrane technology for a wide range of applications within the area of water treatment and purification, from desalination for potable water production, to industrial wastewater treatment, and to the recovery and reuse of municipal wastewaters.

REFERENCES

AlMadani, H.M.N. 2003. Water desalination by solar powered electrodialysis process. *Renewable Energy* 28:1915–1924.

Charcosset, C. 2009. A review of membrane processes and renewable energies for desalination. *Desalination* 245:214–231.

Choi, J.H., Dockko, S., Fukushi, K., and Yamamoto, K. 2002. A novel application of a submerged nanofiltration membrane bioreactor (NF MBR) for wastewater treatment. *Desalination* 146(1-3):413–420.

Choi, J.H., Lee, S.H., Fukushi, K., and Yamamoto, K. 2007. Comparison of sludge characteristics and PCR–DGGE based microbial diversity of nanofiltration and microfiltration membrane bioreactors. *Chemosphere* 67(8):1543–1550.

Cornelissen, E.R., Harmsen, D., de Korte, K.F., Ruiken, C.J., Qin, J.J., Oo, H., and Wessels, L.P. 2008. Membrane fouling and process performance of forward osmosis membranes on activated sludge. *Journal of Membrane Science* 319(1-2):158–168.

Drioli, E., and Curcio, E. 2007. Membrane engineering for process intensification: A perspective. *J. Chem. Technol. Biotechnol.* 82:223–227.

Drioli, E., Curcio, E., Di Profio, G., Macedonio F., and Criscuoli. A. 2006. Integrating membrane contactors technology and pressure-driven membrane operations for seawater desalination: Energy, exergy and cost analysis. *Chemical Engineering Research and Design* 84 (A3):209–220.

Drioli, E., Lagana, F., Criscuoli, A., and Barbieri, G. 1999. Integrated membrane operations in desalination processes. *Desalination* 122:141–145.

Fane, A.G. 2005. Module design and operation. In: *Nanofiltration: Principles and Applications*. Schaefer, A.I., Fane, A.G., and Waite, T.D. (eds.). Elsevier, pp. 67–88.

Phattaranawik, J., Fane, A.G., Pasquier, A.C.S., and Bing, W. 2008. A novel membrane bioreactor based on membrane distillation. *Desalination* 223(1-3):386–395.

Fritzmann, C., Löwenberg, J., Wintgens, T., and Melin, T. 2007. State-of-the-art of reverse osmosis desalination. *Desalination* 216:1–76.

Hagen, K. 1998. Removal of particles, bacteria and parasites with ultrafiltration for drinking water treatment. *Desalination* 119:85–91.

IDA Desalination Yearbook 2010–2011.

Kraume, M., and Drews, A. 2010. Membrane bioreactors in waste water treatment—Status and trends. *Chemical Engineering Technology* 33(8):1251–1259.

Lay, W.C.L., Liu, Y., and Fane, A.G. 2010. Impacts of salinity on the performance of high retention membrane bioreactors for water reclamation: A review. *Water Research* 44:21–40.

Macedonio, F., Curcio, E., and Drioli, E. 2007. Integrated membrane systems for seawater desalination: Energetic and exergetic analysis, economic evaluation, experimental study. *Desalination* 203:260–276.

Macedonio, F., and Drioli, E. 2010. Hydrophobic membranes for salts recovery from desalination plants. *Desalination and Water Treatment* 18:224–234.

Macedonio, F., Drioli, E., Curcio, E., and Di Profio, G. 2009. Experimental and economical evaluation of a membrane crystallizer plant. *Desalination and Water Treatment* 9:49–53.

Macedonio, F., Drioli, E., Gusev, A.A., Bardow, A., and Semiat, R. Efficient technologies for worldwide clean water supply. *Special issue of Chemical Engineering and Processing: Process Intensification.* (Forthcoming).

Mulder, M. 1996. *Basic Principles of Membrane Technology. London:* Kluwer Academic.

Oo, H., Kekre, K.A., Qin, J.J., Tao, G., Lay, C.L., Lew, C.H., Cornelissen, E.R., Ruiken, C.J., de Korte, K.F., and Wessels, L.P. 2008. Osmotic membrane bioreactor: Preliminary pilot study on effects of osmotic pressure on membrane flux and air scouring on fouling. In: *IWA Regional Conference "Membrane Technologies in Water and Waste Water Treatment,"* Moscow.

Peters, T. 2010. Membrane technology for water treatment. *Chemical Engineering Technology* 33(8):1233–1240.

Peter-Varbanets, M., Zurbrugg, C., Swartz, C., and Pronk, W. 2009. Decentralized systems for potable water and the potential of membrane technology. *Water Research* 43:245–265.

Rios, G.M., Belleville, M.-P., and Paolucci-Jeanjean, D. 2007. Membrane engineering in biotechnology: Quo vamus? *TRENDS in Biotechnology* 25(6):242–246.

Van der Bruggen, B., and Vandecasteele, C. 2002. Distillation vs. membrane filtration: Overview of process evolutions in seawater desalination. *Desalination* 143:207–218.

Application of Membrane Separation Technologies to Wastewater Reclamation and Reuse

Peter S. Cartwright
Cartwright Consulting Co., Minneapolis, MN, USA

ABSTRACT

As population growth and industrial and agricultural activities continue to stress the quality of the relatively fixed quantity of available fresh water on this planet, industry is challenged to investigate and develop unique and economical technologies to reclaim and reuse processing wastewater.

The four crossflow pressure-driven membrane technologies of microfiltration, ultrafiltration, nanofiltration and reverse osmosis possess unique properties which favor their use in these applications. Two of these are the facts that they are low energy (the driving force is pressure) and use no chemicals to effect separation.

This paper will define these technologies, explain their operational characteristics and detail the testing requirements for utilization of them in wastewater reclamation and reuse.

INTRODUCTION

Although the total quantity of water on this planet is more or less fixed, its quality is deteriorating, because we have been contaminating it for thousands of years, with little concern for the consequences. The issue that confronts us is the availability of water of sufficient quality. Table 1 is a summary of the world's water resources.

An analogy that may be a bit easier to understand is that if all the world's water were to completely fill a one gallon jug, the fresh water available for use would amount to only about one tablespoon.

Population growth and increased agricultural and industrial activities are contaminating our water supplies, while more stringent regulations and requirements for higher quality water for processing and drinking applications have exacerbated the problems.

The U.N. estimates that over ten million people a year die from drinking polluted water, mostly children.

Today, about 20% of the world's population is without clean water, and it is expected that, without drastic measures, half of the people on this planet will suffer from severe water shortages by 2050.

Across the United States, 39% of water use goes to energy production. Farms use another 40%, and manufacturing an additional 11%. Together, these three sectors use about 300 billion gallons of fresh water every day.

Contamination Issues

The contaminants in water supplies which compromise its quality can be organized into the classes shown in Table 2.

There are not many absolutes in the water treatment industry, but here is one: *it is impossible to make water completely free of all contaminants.* This fact has been recently underscored by the Associated Press article about PCPPs or Endocrine Disruptors found in drinking water

Table 1. Distribution of world water supply (cubic miles)

	Fresh	Saline	Total
Rivers and streams	300		
Freshwater lakes	30,000		
Salt lakes and inland seas		25,000	
Total surface water	**30,300**	**25,000**	**55,300**
Soil moisture and seepage	16,000		
Underground water to ½ mile depth	1,000,000		
Underground water to below ½ mile	1,000,000		
Total ground water	**2,016,000**		**2,016,000**
Glaciers and ice caps	7,000,000		
Oceans		317,000,000	
Total world water supply	**9,046,300**	**317,000,000**	**326,071,300**

Table 2. Water contaminants

Class	Examples
Suspended solids	Dirt, clay, colloidal materials, silt, dust, insoluble metal oxides and hydroxides
Dissolved organics	Trihalomethanes, synthetic organic chemicals, humic acids, fulvic acids
Dissolved ionics (salts)	Heavy metals, silica, arsenic, nitrate, chlorides, sulfates
Microorganisms	Bacteria, viruses, protozoan cysts, fungi, algae, molds, yeast cells
Gases	Hydrogen sulfide, methane, radon, carbon dioxide

supplies throughout the U.S. Undoubtedly, these contaminants have been present for many years, but their concentrations are so low (parts per trillion), that we have only recently been able to measure them.

WATER REUSE

Although still in its infancy, water reuse is growing at an estimated 11% per year in the U.S. Most of the recovered water is from municipal wastewater treatment plants ("reclaimed water") and is used for landscape and agricultural irrigation; however, industrial wastewater reuse is beginning to grow at an even higher rate—over 14%/year, by one estimate.

There are proven technologies available to treat any and all polluted water supplies; it's really a matter of committing financial and engineering resources. For almost all wastewater streams, a comprehensive test is required in order to select the optimum technologies and design the most cost effective water recovery system.

Due to the extreme variation in the specific kind and concentration of contaminants, industrial wastewater reuse requires the most testing and design expertise; however, with the rapidly increasing discharge regulations on both water quality as well as quantity, the incentive to recover and reuse is in place.

The process of treating wastewater and discharging it into a lake, river or aquifer from which drinking water is collected is known as "indirect reuse."

As the paradigm of water reclamation takes hold throughout the world, the concept of "direct reuse," treating wastewater at the source and reusing it directly, will become increasingly common, particularly in residential applications ("graywater reuse"). In many industrial applications, the incoming water has undergone extensive treatment for a particular process, and, overall, it is often more economical to treat this water for reuse than to simply discharge it, particularly as the cost of municipal water continues to increase.

TREATMENT TECHNOLOGIES

The arsenal of treatment technologies available today for industrial and municipal wastewater treatment is extensive. The traditional technologies are listed in Table 3.

A summary of major industrial treatment technologies follows.

As is evident from the previous table, a plethora of treatment technologies is available from removing contaminants from water supplies. For water reuse in most industrial and municipal applications, the most versatile and economical technology platform consists of the four crossflow pressure-driven processes of:microfiltration (MF),ultrafiltration (UF),nanofiltration (NF), andreverse osmosis (RO).

BACKGROUND ON MEMBRANE TECHNOLOGIES

Membrane technologies are based on a process known as "pressure-driven crossflow" filtration, which allows for continuous treatment of liquid streams. In this process, the bulk solution flows over and parallel to the membrane surface, and because the system is pressurized, water is forced through the membrane and becomes "permeate." The turbulent flow of the bulk solution over the surface minimizes the accumulation of particulate matter.

These technologies behave differently than filters in that (with some exceptions) the feed stream is pumped at a high flow rate across the surface of the filter media (membrane), with a portion of this stream forced through the membrane to effect separation of the contaminants, producing the permeate, and the concentrated contaminant remaining in the other stream (concentrate) exits the membrane element on a continuous basis. Figure 1 compares conventional with crossflow filtration.

Crossflow filtration offers the following advantages over traditional filtration technologies:

- Continuous and automatic operation
- Capable of removing contaminants down into the submicron size range
- Usually requires no chemical addition
- Backwashing capabilities
- Generally can operate in turbulent flow conditions
- Systems have a very small footprint

It is important to note that whereas with the media, cartridge and bag filtration technologies, the filtration process must be halted to backwash or replace the medium, crossflow filtration is designed to operate continuously, with the concentrate stream carrying away the contaminants. On the other hand, crossflow filters do become fouled and usually require backwashing operation.

By utilizing surface filters of specific membrane construction, very small pore sizes can be obtained, resulting in two submicron technologies: microfiltration and ultrafiltration.

Microfiltration (MF) is typically used to remove particulate material in the submicron range, most microfiltration devices in use today are designed as cartridge filters in that the entire solution passes through the filter leaving the particulate material behind, either on the filter surface or down inside the filter medium. The microfiltration devices addressed here use the "crossflow" design, which produces two exiting streams: one which has passed through the

Table 3. Traditional treatment technologies

Treatment Technologies	Suspended Solids Removal	Dissolved Organic Removal	Dissolved Salts Removal	Microorganism Removal
Biological processes				
MBR (membrane bioreactor)	X	—	—	X
Activated sludge	X	X	—	X
Anaerobic digestion	X	X	—	—
Bio-filters	—	X	—	—
Extended aeration				
Bio-denitrification	—	L	—	—
Bio-nitrification	X	X	—	—
Pasveer oxidation ditch	X	X	—	X
Chemical processes				
Chemical oxidation				
Catalytic oxidation	X	X	—	X
Chlorination	X	X	—	X
Ozonation	—	L	—	X
Wet air oxidation	X	X	—	X
Chemical precipitation	—	—	X	—
Chemical reduction	—	—	X	—
Ion exchange	—	—	X	—
Liquid-liquid (solvent)	—	—	X	—
Coagulation				
Inorganic chemicals	X	X	—	X
Polyelectrolytes	X	X	—	X
Electolytic processes				
Electrodialysis	—	—	X	L
Electrodeionization	—	—	X	—
Electrolysis	—	—	X	—
Ultraviolet irradiation	—	—	—	X
Extractions				
Incineration				
Fluidized-bed	X	X	—	X
Physical processes				
Carbon adsorption				
Granular activated	X	X	—	—
Powdered	X	X	—	X

(table continues)

Table 3. Traditional treatment technologies (continued)

Treatment Technologies	Suspended Solids Removal	Dissolved Organic Removal	Dissolved Salts Removal	Microorganism Removal
Physical processes (continued)				
Specialty resins	—	L	L	—
Filtration				
Diatomaceous-earth filtration	X	—	—	X
Multi-media filtration	X	—	—	X
Micro-screening	X	—	—	X
Sand filtration	X	—	—	X
Flocculation-sedimentation	X	—	—	X
DAF (dissolved air flotation)	X	X	—	—
Foam separation	X	—	X	—
Membrane processes				
Microfiltration	X	—	—	X
Ultrafiltration	X	X	—	X
Nanofiltration	X	X	L	X
Reverse osmosis	X	X	X	X
Stripping (air or steam)	X	X	—	—
Thermal processes				
Distillation	X	X	X	X
Freezing	—	X	X	—

L = Under certain conditions, there will be limited effectiveness.

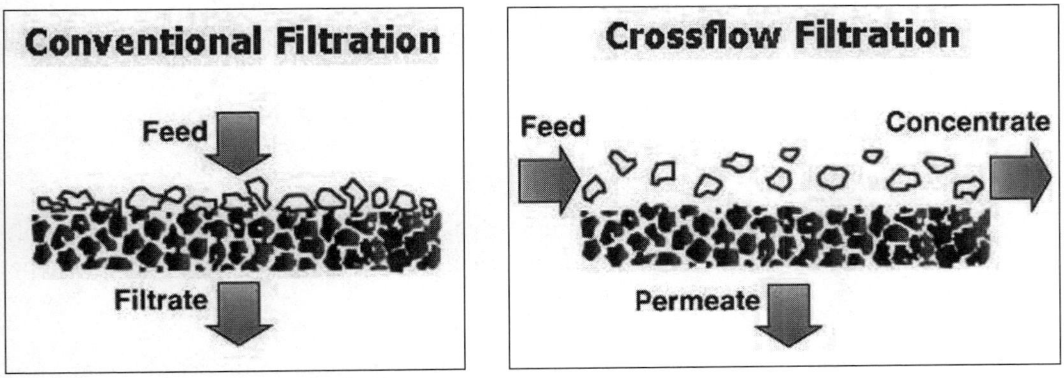

Figure 1. Conventional versus crossflow filtration

membrane media and is purified (permeate), and the other which flows across and parallel to the media surface, continuously removing the contaminants (concentrate).

Generally, microfiltration involves the removal of particulate, or suspended materials ranging in size from approximately 0.10 to 1.0 microns (1,000 to 10,000 angstroms). MF typically operates within a pressure range of 10 to 30 psi (0.68 to 2.0 bar).

Ultrafiltration (UF) is used to separate dissolved, non-ionic materials (macro molecules) typically smaller than 0.10 micron (1,000 angstroms). The removal characteristics of UF membranes can be described in terms of "molecular weight cutoff" (MWCO), the maximum molecular weight of dissolved compounds that will pass through the membrane pores. MWCO terminology is expressed in Daltons. Basically, ultrafiltration is used to remove *dissolved* organic contaminants, while *suspended* solids are removed by microfiltration. UF normally operates in a pressure range of 10 to 100 psi (0.68 to 6.8 bar). UF membranes are available over a wide range of MWCO removal properties, from about 1,000 to over 100,000 Daltons.

MF and UF processes separate contaminants based on a "sieving" process; that is, any contaminant too large to pass through the pore is rejected and exits in the concentrate stream.

Nanofiltration can be considered "loose" reverse osmosis. It rejects dissolved ionic contaminants but to a lesser degree than RO. NF membranes reject a higher percentage of multivalent salts than monovalent salts (for example, 99% vs. 20%). These membranes have molecular weight cut-offs for non-ionic solids below 1000 Daltons.

Reverse osmosis produces the highest quality permeate of any pressure driven membrane technology. Certain polymers will reject over 99% of all ionic solids, and have molecular weight cut-offs in the range of 50 to 100 Daltons.

Both NF and RO membranes reject salts utilizing a mechanism that is not fully understood. Some experts endorse the theory of pure water preferentially passing through the membrane; others attribute it to the effect of surface charges of the membrane polymer on the polarity of the water. Monovalent salts are not as highly rejected from the membrane surface as multivalent salts; however, the high rejection properties of the newer thin film composite RO membranes exhibit very little differences in salt rejection characteristics as a function of ionic valance. As indicated earlier, this difference is significant with NF membranes.

In all cases, the greater the degree of contaminant removal, the higher the pressure requirement to effect this separation. In other words, reverse osmosis, which separates the widest range of contaminants, requires an operating pressure typically an order of magnitude higher than microfiltration, which removes only suspended solids.

Table 4 summarizes the various properties and other features of these technologies.

Device Configurations

To be effective, membrane polymers must be packaged into a configuration commonly called a "device" or "element." The most common element configurations are: Tubular, Hollow (capillary) Fiber, Spiral Wound, and Plate and Frame.

The element configurations are described as follows.

Tubular. Manufactured from ceramics, carbon, stainless steel, or a number of thermoplastics, these tubes have inside diameters ranging from ¼ inch up to approximately 1 inch (6 to 25 mm). The membrane is typically coated on the inside of the tube and the feed solution flows under pressure through the interior (lumen) from one end to the other, with the permeate passing through the wall and collected outside of the tube.

Table 4. Membrane technologies compared

Feature	Microfiltration	Ultrafiltration	Nanofiltration	Reverse Osmosis
Materials of construction	Ceramics, Sintered metals, Polypropylene, Polysulfone, Polyethersulfone, Polyvinylidene fluoride, Polytetrafluoroethyliene	Ceramics, Sintered metals, Polypropylene, Polysulfone, Polyethersulfone, Polyvinylidene fluoride	Thin film composites, Cellulosics	Thin film composites, Cellulosics
Pore size range (micrometers)	0.1–1.0	0.001–0.1	0.0001–0.001	<0.0001
Molecular weight cutoff range (Daltons)	>100,000	1,000–100,000	300–1,000	50–300
Operating pressure range	<30	20–100	50–300	225–1,000
Suspended solids removal	Yes	Yes	Yes	Yes
Dissolved organics removal	None	Yes	Yes	Yes
Dissolved inorganics removal	None	None	20–95%	95–99+%
Microorganism removal	Protozoan cysts, algae, bacteria*	Protozoan cysts, algae, bacteria*, viruses	All*	All*
Osmotic pressure effects	None	Slight	Moderate	High
Concentration capabilities	High	High	Moderate	Moderate
Permeate purity (overall)	Low	Moderate	Moderate-high	High
Energy usage	Low	Low	Low-moderate	Moderate
Membrane stability	High	High	Moderate	Moderate

*Under certain conditions, bacteria may grow through the membrane.

Hollow (Capillary) Fiber. These elements are similar to the tubular element in design, but are smaller in diameter, and are usually unsupported membrane polymers or ceramics. In the case of polymeric capillary fibers, they require rigid support on each end provided by an epoxy "potting" of a bundle of the fibers inside a cylinder. Feed flow is either down the interior of the fiber ("lumen feed") or around the outside of the fiber ("outside-in").

Spiral Wound. This element is constructed from an envelope of sheet membrane wound around a permeate tube that is perforated to allow collection of the permeate. Water is purified by passing through one layer of the membrane and, following a spiral path, flows into the permeate tube. It is by far the most common configuration in water purification applications, but generally requires extensive pretreatment in wastewater applications.

Plate and Frame. Sheet membranes are stretched over a frame to separate the layers and facilitate collection of the permeate, which is directed to a collection tube.

Table 5. Membrane element configuration comparison

Element Configuration	Packing Density*	Fouling Resistance†
Plate & frame	Low	High
Hollow (capillary) fiber	High	High
Tubular	Low	Very high
Spiral wound	Medium	Low

*Membrane area per unit volume.
†Tolerance to suspended solids.

Table 6. Microfiltration and ultrafiltration

Materials of Construction	Device Configuration			
	Hollow Fiber	Tubular	Plate & Frame	Spiral Wound
Polymeric				
PS	X	X	X	X
PES	X	X	X	X
PAN	X	X	X	X
PE	—	X	—	—
PP	X	X	X	—
PVC	—	X	—	—
PVDF	X	X	—	—
PTFE	X	—	X	—
PVP	X	X	—	—
CA	X	—	—	—
Non-polymeric				
Coated 316LSS	—	X	—	None
α-Alumina	—	X	X	None
Titanium dioxide	—	X	—	None
Silicon dioxide	—	X	—	None

PS = Polysulfone
PES = Polyethersulfone
PE = Polyethylene
PP = Polypropylene
PAN = Polyacrylonitrile
PVDF = Polyvinylidene Fluoride
PTFE = Polytetrafluoroethylene
CA = Cellulose Acetate
PVP = Polyvinylpyrrolidone
TF = Thin Film Composite

From the perspective of cost and convenience, it is beneficial to pack as much membrane area into as small a volume as possible. This is known as "packing density." The greater the packing density, the greater the membrane area enclosed in a certain sized device, and generally the lower its cost. The downside of the high packing density membrane elements is their greater propensity for fouling. Table 5 compares the element configurations with regard to their packing densities.

To clarify the membrane materials used for the various element configurations, Tables 6 and 7 are provided.

Table 7. Nanofiltration and reverse osmosis

Materials of Construction	Device Configuration			
	Hollow Fiber	Tubular	Plate & Frame	Spiral Wound
Polymeric				
PS*	—	X	X	X
PES*	—	X	X	X
CA	—	X	X	X
TF	—	X	X	X
Non-Polymeric				
None				

*Base polymer below TF polymer.
PS = Polysulfone
PES = Polyethersulfone
CA = Cellulose Acetate
TF = Thin Film Composite

Table 8. Membrane element cleaning capability

	Membrane Technology				
Element Configuration	MF	UF	NF	RO	Backwashable?
Plate & frame	Yes	Yes	Yes	Yes	No (except for inorganic membrane)
Tubular	Yes	Yes	Yes	Yes	Yes
Hollow fiber	Yes	Yes	Yes	No	Yes
Spiral wound	Yes	Yes	Yes	Yes	No (NF, RO) Yes (MF, UF)

Because of the extreme value of backwashing/backpulsing to minimize the effects of fouling on membrane surfaces, Table 8 categorizes membrane devices with this capability.

SYSTEM DESIGN

Figure 2 is a schematic of a complete membrane processing system (or a single membrane element).

The feed stream enters the system (or membrane element), and as the stream passes along and parallel to the surface of the membrane under pressure, a percentage of the water is forced through the membrane polymer producing the permeate stream. Contaminants are prevented from passing through the membrane based on the polymer characteristics. This contaminant-laden stream exits the membrane system (or element) as the "concentrate" stream, also known as "brine" or "reject."

The permeate rate of a given membrane element cannot be changed without varying the applied pressure or temperature. Recovery, however, can be easily changed by varying the feed flow rate to the element, and this is one of the variables that is controlled by the system designer.

For wastewater treatment and water reuse applications, the minimum recovery is usually no less than 90%.

The relationship between recovery and concentration of solute in the concentrate stream is illustrated by the data in the table and plotted in Figure 3. The concentration effect resulting

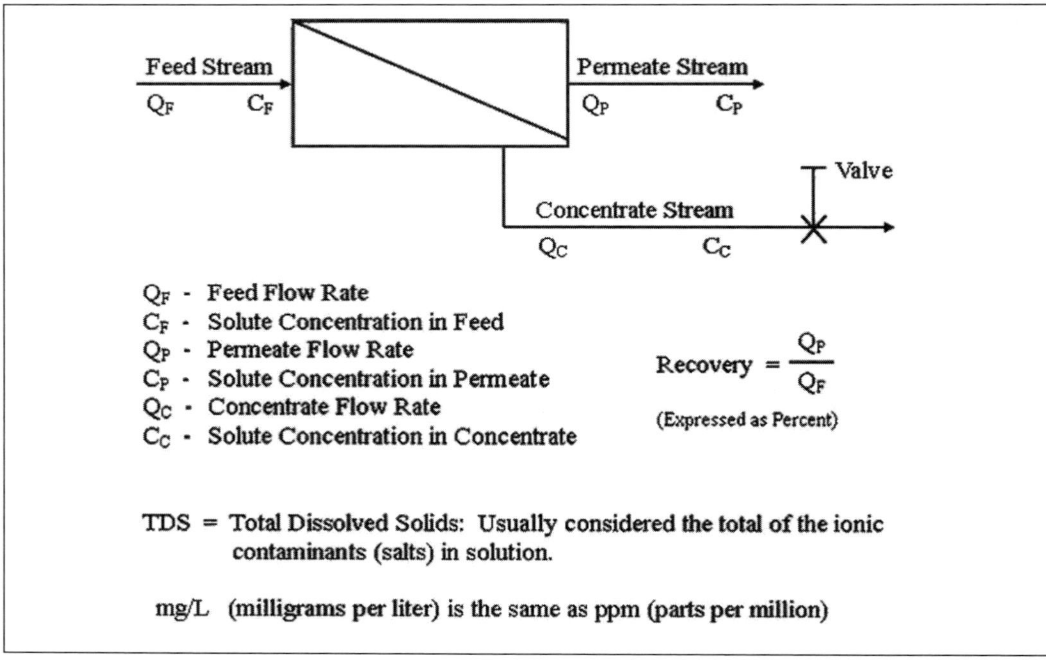

Figure 2. Membrane system schematic

from pumping a certain percentage of the solvent through the membrane is represented mathematically by the term:

$\frac{1}{1 - \text{recovery}}$, also known as "concentration factor" (X).

The advantage of operating systems at high recoveries is that the volume of concentrate is small and the flow rate of the feed pump is smaller; the potential disadvantages are numerous:

- The higher concentration of contaminants is likely to result in fouling. In nanofiltration and reverse osmosis applications, the concentrated salts solution results in high osmotic pressure, requiring a higher-pressure pump and a more pressure tolerant system.
- As higher recoveries reduce the quantity of concentrate to be discharged, the higher concentration of the concentrate stream may present regulatory discharge problems.

MBR TECHNOLOGY

As the newest membrane technology application, and one with huge potential, MBR (membrane bioreactor) technology justifies special mention.

For wastewaters containing biodegradable contaminants, the traditional treatment method is to encourage the use of bacteria to break down the contaminant (bioremediation).

This encouragement can take the form of adding oxygen (in the case of aerobic treatment), providing a mechanical matrix (for bacterial attachment), mixing, and other approaches intended to maximize the metabolic activity of these microorganisms.

MBR offers significant advantages over traditional bioremediation processes, as listed below:

- High-quality effluent, almost free of suspended solids
- The ability to partially disinfect without the need for chemicals

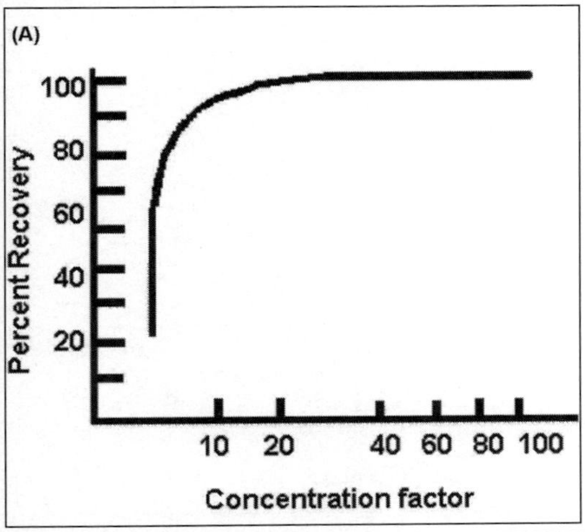

(B)

$$C_c \approx \frac{C_F}{1 - \text{Recovery}} = XC_F$$

$$X = \frac{1}{1 - \text{Recovery}} = \text{Concentration Factor}$$

Percent Recovery	Concentration Factor
33%	1.5
50%	2
67%	3
75%	4
80%	5
90%	10
95%	20
97.5%	40
98%	50
99%	100

Figure 3. (a) Effect of recovery on concentration and (b) effect of recovery on concentration factor

- Complete independent control of HRT (Hydraulic Retention Time) and SRT (Sludge Retention Time)
- Reduced sludge production
- Process intensification through high biomass concentrations with MLSS (Mixed Liquor Suspended Solids) concentrations above 15,000 mg/L
- Treatment of recalcitrant organic fractions and improved stability of processes such as nitrification
- Ability to treat high strength wastes

The membrane device configurations most commonly used today are hollow (capillary) fiber and plate and frame, although tubular and spiral wound devices are becoming more widely used.

The most common biological treatment is aerobic and, typically, air is bubbled into the treatment tank. A very popular approach is to immerse the membrane element in the treatment tank and either allow the hydrostatic head of the solution to provide the driving force or to use a pump to pull the permeate through the membrane (or both). In this case, air bubbles are also directed up over the surface of the membrane, from below (air scouring), in an effort to reduce fouling.

Another design involves pumping water through the membrane system external to the treatment tank, and yet another uses a separate tank for membrane processing downstream of the biological treatment tank. Additional designs and configurations are sure to appear as MBR technology becomes more widely used.

Figure 4 illustrates aerobic MBR applications for both "immersed" and "external" designs.

TESTING

In general, every stream must be tested to develop the following design factors:

- Optimum membrane element configuration
- Total membrane area
- Specific membrane polymer
- Optimum pressure
- Maximum system recovery

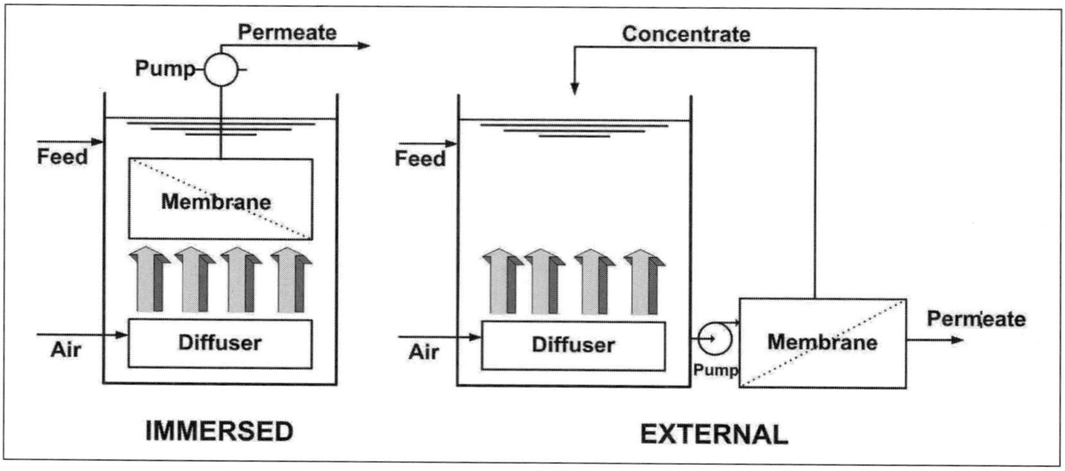

Figure 4. Aerobic MBR applications

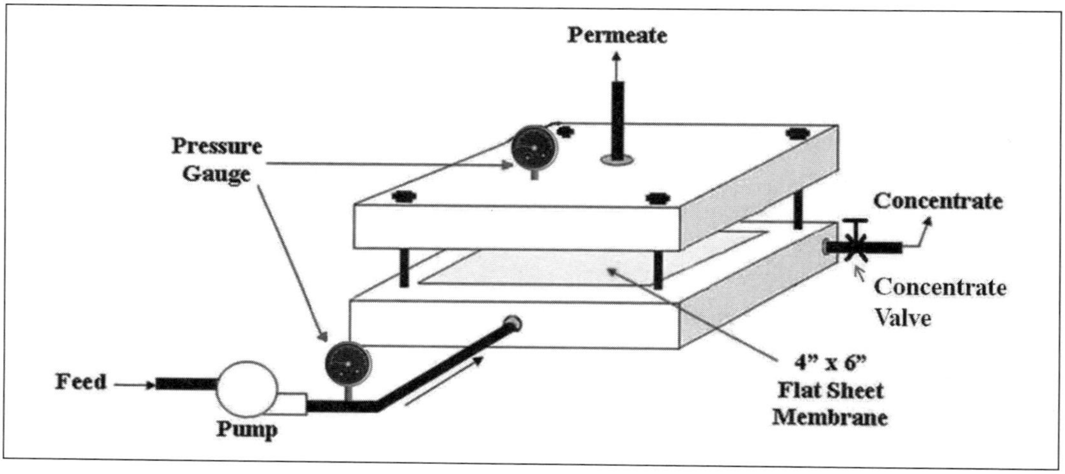

Figure 5. Cell test unit

- Flow conditions
- Membrane element array
- Pretreatment requirements

To generate the necessary design data, several testing options are available.

Cell Testing

A typical cell testing device is illustrated in Figure 5. Cell test devices are available for purchase (or through a consulting engineering firm skilled in the art), which evaluate small sheets of membranes on the stream to be processed. Typically, the sheet is placed between two stainless steel plates, and the test stream pumped across the membrane surface at a selected pressure and flow rate. The permeate is collected and analyzed for degree of separation, possible effect of the stream on the test membrane, and other properties.

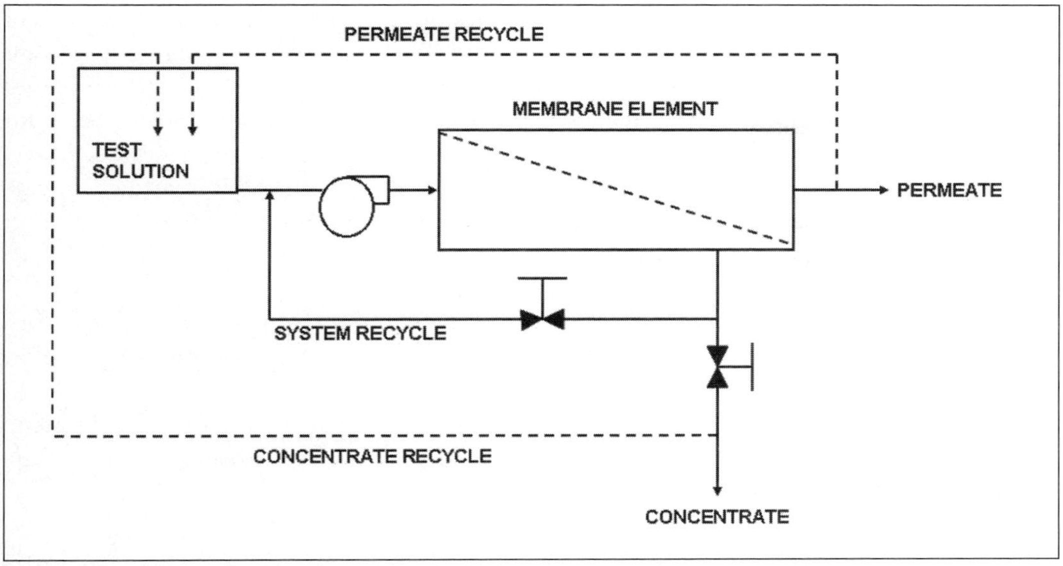

Figure 6. Applications test schematic

The cell test offers a number of *advantages*:

- Only small areas of membranes are needed; excellent for screening membrane polymer candidates.
- Can be run on small volumes of test stream.
- Takes very little time.
- Unit is simple to operate.

The *disadvantages* of this testing approach are:

- Cannot obtain engineering design data.
- Cannot be used for long-term fouling study.
- Is only useful with membranes available as flat sheet.

The cell test approach is useful as an initial step, primarily to select one or more membrane candidates for further evaluation.

Applications Testing

Figure 6 illustrates an applications test schematic. Applications testing utilizes a membrane element in a test unit capable of operating similar to a production unit. Since the data from this testing will be used to scale up the design to full size, it is essential that the membrane element manufacturer supplies an element capable of this scale up.

The applications test equipment should be designed so that very high recoveries can be achieved without compromising the flow rates required to produce turbulent flow, for example. This requires that the pump be capable of not only producing the desired pressure, but also the flow rate to accomplish the minimum crossflow velocity across the membrane surface.

Because the system must be capable of testing at very high recoveries, the concentrate valving must be adjustable to accurately produce extremely low flow rates. This typically involves the assembly of a "valve nest" using micrometer valves. Additionally, the recycle line should be equipped with a diaphragm valve for adjustment of flow and pressure.

The most important feature for application testing equipment is versatility. Different membrane elements have very specific operating parameters, and the equipment must accommodate these. To cover the entire gamut of membrane technologies, two different pieces of application testing equipment are generally required: one for MF and UF, and the other for NF and RO.

The latter must be capable of pressures up to 1,000 psi (68 bar), and it is virtually impossible to find a single pump capable of supplying the flows and pressures required for all four technologies. For MF and UF applications, a variable speed drive centrifugal pump works fine, although the variable speed feature makes it expensive.

Materials of construction are an important consideration in testing considerations: 316L stainless steel is essential for applications requiring pressures in excess of 60 psi (4 bar); below that, schedule 80 PVC is sufficient.

Applications testing is capable of generating complete design data for the full sized system. An applications test can be run on as little as 50 gallons (200L) of test stream, and after setup, can be completed in one hour or less, for each membrane element tested.

A typical applications test is run as follows:

1. To establish "control conditions," high quality water (tap water or water treated with RO or DI) is run into the system at low recovery to minimize any possible contaminant concentration effects. Take data (see Membrane Application Test Data Sheet).
2. Feedwater is then run into the unit set at low recovery, and after stabilization (usually less than 5 minutes), the following data are taken:
   - Pressures
     - Prefilter
     - Primary (feed)
     - Final (concentrate line)
   - Flow
     - Recycle
     - Permeate
     - Concentrate
   - Temperature (recycle)
   - Quality (conductivity)
     - Feed
     - Permeate
     - Concentrate

   The system recovery is then increased incrementally while adjusting the recycle valve to ensure that the correct crossflow velocity is maintained.
3. At the conclusion of the testing, high quality water is again run through the system to determine if the permeate rate or other operating characteristics have been affected.

At each recovery, in addition to the collection of flow and pressure data, analytical samples should be taken for performance evaluation. Of course, the choice of parameters to be measured depends upon the separation goals of the test. It is unusual for system recoveries to exceed 95%; however, that also depends upon the goals of the testing, and it is possible to run a well designed test unit up to 99% recovery.

Once the optimum conditions have been established, such as operating pressure and maximum system recovery, the normalized performance data will enable the test engineer to determine the total membrane area required for the full sized system.

Application testing provides the following advantages and disadvantages:

Advantages

- Fast.
- Provides scale-up data (flow, osmotic pressure as a function of recovery, pressure requirements, etc.).
- Can provide an indication of membrane stability.

Disadvantages

- Does not reveal long term chemical effects.
- Does not provide data on long term fouling effects.

Pilot Testing

Usually this involves placing a test machine (such as that used for the applications test) in the process, operating continuously on a "side-stream" for a minimum of 30 days.

Advantages. Accomplishes all of the functions of the applications test plus provides long term membrane fouling and stability data.

Disadvantages. Expensive in terms of monitoring and time requirements.

CONCLUSIONS

With the exception of the oxygen we breathe, there is no substance more critical to life than water, and no substitute for it.

Many experts feel that there is no other product whose real value so far exceeds its price, and whose price is so often unrelated to its actual cost of production and delivery.

As the world's population continues to grow, as this expanding population tends to relocate to water-short regions, and as climate changes create areas of drought, stress on the quality of our fixed water quantity will become very, very critical. This problem can only be addressed by aggressively and constructively employing such innovative conservation and water reuse.

Solutions are there, but the entire world must give water quality issues high priority and be willing to commit the investments of money, education and commitment to make these solutions a reality.

Removal of Heavy Elements from Aqueous Processes

Lucas Moore, Amir Mahmoudkhani, and Jean Robert Durand
Kemira Oil and Mining, Atlanta, GA, USA

ABSTRACT

Mining is essential to our standard way of life. Whether it is mining for energy (oil, uranium, and coal) or mining for construction and development (metals and minerals), mining affects us all. Mining is also potentially a source of devastation to life because during the mining processes, contamination of the most important and vital fluid, water, is a high probability. Such contamination has become an increasing concern as governmental restrictions begin to tighten, which has made the treatment of such water critical and mandatory. There are methods in practice today for treatment of contaminants such as toxic oxyanions and heavy metals. These methods can often involve many expensive processing steps that may also be limited by variables such as the total dissolved solids, presence of other ions, or the ability to maintain microbial growth. A technology has been developed to successfully reduce cadmium and selenium to a level below the EPA recommendations without the above mentioned limitations on a lab scale.

INTRODUCTION

Oxyanions are negatively charged inorganic compounds with various degrees of oxidation. Examples of oxyanions are CO_x, SO_x, NO_x, PO_x, ClO_2, AsO_x, SeO_x, etc. Heavy metals such as mercury, cadmium, lead, and chromium are toxic metals that can form water-soluble poisonous compounds and deprive a positive biological role. In some cases, the toxicity of these metals can be explained by their ability to imitate the action of essential metals, thus interfering with the function of the body and metabolic processes. The level of toxicity of the heavy metal is generally considered to be dependent on the level of solubility, thus dependent on the ligands associated with the metal. Both oxyanions and heavy metals occur in various concentrations throughout the ecosystem and can be observed in elevated levels, following various mine processes. Such contamination must be treated prior to discharge, but it is also feasible that the contaminant become mobile while in a tailing pond/pit or the initial mine site due to environmental conditions that the contaminants were not exposed to prior to unearthing (Figure 1).

Selenium

Selenium is among the major contaminants in mining aqueous waste streams. It is most commonly observed as selenate, selenite, or selenide in nature (Figure 2). Though selenide (Se^{2-}) has a low toxicity, selenate (Se^{VI}) and selenite (Se^{IV}) are very toxic. Selenate and selenite are generally the forms of selenium found in water. The presence of selenates and selenites in waste water is an immediate problem. If left untreated, selenium will bioaccumulate and pose a threat to all aquatic life downstream. The US National Primary Drinking Water Standard is 50 ppb for selenium and the US National Fresh Water Quality Standard is 5 ppb for selenium.

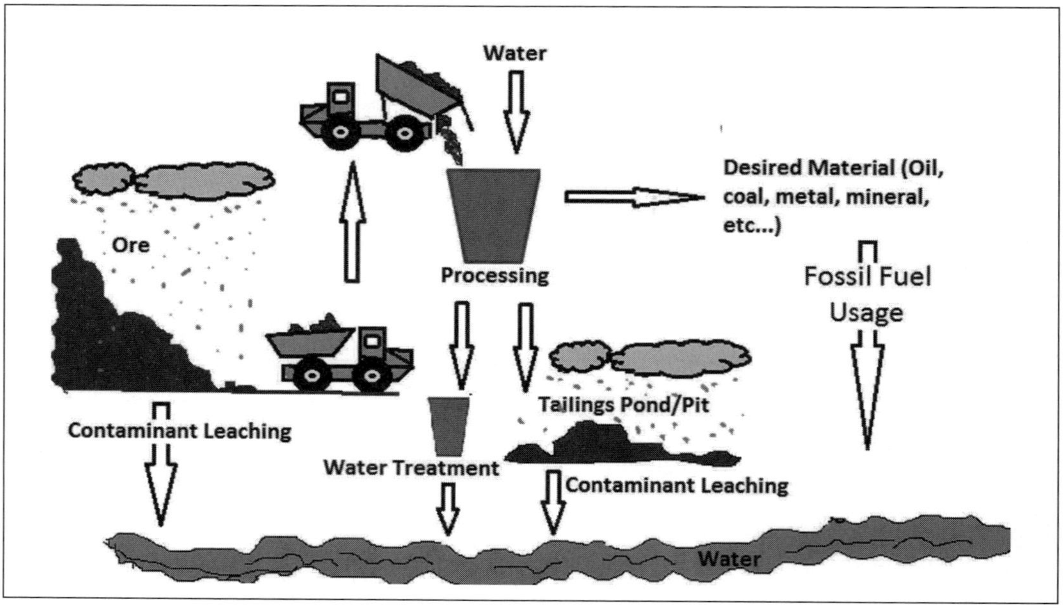

Figure 1. Example of a mine site and possible routes for contamination

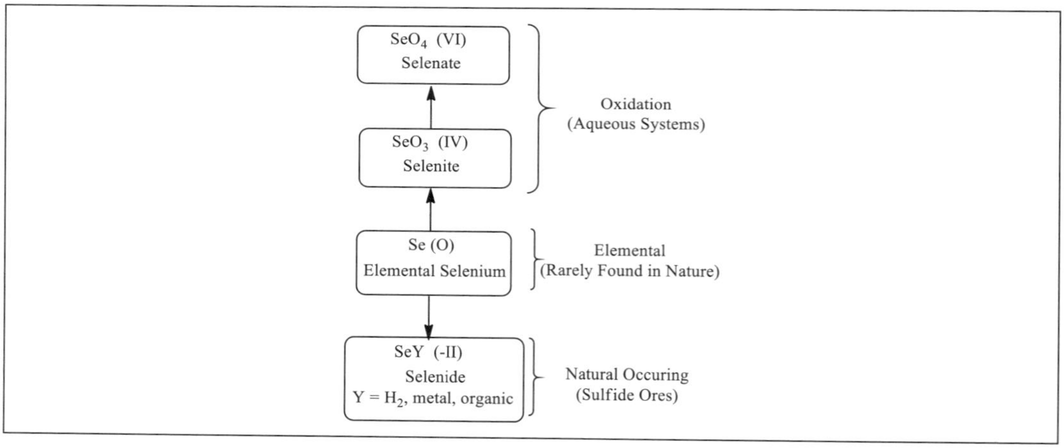

Figure 2. Most common oxidation states of selenium

(EPA 2001 and EPA 2011) The EPA listed selenium in at least 508 of the 1,636 sites listed on the National Priorities List (ATSDR 2011).

Due to the similar chemical nature of selenium and sulfur, selenium is commonly associated with sulfur containing ores such as coal, pyrite, sphalerite, and chalcopyrite. (Adam 2005) Sulfur containing ores are often associated with the mining of copper, nickel, silver, lead, uranium, etc., thus selenium has become a major contamination in processing the above ores. Selenium issues have been reported through the various mining processes throughout the world, from metal smelting processes, coal mining/combustion, to the mining of silver, gold, nickel and phosphate ores (Figure 3). Coal has been reported to have 0.4–24 ppm selenium prior to processing or usage (Lemly 2004).

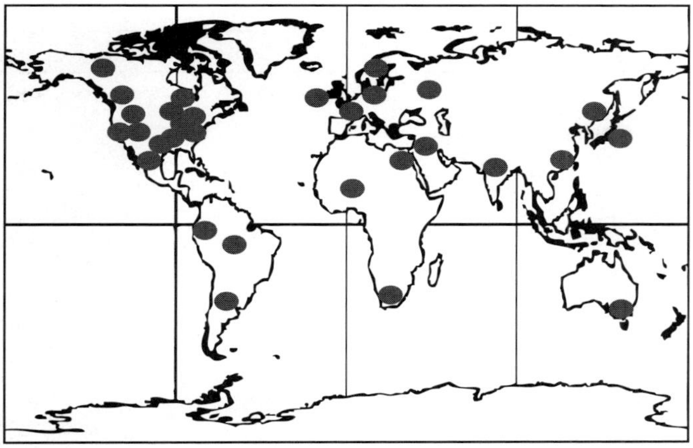

Figure 3. Various sites reported with high levels of selenium contamination, as of 2004 (coal mining/combustion, oil refining, phosphate mining, gold mining, silver mining, nickel mining, metal smelting, and landfill leachate)

Cadmium

Cadmium is a heavy metal that is in the same chemical family as zinc and mercury (group 12) and prefers an oxidation state of +2 and is chemically similar to zinc. Cadmium is typically not found in an elemental pure state; but rather in zinc, lead, and copper ores as an oxide, chloride, or sulfide. The national recommended water quality criteria state that 0.25 ppb is the limit for freshwater. (EPA 2009) The lethal dose of cadmium for a human is ~400 mg. The EPA lists cadmium in at least 1,014 of the 1,669 sites on the National Priorities List (ATSDR 2011). Cadmium's toxicity and the fact that cadmium is a bioaccumulant make it a pollutant of major concern.

The major commercial source of cadmium is as a by-product from zinc production. Some common uses are in plating on steel (anticorrosion), pigments in stabilizers for PVC, electrode material in nickel-cadmium batteries, and in various alloys (Toxic Substance Profile 2011).

Cadmium's major pathway for entering the environment is through volcanic activities, mining and industrial processes, as well as burning of coal. Smelting of non-ferrous metals has been listed as the largest source of cadmium contamination in aquatic environments by human activities (IPCS 2011) (Table 1). Other sources are from the mine drainage water, mine process water, leaching from tailings ponds and the mine sites of these non-ferrous metal mines (Figure 2). Though cadmium only exists at 0.1 ppm in the earth's crust, levels as high as 15 ppm is observed in some cases in sedimentary rocks and marine phosphates. Soils that have elevated levels of cadmium are generally in areas that are also rich in zinc, lead, and copper. Cadmium's high vapour pressure contributes to it toxicity and mobility throughout the environment. During processes such as smelting, cadmium is vaporized, and if not contained, cadmium will oxidize in the environment with either oxygen or other environmental contaminants such as SO_x, NO_x, CO_x, chloride, etc.

Cadmium's similarity to zinc is what leads to its toxicity. Zinc is an essential trace element for life in plants, animals, and many microorganisms. Zinc has many tasks in the functioning of the human body. It has a role in RNA and DNA metabolism, and has been discovered in over 10% of the human proteins. Zinc is also stored in the synaptic vesicles in the brain. Cadmium's chemical similarity to zinc is how cadmium is transported throughout the human body. The kidneys, as well as the human skeletal and respiratory systems are all potential sites

Table 1. Estimates of global cadmium emissions

Source of Cadmium Emissions	Worldwide (tonnes/year)
Natural sources	800
Non-ferrous metal processes	
Mining	0.6–3
Zinc and cadmium	920–4,600
Copper	1,700–3,400
Lead	39–195
Iron and steel production	28–284
Fuel combustion	
Coal	176–882
Oil	41–246
Wood	60–180
Cement manufacturing	8.9–534
Phosphate fertilizer manufacture	68–274
Sewage sludge incineration	3–36

for serious damage during cadmium exposure. Inhalation exposure to cadmium may lead to pulmonary oedema, fibrosis, emphysema, and death, while ingestion may lead to renal tubular dysfunction or tubular cell damage. The major source of human exposure is through the consumption of food or water contaminated with cadmium. The average human consumes 10–40 micrograms of cadmium per day, but consumption can be several hundred micrograms per day in polluted areas.

EXISTING TREATMENT TECHNOLOGIES

Historically, neutralization/precipitation was the method of choice for removing heavy inorganic contaminants. Such methods involve the precipitation of the metal hydroxide by adjusting the pH values with either caustic or lime. For cadmium, the optimal pH needed was higher than 10.5. As the regulation become tighter, i.e., <200 ppb, this method became insufficient. To date, the most common treatment methods are adsorption, coagulation, and media filtration. Examples of media filtration are ion exchange and reverse osmosis (Gaballah 1993; Zhang 2002; Rybock 2009; Mihaylov 1992; Harris 1992).

Adsorption/Media Filtration

These can be as simple as filtering through sand, clay, titanium dioxide, etc. or can be as exotic as filtering through ion exchange resins, activated carbon, or a membrane (reverse osmosis and nanofiltration). Many of these media are commonly used in the water treatment industry. Common problems associated with filtration media include an increased amount of waste, and a potential for fouling or scaling of the membrane. Nitrates, sulfates, and chlorides may also lead to a reduction in the efficiency of oxyanion removal in such a treatment system.

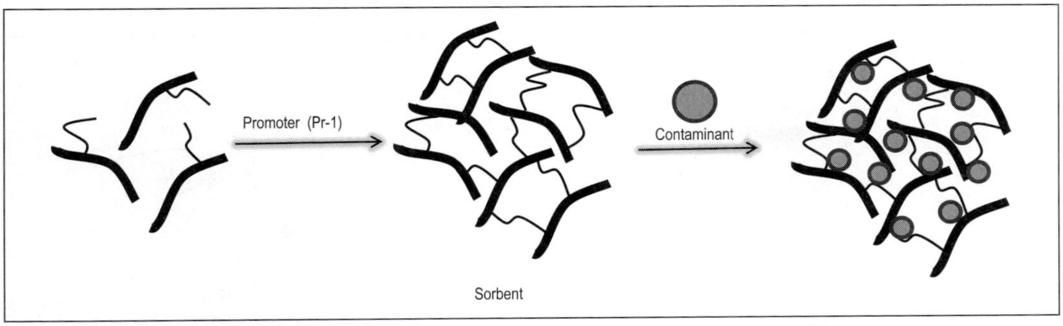

Figure 4. In-situ solidification–chemisorption method

Chemical Treatment

Precipitation and coagulation are the most common forms of chemical treatment associated with heavy metals treatment in process water. A major disadvantage of most chemical treatment possibilities is in the high quantity of chemicals being consumed, thus leading to the need to treat the hazardous solid waste being produced. As with the filtration method, there may also be limitations based on the other contaminants and ions in the process water.

Biotreatment

To date the most cost effective method to remove selenite and selenate contamination from aqueous systems is by using a biological system to reduce selenates to elemental selenium (Cohen 2006; Harrison 2010). As with the previously mentioned methods, there are problems associated with the biomediated reduction. Often the presence of other ions can decrease the effectiveness of the biological systems in reducing the selenates and selenites. Nitrates have been reported as a major problem. If such a problem occurs, a pretreatment step will be needed prior to introduction into the biological treatment area, which will add to the initially high capital expense. The total suspended solids may also limit the efficiency of the selenium treatment.

In-Situ Solidification–Chemisorption Treatment Method

In this work, we are introducing an innovative approach using in-situ solidification–chemisorption method for treatment of contaminated mining process water. A chemically modified inorganic polymeric silica based material is cross-linked in the water stream with the addition of the promoter. The cross-linking produces an insoluble amorphous sorbent with active sites for chemisorption of contaminated oxy- and hydroxyl anionic species, such as selenate or cadmium hydroxides in water, as shown in Figure 4.

In situ cross-linking of inorganic polymer maximizes the number of active sites available for interaction with contaminant, resulting in better use of sorbent material. Due to the irregular nature of cross-linking process, chemisorbed species become encapsulated and immobilized in the resulting amorphous solid mass. The suspended solids precipitate out from solution, and can be separated by gravity settling, filtration or other conventional solid removal methods.

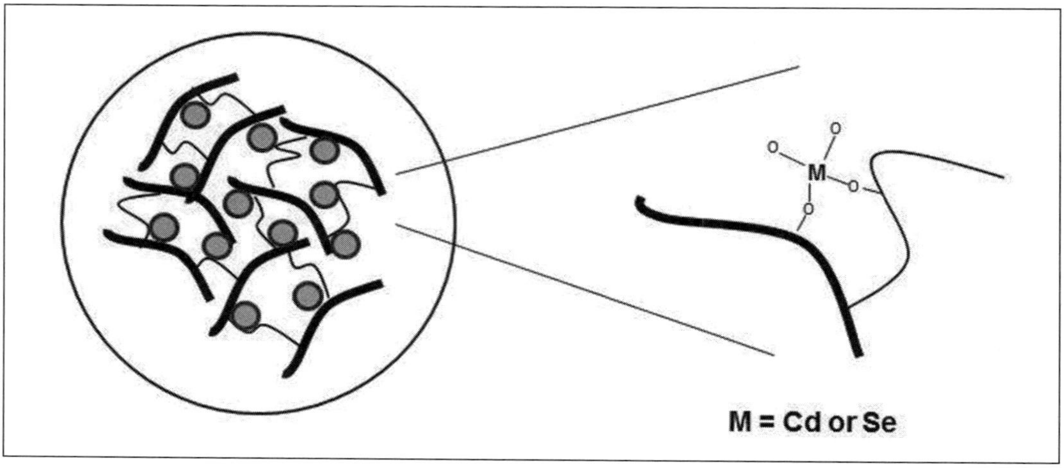

Figure 5. Chemisorption of cadmium or selenium species on inorganic polymeric sorbent

EXPERIMENTAL

Materials

Cadmium chloride ($CdCl_2$) and sodium selenate (Na_2SeO_4) were purchased from Aldrich and used as received with no further purification. Lab-made aqueous solutions containing selenium or cadmium were prepared by dissolution of the above chemicals in city of Atlanta tap water. **Caution!** Cadmium chloride and sodium selenite are extremely toxic and should be handled and disposed according to regulations for toxic substances.

Instruments

In this study, a Thermo Scientific ICP-AES system model iCAP 6200 equipped with a charge injection device (CID) detector and a CETAC ASX-520 autosampler was used for determination of cadmium and selenium species in water samples. Low detection limits (1 ppb for selenium and 0.2 ppb for cadmium) were achieved by pre-concentration of 100 mL aqueous samples. Quantitative elemental analysis of trace elements was conducted on a Bruker S4 Explorer wavelength-dispersive X-ray fluorescence spectrometer. Element distributions of cadmium or selenium before and after treatments were used for qualitative and quantitative analysis of chemisorption of the contaminant species on inorganic polymeric solid sorbent.

RESULTS AND DISCUSSION

Adsorption is the process where the substance is transferred from the liquid phase (solution) to the solid surface. Adsorption involves the inter-phase accumulation/concentration of substances at the surface/interface, and this process occurs at an interface between any two phases such as liquid-solid interface. Chemical adsorption or chemisorption takes place as a result of chemical bond being formed between the solute (dissolved species) and the adsorbent; these bonds are comparable with those leading to the formation of chemical compounds (Figure 5). There are many factors affecting adsorption such as nature of the adsorbent, nature of adsorbate, nature of the solvent, and others. Adsorption processes are capable of removing contaminants if the adsorbent (solid surface) is selected carefully and the solution chemistry is controlled.

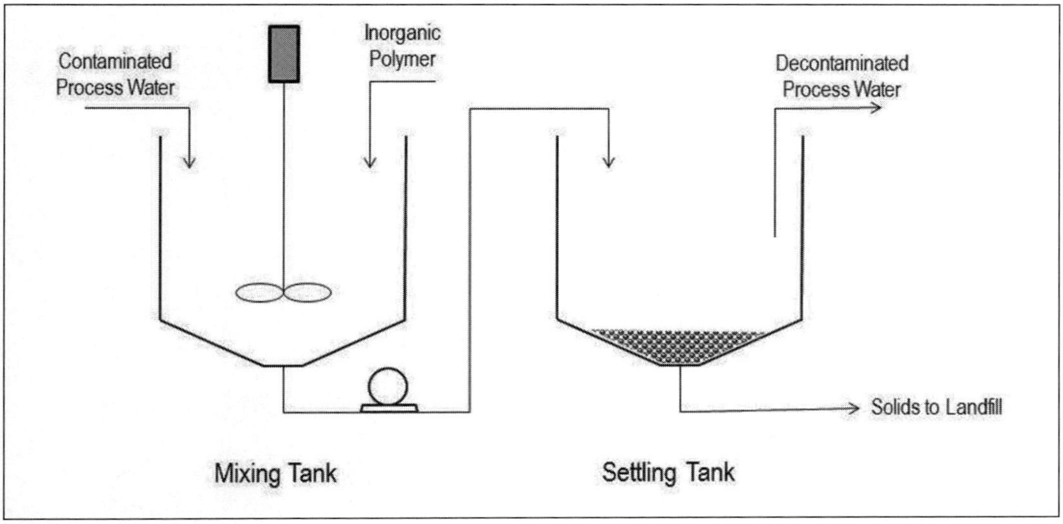

Figure 6. Water treatment in this work

The new water treatment technology developed here is based on in-situ solidification of a sorbent and chemisorption of contaminant species onto the resulting sorbent. An amorphous solid is formed by in-situ cross-linking of a modified silicate-based inorganic polymer. The process, as schematically shown in Figure 6, involves the formation of the sorbent and chemisorption of the contaminated species on active sites of the sorbent, all in a single stage treatment in a continuously stirred (400–500 rpm) mixing tank. Inorganic polymeric system is dosed at 20–1000 ppm to the contaminated water depending on the level of contaminants. After 1–3 hours mixing, the aliquot is transferred into a gravity settling tank to allow for suspended solid precipitation. The clear supernatant is a decontaminated process water, and may be discharged or undergo further treatment if necessary. Faster solid separation may be achieved by filtration and/or use of common organic flocculants such as anionic polyacrylamides. The separated solids may then be safely land-filled.

Treatment of Cadmium Containing Water Samples

Two inorganic polymer systems were evaluated for treatment of cadmium contaminated water samples. Samples were first treated using the sorbent under mixing for 2–3 hours. Resulting suspensions were then left for gravity settling. Filtration of supernatant on a Whatman qualitative filter paper No. 5 (pore size of 2.5 μm) gave a clear, visible solid free water sample for ICP analysis. When this treatment was applied to the removal of cadmium in lab generated water sample, the sorbent alone reduced the cadmium to 17 ppb (for IP-1) and 5.6 ppb (for IP-2) without the need for a promoter (Figure 7). By adding the promoter, the cadmium was further reduced to 1.7 ppb in the case of IP-1 polymer. IP-2 was prepared by slight modification in the IP-1 formulation. The combination of promoter and the polymer systems are labelled as IP-1* and IP-2*, respectively. When flocculation was used for fast settling of suspended solids instead of gravity settling, cadmium level was reduced to 0.8 ppb (for IP-1*) and 0.5 ppb (for IP-2*), indicating the presence of fine solids which may pass through the filter paper at the first trails.

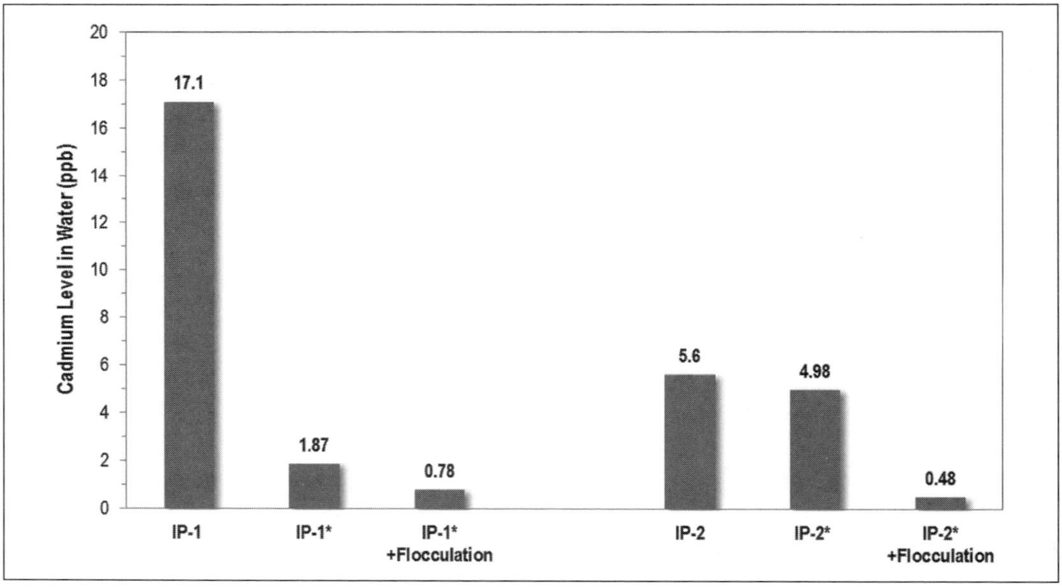

Figure 7. Decontamination of a solution containing 30000 ppb of cadmium using inorganic polymer sorbents

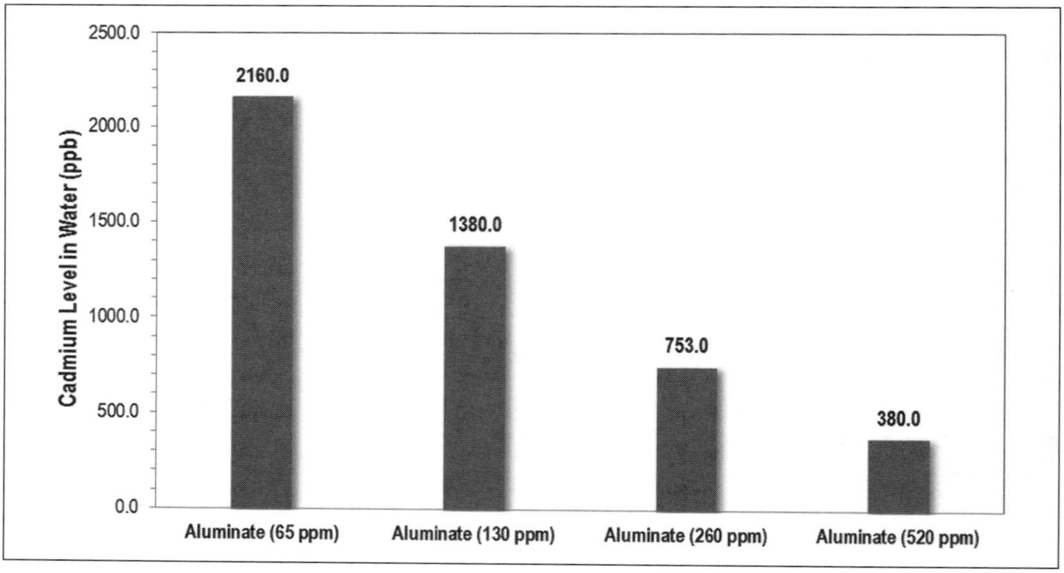

Figure 8. Decontamination of a solution containing 11000 ppb of cadmium using various dosage of sodium aluminate

Comparison of Removal Efficiency of Inorganic Polymer with Aluminate and Ferric Coagulants

The traditional ferrous and aluminate coagulants were also evaluated using lab-generated cadmium water. The ferric treatment did not lead to any reduction in cadmium, but the aluminate coagulation showed a reduction of cadmium levels in contaminated water. As shown

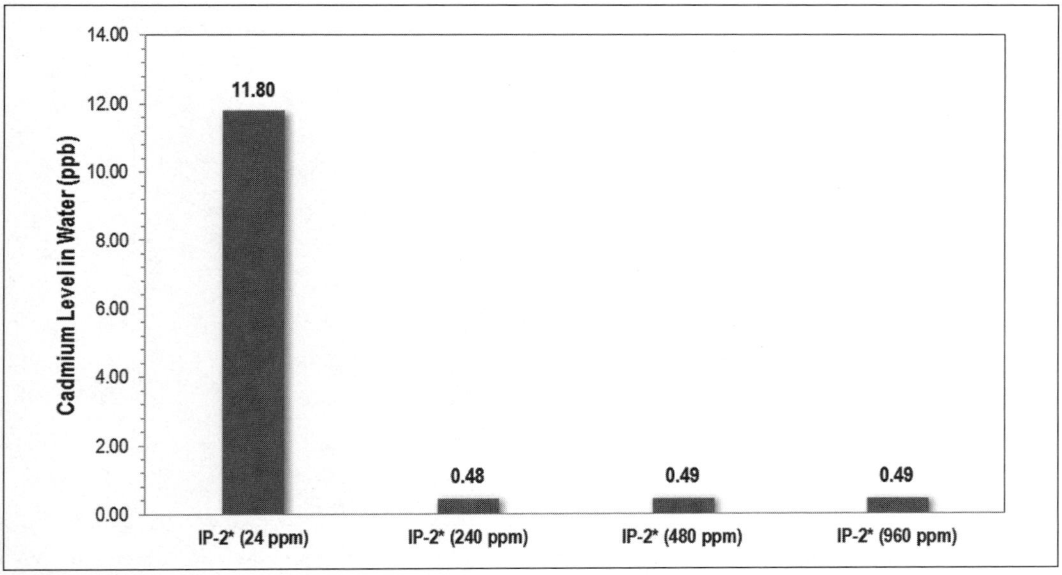

Figure 9. Decontamination of a solution containing 11000 ppb of cadmium using various dosage of inorganic polymer sorbent

in Figure 8, cadmium levels were reduced from 11000 ppb to 380 ppb when 520 ppm of aluminate coagulant was used.

Inorganic polymer systems, however, represent a higher efficiency as cadmium levels reduced from 11000 ppb to 11.80 ppb by using only 24 ppm of the sorbent. Further reduction can be achieved using 240 ppm of the sorbent, resulting in below 1 ppb residual cadmium in the treated water (Figure 9).

Treatment of Selenium Containing Water Samples

During the optimization of this method for treatment of selenate-containing water, a series of inorganic polymeric materials were developed that ranged in their chemical compositions and affinities toward cross-linking in aqueous media (Figure 10). It was found that by using inorganic polymer type 1 (IP-1*), selenium removal efficiency under lab process conditions was more than 99% for treating water samples containing ppm levels of sodium selenate. A reduction in efficiency was observed as the initial selenium-contamination levels were decreased to the lower ppb range (less than 10 ppb); however, some treatments successfully reduced sodium selenate to 3.7 ppb when beginning the treatment with such low levels of selenium.

This treatment method was further optimized by exploring the proper ratio of the best performing sorbent (IP-1) and the sorption promoting chemical (promoter, Pr-1). An improvement in the selenium removal was not observed when the IP-1 was doubled, but the efficiency was increased to >99% when the amount of Pr-1 was doubled, yielding levels of <1 ppb total selenium remaining in the water. The efficiency of the treatment process was monitored via ICP analyses of decontaminated water samples, as well as XRF analyses to evaluate the quantity of absorbed contaminant.

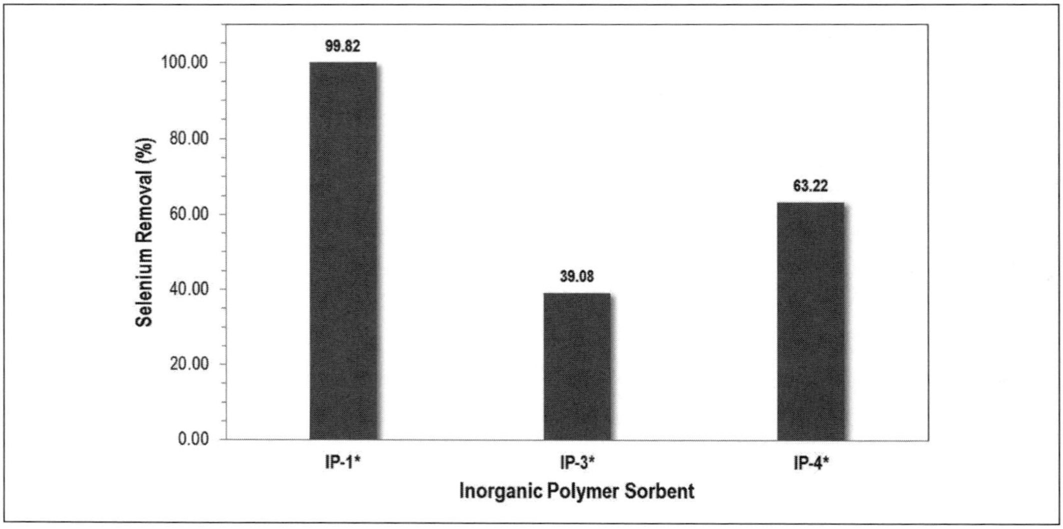

Figure 10. Efficiency of the inorganic polymer systems for selenate removal from water samples

CONCLUSIONS

A new chemisorption technology was developed and evaluated for removing cadmium and selenium from aqueous systems. The contaminant species chemically adsorbed on the active sites within the cavities of the sorbent material, thus efficiently removing the contaminant from the aqueous system. The technology was found to be effective in removing both cadmium and selenate from water samples containing a few ppm of contaminants to levels below 1 ppb. The chemistry of inorganic polymers is diverse, and can be tailored towards total reduction of the contaminants from aqueous media, by the addition of a promoter. Inorganic polymers based on silicates provide an economical and versatile solution for treatment of contaminated mining process waters, within an operational and environment-friendly process.

The chemisorption on inorganic polymers was compared to the traditional coagulant/precipitant methods for removing cadmium from water samples. No cadmium reduction was observed using a ferric system, and an incomplete reduction was achieved via the aluminate treatment. In contrast, the treatment using inorganic polymers offered a significant reduction of cadmium, even at low treatment dosages (24 ppm). Cadmium levels of <1 ppb were achieved at 240 ppm dosage of the inorganic polymer system. Initial screenings revealed that the inorganic polymer systems are not sensitive to high concentration of common cationic and anionic interferences (e.g., Ca^{2+}, Fe^{2+} and SO_4^{2-}) in the process stream. Work is ongoing to evaluate this technology on actual contaminated mine process waters, and the initial data appears promising. We are also conducting leaching experiments to verify whether any contaminant can be released from the adsorbent post-treatment at pH levels below 7. This would allow us to better assess any risk associated with landfilling of the solid waste. Nevertheless, due to the nature of adsorption of toxic species on the inorganic adsorbent, leaching selenate or cadmium from the solid waste is very unlikely at neutral or alkaline conditions.

REFERENCES

Adams, D.J., and Pennington, P. 2005. Selenium and Arsenic Removal from Mining Wastewaters. *Proceedings of the SME Annual Meeting*, Denver, Colorado, February 28–March 2, 2005. Preprint 05-53.

Agency for Toxic Substances and Disease Registry. The Division of Toxicology ToxFAQs. atsdr.cdc.gov/tfacts92.pdf (accessed April 2011).

ATSDR. Agency for Toxic Substances and Disease Registry. The Division of Toxicology and Environmental Medicine (ToxFAQs). atsdr.cdc.gov/tfacts5.pdf (accessed June 2011).

Cohen, R.H. 2006. Use of Microbes for Cost Reduction of Metal Removal from Metals and Mining Industry Waste Streams. *Journal of Cleaner Production, 14,* 1146–1157.

EPA (Environmental Protection Agency). National Recommended Water Quality Criteria for Priority Pollutants. water.epa.gov/scitech/swguidance/standards/current/upload/nrwqc_2009.pdf (accessed June 2011).

Gaballah, I., Goy, D., Kilbertus, G., and Thauront, J. 1993. A New Process for the Decontamination of Industrial Effluent Containing Cadmium Cations. *The Minerals, Metals, & Materials Society,* 43–52.

Harris, T.M., Jones, D.B., Shang, A., Berkenbile, L., and Logsdon III, G. 1992. Recovery of Metals from Wastewater. *The Minerals, Metals, & Materials Society,* 155–163.

Harrison, T., Sandy, T., Leber, K., Srinivasan, R., McHale, J., and Constant, J. 2010. Characterization and Treatment of Selenium in Water Discharged from Surface Coal Mining Operations in West Virginia. *Proceedings of the SME Annual Meeting,* Phoenix, AZ, February 28–March 3, 2010. Preprint 10-142.

IPCS (International Programme on Chemical Safety). The Environmental Health Criteria 134. www.inchem.org/documents/ehc/ehc/ehc134.htm (accessed June 2011).

Lemly, A.D. 2004. Aquatic Selenium Pollution Is a Global Environmental Safety Issue. *Ecotoxicology and Environmental Safety, 59,* 44–56.

Mihaylov, B., and Hendrix, J. 1992. Continuous Precipitation and Separation of Copper, Cadmium, Mercury, and Arsenic. *The Minerals, Metals, & Materials Society,* 81–97.

National Recommended Water Quality Criteria. EPA. water.epa.gov/scitech/swguidance/standards/current/index.cfm (accessed May 2011).

Rybock, J.T., and Anderson, A.L. 2000. Silica Micro Encapsulation: An Innovative Technology for the Treatment of ARD. In *Proceedings of the SME Annual Meeting,* Salt Lake City, Utah, February 28–March 1. Preprint 00-9.

Selenium Treatment/Removal Alternatives Demonstration Project. 2001. Mine Waste Technology Program Activity III, Project 20. EPA/600/R-01/077. Butte, MT: MSE Technology Applications.

Toxic Substance Profile: Cadmium. www.ukmarinesac.org.uk/activities/water-quality/wq8_2.htm (accessed June 2011).

Zhang, B., and Fuerstenau, M.C. 2002. Wastewater Remediation with Cellophane. *Minerals & Metallurgical Processing, 19,* 123–127.

New Electro-Biochemical Reactor for Treatment of Wastewaters

D.J. Adams, M. Peoples, and A. Opara
INOTEC & University of Utah, Department of Metallurgical Engineering, Salt Lake City, UT, USA

ABSTRACT

A number of microorganisms, including undefined microbial consortia, are capable of directly accepting electrode provided electrons in a manner that enhances reduction of terminal electron acceptors; i.e., metals such as arsenic and selenium and various inorganics like nitrates and sulfates. Electrons are provided directly to a system via an electrode and through an electron distribution matrix, using low DC voltage. Providing electrons directly to microbes and reactor environments has a number of advantages over indirect electron provision: adding organic electron donors or hydrogen to bioreactors.

A patented electro-biochemical reactor (EBR) directly provides electrons and an electron receptor environment to microbes and reactors, enhances contaminant transformations, and allows for more precise control of system Eh/pH environments. In conventional treatment systems, nutrients and chemicals provide electrons and electron acceptor environments, in a rather inconsistent manner. Due to the inconsistent availability of electrons and hydrogen within these systems, a large excess of nutrients and chemicals are often required for contaminant transformations and to generate the desired Eh/pH conditions within a portion of the treatment environment.

The EBR provides electrons to microbes and reactor environments using low DC voltage (1–3 volts) in configurations that provide an electron density gradient or a controlled electron density. The provided electrons result in increased microbial contaminant transformation kinetics, reduced retention times, and reductions in the amount of nutrients and chemicals required for contaminant removal. Preliminary tests indicate that EBR technology will remove target contaminants to low ppb levels using ≤½ conventional bioreactor retention times and nutrient concentrations. The applied voltage reduces the use of nutrients, supplies some of the energy (electrons) required for bacterial growth, and supplies energy potential for contaminant transformation, reducing contaminant transformation energies that must be provided for through microbial metabolic processes.

INTRODUCTION

Treatment of various wastewaters is becoming ever more critical due to diminishing water resources, increasing wastewater disposal costs, and stricter discharge regulations that have lowered permissible contaminant levels. The treatment of wastewater for reuse and disposal is particularly important in arid regions of the US and the world. Metal pollutants can be introduced into waters through many activities including mining and mineral processing, abandoned petroleum processing, coal-fired power plants, and agricultural run-off. Over the past decades, appreciation for the role of microbes in the precipitation of minerals and the redox transformation of many elements has grown.

Microbes are now known to be involved in the environmental transformations of many elements. Understanding of the tremendous metabolic diversity of the microbial world has expanded in recent years with the discoveries that microbes can use a variety of elements and compounds, including Fe, Mn, Se, U, As, NO_3^-, CN^-, SO_4^{2-}, and others as terminal electron acceptors in anaerobic respiration. Metal and inorganic removal is highly dependent on the availability of electrons—redox, or Eh/pH, the available oxygen in the system, pH, etc. (Metcalf, 2002, Peoples and Adams 2010, Twidwell, et al. 2000). Microbes and chemicals alter or remove metal and inorganic contaminants by adding and/or removing electrons. It is based on this knowledge that an electrochemical bioreactor was developed to take further advantage of direct electron provision to aid in microbial and biochemical metal and inorganic transformation processes.

Research conducted at Utah Universities since 1995, drawing on related microbial fuel cell studies, has resulted in patented technology harnessing the ability of various microorganisms and undefined microbial consortia to directly utilize electrode provided electrons in a manner that enhances reduction of terminal electron acceptors (Adams, et al., 2009, Adams and Peoples, 2010, Peoples and Adams 2010, Newton, et al., 2008). Electrons are provided directly to a system via an electrode and through an electron distribution matrix, using low DC voltage. Direct electron transfer to cells and reactor systems has a number of advantages over indirect electron provision via addition of organic electron donors or hydrogen to bioreactors. While the mechanisms for direct electron transfer from electrodes to microorganisms are still being investigated, including methods to optimize various patented applications, their benefit is often immediately apparent.

Background

The ability of microorganisms to transfer electrons to electrodes without electron shuttling compounds has been known for about a century (Potter, 1911). Much more recently, acetate was shown to be oxidized to carbon dioxide with direct electron transfer to an electrode as the sole electron acceptor (Bond and Lovley, 2003, Bond et al., 2002). The history of direct electron flow from electrodes to microorganisms is rather short with the first report appearing in 2004 (Gregory et al., 2004). A recent review by Thrash and Coates (2008), indicates that there are a number of investigations on a wide diversity of redox-active molecules that can function as electron shuttles.

The current model for electron transfer to anodes is that of conductive pili (Lovley, 2008, Reguera et al., 2006, Strycharz et al., 2008) and is indicated to proceed through outer-surface *c*-type cytochromes (Holmes et al., 2004, Nevin et al., 2009) required to facilitate electron transfer between the biofilm and the anode surface (Inoue et al., 2010, Kennedy et al., 2007). Additionally, gene expression patterns in current-consuming cells appear to differ from those in current-producing cells (Strycharz et al., 2008). The ability of microorganisms to respire using electrons from an electrode or an electron distribution matrix as a sole or primary electron donor or acceptor makes it feasible for these populations to be more self-sustaining. This mechanism also allows for control of excess microbial growth, thus, reduction of microbial plugging in various treatment system environments resulting from conventional treatment practices using large excesses of nutrients/chemicals.

Electro-Biochemical Remediation

As a historical perspective, microbial reduction of soluble U(VI) to insoluble U(IV) was proposed, but not demonstrated, for immobilizing uranium in subsurface environments (Anderson et al., 2003, Finneran et al., 2002, Gregory and Lovley, 2005) or to concentrate extracted uranium (Phillips et al., 1995). *Geobacter sulfurreducens* reduced U(VI) to U(IV) with an electrode serving as the electron donor (Gregory and Lovley, 2004). Additionally, *Geobacter lovleyi* has been shown to reduce tetrachloroethene (PCE) and trichloroethene (TCE) to *cis*-dichloroethene (*cis*-DCE) with an electrode serving as a sole electron acceptor (Strycharz et al., 2008). Additionally, degradation of complex mixtures of organic compounds has been shown to be enhanced using EBR technology, (Adams–unpublished results, Potter, 1911, Thrash and Coates, 2008).

Electron transfer from electrodes is beginning to receive increased attention; e.g., removal of nitrate in wastewater treatment. Arsenic and nitrate reduction has been examined by Waterland et al., 2010, Oremland and Stolz, 2005, and Virdis et al., 2008. EBR tests show two to ~nine-fold increases in denitrification kinetics in the laboratory (Adams et al., 2009, Adams and Peoples, 2010, Peoples and Adams, 2010, and Newton et al., 2008) and arsenic removal and denitrification in pilot field studies (Adams and Peoples, 2010, Peoples and Adams, 2010). Several studies have reported that mixed microbial cultures were able to denitrify nitrate with an electrode serving as the sole electron donor (Clauwaert et al., 2007, Jia et al., 2008, Park et al., 2005, Virdis et al. 2008). However, in all studies, the microorganisms responsible for this process, the mechanisms of interaction, and denitrification processes warrant further study.

Research conducted at the Utah Universities has shown that directly supplying electrons require less maintenance, monitoring and energy than conventional methods using only organic electron donors. Electrons can be supplied from solar panels, making the use of EBR technology for treatment of waters and contaminated soils an attractive sustainable practice, and pilot-scale tests have been completed and more are underway. Solar-powered and low voltage DC cathodes have been described and deployed at a pilot-scale field site for removal of arsenic and nitrate from contaminated mine waters and large-scale EBR applications are under development.

Though conventional bioreactors have proved successful in an operational sense, the issue of capital costs, treatment residence time, and nutrient costs has dictated their acceptance over other treatment approaches. In conventional systems, a large excess of nutrients/chemicals are required to provide the electrons and electron acceptor environments needed. Moreover, excess nutrients/chemicals are required in conventional systems to adjust reactor chemistry, for microbial growth and contaminant removal, and to compensate for system sensitivity and inefficient and fluctuating electron availability.

The electro-biochemical reactor (EBR) technology overcomes shortcomings in conventional systems by combining electrochemical and biological treatments. It supplies electrons to the reactor and microbes using low *DC* voltage; o*ne volt supplies about a trillion, trillion electrons.* These electrons cost-effectively replace the electrons normally supplied by excess nutrients/chemicals in a controlled manner, effecting a reduction in treatment residence times and nutrient costs; therefore, capital costs. The electrons needed for a full-scale facility can easily be supplied by a small solar grid. The EBR addresses improvements in conventional systems by application of a charge potential that provides electrons to:

1. Stabilize and control the conditions necessary to precipitate metals and reduce inorganics,
2. Reduce or eliminate the amount of nutrients and chemicals required for ORP adjustment,
3. Provide controllable ORP environments that occurs with distance from the electrode rather than the classical and variable ORP gradient produced with excess nutrients/chemicals, and
4. Provide easy microbial access to electrons needed for life functions and transformation of contaminants in aerobic and anaerobic environments.

The EBR technology makes conventional reactors more controllable, efficient, economical, and robust, reducing the reactor size and nutrient/chemical amounts required for effective contaminant transformations. In short, EBR technology starts with the best aspects of proven microbial and chemical systems and takes them to the next level of performance and cost-effectiveness. Both bench and pilot-scale tests indicate that the EBR technology will allow more effective and economical biotreatment of large volume low-contaminant metal and inorganic containing waters as well as provide economic benefits for treatment of high contaminant organic, metal and inorganic waters.

MATERIALS AND METHODS

Microbes

Microbes employed to reduce selenium, arsenic, and nitrate in these tests were site indigenous microbes, isolated, screened for their respective contaminant reductive abilities, grown to high densities, and established within the EBR systems. Nucleic acid based individual microbe and microbial population identification used denaturing gradient gel electrophoresis (DGGE) and terminal restriction fragment (TRF) analysis.

Site Test Waters

Bench scale EBR tests evaluated process waters with a two stage conventional bioreactor system and a single stage EBR. Selenium and nitrate were present in the site water at concentrations of ~4.5 mg/L, nitrate concentrations were adjusted to ~100 mg/L nitrate-N, and the pH was adjusted to ~6.5. Retention times of 12 hours in the EBR and 24 hours in the conventional 2-stage CBR reactor were used. Nutrients were also required at ~1.5× amounts in the CBR system. All reactors were operated under up-flow conditions that approximated plug flow. The EBR system only, was operated at 3 volts DC potential, current was only measurable in the microamp levels. All other conditions in the CBR and EBR stages were identical. All analysis used ICP-MS.

Additional selenium removal tests were conducted at bench-scale using site contaminated waters and a two-stage, up-flow, EBR system; waters contained ~260 mg/L selenium. Because the waters received did not have significant nitrate present, site personnel advised that test waters be spiked with 100 mg/L nitrate-N. The retention time for each EBR stage was between 15 and 24 hours. All analysis used ICP-MS.

Bench scale selenium removal tests were performed with single stage EBR under two different electrode configurations. Tested waters contained 550–1000 ppb selenium. Retention times varied between 17 and 24 hours. All analysis used ICP-MS.

Pilot-scale tests. A two-stage EBR system was operated under conditions that approximated plug flow (up flow); total retention times varied from 10 to 18 hours or flow rates from

~0.5 gpm to ~0.8 gpm. Influent arsenic concentrations were ~750 mg/L and nitrate concentrations averaged about 24 mg/L nitrate-N. Pilot-scale tests were initiated in summer and finished in late fall with water temperatures ranging from ~23°C to ~9.5°C. Analysis was completed by a University certified laboratory and a third party lab.

Nutrients

In both the selenium and arsenic tests, a laboratory culture media, Trypticase Soy Broth at 15 to 30 g/L or a balanced molasses, urea (>0.25g/L to 2.5g/L), yeast extract (>0.25 g/L to 1.0 g/L), and phosphate (0.025g/L to 1.0 g/L) mixture was used. Nutrient concentrations were varied to examine nutrient components of C:N:P:S, etc., for growth of new microbial cells and were balanced to microbial cellular levels. Higher nutrient levels were used to establish the biofims to their desired density. Lower nutrient levels were used once the EBR and microbial populations reached more stabilized operating conditions. Nutrients were added on a daily basis by mixing the nutrient into an appropriate amount of pH adjusted test water and pumping them into each EBR separately over a five minute time period, then returning to normal flow rates.

RESULTS AND DISCUSSION

Microbes

Many species of microorganisms can affect the reactivity and mobility of metals and inorganics, and can be used to remove these contaminants from waters. The microorganisms used are naturally-occurring INOTEC/University repository microbes and site microbial consortia; not genetically engineered or modified microbes. The resulting products of the processes are elemental selenium, elemental arsenic and arseinc sulfides; nitrates and nitrites are converted to nitrogen gas.

Microbes present in the arsenic EBR included *Desulfobacterium sp.*, *Desulfotomaculum sp.*, *Shewanella sp.*, *Bacillus sp.*, and several uncultured (unidentified) microbes. Bacteria present in the selenium EBR included *Pseudomonas sp.*, *Desulfobacterium sp.*, *Shewanella sp.*, *Bacillus sp.*, and several uncultured (unidentified) microbes. The unidentified microbes were shown through site population analysis to be site indigenous microbes (data not presented). Microbial populations were examined after biofilm establishment using site waters containing indigenous microbes. Microbial density on the EBR support materials estimated through assays at $\sim 2\times 10^{11}$/g.

Voltage, ORP, and Nutrients

The voltage required to increase contaminant transformation efficiencies varies with the bioreactor's microbial support materials, water chemistry, and microbes. Supplied voltage provides electrons at the bacterial surface and a readily available supply of elections to the bacterial-contaminant-surface environment that lowers the microbial contaminant interaction and transformation energy requirements. Providing electrons at the microbial support surface interface allows a more controlled oxidation-reduction potential (ORP) environment or gradients for more effective removal of single or multiple contaminants in a single bioreactor.

Microbes in the EBR system interact with the electrode through direct contact, mediating the transfer of electrons via conductive pilli and microbial surface interactions. Interaction can also occur through energy shuttle compounds that move energy to both electrode and

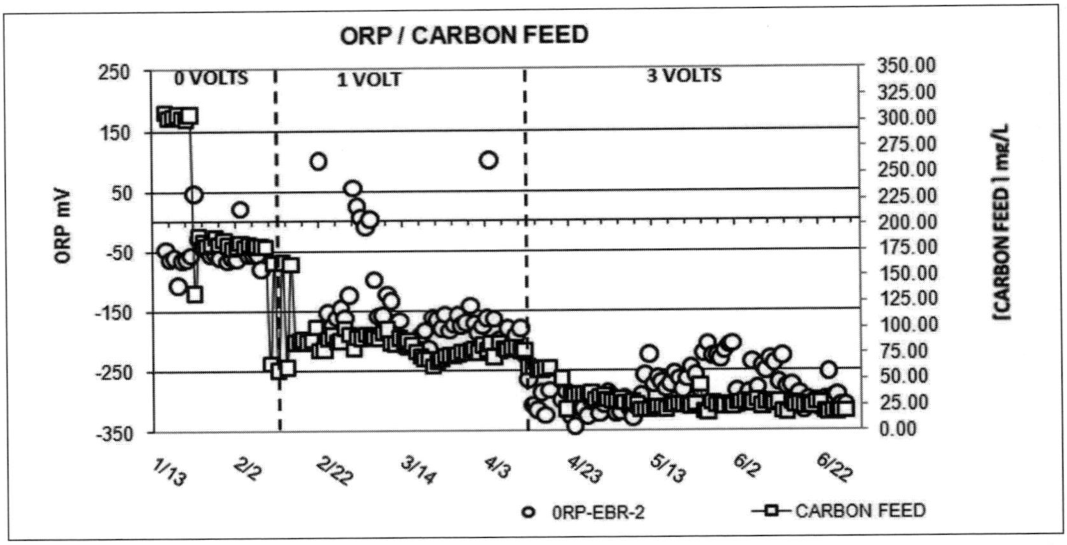

Figure 1. Relationships between voltage, ORP, and carbon input measurements made in a two-stage EBR at the end of the second stage

non-electrode and conductive matrix bound microbes throughout the EBR system. The applied voltage potential assists in controlled system ORP in a manner that eliminates some of the need for excess nutrients to lower ORP; thus reducing nutrient costs. Readily available electrons supply some of the energy (electrons) required for bacterial growth and contaminant transformation, again reducing nutrient costs. Figure 1 shows typical ORP measurements made in a two-stage EBR and a relationship with voltage and nutrient (carbon) amount. Note the stability of ORP obtainable within the EBR system at 3 volts and the lower amounts of nutrient required to achieve the desired ORP.

Bench-Scale Selenium Removal Tests

Figure 2 shows the results of bench-scale tests using mining process water containing ~4.5 mg/L selenium and spiked nitrate levels of ~100 mg/L. The figure shows comparisons with a two stage conventional bioreactor (CBR) system and a single stage EBR. Retention times were 12 hours in the EBR and 24 hours in the conventional 2-stage CBR reactor. Nutrient amounts were also ~1.5× in the CBR system. All reactors were operated under up-flow conditions that approximated plug flow. The EBR only, was operated at 3 volts DC. All other conditions in the CBR and EBR stages were identical. The EBR effluent was below 0.030 mg/L while the two state CBR produced selenium effluents averaging >0.060 mg/L. Monitoring points within the EBR indicated that nitrates were also completely removed in the first few inches of the system.

Figure 3 shows the results of a second bench-scale test using a two stage, up-flow, EBR system to treat waters containing ~260 mg/L selenium. Because the waters received did not have significant nitrate present, site personnel advised that test waters be spiked with 100 mg/L nitrate-N. The retention time for each EBR stage was ~15 hr for a total retention time of 30 hr. The EBR system produced final selenium effluents averaging ~13 µg/L. Effluents from EBR-2 ranged from a high of 55.4 µg/L to a low of 2.8 µg/L. If the high and low selenium effluent readings were discarded for the 3 volt test period, the average selenium effluent from

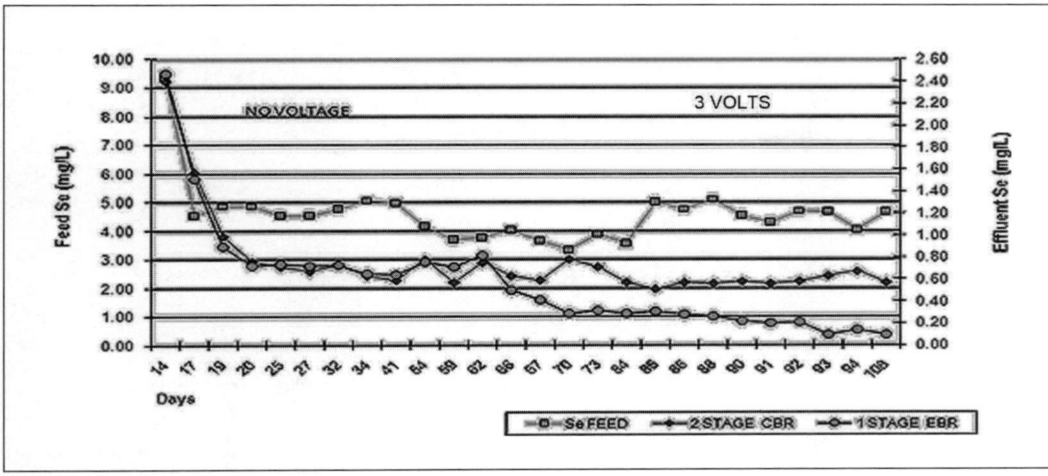

Figure 2. Comparison of treatment of mining process water containing selenium using two bioreactor types. A two-stage conventional bioreactor (CBR), configured identically to a single-stage EBR, but without applied potential, using twice the retention time, and ~1.5 times the nutrient amount as the single-stage EBR.

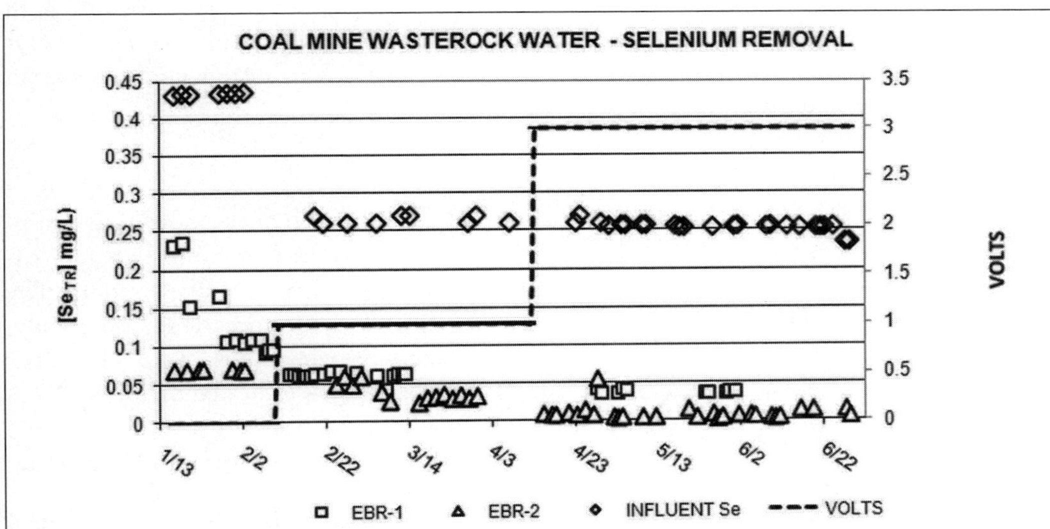

Figure 3. Two-stage EBR treating mining influenced wastewaters containing selenium. EBR stages were compared with applied voltage and selenium removal.

EBR system drops to ~11µg/L. Nitrate in the EBR system was removed in the first few inches of the EBR 1st stage.

In mining solutions, for example those shown in Figures 2 and 3, spiked with 100 mg/L nitrate-N nitrate was at zero at the first monitoring point in the EBR. Once equilibrated, EBR denitrification rates reached >95 mg nitrate-N per hour.

Additionally, a mine process water containing 550–1,000 ppb selenium was tested using EBR with different electrode configurations. Multi, where eight electrodes evenly spaced

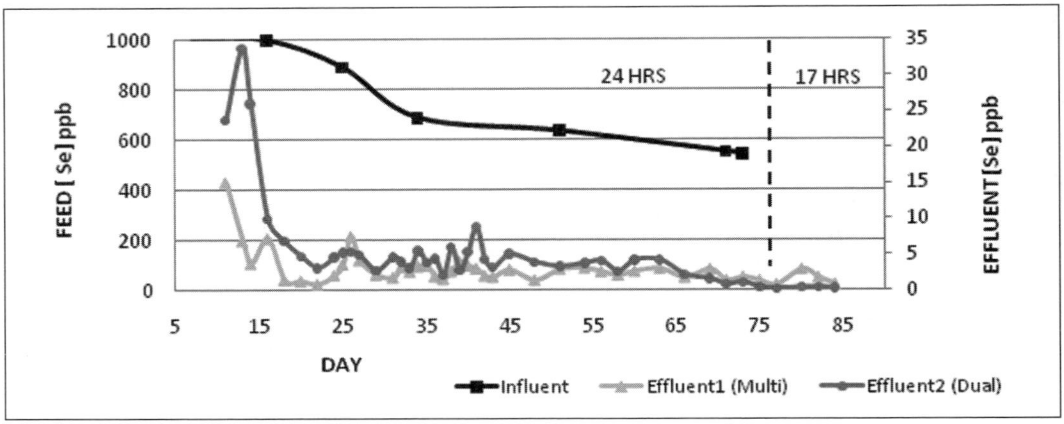

Figure 4. Process mine water treated in two different EBR configurations and with two retention times

over the length of the EBR and connected to a power supply providing a 3 V potential, and dual, where two electrodes, one at the top and one at the bottom of the EBR also providing 3 V potential. Each EBR was operated as single-stage reactor with 24 hours retention time, Figure 4. Upon reaching steady state, >99.8% of selenium was being removed using both configurations; effluent selenium concentrations were <1 ppb. Initial results with a 17-hour HRT show similar results. Optimal EBR electrode configurations are dependent on selenium concentration and water chemistry.

Pilot-Scale Tests—Arsenic Removal

Pilot-scale tests were conducted on site using EBR components configured within a shipping container. The two stage EBR pilot system was operated under conditions that approximated plug flow (up flow) using total retention times varying between 10 to 18 hours; flow rates from ~0.5 gpm to ~0.8 gpm. The light colored circles in Figure 5 indicate replicate analysis splits, with the submitted samples shown in light diamonds being diluted ½. This analysis control check was completed because University laboratory analysis indicated between 10 and 20 ppb arsenic concentrations; significantly lower than those reported by a third party lab. However, the third party lab analysis met site arsenic treatment goals. Influent NO_3 was about 24 mg/L and was removed in the first EBR stage.

Note that after about three days, the system arsenic effluent in EBR1 increases during flow rates of ~0.8 gpm. However, during the same test periods, the efficiency of EBR2 increased, maintaining the final average arsenic effluent at about 15 ppb (University analysis) or ~40 ppb (third party analysis). Third party analysis met target discharge goals.

Pilot-scale tests were conducted through summer and late fall covering a >15°C water temperature drop. EBR system maintained a consistent metal and nitrate removal throughout the testing period, Figure 6.

Thus far, nine fully successful bench scale tests on various waters have been completed for mining companies to treat both process and wastewaters from hard rock mines and coal mines containing selenium and nitrate, arsenic and nitrate, and mercury. A pilot-scale testing for selenium removal from mining waters has also been completed successfully.

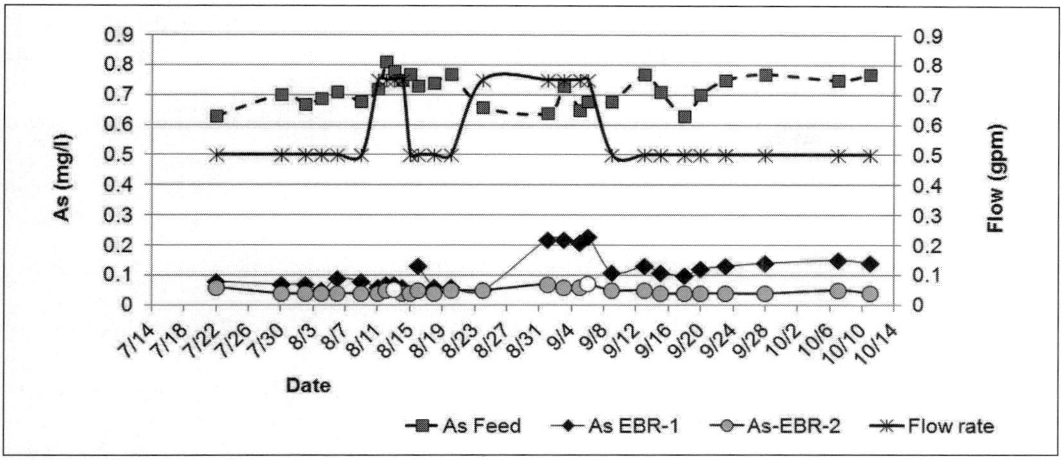

Figure 5. Arsenic removal in a pilot-scale EBR system operated on-site using 3 volts and a total retention time of 10 to 18 hours

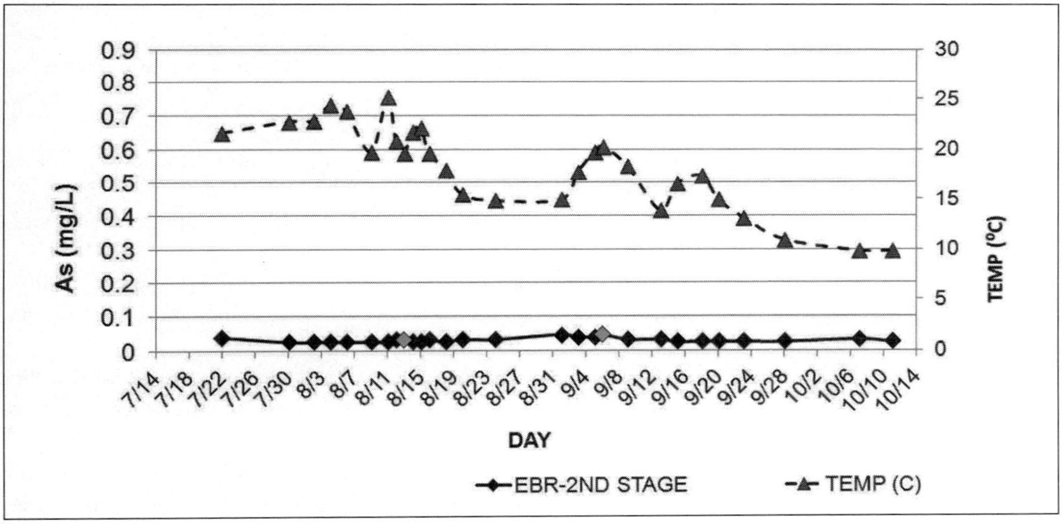

Figure 6. Relationship between temperature (°C) and arsenic effluent levels during on-site pilot-scale testing

CONCLUSIONS

The EBR has been shown to be effective in removal of selenium, arsenic, nitrate as well as other metals and inorganics.

- The EBR system provides a more precise system control; better controlled ORP, produces a more robust biofilm, and increases contaminant transformation/removal kinetics and efficiency.
- EBRs are capable of removing target contaminants to low ppb levels.
- From the results presented and for all tests conducted to date, EBRs show ≥40% lower capital costs and ≥40% less operational nutrient/reagent costs than conventional bioreactors.

- The EBR system produces more controllable and stable bioreactor environments.
- Because of the low DC voltage potential used, power requirements for a full-scale facility can be supplied by a small solar grid.
- EBR technology starts with the best aspects of proven microbial and chemical systems and takes them to the next level of performance and cost-effectiveness.

REFERENCES

Adams, D.J., Peoples, M., Newton, N., and Nanduri, M. 2009. *Electrobiochemical Reactor: Removal of Metals, Nitrate and BOD. Thatcher, AZ: INAP.*

Adams, D.J., and Peoples, M. 2010. New electro-biochemical reactor for removal of selenium, arsenic, and nitrate. Paper/Presentation. Sydney, NS: IMWA.

Adams, D.J., Diaz, X., Miller, J.D., and Chatwin, T. 2005. Arsenic removal from contaminated waters. San Francisco, CA: International TMS Meetings.

Bond, D.R., and Lovley, D.R. 2003. Electricity production by *Geobacter sulfurreducens* attached to electrodes. *Appl. Environ. Microbiol.* 69:1548–1555.

Bond, D.R., Holmes, D.E., Tender, L.M., and Lovley, D.R. 2002. Electrode-reducing microorganisms that harvest energy from marine sediments. *Science* 295:483–485.

Clauwaert, P., Rabaey, K., Aelterman, P., De Schamphelaire, L., Pham, T.H., Boeckx, P., et al. 2007. Biological denitrification in microbial fuel cells. *Environ. Sci. Technol.* 41:3354–3360.

Finneran, K.T., Anderson, R.T., Nevin, K.P., and Lovley, D.R. 2002. Potential for bioremediation of uranium contaminated aquifers with microbial U(VI) reduction. *Soil Sediment Contam.* 11:339–357.

Gregory, K.B., and Lovley, D.R. 2005. Remediation and recovery of uranium from contaminated subsurface environments with electrodes. *Environ. Sci. Technol.* 39:8943–8947.

Gregory, K.B., Bond, D.R., and Lovley, D.R. 2004. Graphite electrodes as electron donors for anaerobic respiration. *Environ. Microbiol.* 6:596–604.

Holmes, D.E., Bond, D.R., O'Neilf, R.A., Reimers, C.E., Tender, L.R., and Lovley, D.R. 2004. Microbial communities associated with electrodes harvesting electricity from a variety of aquatic sediments. *Microb. Ecol.* 48:178–190.

Jia, Y.-H., Tran, H.-T., Kim, D.-H., Oh, S.-J., Park, D.-H., Zhang, R.-H., and Ahn, D.-H. 2008. Simultaneous organics removal and bio-electrochemical denitrification in microbial fuel cells. *Bioprocess Biosyst. Eng.* 31:315–321.

Kennedy, J.P., Richins, B.C., and Adams. D.J. 2007. Arsenic binding proteins for removal of arsenic from solution. Presented/Published in the Environmental and Subsurface Science Symposium, Bioremediation and Biotechnology.

Lovley, D.R. 2008. The microbe electric: Conversion of organic matter to electricity. *Curr. Opin. Biotechnol.* 19:564–571.

Metcalf & Eddy, Inc., Tchobanoglous, G., Burton, F., and Stensel, H.D. 2002. *Metcalf & Eddy, Wastewater Engineering Treatment and Reuse*, 4th ed., pp. 616–700.

Newton, N.N., Adams, D.J., and Miller, J.D. 2008. *Biotreatment of Cyanide and Arsenic in Drinking Water and Mine Waste Streams.* Littleton, CO: SME.

Oremland, R.S., and Stolz, J.F. 2005. *Arsenic, microbes and contaminated aquifers. Short Survey Trends Microbiol. 13(2):*45–49.

Park, H.I., Kim, D.K., Choi, Y.-J., and Pak, D. 2005. Nitrate reduction using an electrode as direct electron donor in a biofilm-electrode reactor. *Process. Biochem.* 40:3383–3388.

Peoples, M., and Adams, D.J. 2010. Electro-biochemical Reactor for Removal of Metals and Nitrates. Poster Presentation. SME, Phoenix, AZ.

Phillips, E.J.P., Lovley, D.R., and Landa, E.R. 1995. Remediation of uranium contaminated soils with bicarbonate extraction and microbial U(VI) reduction. *J. Ind. Microbiol.* 14:203–207.

Potter, M.C. 1911. Electrical effects accompanying the decomposition of organic compounds. *Proc. R. Soc. London B* 84:260–276.

Reguera, G., McCarthy, K.D., Mehta, T., Nicoll, J.S., Tuominen, M.T., and Lovley, D.R. 2005. Extracellular electron transfer via microbial nanowires. *Nature* 435:1098–1101.

Reguera, G., Nevin, K.P., Nicoll, J.S., Covalla, S.F., Woodard, T.L., and Lovley, D.R. 2006. Biofilm and nanowire production leads to increased current in *Geobacter sulfurreducens* fuel cells. *Appl. Environ. Microbiol.* 72:7345–7348.

Strycharz, S.M., Woodard, T.L., Johnson, J.P., Nevin, K.P., Sanford, R.A., Loeffler, F.E., and Lovley, D.R. 2008. Graphite electrode as a sole electron donor for reductive dechlorination of tetrachloroethene by *Geobacter lovleyi. Appl. Environ. Microbiol.* 74:5943–5947.

Thrash, J.C., and Coates, J.D. 2008. Review: Direct and indirect electrical stimulation of microbial metabolism. *Environ. Sci. Technol.* 42:3921–3931.

Twidwell, L.G., McCloskey, J., Miranda, P., and Gale, M. 2000. *Technologies and potential technologies for removing selenium from process and mine wastewaters.* In *Proceedings of Minor Elements*. Littleton, CO: SME. p. 53.

Virdis, B., Rabaey, K., Yuan, Z., and Keller, J. 2008. Microbial fuel cells for simultaneous carbon and nitrogen removal. *Water Res.* 42:3013–3024.

Waterland, R., Adams, D.J., Cota, T., and Corbetta, P. 2010. In situ denitrification of spent ore at the Wharf mine in Lead, South Dakota. SME Annual Meeting, Phoenix, AZ.

Kinetics and Thermodynamic Studies on Biosorption of Heavy Metals Using *Ecklonia Maxima*

A.B. Marshall and D.I.O. Ikhu-Omoregbe
Dept. of Chemical Engineering, Cape Peninsula University of Technology, Cape Town, South Africa

ABSTRACT

Biosorption is the property of certain microbial biomaterials to bind and concentrate heavy metals from very dilute aqueous solutions. It is a relatively new technology for the treatment of industrial wastewater, therefore information is not readily available regarding the mechanism of the process. *Ecklonia Maxima* is a brown seaweed found along the west coast of South Africa between Cape Agulhus and Cape Columbine. A previous study conducted by Aldrich et al. proved that *Ecklonia Maxima* was capable of removing divalent metals such as Cu, Pb and Cd from dilute synthetic solutions. *Ecklonia Maxima* was utilized in the removal of Mn^{2+} and Fe^{2+} from synthetic solutions. The optimum pH for Mn^{2+} removal was found to be 5.8 and the optimum pH for Fe^{2+} was 4.8. The biosorption process was best described by the pseudo-second-order kinetic equation for both metals, and the Freundlich isotherm was the best fit for the Mn^{2+} saturation data.

INTRODUCTION

Heavy metals, which are released from industrial, mining and agricultural activity, are not biodegradable. This can be detrimental to a variety of living species due to the accumulation of heavy metals throughout the food chain. The removal of these heavy metals is considered a high priority, particularly in light of more stringent environmental legislation (Volesky, 2003; Baysal et al., 2003).

The conventional techniques used to remove heavy metals from waste effluents include precipitation, coagulation, reduction and membrane processes, ion exchange and adsorption. However, the applicability of such processes is often limited by technical and/or economic constraints. For example, precipitation processes cannot meet the metal concentration limits required by regulatory standards and produce wastes that are difficult to treat. On the other hand, ion exchange and adsorption processes are comparatively effective but require expensive adsorbent materials and challenging plant management, especially when the heavy metal concentrations are in the range 10–100 mg/L (Goyal et al., 2003; Kapoor et al., 1995). There is a real need to find a cost-effective alternative for the removal of heavy metals from wastewater.

Water resources in South Africa are multi-functional. The same water source is used in industrial and agricultural applications. It is also used by the general population—in the case of informal settlements usually before the water is treated by the municipal treatment plants. It is therefore important that effluents from industrial and agricultural applications are treated before it is discharged into rivers and estuaries used by the general population. The quality of water resources in South Africa is also steadily declining due to the increase in urbanisation and industrialisation. Population increases in South Africa are also expected to lead to an increase in agricultural development, which will then lead to an increased demand for irrigation water (Jackson et al., 2009).

Current methods for the removal of iron and manganese include precipitation which should always be followed by a secondary treatment, reverse osmosis which is very expensive and ion exchange. Rural communities in South Africa may not be able to afford these treatment processes and also do not have the technological advances necessary to effectively implement these processes. An alternative method to treat contaminated water is required which is cost effective and easy to implement.

Passive biosorption is a property of certain types of inactive/dead microbial biomasses to bind and concentrate heavy metals from very dilute aqueous solutions. The biomass acts as an ion exchanger of biological origin, where the biomass sequesters heavy metals from industrial effluents, or recovers precious metals from processing solutions.

Seaweed is abundantly available along the South African coastline. Preliminary tests show that seaweed is successful in removing divalent metals. However, very little is known about the mechanism of the biosorption process which makes it difficult to predict the performance of the biosorption process. An understanding of the mechanism of biosorption could facilitate the design of a water treatment process requiring very little technological knowledge which can be implemented in any rural area in South Africa where it is needed.

MATERIALS AND METHODOLOGY

Materials

The stock metal solutions for Iron(II) and Manganese(II) were made up of metal standards (atomic absorption standard solutions from Merck) diluted with distilled water to the required concentration of 1000ppm. Analytical grades of HCl and NaOH (Merck) were used to adjust the pH of the solution.

The biomaterial, *Ecklonia Maxima*, was dried and ground to the correct size of 1.0–1.2mm diameter. The biomaterial was activated at a pH of 5 using $CaCl_2$, to encourage heavy metal sorption onto the biomaterial surface.

Kinetics

Batch kinetic experiments were performed at the ambient room temperature (between 20–24°C) using a 1-liter baffled container at a range of temperatures using an overhead stirrer with a baffle attachment at a speed of 350–400rpm. The temperature was kept constant by submerging the container in a water bath. The biomaterial load was approximately 1g/1000mLmetal solution at an initial heavy metal concentration of 100ppm. Periodic 1 mL samples were taken over a period of 360 minutes (at least 12 samples), and diluted as required so that analysis could be completed on the Atomic Adsorption Spectrophotometer. Diluted samples were analyzed for heavy metal content of either Fe(II) or Mn(II) only. The metal uptake by the biomaterial was determined from the difference between the initial and final metal concentrations. Only single metal species experiments were performed.

Pseudo 1st-order and 2nd-order models were used to describe the biosorption curve. The linear pseudo-first-order equation (Ozacar & Sengil, 2003; Xie et al., 2008) is given by:

$$\log(q_{e,1} - q_t) = \log q_{e,1} - k_1 t/2.303 \tag{1}$$

where q_t and q_e are the amounts of metal ions adsorbed at time t and at equilibrium (mg/g), respectively and k_1 is the rate constant of the pseudo-first-order adsorption process (min^{-1}).

The linear pseudo-second-order equation (Ozacar & Sengil, 2003) is given by:

$$t/q_t = 1/(k_2 q_{e,2}^2) + t/q_{e,2} \quad (2)$$

where k_2 is the equilibrium rate constant for the pseudo-second-order biosorption ($g \cdot mg^{-1} \cdot mm^{-1}$).

Saturation Capacity

Heavy metal solutions were prepared at various concentrations—starting at 2ppm and increasing until 256ppm. The biomaterial load was 1g/1000mL of heavy metal solution. Contact was allowed for 24 hours between the synthetic heavy metal solution and the biomaterial under gentle stirring. The final saturation capacity samples were measured for solution pH, filtered (Schleider & Schuell filter paper), and analyzed for ion content. The loading of the sorbents on the biomaterial was determined from the difference of metal ion concentrations in the initial and final solutions.

The Langmuir and Freundlich isotherms were used to simulate the adsorption of the Mn^{2+} and Fe^{2+} by the seaweed biomaterial.

The linearized form of the Langmuir isotherm used is given by (Xie et al. 2008):

$$C_e/q = 1/(b.q_e) + C_e/q_e \quad (3)$$

where C_e is the equilibrium concentration (mg/L), and b is related to the adsorption energy according to the Arrhenius equation.

The linearized form of the Freundlich isotherm is given by (Xie et al. 2007; Tewari et al., 2005):

$$\ln q = \ln k_f + (1/n) \ln C_e \quad (4)$$

where k_f is a constant of adsorption capacity and n is the intensity of adsorption.

Thermodynamics

Thermodynamic properties of the biosorption process were estimated from the kinetics and saturation data obtained.

Free energy change was based on Gibbs equation, which was derived from the applicable sorption isotherm. If the adsorption was in accordance with the Freundlich isotherm, the equation would be (Xie et al. 2008):

$$\Delta G^\circ = -nRT \quad (5)$$

where n is the coefficient of the Freundlich isotherm, R is the universal gas constant (kJ/kg·K), and T is the absolute temperature (K).

Enthalpy change was estimated from a plot of ΔG° vs. T according to the following equation:

$$\Delta G^\circ = -\Delta S(T) + \Delta H \quad (6)$$

The enthalpy change was estimated from the intercept and the entropy change from the slope of the straight line (Babarinde et al. 2007).

RESULTS AND DISCUSSION

Effect of pH

Ecklonia Maxima was activated using $CaCl_2$ at a pH of 5.0. The activation of biomaterials using Ca^{2+} was proposed by Kratochvil and Volesky (1998) because the Ca^{2+} keeps the alginate locked inside the biomass. The experiments were performed at pH values of 5.0 to 5.5 for Manganese (II) Sulphate. It was difficult to constitute the Iron (II) sulphate metal at the same pH level due to the solubility of the Iron(II). Micro-precipitation was observed at pH values of 2.8–3 using Iron (II) sulphate salts. An Ammonium Iron (II) Sulphate salt was used instead, allowing the pH to be increased to pH range 3.8 to 4.5 for the Fe^{2+} solution.

Figure 1 shows that the biosorption capacity increased with an increase in pH. Similar observations were made for Fe^{2+} (only Mn^{2+}results shown). This agrees with reports that biosorption is strongly pH sensitive (Yin et al., 1999; Matheickal & Yu, 1996; Fourest & Roux, 1992). Sorption increased between the pH of 5.0 to 5.8, at which point a plateau was reached. Feng et al. reported in 2004 that an optimal pH of 5.5 was required for the biosorption of heavy metals such as Cd(II) and Co(II). It is also consistent with the findings of Omar (2008) who stated that the maximum capacity is reached at pH values around the 5.0 mark depending on the activation process used as well as the heavy metals removed by the biosorbent.

Generally, higher pH values result in higher metal cation uptakes due to the lowered metal solubility (Davis et al., 2000; Wang & Chen, 2006; Jha et al., 2009).

Kinetics Analysis

Manganese and iron biosorption are characterized by two distinct periods in the sorption process during kinetic experiments. The first is a period of rapid and continuous sorption leading to rapid decrease of metal ions in solution. This is followed by a period of slower sorption, leading to a dynamic equilibrium between the ions in solution and those absorbed by the biomaterial. This is clearly seen in Figures 2 and 3.

Figure 2 shows the capacity of biosorption for manganese as a function of time. During the first 50 minutes, there is a rapid increase in metal removal, with nearly 65% recovery of Mn^{2+}. Sorption equilibria was reached within 150 minutes for all temperature profiles investigated. Sorption uptake seems to be highest at 30°C.

Figure 3 shows the same initial rapid uptake of metal ions, but this is followed by a continued upward trend of the curves. It seems to indicate that metal uptake was still ongoing, even after 300min, and that sorption had not been reached. A fixed cell biomaterial such as the seaweed, offers a finite number of surface binding sites, therefore initial uptake of heavy metals would be expected to show saturation kinetics at high total metal ion concentrations (Sag & Kutsal, 2000). The fact that the biomaterial was still taking up metal ions would suggest that there were still available binding sites within the biomaterial structure.

Pseudo-first- and second-order equations were developed based on the sorption capacity of the solid phase. The results for the first order kinetics are detailed in Table 1. The correlation coefficients are very low, and biosorption probably does not follow first order kinetics.

Table 2 details the results obtained for the pseudo-second-order kinetics. The R^2 values are all above 0.9, indicating a good fit of the data to the equation used. There is a good correlation between the calculated and experimental capacity values. The Fe^{2+} capacity values are generally higher than those for the Mn^{2+}, with the highest amount of Fe^{2+} removed at 45°C and 30°C for Mn^{2+}.

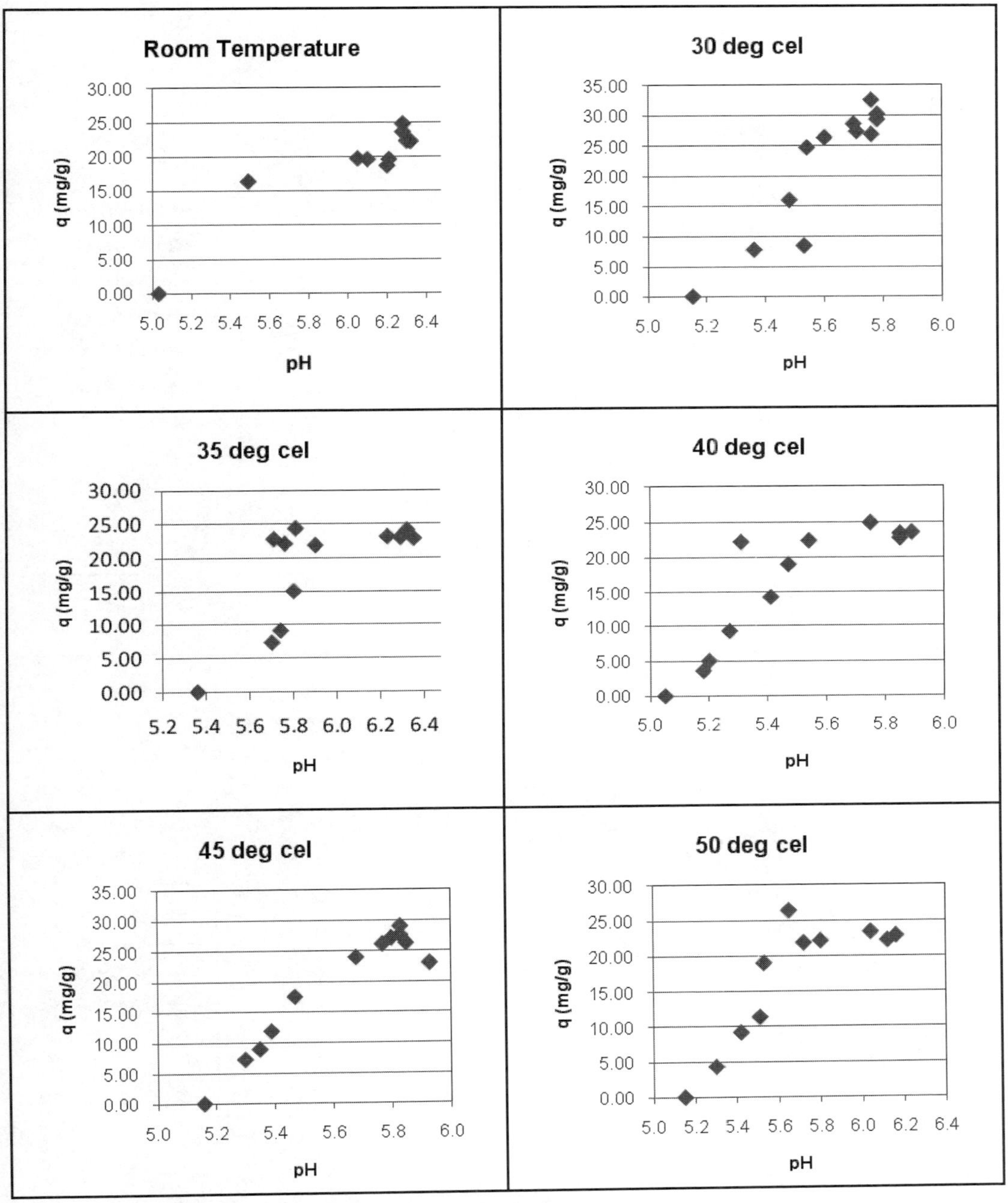

Figure 1. Individual graphs for Mn^{2+}–pH vs. capacity

Adsorption

From Table 3 it is seen that the maximum removal of Mn^{2+} takes place at the elevated temperatures. A maximum capacity of 69.86mg/g is calculated at 45°C, which is approximately 20mg/g lower than the calculated room temperature capacity. This value is still significantly lower than the maximum calculated value of 119.76mg/g for the Fe^{2+}. The maximum value for Fe^{2+} is calculated at room temperature (22°C), and tends to decrease with an increase in temperature.

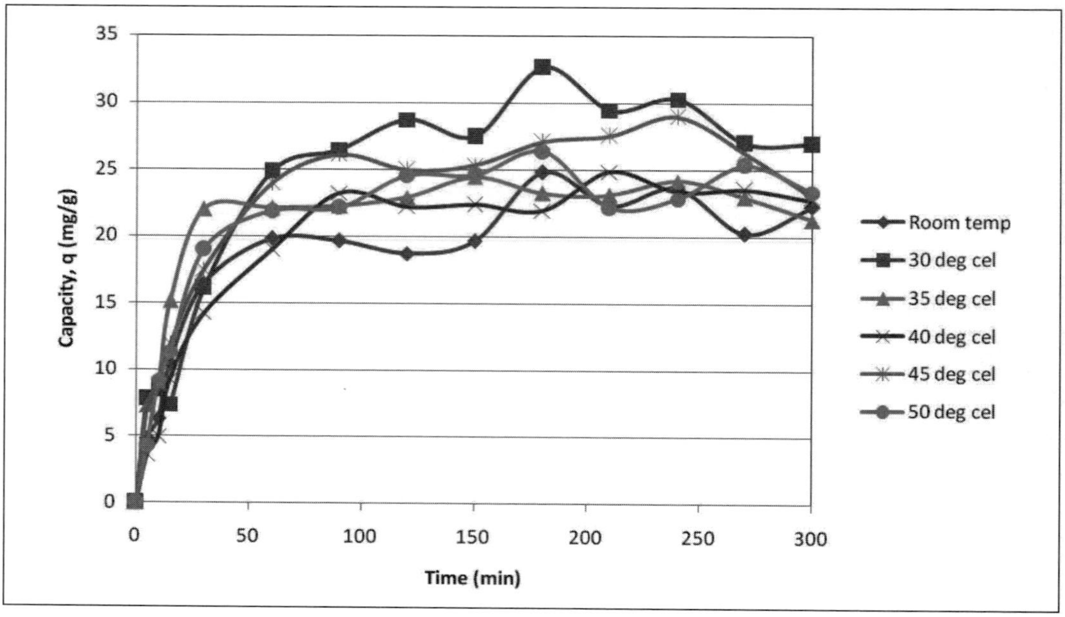

Figure 2. Capacity vs. time for Mn^{2+}

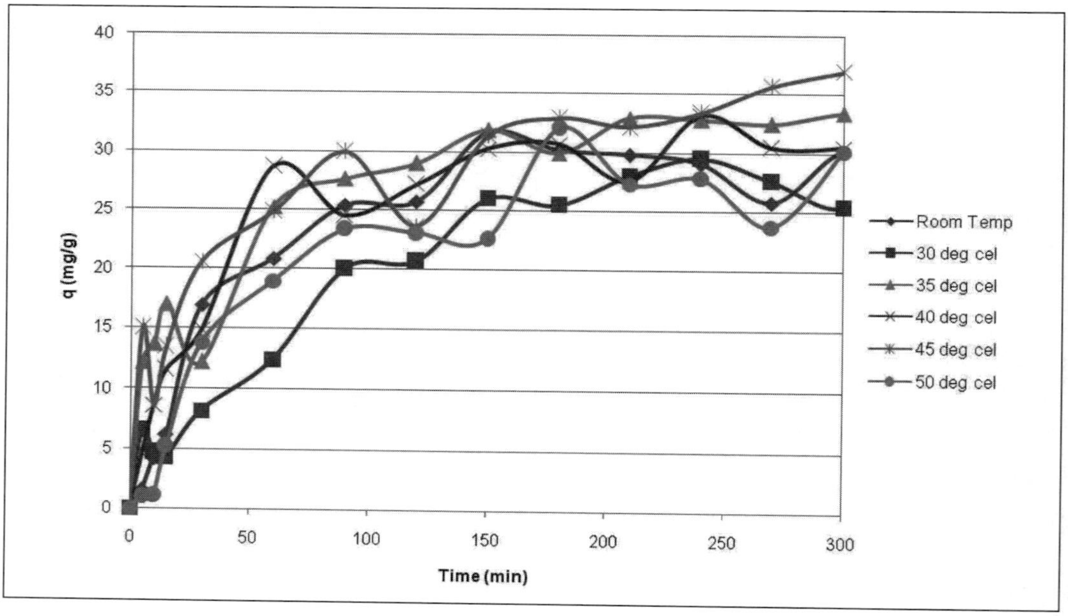

Figure 3. Capacity vs. time for Fe^{2+}

Data was fit to both the Langmuir and Freundlich isotherms. Based on the correlation coefficients as well as the calculated n values, Mn^{2+} is described by the Freundlich isotherm. Fe^{2+} seems to be described by the Langmuir and the Freundlich isotherms, depending on the operating temperature. However, both isotherms tend not to describe the Fe^{2+} biosorption completely.

Table 1. Pseudo-first-order kinetics

	Mn^{2+}		Fe^{2+}	
Temperature	k_1 $(min)^{-1}$	R^2	k_1 $(min)^{-1}$	R^2
Room temp.	0.1819	0.816	0.2464	0.826
30°C	0.1750	0.737	0.2279	0.831
35°C	0.1888	0.586	0.3039	0.846
40°C	0.3063	0.466	0.2164	0.825
45°C	0.2303	0.785	0.2372	0.907
50°C	0.1911	0.695	0.2026	0.846

Table 2. Pseudo-second-order kinetics

	Mn^{2+}				Fe^{2+}			
Temperature	k_2 $(\times10^{-3})$ g/mg·min	q_e (cal) mg/g	q_e (exp) mg/g	R^2	k_2 $(\times10^{-3})$ g/mg·min	q_e (cal) mg/g	q_e (exp) mg/g	R^2
Room temp.	2.6931	23.810	23.6	0.983	0.56201	35.714	30.20	0.924
30°C	1.9800	30.303	29.9	0.976	0.51487	33.333	29.55	0.921
35°C	11.9506	22.727	23.7	0.991	1.35406	35.714	33.38	0.985
40°C	2.0075	25.000	24.3	0.988	1.18785	34.483	33.18	0.984
45°C	4.6431	26.316	26.6	0.977	0.99705	38.462	37.01	0.984
50°C	3.9929	24.390	24.7	0.988	0.82277	32.258	32.04	0.933

Table 3. Simulated results of biosorption

	q_{max} (mg/g)		Freundlich Constants				Langmuir Constants			
Temperature	Mn^{2+}	Fe^{2+}	n Mn^{2+}	R^2	n Fe^{2+}	R^2	b Mn^{2+}	R^2	b Fe^{2+}	R^2
Room temp.	49.90	119.76	0.904	0.990	0.619	0.927	1.918	0.939	4.146	0.891
30°C	59.88	59.88	0.835	0.983	1.140	0.795	2.319	0.869	5.264	0.488
35°C	39.92	49.90	0.724	0.991	0.745	0.821	2.370	0.982	6.098	0.983
40°C	49.90	39.92	0.674	0.978	1.510	0.536	2.586	0.843	4.848	0.686
45°C	69.86	52.84	0.705	0.992	1.245	0.823	2.527	0.924	2.820	0.886
50°C	69.86	99.80	0.759	0.996	1.915	0.796	2.487	0.925	3.255	0.378

Generally speaking:

- If $1/n = 0$ the partition between two phases does not depend on concentration.
- If $1/n < 1$ the isotherm follows normal a Langmuir type isotherm and is not described by the Freundlich isotherm.
- If $1/n > 1$ it is an indication that co-operative sorption involving strong interactions between the molecules and adsorbate is taking place (Munir et al., 2010).

The values of $1/n$ for Mn^{2+} range between 1.106 and 1.484, clearly indicating a linear fit. The values of $1/n$ for Fe^{2+} range between 0.522 and 1.6155, with only room temperature and 35°C being described by the Freundlich isotherm. The Freundlich isotherm equation is

Table 4. Thermodynamic data

	ΔG°		
Temperature	Mn^{2+}	Fe^{2+} (Freundlich)	Fe^{2+} (Langmuir)
Room temp.	–2.22	–1.5196	–3.488
30t	–2.105	–2.8725	–4.1839
35°C	–1.854	–1.9081	–4.6297
40°C	–1.755	–3.9309	–4.1077
45°C	–1.866	–3.2925	–2.7414
50°C	–2.041	–5.1445	–3.1694

an empirical equation based on a heterogeneouos surface. This would suggest that the binding sites are not equivalent and/or independent (Tewari et al., 2005). A value of n larger than one means that the system is more heterogeneous, which usually results in a non-linearity of the adsorption isotherm (Lesmana et al., 2009). If n falls within the range 1–10, it indicates favourable biosorption of heavy metals (Jha et al., 2009) through multiple forms of adsorption.

Thermodynamics Analysis

Thermodynamic parameters were calculated based on the kinetics results analysis as well as the isotherm analysis. The values of Gibbs free energy was calculated based on the Freundlich isotherm for Mn^{2+}, and on the Langmuir isotherm parameters for Fe^{2+}. A negative value for $\Delta G°$ confirms the feasibility of the process and the spontaneous nature of metal ion adsorption with a high preference of metal ions for the biosorbent (Baysal et al., 2009). Values below 8 would indicate a physical process rather than a chemical process (Sari et al. 2008). For Mn^{2+}, the free energy values increase positively with an increase in temperature, which implies that the spontaneity of the biosorption process reduces with an increase in temperature. The opposite it noted for the Fe^{2+}. See Table 4.

ΔH° was determined from the intercept of the plot of $\Delta G°$ vs. T. The values were calculated as –12.9 J/mol for Fe^{2+} and –4.963 J/mol for Mn^{2+}. The negative values indicate that the process is exothermic. ΔS° was determined as –0.029 J/mol for Fe^{2+} and –0.009 for Mn^{2+}

CONCLUSION

The biosorption process is pH sensitive. The optimum pH for Mn^{2+} removal is 5.8 and 4.8 for Fe^{2+}. The maximum capacity of heavy metals removed was approximately 120mg/g for Fe^{2+} and 70mg/g for Mn^{2+}. The pseudo-second-order kinetics system best described the data for both the Mn^{2+} and Fe^{2+}. The Freundlich isotherm best described Mn^{2+} biosorption, but the biosorption of the Fe^{2+} was described to a certain extent by both the Langmuir and the Freundlich isotherms.

The negative values for $\Delta G°$ indicate that the adsorption of both iron and manganese could take place spontaneously. The negative values for $\Delta H°$ indicate that the sorption process is exothermic, and the negative values of $\Delta S°$ indicates a high level of reversibility in the biosorption process. As a result the process is better carried out at lower temperatures.

REFERENCES

Adesola Babarinde, N.A., Oyesiku, O.O., and Dairo, O.F. 2007. Isotherm and thermodynamic studies of the biosorption of copper (II) ions by *Erythrodontium barteri*. *International Journal of Physical Sciences* 2(11):300–304.

Baysal, Z., Cinar, E., Bulut, Y., Alkan, H., and Dogru, M. 2009. Equilibrium and thermodynamic studies on biosorption of Pb(II) onto *Candida albicans* biomass. *Journal of Hazardous Materials* 161:62–67.

Davis, T.A., Volesky, B., and Vieira, R.H.S.F. 2000. Sargassum seaweed as biosorbent for heavy metals. *Water Research* 34(17):4270–4278.

Feng, D., and Aldrich, C. 2004. Adsorption of heavy metals by biomaterials derived from the marine algae Ecklonia Maxima. *Hydrometallurgy* 73:1–10.

Fourest, E., and Roux, J.C. 1992. Heavy metal biosorption by fungal mycelia by-products: Mechanism and influence of pH. *Applied Microbial Biotechnology* 37:399–403.

Goyal, N., Jain, S.C., and Banerjee, U.C. 2003. Comparative studies on the microbial adsorption of heavy metals. *Advances in Environmental Research* 7:311–319.

Jackson, V.A., Paulse, A.N., Odendall, J.P., and Khan, W. 2009. Investigation into the metal contamination of the Plankenburg and Diep rivers, Western Cape, South Africa. *Water SA* 35(3):289–299.

Jha, B., Basha, S., Jaisware, S., Mishra, B., and Thakur, M.C. 2009. Biosorption of Cd(II) and Pb(II) onto brown seaweed, *Labophora variegate* (Lamouroux): Kinetic and equilibrium studies. *Biodegradation* 30:1–13.

Kapoor, A., and Viraraghavan, T. 1995. Fungal biosorption—An alternative treatment option for heavy metal bearing wastewaters: A review. *Bioresource Technology* 53:195–206.

Kratochvil, D., and Volesky, B. 1998. Advances in the biosorption of heavy metals. *Trends in Biotechnology* 16:291–300.

Lesmana, S.O., Febriana, N., Soetaredjo, F.E., Sunarso, J., and Ismadji, S. 2009. Studies on potential applications for the separation of heavy metals from water and wastewater. *Biochemical Engineering Journal* 44:19–41.

Matheickal, J.T., and Yu, Q. 1997. Cu(II) binding by marine alga E*cklonia radita*. *Water Science Technology* 18:25–34.

Munir, K., Yusuf, M., Noreen, Z., Hameed, A., Hafeez, F.Y, and Faryal, R. 2010. Isotherm studies for determination of removal capacity of bi-metal (Ni and Cr) ions by *Aspergillus niger*. *Pakistani Journal of Botany* 42(1):593–604.

Ozacar, M., and Sengil, I.A. 2003. Adsorption of reactive dyes on calcined alunite from aqueous solutions. *Journal of Hazardous Materials* 98:211–224.

Sag, Y., and Kutsal, T. 2000. Determination of the biosorption activation energies of heavy metal ions on *Zoogloea ramigera* and *Rhizopus arrhizus*. *Biochemical Engineering Journal* 6:145–151.

Sari, A., Mendil, D., Tuzen, M., and Soylak, M. 2007. Biosorption of Cd(II) and Cr(III) from aqueous solution by moss (*Hylocomium splendens*) biomaterials: Equilibrium, kinetic and thermodynamic studies. *Chemical Engineering Journal* 144:1–9.

Tewari, N., Vasudevan, P., and Guha, B.K. 2005. Study on biosorption of Cr(IV) by *Mucor hiemalis*. *Biochemical Engineering Journal* 23:185–192.

Volesky, B. 2003. Biosorption process simulation tools. *Hydrometallurgy* 71:179–190.

Wang, J., and Chen, C. 2006. Biosorption of heavy metals by *Saccharomyces Cerevisiae*: A review. *Biotechnology Advances* 24:427–451.

Xie, S., Yang, J., Chen, C., Zhang, X., Wang, Q., and Zhang, C. 2008. Study on biosorption kinetics and thermodynamics of uranium by *Citrobacter freudli*. *Journal of Environmental Radioactivity* 99:126–133.

Flotation of Chalcopyrite in Water Containing Bacteria

Wenying Liu, Chris Moran, and Sue Vink
University of Queensland, Brisbane, Queensland, Australia

ABSTRACT

The use of multiple sources of water supply and introducing water reuse in mineral processing have been part of the operating philosophy for the minerals industry to save freshwater and reduce mine water discharge. A consequence of implementing these practices is that water quality can be compromised, which in turn can impact flotation performance. A number of studies have shown that dissolved abiotic water constituents can have different effects on flotation performance depending on the operating conditions and reagent regimes. However, far less is known about the impacts and processes associated with biotic water constituents, including microorganisms. In this study, we investigated the effect of bacteria (*E. coli*) on mineral (chalcopyrite) flotation performance using both microflotation and scale-up batch flotation tests. Microflotation results indicated that the presence of *E. coli* in water had an adverse effect on the flotation of chalcopyrite. The magnitude of the adverse effect increased with the increase of bacterial cell concentration. The adverse effect could be mitigated to different extents by increasing the collector concentration depending on the bacterial cell concentration. The adverse effect was also observed in batch flotation tests. Froth images taken showed that the bubble size and bubble load were affected by the presence of the bacteria.

INTRODUCTION

Water represents a key component across the minerals industry, including exploration, mining, mineral beneficiation and closure. The industry as a whole is running under a general context of global water challenges, such as limited freshwater availability (Peters and Meybeck 2000; Ridoutt and Pfister 2010), environmental pollution from the mining industry (Gray 1997; Johnson et al. 2002), and community concerns (Jenkins 2004; Kapelus 2002). These challenges are motivating the mining industry to save freshwater and minimise mine water discharge into local streams. Research and analysis of mine water management has led to the conclusion that the use of multiple sources of water supply and introducing water reuse into the site water system could be rational water management practices to realise these reductions (Cote et al. 2007). These practices are common in minerals extraction and concentration activities, e.g., utilization of saline water in coal mining (Turek et al. 2005), reusing water in gold cyanidation (Bahrami et al. 2007) and sulphide ore flotation (Rao and Finch 1989). However, as the efficiency of a water system improves on a site there is an increased tendency for water quality to vary in response to different physical and physico-chemical processes occurring in different parts of the site water system, which can in turn influence the efficiency of mineral separation processes, especially froth flotation (Broman 1980; Rao and Finch 1989). The main water constituents can be broadly grouped into two categories: abiotic, meaning not alive (e.g., inorganic metal ions), and biotic, meaning of or related to life (e.g., organics, microorganisms). The mechanism of how a particular water constituent influences flotation performance is different for each mineral and reagent combination.

In some cases, abiotic water constituents can impact adversely on flotation performance by decreasing the recovery of the target mineral and/or causing poor selectivity between the target mineral and gangue minerals. For example, metal ions (e.g., Fe^{2+}/Fe^{3+}, Zn^{2+}) can decrease flotation by precipitation of hydrophilic metal hydroxides on the mineral surfaces at a particular pH and metal concentration forming a barrier for collector adsorption and subsequent bubble attachment (Celik and Somasundaran 1986; Hoover 1980). Metal ions (e.g., Pb^{2+}, Fe^{2+}, Cu^{+}) can also dilute the concentrate or decrease the grade in varying degrees by inadvertent activation of the unwanted minerals (Chandra and Gerson 2009; Seke and Pistorius 2006). In other cases, positive impacts of abiotic water constituents on flotation performance have been observed. For example, investigation of oil agglomeration of coal in inorganic salt solutions showed that coal recovery increased markedly as salt concentration (NaCl) was raised due to the compression of electrical double layer in the presence of salts (Yang et al. 1988). In high salt concentrations gas bubbles are finer, which may result in reduction of reagent consumption. Pugh et al. (1997) examined the influence of aqueous solutions of a number of inorganic electrolytes on the flotation performance of fine graphite particles. The increase in flotation performance in the presence of the electrolytes was attributed to the generation of higher concentrations of small non-coalescing bubbles and also a reduction in the electrostatic repulsion between particles and bubbles.

There is also some research on the impact of biotic water constituents on mineral flotation, e.g., organics, which can interfere with the flotation process by acting as surfactants, dispersants or flocculants. For example, residual reagents such as xanthate and their decomposition products retained by water reuse can adsorb unselectively on most sulphides and reduce flotation selectivity (Seke and Pistorius 2006). However, in some cases retention of some reagent through water reuse is advantageous as it can lower reagent consumption. One example is that the dosage of amine collector in the reverse flotation of iron ores was reduced up to 50% by using water recovered from tailings pond without compromising the flotation performance (Stapelfeldt and Lima 2001).

The literature review shows that most of the current research has investigated how abiotic water constituents influence flotation performance. In terms of flotation with traditional chemical reagents, literature emphasises the potential importance of biotic water constituents, particularly bacteria, in mineral separation processes because flotation plants can provide ideal conditions for microbial growth (Levay et al. 2001). Bacteria can also be introduced into the water system from multiple water sources, especially under the context that treated sewage effluent with a diversity of bacteria is now being considered as an alternative source of water supply for mine sites (Slatter et al. 2009). The presence of bacteria in water could cause problems to froth flotation. For example, a mine site located in Peru experienced flotation problem possibly caused by a high concentration of bacteria in river water. Flotation with water containing bacteria involves complex interactions among biological (biotic water constituent), chemical (flotation reagent) and geological materials (mineral), but there is a lack of understanding regarding how the flotation process is affected by the presence of bacteria. Motivated by the existing work, in this study, we investigated the effect of bacteria (*E. coli*) on the flotation of pure mineral (chalcopyrite) and tested the conventional scenario of increasing the collector concentration to mitigate the adverse effect caused by the bacteria using microflotation. Batch tests were carried out to investigate whether the effect of *E. coli* still exists with frother addition. Froth images of batch tests were taken to study the effect of bacteria on the appearance of froth. The chalcopyrite mineral was selected for its high purity so that the floatability characteristics of an essentially pure material in the presence of the bacteria could be determined without the interference of various mineral impurities. *E. coli* is used in this study because it is one kind of the main bacteria in sewage effluent and the river water used on the mine site in Peru, and it is relatively safe to work within laboratories.

EXPERIMENTAL

Materials

Mineral Samples

Chalcopyrite sample was purchased from GEO Discoveries, Australia. The sample was crushed using a rolls crusher. The product from the crusher was manually screened to collect –2.8+0.6 mm size fractions. The collected sample was sealed in polyethylene bags and kept in a cold room. Just prior to use, the sample was ground with a pulveriser and screened to produce –150+53 µm particles for microflotation experiments. The same pulveriser was used to grind 100 g of chalcopyrite to achieve a D_{80} of 150 µm for batch tests. Flotation was started immediately after grinding and screening to minimise oxidation and other contamination effects in the sample.

Bacterial Strain

A pure culture of *E. coli* strain K12 was cultured in LB medium consisting of tryptone (10 g/L), yeast extract (5 g/L) and sodium chloride (10 g/L). The culture was grown in flasks shaken at 37°C for 16 hours. The cells were harvested by centrifugation (eppendorf centrifuge 5810R) at 4000 × g for 15 min. The cell pellets were washed three times with 1× PBS buffer solution, suspended in the same solution and then stored at 4°C until use in flotation experiments. The bacterial cell concentration was measured by optical density (OD) using nanodrop ND-1000 spectrophotometer (Thermo Scientific), where 1 OD equals 1×10^9 cell/mL.

Apparatus and Procedures

Microflotation Experiments

Microflotation experiments were conducted with University of Cape Town microflotation unit (Bradshaw and Connor 1996). Despite repeated screening, the particle size fraction used still contained a significant amount of very fine poorly floating particles which had to be removed before flotation. For each microflotation experiment, 3 g of the –150+53 µm size fraction was placed in a 250 mL beaker and mixed with 150 mL of distilled water. The beaker was placed in the ultrasonic bath for exactly 1 min. Fine poorly floating particles were removed by decanting the suspension. The desliming procedure was repeated for an additional two times. The deslimed chalcopyrite was transferred to the microflotation unit. *E. coli* were added into the flotation unit to different target concentrations. The slurry was conditioned for 2 min to allow the bacteria to interact with the mineral. Then sodium ethyl xanthate collector was added to render the mineral surface hydrophobic. The slurry was conditioned for another 2 min. The total volume of the pulp at the conditioning stage was 150 mL. The pulp pH was kept at 6–7. The pulp was stirred continuously with a peristaltic pump at a speed of 75 rpm. After conditioning, distilled water was added into the flotation unit to a final volume of 350 mL. Air bubble produced through a syringe at the bottom of the microflotation unit was immediately provided to the unit. Air flow rate was maintained at 18 mL/min for 90 s of flotation time. The flotation concentrate and tail were collected on top and at the bottom of the unit and filtered. The products were dried at room temperature and weighed for the analysis. All tests were carried out in triplicates to ensure reproducibility. The standard errors were calculated and used as error bars. The chalcopyrite flotation yield was calculated as: yield=weight of concentrate/(weight of concentrate + weight of tail).

Batch Tests

For each batch test, 100 g of the D_{80} of 150 μm size fraction was placed in an 800 mL beaker and mixed with 500 mL of distilled water to form slurry. The slurry was transferred to 1.5 L batch flotation cell. *E. coli* were added into the flotation cell to different target concentrations. Distilled water was added into the flotation unit to a final volume of 1.5 L. The slurry was conditioned for 2 min to allow the bacteria to interact with the mineral. Then sodium ethyl xanthate collector was added to a final concentration of 200 g/t. The slurry was conditioned for 2 min, followed by the addition of frother methyl isobutyl carbinol (MIBC) to a final concentration of 200 g/t. The slurry was conditioned for another 1 min. The pulp pH was kept at 6–7. The pulp was stirred continuously with an impeller installed at the bottom of the flotation cell at a speed of 750 rpm. Air produced at the bottom of the cell was immediately provided at a flow rate of 5 L/min. The concentrate was scraped at 1, 2, 4, 8 min. The concentrate and tail were dried in the oven at 70°C overnight and weighed for the analysis. All tests were carried out in triplicates to ensure reproducibility. The standard errors were calculated and used as error bars. The chalcopyrite flotation yield was calculated the same way as in the microflotation. Images were taken by the digital camera installed over the flotation cell to monitor froth colour and bubble size.

RESULTS AND DISCUSSION

Microflotation

Effect of E. coli *Concentration*

Prior to investigation of the effect of *E. coli*, baseline tests in the absence of *E. coli* were carried out for microflotation. Figure 1 shows the chalcopyrite flotation yield as a function of the collector concentration. As expected, the flotation yield increased with the increase of the collector concentration up to the saturation point of 0.6 mg/L after which further increase of the collector concentration did not increase the flotation yield. Therefore, the collector concentration of 0.6 mg/L was used in the experiment investigating the effect of *E. coli*.

Four different concentrations of *E. coli* used in the microflotation were chosen by reference to the bacterial cell concentration in typical sewage effluent and the river water used at the mine site in Peru. Figure 2 shows that the flotation yield of chalcopyrite decreased when *E. coli* cell concentration was higher than 10^5 cell/mL. The magnitude of the adverse effect of *E. coli* on chalcopyrite flotation was dependent on the *E. coli* cell concentration. Increasing *E. coli* cell concentration resulted in a greater reduction in chalcopyrite flotation yield.

The observed adverse effect of *E. coli* on the flotation of chalcopyrite seems to be due to the direct microbial attachment and/or indirect interactions of bacteria and mineral, with organic carbon compound secreted by the bacteria acting as surface active agents. Before adhering to the particle surface, the bacteria must first establish contact with them. Because the slurry in the flotation unit is under a turbulent state, transporting of the bacteria from bulk liquid to the vicinity of the solid-water surface in a turbulent flow should be driven by fluid dynamic forces (Characklis 1981). After that, the bacteria and/or the organic carbon compounds secreted

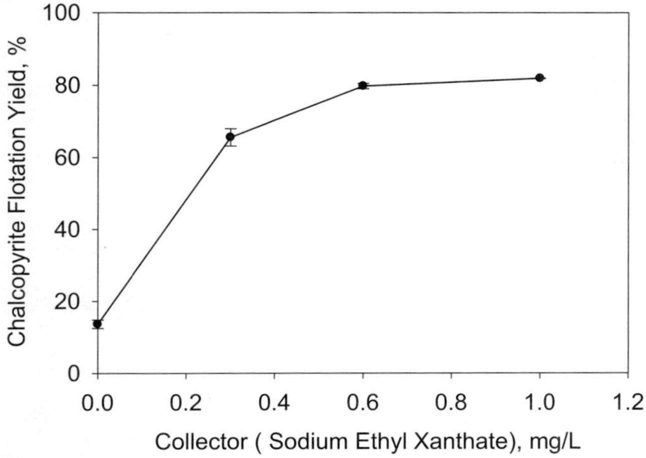

Figure 1. Chalcopyrite flotation yield as a function of the collector concentration in the absence of *E. coli*

by the bacteria have the possibility to attach to the mineral surface. Both types of interaction can bring out surface modifications leading to poor flotation performance. However, that the bacteria cell or the substances excreting by the bacteria or both are responsible for the poor flotation performance is not clear. There is a need to carry out further experiments to study the mechanisms of attachment of microbes and microbial products to mineral surface.

Mitigation of the Negative Effect by Increasing Collector Concentration

The adverse effect of *E. coli* on chalcopyrite flotation has been observed in laboratory experiments. A question of interest would be whether adjusting the collector concentration can mitigate the negative effect as mine sites would normally do to deal with water quality problems in flotation. Figure 3 shows the chalcopyrite flotation yield at different *E. coli* and the collector concentrations. In general the flotation yield increased as the collector concentration was increased. The effect of increasing collector concentration on the flotation yield was more pronounced at lower bacterial cell concentrations. At lower cell concentrations, a higher flotation yield was achieved as the collector concentration was increased. At higher cell concentrations, the flotation yield did not increase as the collector concentration was raised. Under the condition of a high bacteria concentration, an extremely high concentration of the collector would possibly be able to improve the flotation performance to some extent. This would result in increased costs associated with reagents for mine sites.

Batch Tests

Effect of E. coli *Concentration*

Microflotation is a fairly well accepted method of testing mineral floatability despite some shortcomings particularly in view of no addition of frother. As a result, batch tests were conducted to see whether the adverse effect of *E. coli* still exists at a scale-up level with frother addition and whether the presence of bacteria can affect the froth layer properties. Figure 4 shows the chalcopyrite flotation yield as a function of time at different *E. coli* concentration in batch

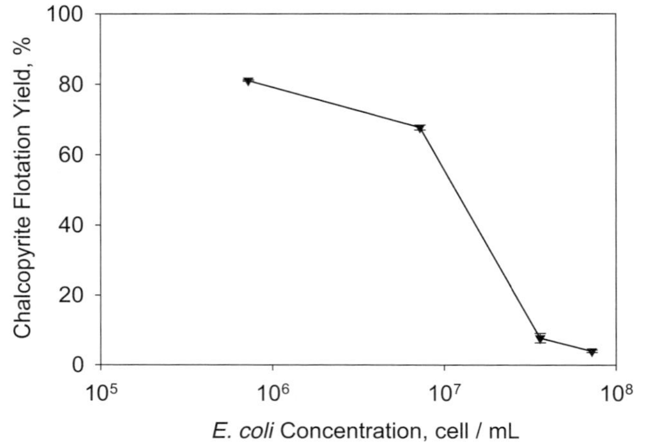

Figure 2. Effect of *E. coli* cell concentration on the flotation of chalcopyrite

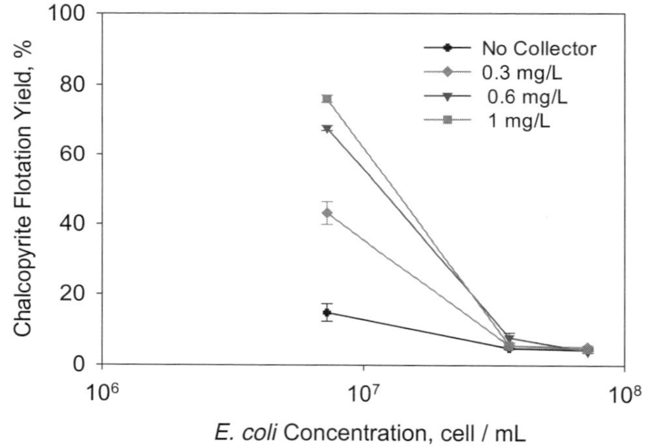

Figure 3. Effect of *E. coli* cell concentration on the flotation of chalcopyrite at different collector concentrations

tests. For all concentrates collected at the same time interval, increase of *E. coli* concentration resulted in reduction of the chalcopyrite flotation yield.

Change of Froth Appearance

To explore the effect of *E. coli* on froth structural features, froth images were taken at different *E. coli* concentrations. Froth images collected (Figure 5) showed that the appearance of froth layer, such as bubble size and froth colour, was different depending on *E. coli* concentrations. Further image processing could provide more information on the bubble size distribution and bubble load.

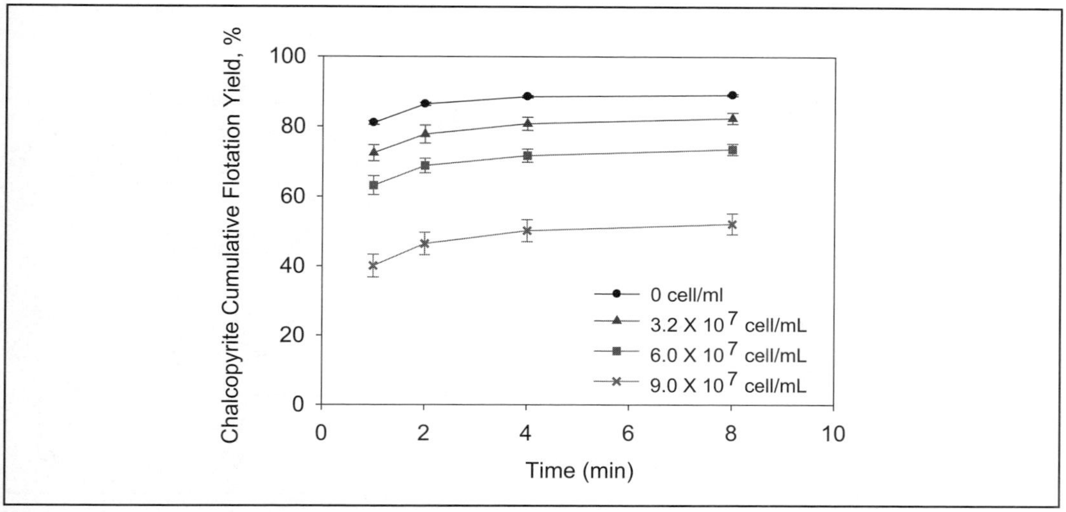

Figure 4. Effect of *E. coli* cell concentration on the flotation of chalcopyrite as a function of time

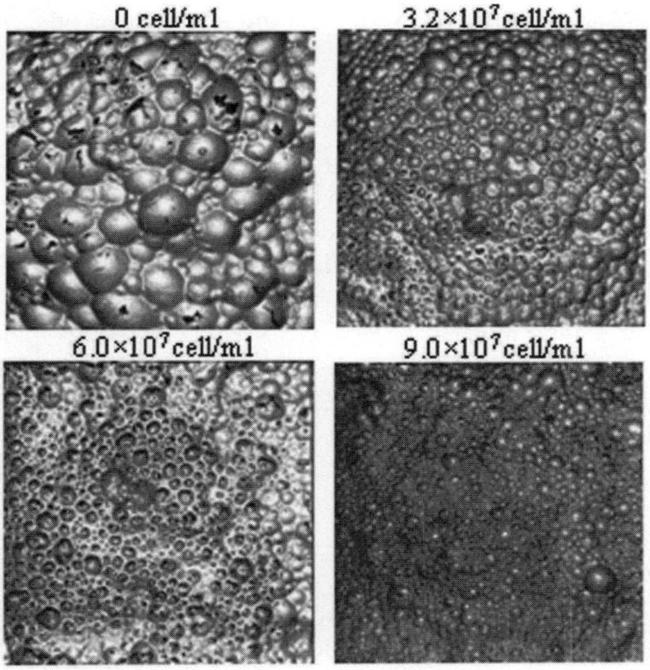

Figure 5. Froth images taken at different *E. coli* concentrations

Comparison of Microflotation with Batch Test

Comparison of microflotation and batch test results is presented in Figure 6. At the same bacterial cell concentration, the reduction of chalcopyrite flotation yield was found to be much higher in microflotation compared with batch tests, which could be attributed to the differences of microflotation and batch tests in a number of aspects, such as pulp density, frother addition and fluid dynamic conditions. Although the magnitude of the adverse effect of the

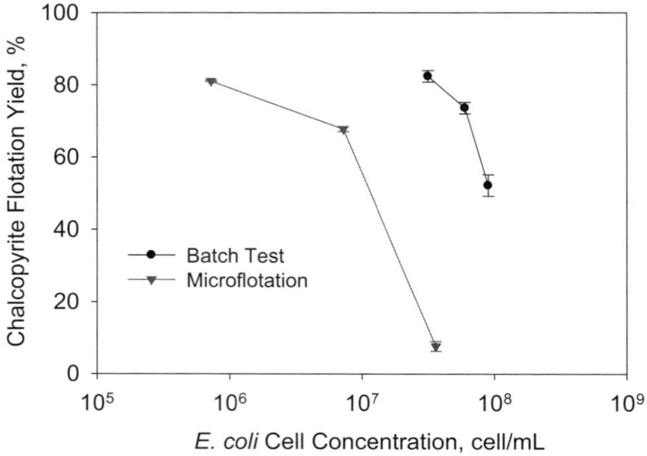

Figure 6. Comparison of microflotation and batch test

bacteria was different under these two conditions, the trends were similar. Therefore, microflotation results are valid and can be used to represent the adverse effect of *E. coli* on chalcopyrite flotation.

CONCLUSIONS

The adverse effect *E. coli* in water on chalcopyrite flotation has been observed in both microflotation and batch flotation tests. The magnitude of the adverse effect was greater when the *E. coli* cell concentration was raised. The adverse effect can be partly mitigated by increasing the collector concentration depending on the bacteria concentration. The presence of bacteria in water also affected the froth properties in batch tests.

The interactions among mineral, reagent and bacteria are compounded by an already-very-complex surface chemistry in a flotation system even in the absence of bacteria. Therefore, the questions of interest in the further research are what the substances adhering to the mineral surface are, bacteria cell, the chemicals excreting by the bacteria or both, and what the consequences of the adhesion are. Following the existing experimental results obtained, we will do further work to better understand the interaction mechanisms of bacteria and mineral.

ACKNOWLEDGMENT

Financial support for this study from ARC-Linkage project (#LP0883872) "Impact of recycled and low quality process water on sustainable mineral beneficiation practices," part of AMIRA project P260E, is gratefully acknowledged.

REFERENCES

Bahrami, A., Hosseini, M., and Razmi, K. 2007. An investigation on reusing process water in gold cyanidation. *Mine Water Environ.*, *26*(3): 191–194.

Bradshaw, D.J., and Connor, C.T. 1996. Measurement of the sub-process of bubble loading in flotation. *Miner. Eng.*, *9*(4): 443–448.

Broman, P.G. 1980. Water reuse at sulphide ore concentrates in Sweden: Practice, experience and current development. In *Complex Sulphide Ores*. Edited by M.J. Jones. London: Institution of Mining and Metallurgy.

Celik, M.S., and Somasundaran, P. 1986. The effect of multivalent ions on the flotation of coal. *Sep. Sci. Technol., 21*(4): 393–402.

Chandra, A.P., and Gerson, A.R. 2009. A review of the fundamental studies of the copper activation mechanisms for selective flotation of the sulfide minerals, sphalerite and pyrite. *Adv. Colloid Interface Sci., 145*(1-2): 97–110.

Characklis, W.G. 1981. Bioengineering report: Fouling biofilm development—A process analysis. *Biotechnol. Bioeng., 23*(9): 1923–1960.

Cote, C.M., Moran, C.J., and Hedemann, C.J. 2007. Evaluating the costs and benefits of salt management strategies at mine sites using a systems model. *26*(4): 229–236.

Gray, N.F. 1997. Environmental impact and remediation of acid mine drainage: A management problem. *Environ. Geol. (Heidelberg, Ger.), 30*: 62–71.

Hoover, M.R. 1980. Water chemistry effects in the flotation of sulphide ores—A review and discussion for molybdenite. In *Complex Sulphide Ores*. Edited by M.J. Jones. London: Institution of Mining and Metallurgy.

Jenkins, H. 2004. Corporate social responsibility and the mining industry: Conflicts and constructs. *Corporate Social Responsibility and Environmental Management, 11*(1): 23–34.

Johnson, C.A., Leinz, R.W., Grimes, D.J., and Rye, R.O. 2002. Photochemical changes in cyanide speciation in drainage from a precious metal ore heap. *Environ. Sci. Technol., 36*: 840–845.

Kapelus, P. 2002. Mining, corporate social responsibility and the "community": The case of Rio Tinto, Richards Bay Minerals and the Mbonambi. *J. Bus. Ethics, 39*(3): 275–296.

Kitchener, J.A. 1984. Surface forces in flotation—A critique. In *Principles of Mineral Flotation: The Wark Symposium*. Edited by M.H. Jones and J.T. Woodcock. Victoria: The Australian Institute of Mining and Metallurgy.

Levay, G., Smart, R.S.C., and Skinner, W.M. 2001. The impact of water quality on flotation performance. *J. S. Afr. Inst. Min. Metall.*, 69–75.

Peters, N.E., and Meybeck, M. 2000. Water quality degradation effects on freshwater availability: Impacts of human activities. *Water Int., 25*(2): 185–193.

Pugh, R.J., Weissenborn, P., and Paulson, O. 1997. Flotation in inorganic electrolytes; the relationship between recover of hydrophobic particles, surface tension, bubble coalescence and gas solubility. *Int. J. Miner. Process., 51*(1-4): 125–138.

Rao, S.R., and Finch, J.A. 1989. A review of water re-use in flotation. *Miner. Eng., 2*(1): 65–85.

Ridoutt, B.G., and Pfister, S. 2010. Reducing humanity's water footprint. *Environ. Sci. Technol., 44*(16): 6019–6021.

Seke, M.D., and Pistorius, P.C. 2006. Effect of cuprous cyanide, dry and wet milling on the selective flotation of galena and sphalerite. *Miner. Eng., 19*(1): 1–11.

Slatter, K.A., Plint, N.D., and Cole, M. 2009. *Water management in Anglo Platinum process operations: effects of water quality on process operations.* Paper presented at the International Mine Water Conference, Pretoria, South Africa.

Stapelfeldt, F., and Lima, R.M.F. 2001. *Recycling of process water containing amines in the reverse flotation of iron ores.* Paper presented at the International Mine Water Association Symposium, Belo Horizonte, Brazil.

Turek, M., Dydo, P., and Surma, A. 2005. Zero discharge utilization of saline waters from "Wesola" coal-mine. *Desalination, 185*(1-3): 275–280.

Yang, G.C.C., Markuszewski, R., and Wheelock, T.D. 1988. Oil agglomeration of coal in inorganic salt solutions. *Int. J. Coal Prep. Util., 5*(3): 133–146.

Relationships Observed in Rock Pile Microbial Populations

D.J. Adams
Department of Metallurgical Engineering, University of Utah, Salt Lake City, UT, USA

ABSTRACT

Rock pile samples were collected for microbial analysis and evaluation from the same zones as samples undergoing geochemical and geophysical characterization. In general, the samples were collected based on visual and geophysical variations observed within the geochemical and geophysical sample boundaries. Microbial analysis indicated that *Acidithiobacillus* and *Leptospirillum sp.* were common across the mine site environment, but that specific species often dominated different geological zones and specific geochemical locals within these zones. *Acidithiobacillus* and *Leptospirillum species* were usually accompanied by heterotrophic, acidophilic, and *Archaea species.*

The microbial populations present in the rock pile were potentially capable of numerous reactions including a number of dissolution and/or cementation type reactions as well as pyrite oxidation when conditions favor these reactions. Some of the more important processes caused or influenced by microorganisms include break-up of particles, growth of microorganisms in micro-cracks or fractures, various chemical processes, production of organic acids and chelators, pyrite oxidation, oxidization of nitrogen-containing organic materials and formation of nitric and other inorganic acids. Microbes influence the moisture, pH, and redox regimes in microcosms. As microorganisms and microbial products participate in and accelerate many geochemical reactions within the rock pile, they use various minerals present as electron donors or terminal electron acceptors. Microbes can accelerate rock pile chemistry changes, and respond to rock pile chemistry. Various relationships between geochemical parameters and microbial populations will be presented.

INTRODUCTION

The rock pile for this study ranged in elevation from 6,800 ft to 10,000 ft. It is located near the margin of a caldera in a highly mineralized region. The mine rock piles extend to heights approaching 1,600 ft with slope angles in the range of 36°. Natural failures within alteration scars have produced debris flows in the vicinity of the mine rock piles and shallow failures have been noted in the rock piles.

Site rock pile materials contain trace amounts of sulfides other than pyrite, such as chalcopyrite, galena, and sphalerite. Other than pyrite and the iron fraction of the chalcopyrite, oxidation of these sulfides will not produce significant acid. However, it has been noted that considerable heat is being generated in one or more site wells, rock piles, and vent areas. Additional minerals present at the site, such as a large amount of gypsum and carbonate minerals (calcite and dolomite) may also contribute to weathered rock pile stability or instability. Samples collected for microbial analysis examine populations present within the rock piles (using trenches), surrounding rock outcrops, soils, and surface crust materials.

This document presents the general microbial analysis of the mine site and the rock pile coded "GHN." During re-grading of the GHN rock pile, trenching was completed to examine rock pile geophysical and geochemical make-up; at the same time, paired samples were selected, under direction of the site geologist, and collected for microbial analysis. The objectives of the microbial data collection and analysis include: (1) Analysis of collected GHN trench materials and other broad mine site samples to gather site and rock pile microbial population profile data. Examination of microbial population characteristics and possible correlations with various sample geochemical parameters. (2) Quantification of microbes in terms of heterotrophs, sulfate reducers, *Acidithiobacillus/Leptospirillum* sp., and acidophilic microbes to speculate on the microbial potential for weathering and/or mineralization.

Background

Pyrite is oxidized by oxygen and/or ferric iron and microbial activity can accelerate this oxidation rate. This document covers macro-environments like geological units and mine surface areas to micro-environments making up geological units and interfaces and the cracks within rocks. A rate law derived by Williamson, et al., 1994, indicates the abiotic rate of pyrite oxidation increases with increasing oxygen concentration and increases slightly as pH decreases. The rate of sulfide mineral oxidation increases as pH decreases into a range conducive to bacterial mediation of ferrous iron oxidation (Nordstrom, 1982, Kleinmann, etal,1981,Singer and Stumm, 1970). Microbial pyrite oxidation rates begin to exceed chemical oxidation rates at around pH 3.5–4 and at pH 2 can be several orders of magnitude greater (Kelley and Tuovinen, 1988). Both abiotic and biotic rates tend to increase with increasing temperature. The various lithologies piled at the mine site also contain carbonate minerals (e.g., calcite and dolmite) and silicate minerals (e.g., plagioclase, K-feldspar, hornblende, biotite, augite, chlorite, epidote).

Microbiology

Acid is typically formed by the oxidation of pyrite (FeS_2), containing reduced iron and sulfur species. When solid-phase pyrite in rock is exposed to oxygenated water (Johnson, et al. 1993), the following chemical reaction occurs:

$$FeS_2 + 3.5O_2 + H_2O \rightarrow Fe^{2+} + 2SO_4^{\ 2-} + 2H^+$$

As a result of this oxidation, the concentration of both reduced iron and acid in drainage waters increases significantly. In addition, the simultaneous presence of reduced iron and oxygen promotes the growth of acidophilic, autotrophic iron- and sulfur-oxidizing bacteria. These include *Acidithiobacillus species, Acidiphilium sp., Leptospirillum sp.*, and others such as *Sulfolobus acidocaldarius,* and the colorless sulfur-oxidizing bacteria (Evangelou and Zhang, 1995).

Although these and other bacterial species, including acid tolerant heterotrophs, may play an important role in mineral leaching, *Thiobacillus ferrooxidans (*now known as *Acidithiobacillus,* and is referenced in this document as such, except where the material is referenced from a source that refers to this microbe as *T. ferrooxidans.*) has often been used as a model bacterium for studying iron oxidation (Wichlacz and Unz, 1981). It is widely believed that *T. ferrooxidans* and other iron oxidizers act as a catalyst for acid production by constantly oxidizing reduced iron (Fe^{2+}) to ferric iron (Fe^{3+}) which then acts as a chemical oxidant of pyrite, thereby liberating more reduced iron, sulfate, and large amounts of acidity as shown in the following two reactions (Harrison, 1984).

$$4Fe^{2+} + O_2 + 4H+ \rightarrow 4Fe^{3+} + 2H_2O$$

$FeS_2 + 14Fe^{3+} + 8H_2O \rightarrow 15Fe^{2+} + 2SO_4^{2-} + 16H^+$

Oxidation of Fe^{2+} serves as the rate-limiting step in biologically mediated acid production (Singer and Stumm, 1970). Because bioavailable organic carbon tends to be relatively scarce in most mine rock piles, especially those with acid drainage, *T. ferrooxidans* and other iron oxidizers often compete quite well with other bacterial species for oxygen and nutrients. From analysis of various other acid producing rock piles and their associated waters, it is expected that the site microbes involved in the oxidation of sulfide to elemental sulfur belong principally to the genera *Thiobacillus, Leptospirillum, Sulfolobus,* and the colorless sulfur-oxidizing bacteria, with *Thiobacillus sp.* most likely being the dominant genus (Evangelou et al., 1995, Lapakko and Antonson, 2002, Berner, 1981).

Sulfur bio-oxidation takes place from a pH of 8.5 to 1.9; however, the species involved will vary and is somewhat dependent on rock pile geochemistry, moisture, etc., oxidizing the sulfur to sulfate. Although *A. ferrooxidans* is often considered to be the most important member of a rock pile population in terms of acid generation, mixed bacterial cultures can be more efficient at mineral decomposition than is *A. ferrooxidans* alone. Additionally, *A. ferrooxidans* seldom if ever grows as a pure culture. Most environments at a pH $\geq$3.0 contain a consortium of heterotrophic bacteria, and a variety of other acidophilic, chemolithotrophic, sometimes sulfate reducing bacteria. Various groups of bacteria are able to rapidly attack a variety of minerals under the right environmental conditions. A "succession" or change in species population domination takes place as the pH of the rock pile environment is lowered by the production of sulfate. Similar types of microbial succession can be observed with relationship to site and/or rock pile temperatures, moisture, geochemistry, and other environmental parameters.

All microbial species require nutrient components, carbon, nitrogen, phosphate, sulfur, and a number of other elements and vitamins, in certain ratios. Since nutrient components are required in specific ratios for microbial growth, energy, and contaminant transformation, any nutrient component can be limiting. A limiting nutrient component will affect the utilization of the other nutrient components and the transformation, sequestration, or mobilization of the minerals present. Acid generated by microbial actions of growth or pyrite oxidation will increase dissolution of other minerals present.

Site Microbial/Geochemical Environments

Oxic environments:

- Dissolved oxygen > 30 μM
- Mn^{2+} below detection

Suboxic environments:

- Dissolved oxygen > 1 μM, < 30 μM
- Fe^{2+} below detection
- Mn^{2+} detectable

Anoxic environments:

- Dissolved oxygen < 1 μM)
- Anoxic – Sulfidic
- H_2S > 1 μM
- Anoxic – Non-sulfidic
- H_2S < 1 μM

These environments are based on measured dissolved oxygen and H_2S and strongly tied to redox reactions (Ehrlich, 1990, Schrenk et al., 1998, Baker and Banfield, 2006). Concentrations of ions, Eh, and pH have major effects on both chemical reactions and microbial formation of authigenic minerals.

Iron oxyhydroxides generated by pyrite oxidation, as well as other reaction products, can aid in the cementation of soil and sand materials. Formation of these types of aggregates would tend to increase rock pile stability. Ferricretes represent iron transported by reducing acid waters and precipitated as hematite and goethite at higher Eh (above 0.4) and/or pH (above 6) more oxidizing, less acidic conditions. These conditions represent major shifts in redox chemistry and possibly in the site microbiology. Both weathering and microbial actions can influences slope stability both positively and negatively.

Microbiology and Reduction Reactions

In many rock pile environments, particularly hard rock mines because there is a shortage of carbon, there is often not a well defined sequence of reduction or redox reactions affecting the inorganic species present in the system. For each of the major redox elements there is a specific range of pE for the initiation of the reduction of the element. Microorganisms use oxidized forms of Fe, S, Mn, and some metal oxyanions as terminal electron acceptors. Nitrate respiration below pE 8 produces products that include NO_2^-, N_2, N_2O, NH_4^+. Solid phase manganese reductions can occur in the presence of NO_3^-. However, solid phase iron reduction, does not occur in the presence of NO_3^- or O_2 and the Mn:Fe ratio is an indication of whether a denitrifying environment is present. Anaerobic microbes produce products that include H_2S, HS^-, $S_2O_3^-$. Changes in pE induce both chemical reduction sequences as well as sequences in microbial ecology.

MATERIALS AND METHODS

General and specific standard operational procedures (SOPs) were developed for collection and microbial analysis of rock pile samples. These SOPs were adjusted as on-site conditions and laboratory culturing and characterization of diverse microbial populations required. Microbes were examined via classical and nucleic acid methods at the University of Utah. Enrichment and isolations were completed to characterize the Bacteria and Archaea at the mine site. Samples were collected from a number of different mine site and rock pile environments for correlation analysis, including:

- General site and pit locations
- Roads and road cuts
- Pits, and trenches (GHN site)—surface and rock pile samples with depth
- Vents, wells, and streams

In general, samples were used as inoculum in a variety of different media, which were incubated under aerobic and microaerophillic, and limited anaerobic conditions. Metabolic and biochemical characterizations were carried out, including assessment of growth rates, and their range of metabolic capability. Some new and unknown bacteria were isolated.

Microbial Sampling and Analysis Approach

Many factors affect the microbial community biomass, structure and diversity, and activity as shown in the first level boxes in Figure 1. Level two and three boxes show the techniques being utilized to provide information on the community culturable and non-culturable microbial numbers and general characteristics, bacterial population or community structure, and to provide data needed to establish biogeochemical relationships. These methods of microbial

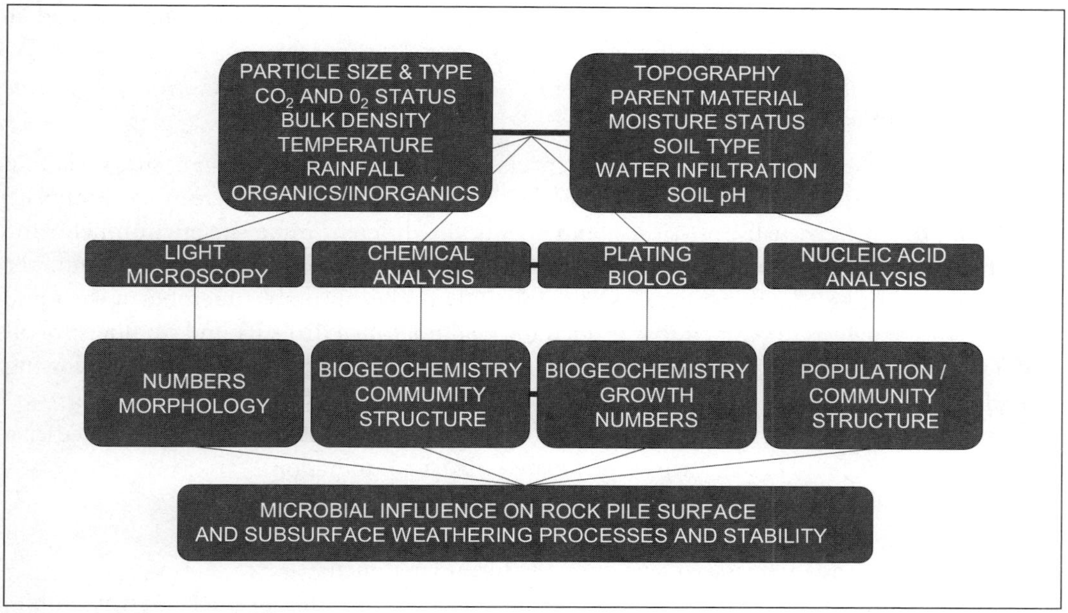

Figure 1. Overview of factors affecting microbial community structure, methods of analysis to assess the microbial population that incorporate quality control between samples and over time, and methods to assess the culturable and non-culturable microbial populations

data collection also provide the capability for quality control between samples, sample replicates, and samples analyzed over time.

Sampling and Sample Storage

Defining the site microbiology entails collecting, isolating, and characterizing microbes from samples collected at specific sites, with specific lithologies and geochemistry. In order to evaluate the microbial influence on the rock pile stability, microbiological samples from site rock pile trenches, drill cuttings, hot spots, associated soils and rock outcrops, crust materials and rock pile vent areas were collected.

The site microbial analysis and subsequent speculation on microbial roles is based on a total of 148 paired samples that included 15 duplicate samples, and sample sets across two geological units. Site materials were aseptically collected into sterile propylene screw-capped, plug-sealed tubes. Microbiological and geochemistry analysis were not from the same sample, but from the same sample zone.

GHN trench samples were collected within described geologic units indicating different zones encountered in the rock pile with depth. Freshly cut ~3' benches were used to describe the different zones through the depth of the rock pile during re-contouring (McLemore, 2007). Samples were collected from freshly trenched and newly exposed surfaces using gloved hands and an alcohol cleaned stainless steel trowel or spoon, filling the tubes so as to eliminate as much air space as possible and to preserve the sample moisture content. Filled tubes were stored in a cooled ice chest and transferred to a refrigerator at ~4°C as soon as possible after sampling. Once at the laboratory, one sample from each set was frozen at –80°C; the other paired sample remained at 4°C for various analyses. Frozen samples were used for checks to ascertain any decrease in viability and for random reproducibility checks of samples stored at 4°C and as a repository of samples for additional analysis.

Media Development and Analysis

Media was developed and optimized specifically for culture of general site rock pile microbes falling into general microbial categories. This was completed by screening composite samples representing visually, physically, and chemically different mine site environments for microbial growth and diversity. This nutrient/media screening was used to reduce the number of different culture media combinations used for study of the culturable microbes at the mine site. Briefly, media was developed that yielded optimal microbial diversity and numbers for all initial samples and microbes examined. Screening included replicate samples and the following procedures:

- Metabolic profile analyses on environmental, gram positive, and gram negative BIOLOG™ plates and select formulated media combinations
- Samples at pH values of 3.5 to 7.0
- Testing in media and saline solutions with various trace elements and screening in various standard, classical, and specially formulated media
  - Comparing site results on specific media with cultivation of stock control samples from American Type Culture Collection, laboratory stock cultures, and various control samples
    - *Acidithiobacillus ferrooxidans, A. thiooxidans, Leptospirillium sp.,* a general acidophilic microbial mix, several sulfate reducing bacteria, *and* a general heterotrophic microbial mix
- Replicate plating on solid media and comparative triplicate MPN analysis in liquid samples of all media types
- Nucleic acid profiling Denaturing Gradient Gel Electrophoresis (DGGE) of site and control microbes (The sequence amplification of DGGE bands was completed by University of Utah DNA analysis services and included primers for Bacteria and Archaea.)

RESULTS AND DISCUSSION

Microbial Culturing and Characterization

The media and analysis of bacteria in the rock pile system was designed around bacterial metabolism rather than specific bacteria; however, specific types of bacteria were also identified. Analysis is looking at bacteria that perform specific biochemical functions or transformations rather than just focusing on specific bacteria of a particular genus and species. Bacteria can be thought of as small chemical factories; under the right conditions, many different genus and species of bacteria can perform similar geochemical transformations. This is stated because the methods of analysis determine the types and relationships of the information obtained.

Microbial enumeration and characterizations have defined the site microbiology in both general and specific characteristics. Variations of stock media were tested over a broad pH range and numerous minor adjustments in component levels to determine the optimal pH and component concentration for use of the media types selected with site samples. This was assisted by evaluation of 10 visually different site wide sample combinations to determine how effective the different media would be to maximize the successful culture of site microbes.

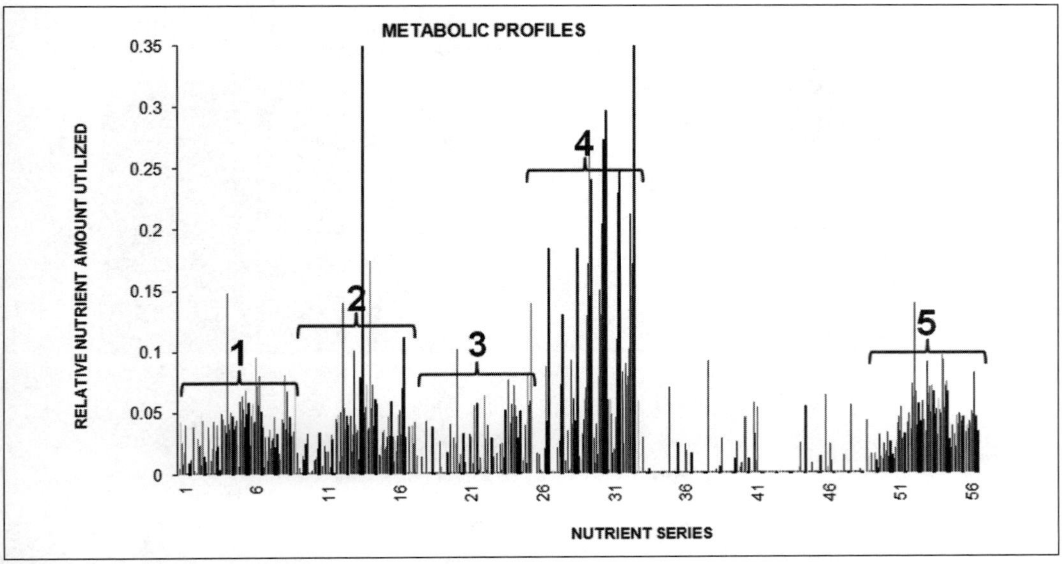

Figure 2. Metabolic profiles of 10 site sample combinations collected across zones shown in Figure 3, indicating the five metabolic profiles of the 56 nutrient combinations tested that were targeted for cultivation of site rock pile microbes

Five main metabolic groupings, from a total of 56 nutrient media screened, Figure 2, were used to adjust culture media content in an attempt to cultivate as many different microbes from site samples as possible. This data was useful in reducing the number of different culture media combinations required for this effort and was confirmed through culture of known environmental sample controls. Control samples included *Acidithiobacillus ferrooxidans, A. thiooxidans, A. denitrificans, Leptospirillum,* a general acidophilic microbial mix, several known sulfate reducing bacteria, and various laboratory repository rock pile cultures, including heterotrophs.

In the microbial population analysis, based on metabolism profiles, it should be noted that statistical analysis of the populations indicates that the acidophilic microbes may be a subset of the heterotrophic microbes that grow well at lower pH ~3.5. However, this was contradicted with observances that acidophilic microbes are definitely the predominant microbes in some samples and heterotrophs in others rather than both being high in many of the same samples.

Acidophilic and heterotrophic populations were shown to be associated with different microbial population profiles and therefore could indicate geologic zones where a succession of microbial populations is taking place. In these areas, higher concentrations of *Acidithiobacillus/Leptospirillum sp.* are associated with the higher concentrations of acidophilic microbes than with higher concentrations of heterotrophic microbes. All other microbial metabolic types, *Acidithiobacillus/Leptospirillum sp.*, *Archaea sp.*, and SRB were all shown to be distinctly different metabolic types and significantly different at the 99% or higher confidence level.

General Site Microbial Analysis

A. ferrooxidans has often been used as the type bacterium for modeling iron and pyrite oxidation and acid generation. In this rock pile, it was common to find combinations of different microbial types (with different metabolic characteristics), present even at lower pH ranges. Both the rock pile trench samples and the general site microbial profiles were similar

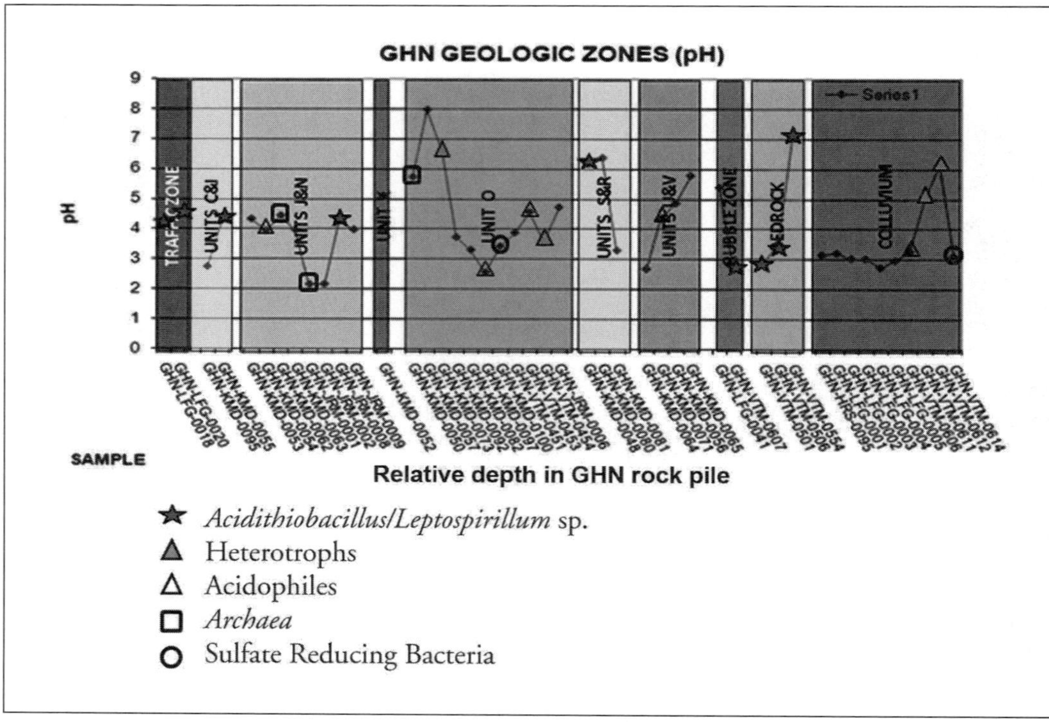

Figure 3. Example GHN trench pH and sample profiles with geologic units described to indicate the different zones encountered in the rock pile as it was trenched in ~3' benches during re-contouring (McLemore, 2007). Locations with dominant, ≥50% of a single metabolic type, populations.

and have similar population profile ratios; *Acidithiobacillus/Leptospirillum sp.* > heterotrophs ≥ Acidophilic microbes ≥ *Archaea* > SRB populations.

Out of a total of 133 different samples examined, 81 samples had dominant populations. In about ~36% of the dominant population samples, *Acidithiobacillus/Leptospirillum sp.* were the predominant microbes present; 50% or more of the total microbial number present. In about 32% of the samples, either heterotrophs or acidophilic microbes were dominant. Acidophiles are organisms that can withstand and even thrive in acidic environments where the pH values range from 1 to 5 and include types of *Bacteria* and *Archaea* that are found in a variety of acidic environments. It has only been recently that the importance of acidophiles and their role in acid mine drainage has been recognized.

From the results obtained, it is thought that the acidophilic and heterotrophic bacterial species may play an important role in both mineral leaching, dissolutions and transformations that provide opportunities for re-precipitation and crust formation. They probably play important roles on both sides of the mineral leaching and re-precipitation event zones within a rock pile environment.

Archaea microbes were dominant in about 9% of the samples and were noticeably present in populations where *Acidithiobacillus/Leptospirillum sp.* dominated, but not the other way around. Sulfate reducers were dominant in ~4% of the samples and were all found in GHN trench samples, rather than in general site samples.

Specific environments within the GHN rock pile were more favorable for particular microbial types; note that the predominant metabolic types occur mainly in the GHN site. This may be explained through the observation of the different geologic zones and the zones of increasing oxidation as samples ranged from inside the rock pile to the rock pile surface, or traffic zone. GHN microbial analysis and speculation as to microbial roles in different geological units is supported by a total of 45 samples. However, some zones have as few as one sample. GHN defined geological units are based on trench samples and the pH profile of the GHN geological units is shown in Figure 3.

In general, the proportions of microbial types appear to vary depending on the site geology, geochemistry, and the specific site conditions. Samples were analyzed in most geologic zones that ranged from a pH of ~3.5 to pH ~8.1. While many samples have *Acidithiobacillus/ Leptospirillum sp.*, they were not the dominant microbes in all samples. A number of samples contained dominant populations of heterotrophs, acidophiles, sulphur-reducing and other sulphur-oxidizing bacteria.

Gram-positive iron and sulfur-oxidizing bacteria related to *Sulfobacillus sp.* were also indicated in some of the samples, but were not positively identified. Acidophilic heterotrophic microbes, such as *Acidiphilium sp.* were also indicated in a number of samples. From the data gathered, heterotrophic microbes, acidophilic microbes, and *Archaea sp.* are believed to assist the establishment and growth of iron-oxidizing bacteria like *A. ferrooxidans* and the *Leptospirillum sp.* This is thought to be due to their ability to mobilize essential nutrients or to remove inhibitory substances. How much these interactions contribute to overall mineral mobilizations or sequestrations in practice is still unclear.

In combined samples examined for microbial population profiles it was shown that denitrifyers were present. Denitrification is a fairly common trait in many microbial genera and with the past presence of blasting compounds; the presence of denitrifyers was expected. Molds were observed in a number of site surface and near surface samples, which was expected, and can contribute to microenvironment changes.

Low level correlations within the site samples and the geochemical analysis conducted on similar site samples were present, but due to small sample sizes ranging from ≤35 per factor, are not discussed. With larger sample sets, correlations may have been interesting as the low level correlations hinted at microbial succession in locations at the site. Interesting low level correlations include the association of *Archaea sp.* with *Acidithiobacillus/ Leptospirillum sp.* The occurrence of sulfate reducers with heterotrophs, and the occurrence of *Archaea type sp.* with acidophilic microbes. A correlation of heterotrophs and SRB with the finer materials was also hinted at, while the gravel appeared to have a more diverse population.

Nucleic Acid Profiling

It is now accepted that the vast majority of microbes have not yet been isolated, identified or characterized. This is largely due to a lack of knowledge of how these organisms survive and grow in natural habitats. When one also considers that a bacterium is often part of a larger more complex community or ecosystem with possible co-dependence on other members, it is understandable why traditional culturing techniques fail to accurately reflect the large microbial diversity in an environmental sample.

A microbial study based on only the culturable microbes can be constrained with respect to viewing the microbial diversity. This is because most microbes defy cultivation by standard methods; it is difficult to obtain a representative picture of total microbial diversity. However, molecular tools and perspectives based on gene sequences were found to be problematic outside a clean room environment due to low numbers of microbes, and small amounts of extractable nucleic acid in the drier mine site environments. In this setting and for these purposes, the culturable microbial types may provide a better basis for evaluation of microbial involvement and their contribution to potential reactions involved in rock pile weathering and stabilization.

Extraction of genomic DNA was attempted from both cultured and uncultured soil microorganisms; isolation directly from the soil sample. Although all commercially available extraction methods were tried, only DNA preps from cultured microbes were mostly successful. Cell lysis using high salt (NaCl), lysozyme, proteinase K, and SDS at high temperature on some soil samples yielded substantial amounts of DNA. However, impurities were sufficient to inhibit PCR amplification of the 16S rDNA, even in the presence of the PCR inhibitors PVP40 and PTB.

Some of the general site microbial species identified included various species within the following genera: *Acidithiobacillus, Leptospirillum, Pseudomonas, Bacillus, Desulfobaterium, Desulfotomaculum*, and *Desulfobacter.* A number of site microbes selected for sequence analysis were unable to be identified and were sometimes quite unique. For example, SRB in both GHN-KMD-0096 and GHN-KMD-0082 sample materials reduced sulfate at extreme rates, >100 times faster than reported in the literature, and grew well at pH's ranging from <3.5 to 7.5.

CONCLUSIONS

- In general, the proportions of microbial types appear to vary depending on the site geology, geochemistry, and the specific site conditions.
- While many samples have *Acidithiobacillus sp., Acidithiobacillus sp.* was not dominant in all samples. At this site, it was common to find combinations of different microbial types (different metabolic types), present even at lower pH ranges.
  - *Acidithiobacillus/Leptospirillum sp.* were the dominant microbes present; 50% or more of the microbial number present, in about ~36% of the samples.
  - Heterotrophs or acidophilic microbes were dominant in about 32% of the samples.
  - Possible *Archaea* type microbes are dominant in about 9% of the samples.
  - Sulfate reducers are dominant in ~4% of the samples.
  - Several new and unknown microbes were found.
  - No microbial growth was obtained in samples cultured at higher temperatures, i.e., 60°C; only limited samples were collected from higher temperature zones.
- Both the GHN rock pile and the general site microbial profiles are similar, but not identical, and have similar population profile ratios; *Acidithiobacillus/Leptospirillum sp.* > heterotrophs ≥ Acidophilic microbes ≥ *Archaea* > SRB populations.
- The GHN site has had acid generating sites in the past, there were higher acid zones at the time the trenches were sampled, the microbial profiles indicate microbial involvement in these events, and there appear to be acid generating zones within other site rock piles presently.
  - The GHN microbially-mediated acid producing zones noted in trench samples were toward the outside, the oxidized portion, of the rock pile.

 - Because the general site microbial profiles are quite similar to the GHN rock pile, this indicates that the type of acid producing zones observed in GHN are probably occurring in other site rock piles.
- From the results obtained in these studies, it is thought that the acidophilic and acidophilic heterotrophic bacterial species may play an important role in both mineral leaching, dissolutions and transformations that provide opportunities for re-precipitation and crust formation, on the surface and within selected rock pile environments. They probably play important roles in the succession of different microbial types within the rock pile environments.
  - All microbes produce some form of acids during growth cycles and require the minerals present in the various samples for life functions.
  - Microbes present can produce organic chelators that can remove metallic ions from rocks and others can form accretion deposits.
  - Microbes can act as precipitation nuclei.
  - As microorganisms and microbial products participate in and accelerate many geochemical reactions within the rock pile, they use various minerals present as electron donors or terminal electron acceptors.
  - Microbes can accelerate rock pile chemistry changes, respond to rock pile chemistry, and be a main driving component of rock pile chemistry changes.
- In rock piles, microbial populations can often be at low levels due to lack of moisture, nutrients, or other appropriate environmental conditions.

ACKNOWLEDGMENTS

The author thanks B. Richins, J. Eshler, and J. Kennedy for their contribution to this paper.

REFERENCES

Baker, B.J., and J.F. Banfield. 2006. Microbial Communities in Acid Mine Drainage. *FEMS Microbiology Ecology* 44(2):139–152.

Berner, R.A. 1981. A New Geochemical Classification of Sedimentary Environments. *J. Sed. Petrol.* 51(2):359–365.

Ehrlich, H.L. 1990. *Geomicrobiology*, 2nd ed. Marcel Dekker.

Evangelou, V.P., and Y.L. Zhang. 1995. Pyrite Oxidation Mechanisms and Acid Mine Drainage Prevention. *Crit. Rev. Environ. Sci. Tech.* 25(2):141–199.

Harrison, A.P., Jr. 1984. The Acidophilic Thiobacilli and Other Acidophilic Bacteria That Share Their Habitat. *Ann. Rev. Microbiol.* 38:265–292.

Johnson, D.B., M.A. Ghauri, and S. McGinness. 1993. Biogeochemical Cycling of Iron and Sulfur in Leaching Environments. *FEMS Microbiol. Rev.* 11:63–70.

Kelley, B.C., and O.H. Tuovinen. 1988. Microbial Oxidation of Minerals in Mine Tailings. In *Chemistry and Biology of Solids Waste: Dredged Material and Mine Tailings*, pp. 33–53.

Kleinmann, R.L.P., D.A. Crerar, and R.R. Pacelli. 1981. Biogeochemistry of Acid Mine Drainage and a Method to Control Acid Formation. *Min. Eng.*

Lapakko, K.A., and D.A. Antonson. 2002. Drainage pH, Acid Production, and Acid Neutralization for Archean Greenstone Rock. In *Proc. 2002 SME Annual Meeting*, February 25–27, Phoenix, AZ.

McLemore, V. 2007. Simplified Conceptual Geologic Model (personal communication).

Schrenk, M.O., K.J. Edwards, R.M. Goodman, R.J. Hamers, and J.F. Banfield. 1998. Distribution of *Thiobacillus ferrooxidans* and *Leptospirillum ferrooxidans*: Implications for Generation of AMD. *Science* 279(5356):1519–1522.

Nordstrom, D.K. 1982. Aqueous Pyrite Oxidation and the Consequent Formation of Secondary Iron Minerals. In *Acid Sulfate Weathering*. Edited by K.A. Cedric, D.S. Fanning, and I.R. Hossner. *Soil Sci. Soc. Amer.* Spec. Pub. 10, pp. 37–56.

Singer, P.C., and W. Stumm. 1970. Acid Mine Drainage: The Rate Determining Step. *Science* 167:1121–1123.

Wichlacz, P.L., and R.F. Unz. 1981. Acidophilic Heterotrophic Bacteria of Acidic Mine Waters. *Appl. Environ. Microbiol.* 41:1254–1261.

Williamson, M.A., and J.D. Rimstidt. 1994. The Kinetics and Electrochemical Rate-Determining Step of Aqueous Pyrite Oxidation. *Geochim. Cosmochim. Acta* 58:5443–5454.

Mining Effluent Reutilization in Biomass Production

E. Martínez and M.J. García
AGQ Labs And Technological Services, Burguillos, Seville, Spain

ABSTRACT

The great environmental, economical, mediatic and social impact of mining effluents makes urgent new sustainable solutions. One of them is the reutilization in an agronomic controlled woody energy crops production.

This procedure permits to reuse acidic waters and other mining effluents to successfully develop plantations with potential energy use while to recover degraded soils near from the productive activity.

The authors of this work believe in synergy between energetic crops and mining industry provides a new source of energy and a solution for effluents minimization.

Based on a great knowledge of agronomic process and by application of advance techniques for water and soils treatment, an environmental concern can be solved and also goods production achieve. It can be a biomass fuel production, or a landscaping or an environmental restoration.

INTRODUCTION

Mining is one of main primary industries in our days. The continuous demand of metals is compensated by a huge throughput on mineral processing facilities. Although its relevance in world economy, meanly in development countries, it has a negative global image by the important impact that produces. New trends are trying to change social perception and great advances are reaching in environmental issues. Concerns about sustainability and social responsibility are changing the performance in 21st century mining (Jenkins and Yakovleva 2006).

An innovative approach is presented in this paper to value mining effluents through its use in energy crop plantations. The objective is to change environmental negative effects of mining effluents, into a solution with high potential benefits: biomass fuel production.

After a prior chemical modification of mining effluents composition, it can be used as irrigation solutions in woody energy crops, with the objective of maximum consumption without negative effects in the environment.

By means of the foliar (leaf) analysis, and knowing as well the chemical composition of the soil at different depth levels of the profile to root zone, and supposing that it is a result of the interaction that the roots activity and the soil have on the applied irrigation solution, we can estimate and evaluate the reaction and behavior of the inputs (water quality, fertilizers, additives, chelates, etc.).

In Spain and Europe the interest in bioenergy production, from forest residues and energy crops, has grown exponentially in the last years. Biomass plays an important role as a renewable energy because it can be considered as a carbon neutral energy source.

The proposed process will provide an exhaustive continuous monitoring, certifying the sustainability of the process; allowing to undertake corrective measures, key to avoid

Table 1. Tinto River quality characterization in the proximity of former mining operations

	Concentration	Units
pH	2.57	ud
C.E.	3208	microS/cm
As	0.295	mg/L
Cd	0.528	mg/L
Cu	19.12	mg/L
Fe	324.5	mg/L
Mn	16.17	mg/L
Ni	0.277	mg/L
Pb	0.078	mg/L
Zn	21.7	mg/L
SO_4	2118	mg/L

contamination or alterations over the plant, as well as managing the plantation in the most optimal manner.

MINING EFFLUENTS

Great innovations in environmental management with new technologies are continuously reporting (Warhurst and Bridge 1996).

Although other effluents have been used to woody energy crops (Zalesny et al. 2009) this work has been focused in mining effluents.

Different effluents and acid mines drainage (AMD) can be employed in our process, meanly contaminated by heavy metals and low or high pH. Some demo facilities are running in different locations nowadays, confidential agreements not allow us to publishing.

At the moment, the most inconvenient and limiting constituents for its application are Boron and Chloride.

In present work we have studied the contaminates waters, by ancients and former mining operations, of Tinto River located in Huelva, a province in the south of Spain near the border with Portugal. Table 1 shows a typical AMD characterization from the river.

The pH values are low and quite stable along the river due to hydrolysis equilibrium produced by ferric iron in water solution. Heavy metals are in high concentrations, and only few species of microorganism (bacteria and algae) can be found in those waters.

River waters are not suitable for its use with agriculture purposes, moreover in some cases have produced deforestation and negative effects in agronomic lands when flooding (Nieto et al. 2007).

The quality of river water remains constant throughout the year, only few differences were found between winter, and summer period.

ENERGETIC CROPS AND BIOMASS PRODUCTION

Potential biomass species include some of over 700 *Eucalyptus* species. *Eucalyptus* is the most valuable and widely planted hardwood in the world (Rockwood et al. 2008).

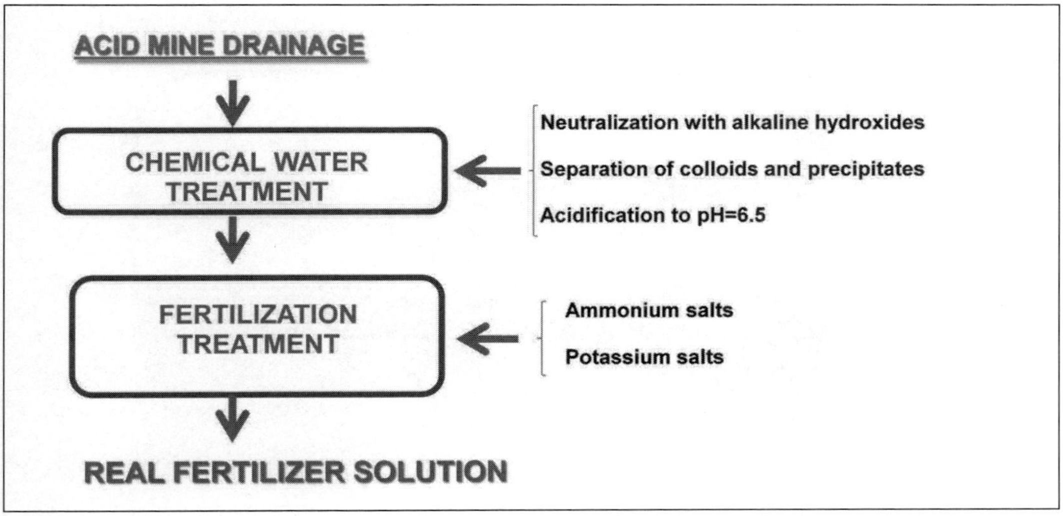

Figure 1. Flowsheet of the treatment

Table 2. Treated AMD

	Concentration	Units
pH	6.50	ud
C.E.	3,150	microS/cm
SO_4	1,440	mg/L
Cl^-	130	mg/L
NO_3^-	30	mg/L
$PO_4H_2^-$	70	mg/L
Ca^{2+}	440	mg/L
Mg^{2+}	122	mg/L
Na^+	40	mg/L
K^+	0.3	mg/L
Cu^{2+}	0.5	mg/L
Zn^{2+}	0.8	mg/L
B	0.2	mg/L

The present study employees the two main species: *Eucalyptus globulus* and *Eucalyptus camaldulensis*, often used in Huelva (SW Spain) in the cellulose, wood and energy production industries. Normal values for gross calorific value (MJ/kg) are close to 20, and ash content less than 2 wt.%. Both species have short rotation coppice and offer high output per hectare.

ACID MINE DRAINAGE TREATMENT

As was mentioned before, those acidic waters are not possible to use for irrigating, so a previous chemical treatment was necessary for crops applications.

After a laboratory development the chemical and physical requirements were fixed and the variables optimized, the process shown in Figure 1 was applied (see also Table 2).

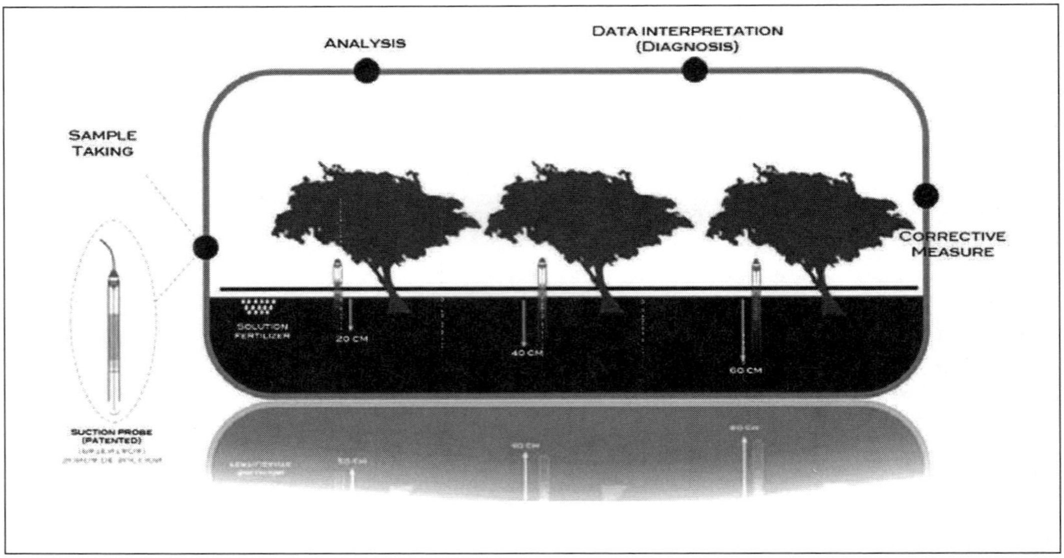

Figure 2. Nutritional monitoring system

The equipment for water treatment is as easy as follows:

- A stirred tank for alkaly addition in a previous neutralization stage.
- A filtration continuous system to eliminate solid residues. Mainly metallic hydroxides.
- Another stirred tank for acidification to pH=6,5 with HNO_3 and H_3PO_4.
- One automatic feed system for fertilizers, and other nutrients for the plants.

During all tests, an automatic control system, including close loops for pH adjustments, was used to fix the demands of water and nutrients by the plant along the phenological cycle. Another aim was to enhance the plants production and quality of final product in the specific crop, overcoming any adversities: salinity, ($SO_4^{=}$), (B), avoiding any environmental risks.

EXPERIMENTAL PLANTATION

A research plot area of 1 hectare was located close to our main technological center, in Burguillos (Sevilla, SW Spain), 37°35'41"N 5°58'55"W (see Figures 2–6). Each line of plants was equipped with a series of irrigation emitters with a designed flow of 1.6 L/h and spaced 1 m apart.

Soils are very calcareous, alkalines (pH: 8–8.4), with low organic matter content (<1.5%).

Suction probes at different depths were installed, constituting a tool to know what happens in the raticular zone of trees, in order to estimate the availability of the nutrients in the soil at any time (amounts and balances), as well as the rate of absorption of the fertilizers and the conditions of absorption (pH, humidity and salinity), being so the result due to interactions produced in the fertilizer-soil-plant dissolution process.

Suction probes will also keep great control over any possible leaching, providing means to reduce/avoid/overcome/identify any environmental risks.

Chemical composition of the soil solutions at different depth of root-zone profile, together with foliar (leaves) analytical results, and permits optimize irrigation by zones of the AMD treated and fertilizer (fertilizer solution).

With those samples (8 samples every 5 months) we are able to monitor:

- Real Fertilizer Solution Analysis and its evolutions along the test.

Figure 3. Crops area beside AGQ Technological Center

Figure 4. Early plantation stages

- Soil Solutions Analysis at three different depths, and availability of each nutrient. Also is a global view of synergies, antagonisms, and hydraulic gradient.
- Foliar composition evolution, monitoring carential levels in advance, always sampling around 20 trees—always the same—close to the suction probes.

A physiological plant answer to the applied fertilizer solution, with a dynamic and real balance of ions and water through root-zone can help us optimize the plants development of the agronomic project (yield and quality), and minimize environmental and economic impact.

Figure 5. Crops development

Figure 6. Plants evolution, and suction probes details

Table 3. Sampling program results in May 2010

May-10	pH	CE mS/cm^2	$H_2PO_4^=$ (mg/L)	Cl^- (meq/L)	$SO_4^=$ (meq/L)	NO_3^- (meq/L)	NH_4^+ (meq/L)	Ca^{2+} (meq/L)	Mg^{2+} (meq/L)	Na^+ (meq/L)	K^+ (meq/L)	B (meq/L)
Real fertilizer solution	6,70	3050	68,1	3,86	28,86	4,86	2,27	21,87	10,45	1,80	1,83	0,18
S. Probe at 20 cm	7,65	3332	28,9	4,98	24,52	3,43	0,66	24,33	11,50	2,21	1,08	0,21
S. Probe at 40 cm	7,80	3550	17,9	7,20	24,91	2,21	<0,14	25,43	11,25	3,96	0,71	0,25
S. Probe at 60 cm	8,20	3890	8,2	10,81	24,98	0,53	<0,14	28,30	11,30	6,28	0,22	0,23

Table 4. Sampling program results in September 2010

September-10	pH	CE mS/cm^2	$H_2PO_4^=$ (mg/L)	Cl^- (meq/L)	$SO_4^=$ (meq/L)	NO_3^- (meq/L)	NH_4^+ (meq/L)	Ca^{2+} (meq/L)	Mg^{2+} (meq/L)	Na^+ (meq/L)	K^+ (meq/L)	B (meq/L)
Real fertilizer solution	6,40	3250	70,1	3,86	31,22	6,54	3,67	20,45	11,21	1,80	2,25	0,28
S. Probe at 20 cm	7,45	3112	38,9	4,78	30,12	4,53	0,45	23,42	10,30	2,11	1,12	0,21
S. Probe at 40 cm	7,50	3093	27,9	5,20	29,91	1,75	<0,14	24,56	10,98	2,87	0,54	0,22
S. Probe at 60 cm	7,52	2920	0,2	6,82	28,15	0,76	<0,14	28,15	11,30	3,54	0,10	0,20

Table 5. Sampling program results in May 2011

May-11	pH	CE mS/cm^2	$H_2PO_4^=$ (mg/L)	Cl^- (meq/L)	$SO_4^=$ (meq/L)	NO_3^- (meq/L)	NH_4^+ (meq/L)	Ca^{2+} (meq/L)	Mg^{2+} (meq/L)	Na^+ (meq/L)	K^+ (meq/L)	B (meq/L)
Real fertilizer solution	6,20	3120	72,1	4,20	32,25	8,02	5,11	21,35	11,05	1,72	2,00	0,29
S. Probe at 20 cm	7,22	3200	41,3	4,61	31,05	6,11	2,11	24,33	10,30	2,11	1,58	0,22
S. Probe at 40 cm	7,40	3505	25,5	5,20	30,08	2,50	<0,14	25,01	11,12	2,28	0,82	0,24
S. Probe at 60 cm	7,75	3795	0,4	5,58	36,12	2,27	<0,14	27,65	11,87	2,94	0,31	0,21

RESULTS

Treated water was used as Real Fertilizer Solution with an irrigation flow average of 10.6 m^3/h/Ha. Weather conditions and plants behavior make the different in flows along the test duration. After consolidated plants growth and when fertilizers began to be added (ammonium nitrate and potassium nitrate), three sampling program was conducted.

Tables 3–5 show the results obtained in May and September 2010, and in May 2011. In all cases, the fertilizer solutions and de aqueous solutions at three different depths were sampled an analyzed.

High consumption of nitrogen forms, phosphorus as phosphate and potassium is observed. A concentration of chloride and sodium is produced and this can be affect plants growth, although a rate of 0.7 mm of diameter per day was obtained.

In order to prevent limitation effects the irrigation flow rate was increased.

In this sampling no high salinity concentration is found, chloride and sodium levels at 60 cm depths are lower than in the previous sampling. Phosphorus and nitrogen were taken by plants till total consumption, and a growth rate of 1 mm of plants diameter per day was reported.

After 2010–2011 winter another sampling program was conducted, during that period the principal differences were about the rains and low temperature, but at the beginning of spring season the amount of nitrogen was increased in real solution fertilizer.

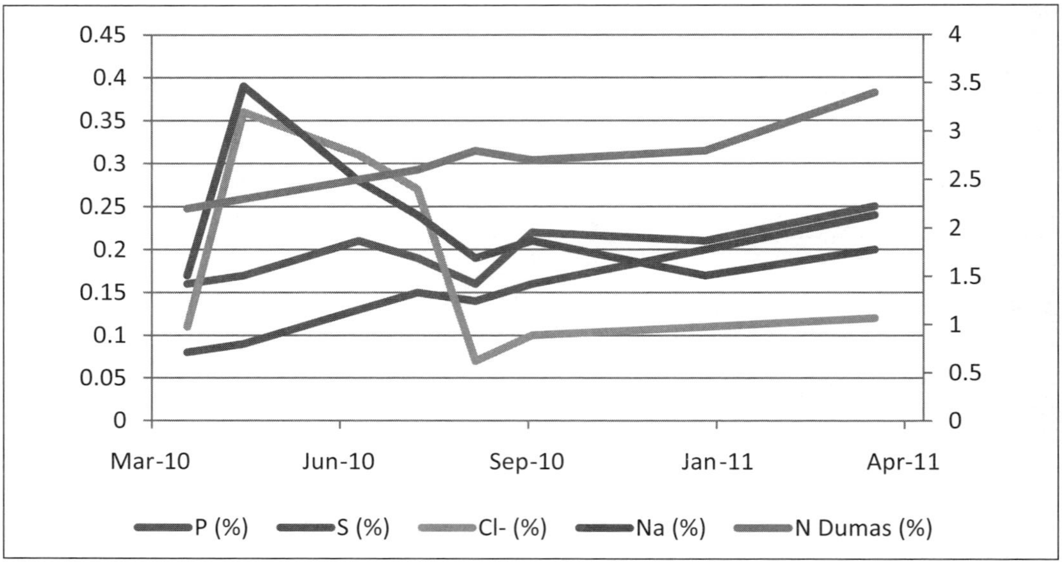

Figure 7. Foliar analysis, elements concentrations in leaves

A continuous growth was observed with rates of 1.6 mm of diameter per day. The consumption of nitrogen, phosphorus and potassium, was almost total, and no effects of salinity was found.

At the same time a continuous and detailed program of foliar analysis was completed; in Figure 7 the elements concentrations in leaves are represented.

From all the previous data some comments are pointed out:

- High plants absorption, by means of consumption, of main nutrients ions (N, P, K). As results a great growth rate was achieved.
- A relevant water limitation was found in the first sampling, with no absorption effect of ions, especially Cl^- and Na^+. An increase of salinity due to the low irrigation flow, and as a consequence, a minor metabolic activity, was found.
- In the second sampling, salinity problem is minimized after increasing the flow coefficient. It is also observed less concentration of toxic elements, Cl and Na, in foliar analysis.
- In all cases the chemical composition of solution at high depths is not pollutant and has no environmental restriction.

CONCLUSIONS

Acidic chemical water treatment is suitable as initial stage for preparing a fertilizer solution to use in woody energy crops. By means of an agronomic sophisticated process this fertilizer solutions can be used as nutrient for eucalyptus species.

Eucalyptus plants growth without problems and productivity are closely tied to water (real solution fertilizer) use.

The knowledge of what happens in the soil-plant-water system, allows us to know physiological and plant answer to the applied fertilizer solution, and availability and real balance of ions and water through root-zone. With a result of optimization of agronomic results (yield and quality), and minimize environmental impact

This control also contributes for optimization in the use of fertilizers, and efficiency in production cost.

A new viable processes to convert mining effluents to woody energy crops can be applied.

REFERENCES

1. Jenkins, H., and Yakovleva, N. 2006. Corporate social responsibility in the mining industry: Exploring trends in social and environmental disclosure. *Journal of Cleaner Production* 14(3-4):271–284.
2. Nieto, J.M., Sarmiento, A.M., Olías, M., Canovas, C.R., Riba, I., Kalman, J., and Delvalls, T.A. 2007. Acid mine drainage pollution in the Tinto and Odiel rivers (Iberian Pyrite Belt, SW Spain) and bioavailability of the transported metals to the Huelva Estuary. *Environment Internacional* 33(4):445–455.
3. Rockwood, D., Rudie, A., Ralph, S., Zhu, J., and Winandy, J. 2008. Energy product options for *Eucalyptus* species grown as short rotation woody crops. *International Journal of Molecular Sciences ISSN 1422-0067.*
4. Warhurst, A., and Bridge, G. 1996. Improving environmental performance through innovation: Recent trends in the mining industry. *Minerals Engineering* 9(9):907–921.
5. Zalesny, R.S., Jr., Wiese, A.H., Bauer, E.O., and Riemenschneider, D.E. 2009. Ex situ growth and biomass of Populus bioenergy crops irrigated and fertilized with landfill leachate. *Biomass Bioenergy* 33:62–69.

Effect of Cations, Anions, and Ionic Strength on the Flotation of Pentlandite–Pyroxene Mixtures

N.J. Shackleton, V. Malysiak, and D. De Vaux
Anglo American Research, Crown Mines, South Africa

N. Plint
Anglo American Platinum, Marshalltown, South Africa

ABSTRACT

This study examines the surface alteration and flotation behaviour of synthetic pentlandite in mixture with pyroxene, in the presence of process water containing various ionic concentrations. Ionic strength was also investigated using sodium chloride to modify the water ion concentration. It is proposed that ions found in process waters passivate the mineral surfaces and inhibit collector adsorption which affects the floatability of pentlandite.

ToF-SIMS and XPS analyses were used to characterise the mineral surface as well as the extent of surface alteration. Microflotation was used to determine the hydrophobicity of the synthetic minerals.

The results have shown that there seems to be a synergistic effect between an ionic strength of ≥0.1 and ion concentration in process water. However, the ions in the process water play a more significant role compared to ionic strength in terms of mineral surface alteration and floatability of the mineral mixtures evaluated.

INTRODUCTION

It is well established that the quality of process water has a significant influence on flotation recovery and selectivity, especially when the water is recycled, and the concentration of the ions increases (Levay et al., 2001). Many in-house studies have been carried out to determine which ions play a detrimental role during Platinum Group Element (PGE) mineral flotation and to possibly determine threshold values. The results obtained do not specifically identify detrimental ions but show that waters with varying cation and anion concentrations result in varying flotation responses. Currently only nitrate and ammonium have been identified as problem ions which negatively affect the flotation response of the PGE minerals. The fundamental studies that have been conducted have shown that certain ions, viz. calcium, play a detrimental role in the flotation of pentlandite and pyroxene mixtures when compared to the results obtained using a simple electrolyte (Malysiak, 2003).

During the study described in this paper, synthetic water, containing key ions typically found in process water, was prepared in order to determine the effect of varying the concentration of cations and anions on the floatability and selectivity of pentlandite–pyroxene mixtures in the presence and absence of SIBX at pH 9. Ionic strength was also varied using ions which have a minimal effect on flotation to de-couple ionic strength from the cations and anions present in solution. Microflotation and ToF-SIMS (Time of Flight Secondary Ion Mass Spectrometry) were used as tools in determining the effect of surface alteration on the flotation response.

The objective of the study is to obtain a fundamental understanding of the floatability of pentlandite and pyroxene mixtures by measuring the effect of parameters such as collector adsorption, ionic activation, and the distribution of ions on the mineral surfaces and relating these to the microflotation behaviour of the samples.

EXPERIMENTAL METHODS

Minerals

Natural pyroxene and synthetic pentlandite [$(Fe,Ni)_9S_8$] was used during the study. Pentlandite was synthesised at Anglo Research (AR) using a method developed by Johnson Matthey. After hydrogen de-sorption, reduced iron (11g) and reduced nickel (11.86g) were mixed with sulphur flake (11.41g) and transferred to a quartz ampoule. The evacuated, sealed ampoule was stage heated in a furnace to 1150°C for 2 hours then cooled to ambient temperature. The final product was sent for mineralogical analysis to verify the homogeneity of the sample.

Natural pyroxene from Amandelbult Section was crushed to 2mm and selected by hand picking.

Mineral samples were stored under argon in a freezer and freshly ground in an agate mortar just prior to each experiment. The products were screened to obtain size fractions of +25µm –53µm for pentlandite and +53–106µm for pyroxene; these fractions were combined in a 1:1 ratio of pentlandite to pyroxene for both the microflotation and surface analysis testwork.

Reagents

Purified sodium isobutyl xanthate (SIBX) powder was obtained from SENMIN and used as the collector at a dosage of 5.00E–05M. Other chemicals were of analytical grade quality. Sodium carbonate (0.1M) and hydrochloric acid (0.1M) were used for pH adjustment.

Synthetic Water Composition

During the study, deionised water (DI) and synthetic process water, which contains similar amounts of key ions typically found in process water used on PGM concentrators, was used for the microflotation tests, XPS and ToF-SIMS analyses. Variations in the concentration of the cations and anions (synthetic water, 5× synthetic water and 10× synthetic water) were also evaluated (Table 1). Table 2 shows the ionic strength trials where ionic strength was varied using sodium chloride solutions with a concentration of 2.5g/L and 25g/L.

Time of Flight Secondary Ion Mass Spectrometry

Surface analysis of the minerals was carried out using a PHI TRIFT IV ToF-SIMS instrument operating in the static SIMS regime. A pulsed primary ion beam bombards the sample surface, causing the emission of atomic and molecular secondary ions. A small percentage of the secondary ions are charged and can therefore be extracted by an electric field into a mass spectrometer. The mass spectra are recorded by measuring the time difference between pulsing the primary ion gun and the arrival of secondary ions on a fast dual microchannelplate detector at the spectrometer, by means of a multistop time-to-digital converter. ToF-SIMS analysis was carried out on stirred trial products (samples were conditioned in the desired solution at pH 9 for 1 minute prior to reagent addition; a 2 minute conditioning period was used for SIBX).

Table 1. Synthetic water—cation and anion concentrations

	Deionised Water	Synthetic Water	5× Synthetic Water	10× Synthetic Water
Cation/Anion	ppm			
Ca^{2+}	29	80	400	800
Mg^{2+}	33	80	400	800
Na^{+}	87	135	675	1350
Cl^{-}	143	270	1350	2700
SO_4^{2-}	110	250	1250	2500
NO_3^{-}	88	135	675	1350
CO_3^{2-}	16	40	200	400
TDS	506	990	4950	9900
Ionic strength	0.012	0.025	0.13	0.25

Table 2. Synthetic water—variations in ionic strength

Cation/Anion	2.5 g/L NaCl, ppm	25 g/L NaCl, ppm
Na^{+}	984	9,835
Cl^{-}	1,516	15,165
TDS	2,500	25,000
Ionic strength	0.043	0.43

The samples were filtered and washed with deionised water (conductivity 0.7μmScm^{-1}), adjusted to pH 9, to remove any physically attached ions. All samples were dried in an argon atmosphere at ambient temperature, screened and stored under argon in a freezer.

Throughout the study a 30kV, 100μm-Au_1 bunched cluster beam with charge compensation was used. Five areas (300×300μm) of each sample were imaged and analysed for K, Si, Na, Ca, Mg, Al, Ni, Fe, and Cu during positive ion analysis and S, O, OH, and xanthate during negative ion analysis. The data obtained were evaluated using Statistica. The intensities obtained are normalised for the elements of interest and presented as a percent normalised yield.

Microflotation Testwork

A microflotation cell (volume 250cm^3) was used to determine the flotation response of pyroxene and pentlandite–pyroxene mixtures. The cell consists of a conical tapered cylindrical tube with air introduced through a needle at the base of the cell (Wesseldijk et al., 1999). Mineral loaded bubbles rise through the cell and are deflected off the cone at the top of the cell, after which they burst, resulting in the minerals dropping into the concentrate launder. The testwork involved taking a 2g sample (1g of pentlandite (+25–53μm) and 1g of pyroxene (+53–106μm)) which was added to 250mL of the required solution, which had been adjusted to pH 9. The desired pH was maintained throughout the flotation by adjusting with either sodium carbonate or hydrochloric acid. Concentrates were collected at time intervals of 2, 5, and 10 minutes. The floated and non-floated fractions were allowed to dry in air, screened and weighed. The conditioning period for SIBX was 2 minutes.

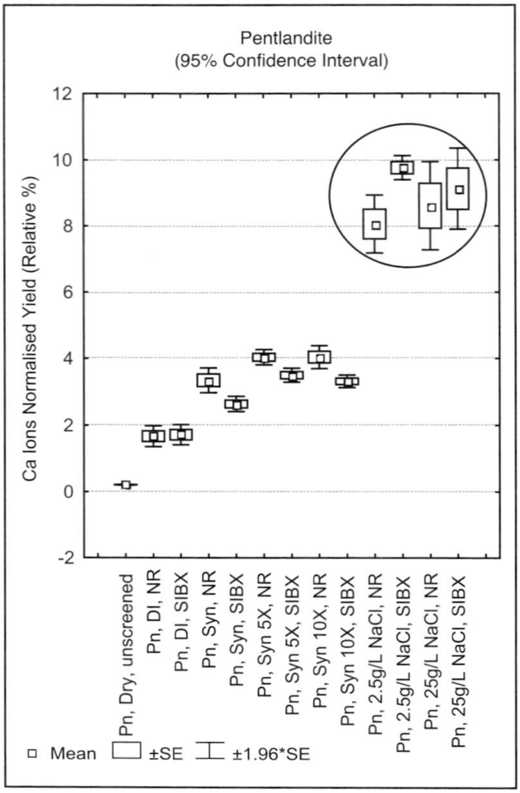

Figure 1. Calcium ions normalised yield relative percent for pentlandite

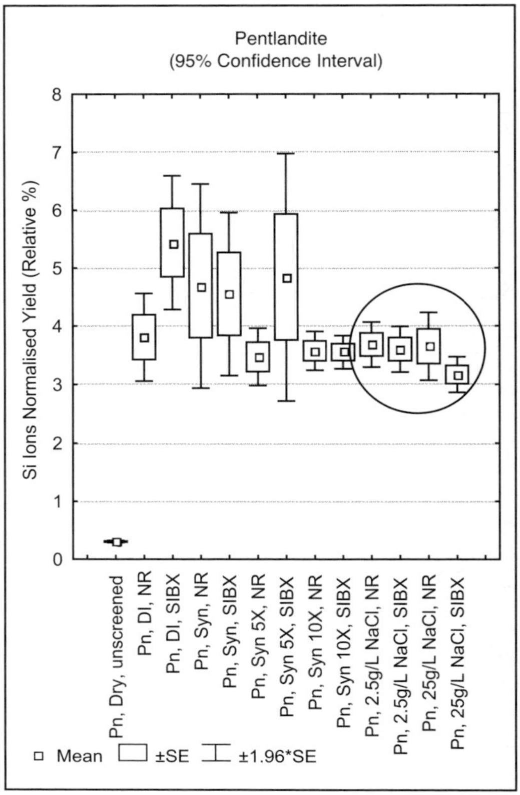

Figure 2. Silicon ions normalised yield relative percent for pentlandite

RESULTS

The figures and tables in this section show data for either pentlandite or pyroxene on their own but the interpretation should always consider that the testwork was carried out as pentlandite–pyroxene mixtures and not as single minerals. Therefore dissolution products from either pentlandite or pyroxene may contribute to the surface alteration of both minerals and affect floatability.

Surface Analysis

ToF-SIMS analysis is a well-established technique in determining the occurrence of atomic/molecular species on the surface of mineral samples. It must be noted that the ToF-SIMS values are relative and not absolute. Figures 1 to 4 show the calcium, silicon, aluminium and magnesium ions normalised yield obtained for pentlandite and Figure 5 shows the calcium ions normalised yield obtained for pyroxene.

It is clear that calcium, silicon, aluminium and magnesium ions adsorb onto the pentlandite mineral surface thereby masking the nickel-iron mineral surface. These passivation layers can hinder xanthate adsorption which would negatively influence the pentlandite recovery. It should be pointed out that as the concentration of calcium ions in solution increases an

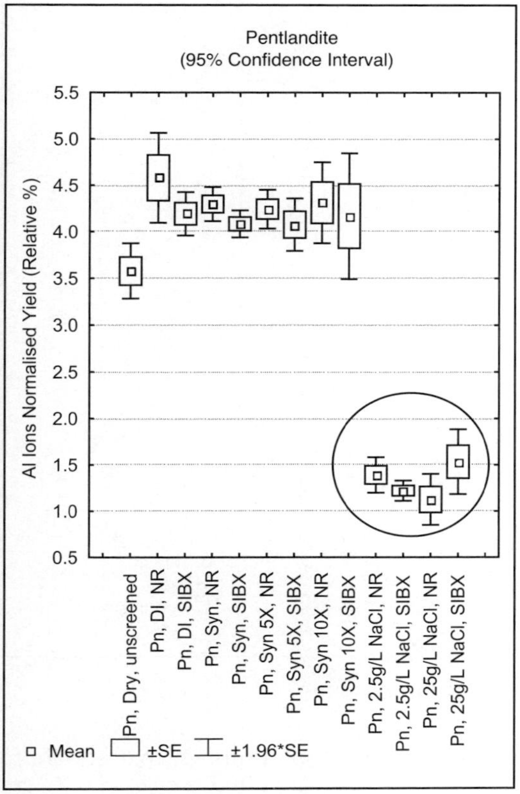

Figure 3. Aluminium ions normalised yield relative percent for pentlandite

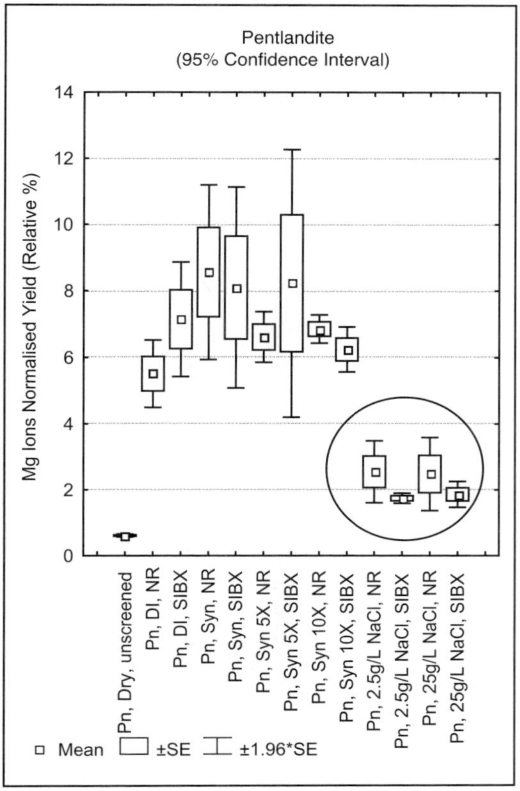

Figure 4. Magnesium ions normalised yield relative percent for pentlandite

increase in calcium ion adsorption is observed on the pentlandite mineral surface. However, a maximum calcium adsorption is reached at a calcium solution concentration of around 400 ppm (5× synthetic water trials).

The ionic strength testwork (circled) has shown a significant increase in calcium ion adsorption on pentlandite surfaces in the presence of sodium chloride (Figure 1). The aluminium and magnesium ion surface concentration is significantly lower on pentlandite surfaces compared to the DI and synthetic water trials (Figures 3 and 4). The aluminium surface concentration is also significantly lower than that observed on the dry untreated pentlandite mineral surface which is probably due to a masking effect from the resultant Ca adsorption. The higher calcium adsorption for the NaCl trials compared to the synthetic water trials is probably due to minimal competition from other ions in solution and a higher affinity of calcium for both pentlandite and pyroxene surfaces (Figures 1 and 5).

Figures 6 and 7 show the xanthate (SIBX) and Figures 8 and 9 show the sulphur ions normalised yield obtained for pentlandite and pyroxene, respectively.

The results show that for pentlandite there is xanthate adsorption when compared to the corresponding no reagent trials (Figure 6). The xanthate adsorption increases as the cation and anion concentrations increase in the synthetic water but does not increase further in the 10× synthetic water when compared to the 5× synthetic water trial; again suggesting that a threshold value has been reached. The baseline for xanthate detection on pentlandite surfaces after treatment in solution increased when compared to the dry untreated sample. This may be due

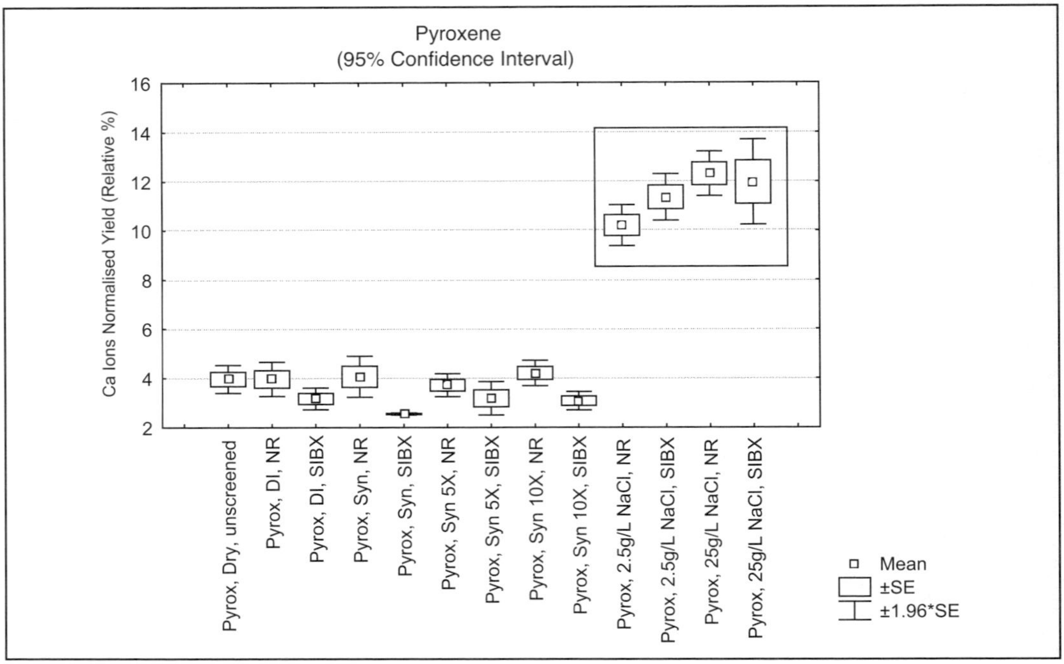

Figure 5. Calcium ions normalised yield relative percent for pyroxene

to the interference of other elemental and molecular species in the solutions which absorb onto the mineral surface.

The ionic strength testwork (shown in the rectangles) has shown a significant increase in xanthate ion adsorption for pentlandite in the presence of sodium chloride especially at the 25g/L NaCl concentration (Figure 6).

The results for xanthate adsorption onto pyroxene surfaces, although not significant, show a trend of increased adsorption as the cation and anion concentrations in solution increases with higher values obtained for the NaCl trials (Figure 7). The sulphur ions results, which may infer xanthate adsorption, show a significant difference in the surface coverage for the sodium chloride trials compared to the synthetic water trials for both pentlandite and pyroxene (Figures 8 and 9). This would most likely result in a higher flotation response for both minerals. The higher xanthate/sulphur adsorption on both pentlandite and pyroxene may be due to higher iron/nickel species and/or sodium adsorption on the mineral surface.

Microflotation

Microflotation testwork was performed with the aim of linking the trends observed from the surface analysis experiments to floatability. To evaluate the contribution of entrainment and natural floatability, flotation tests were conducted without any reagent addition.

Effect of Cation and Anion Concentrations

The objective of these trials is to determine the effect of varying the concentration of cations and anions on the floatability and selectivity of pentlandite–pyroxene mixtures in the presence and absence of SIBX at pH 9. Deionised water, synthetic water, 5× synthetic water

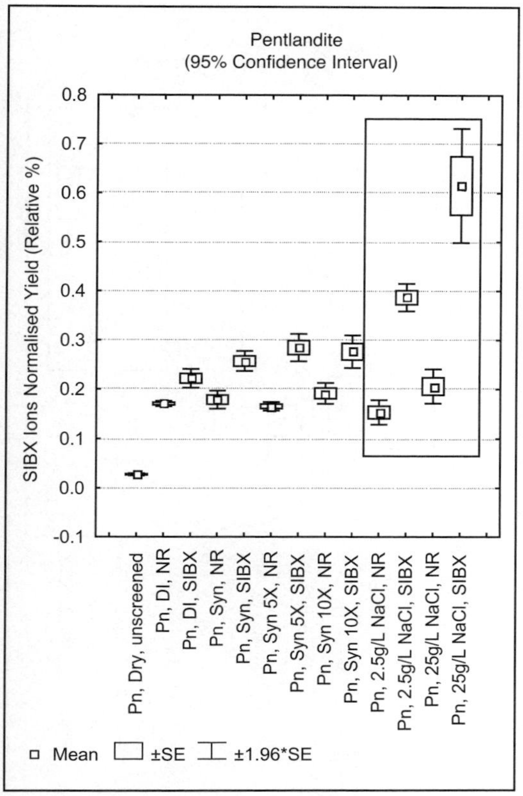

Figure 6. Xanthate ions normalised yield relative percent for pentlandite

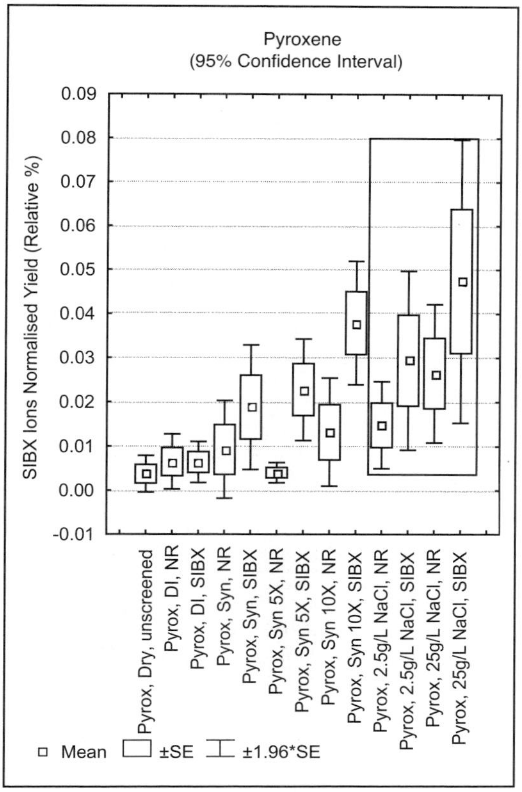

Figure 7. Xanthate ions normalised yield relative percent for pyroxene

and 10× synthetic water were evaluated during the testwork. The 5× synthetic water concentration was chosen as it is similar to the amount of ions observed in the Rustenburg waters and the 10× synthetic water was chosen to possibly determine threshold values of the ions present with respect to flotation recovery. Figures 10 and 11 show the recovery–time curves obtained for pentlandite and pyroxene minerals, respectively.

The microflotation data has shown that in the absence of reagents the highest pentlandite recovery (~26%) was obtained in deionised water (Figure 10). All the synthetic water concentrations gave similar pentlandite recoveries (~14%). In the presence of xanthate the highest pentlandite recovery (~74%) was obtained in the 10× synthetic water concentration, while the synthetic and 5× synthetic water concentrations gave a pentlandite recovery of ~67%. DI water gave the lowest pentlandite recovery (~55%).

Pyroxene recovery in the absence of reagents was lowest (~2%) in the 10× synthetic water concentration, while the highest pyroxene recovery was achieved in deionised water (~4%). In the presence of xanthate the highest pyroxene recovery (~17%) was obtained using synthetic water, while the DI, 5× synthetic and 10× synthetic water concentrations gave similar pyroxene recoveries (~8%). This indicates that ions present in process water play a significant role in the floatability of pentlandite and pyroxene.

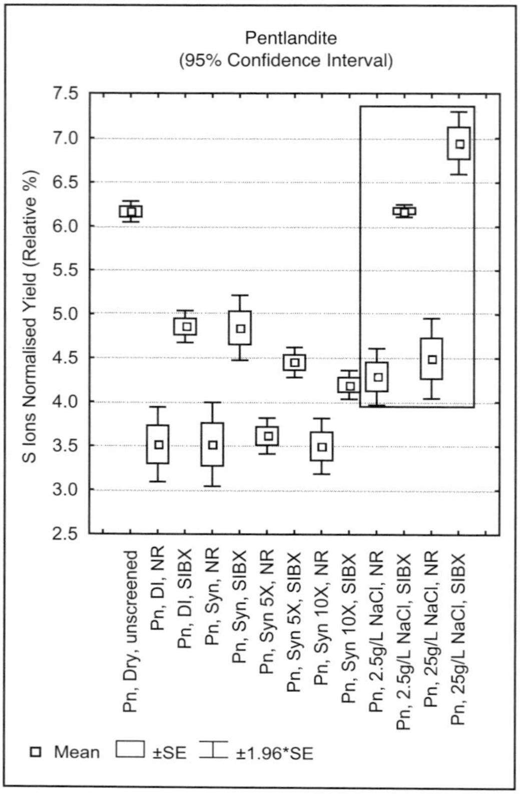

Figure 8. Sulphur ions normalised yield relative percent for pentlandite

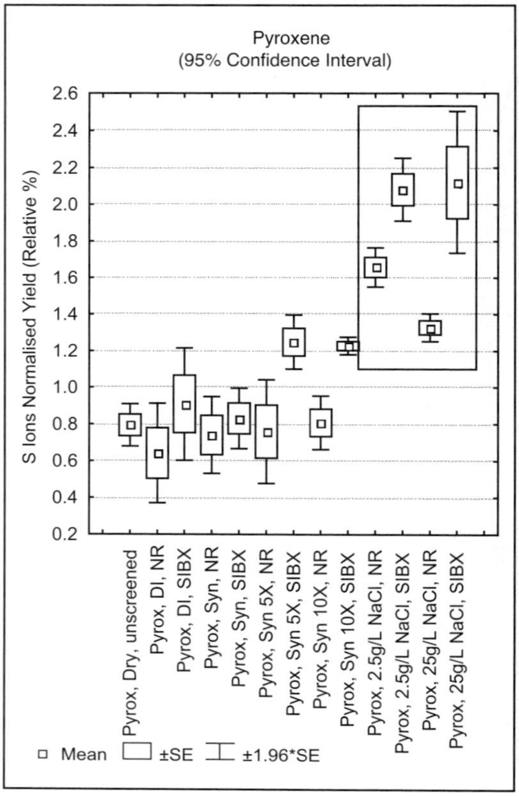

Figure 9. Sulphur ions normalised yield relative percent for pyroxene

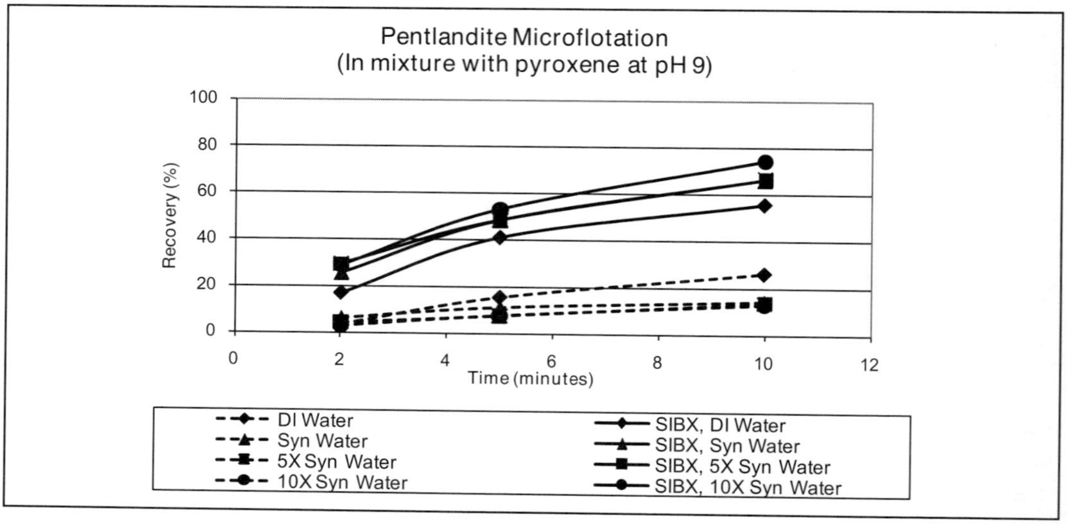

Figure 10. Pentlandite recovery–time curves at pH 9 comparing no reagents and SIBX in DI water and various synthetic water concentrations

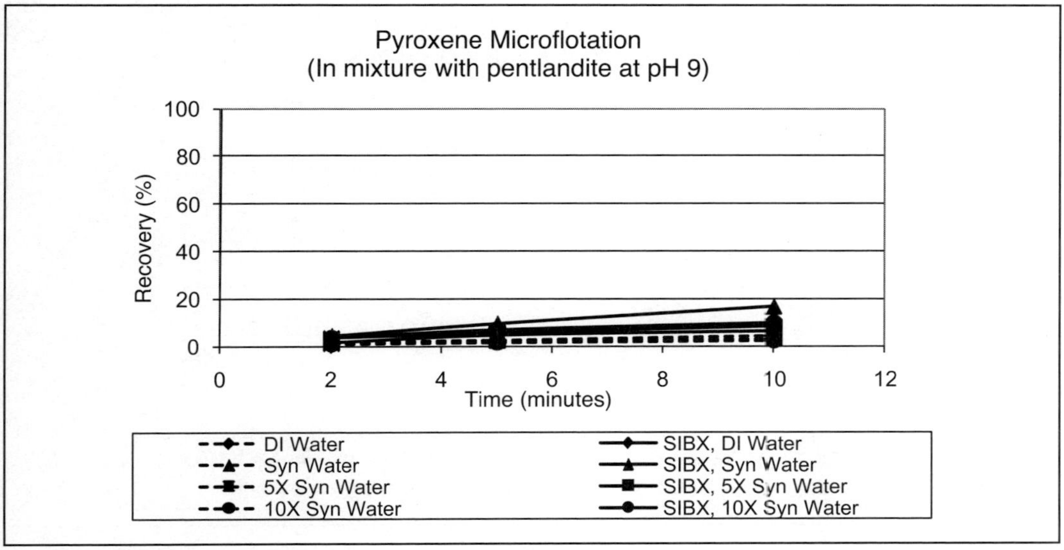

Figure 11. Pyroxene recovery–time curves at pH 9 comparing no reagents and SIBX in DI water and various synthetic water concentrations

Effect of Ionic Strength

Figures 12 and 13 show the recovery–time curves obtained for pentlandite and pyroxene minerals, respectively. These graphs compare the results obtained for pentlandite–pyroxene mixtures in the absence of reagents and with xanthate addition using waters with various ionic strengths at pH 9. Ionic strength was varied, using Na and Cl ions to de-couple ionic strength from the cations and anions present in solution.

The microflotation data shows similar pentlandite (~43%) and pyroxene (~12%) recoveries in the absence of reagents irrespective of the ionic strength. These recoveries are much higher than those obtained for the synthetic water concentration trials. In the presence of xanthate, pentlandite recoveries of ~74% were obtained, which is similar to the results obtained for the 10× synthetic water trial. However, the pyroxene recovery was significantly enhanced irrespective of the ionic strength when compared to the cation and anion concentration trials. This indicates that ionic strength plays a role in the flotation recovery of pentlandite and pyroxene.

DISCUSSION

The key objectives of this study were to determine:

1. Whether the concentration of cations and anions in process water plays a role in surface alteration and floatability of the value and gangue minerals.
2. Identify which ions affect surface alteration and floatability of both the value and gangue minerals.
3. Determine possible thresholds values of the ions which affects surface alteration and floatability of both the value and gangue minerals.
4. Whether the ionic strength plays a role in surface alteration and floatability of the value and gangue minerals.
5. Whether there is a synergistic effect between the ions present in process water and ionic strength.

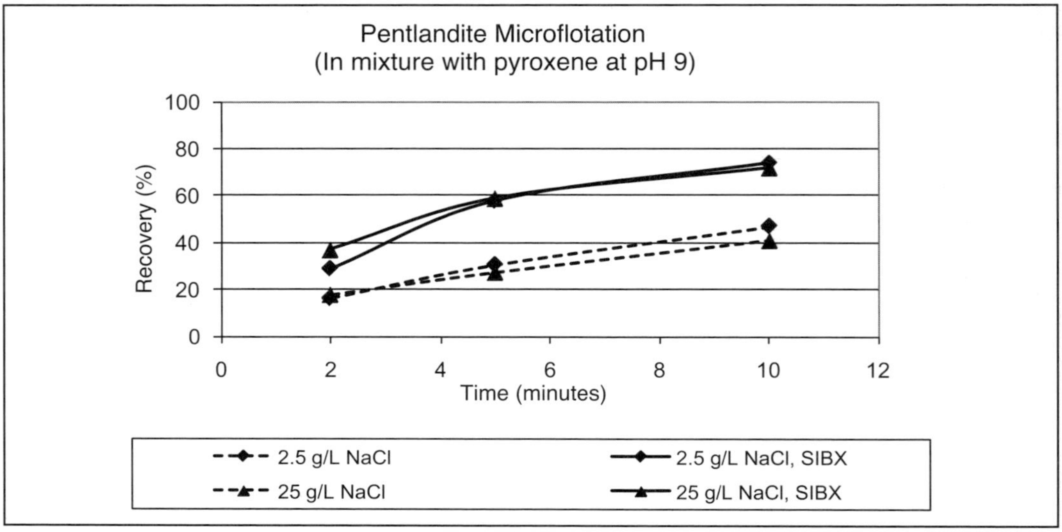

Figure 12. Pentlandite recovery–time curves at pH 9 comparing no reagents and SIBX in 2.5 g/L and 25 g/L NaCl solutions

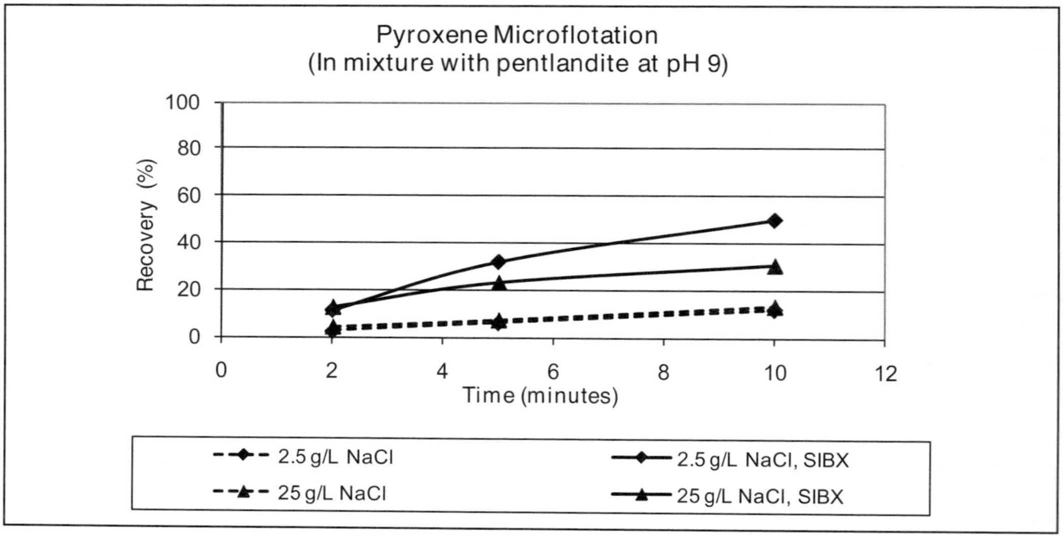

Figure 13. Pyroxene recovery–time curves at pH 9 comparing no reagents and SIBX in 2.5 g/L and 25 g/L NaCl solutions

When comparing the results obtained from the deionised and synthetic water trials it is clear that the presence of key ions typically found in process water reduces the pentlandite recovery compared to those achieved in deionised water in the absence of xanthate (Figure 10). When xanthate is added to the system then the recovery of pentlandite increases as the ion concentration in the water increases. Hodgson and Agar (1989) concluded that xanthate chemisorption onto the pentlandite surface takes place on the nickel sites. Based on this observation, one possible mechanism contributing to the higher pentlandite flotation response could be a higher number of nickel sites available for xanthate adsorption.

The ToF-SIMS results have shown the adsorption of calcium, silicon, aluminium and magnesium ions on the pentlandite surface (Figures 1 to 4). Calcium in particular is readily adsorbed onto the mineral surfaces and increases as the ion concentration increases; a maximum adsorption is reached at a calcium concentration of around 400 ppm in solution (5× synthetic water trials). The surface concentrations of calcium, silicon, aluminium and magnesium ions did not seem to hinder the pentlandite floatability at the concentrations evaluated.

In terms of pyroxene recovery; it has previously been shown (Shackleton, 2003) that the higher pyroxene recovery observed in a mixture with pentlandite is attributed to a higher degree of inadvertent activation by Ni^{2+} species and the same is true for this study (Figures not shown). Copper ions were also observed on the pyroxene mineral surfaces due to occlusions of chalcopyrite in the natural pyroxene mineral. Xanthate anions would adsorb onto the positively charged nickel and copper sites due to electrostatic attraction. The lower floatability of pyroxene observed at the higher ion concentrations (5× and 10× synthetic water) could be associated with a lower concentration of pentlandite dissolution products in solution due to the formation of a calcium passivation layer on pentlandite surfaces. Smart (1991) has also shown that xanthate has a cleaning action on gangue mineral surfaces which may take the form of detachment of colloidal particles, which inadvertently activate the pyroxene surface, from these surfaces. Cleaning of the gangue mineral surfaces is expected to assist in selectivity.

Although suggestions have been made regarding the mechanisms for the flotation response of pentlandite and pyroxene in process waters with varying ion concentrations the comparison is less straightforward due to the difference in ionic strength especially for the 10× synthetic water trials. One of the factors that could also contribute to a higher recovery is an increase in ionic strength. It is well-documented by Blake and Ralston 1985, that at an ionic strength of 0.1, the bubble size decreases and bubble density increases per unit volume. For a particle to adhere to a bubble the following events must occur; (1) thinning and rupture of the intervening film of liquid between particle and bubble; and (2) expansion of the three-phase line of contact to form a wetting perimeter. The sum of these two events is the induction time and must be less than the contact time if flotation is to occur. An increase in ionic strength leads to a decrease in induction time and an increased flotation rate together with a decrease in the critical hydrophobicity required for initiating flotation. In addition, the influence of ionic strength affects the attachment efficiency by a more rapidly thinning thin liquid film at a higher salt concentration due to an increased rate of drainage at the boundary ring (Hewitt et al., 1974). An increased ionic strength is also advantageous in recovering weakly hydrophobic particles (Crawford and Ralston, 1988). However, for this study an increase in recovery is only observed for the synthetic water trial (IS of 0.025) compared to the DI water (IS of 0.012) trial in the absence of reagents without any further increase in recovery as ionic strength increases to 0.13 for the 5× synthetic water and 0.25 for the 10× synthetic water. This suggests that ionic strength does not play a role or that the ions present in solution counteract the effect of ionic strength. In the presence of xanthate, an increase in pentlandite recovery is observed as the ionic strength increases which may purely be due to the induced hydrophobicity introduced by the collector as discussed earlier which overcomes the effect of the ions present in solution. It is also evident that the ions present in the synthetic water trials depress the pyroxene recoveries thereby introducing selectivity between the value and gangue minerals.

In order to de-couple ionic strength from the cations and anions present in solution the ionic strength was varied using ions, viz. NaCl. Two sodium chloride concentrations were evaluated, namely, 2.5g/L (IS of 0.043) and 25g/L (IS of 0.43). The microflotation results show that a much higher recovery of pentlandite and pyroxene was obtained using the NaCl

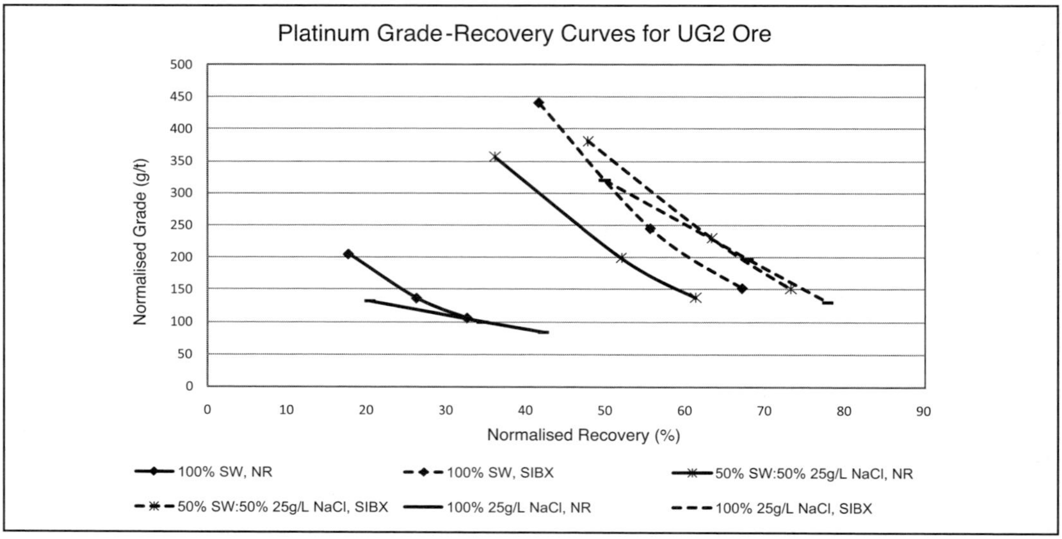

Figure 14. Normalised Pt grade–recovery curves comparing the effect of varying sodium chloride concentrations in the absence and presence of xanthate (SIBX)

solutions compared to both the DI and synthetic water trials irrespective of cation and anion concentrations in the absence of xanthate (Figures 12 and 13). In the presence of xanthate the results show that selectivity is reduced as the pyroxene recovery increased significantly (~38%) compared to the recovery obtained using an equivalent ionic strength solution, viz. synthetic water trial. When comparing the 2.5 and 25g/L NaCl trials, the pyroxene recovery is reduced by ~20%. As observed for the higher ion concentration trial (10× synthetic water), the reduction in pyroxene recovery may be due to an increase in the concentration of ions present in solution which now begin to depress the pyroxene recovery as both the 25g/L NaCl and 10× synthetic water trial have similar ionic strength. Thus it can be concluded that ionic strength plays a role in pentlandite floatability but only when the ionic strength is in the order of 0.1 or higher as seen for the 10× synthetic water ion concentration (IS of 0.25). However, there seems to be no effect of ionic strength for pyroxene when there are minimal concentrations of competing ions which would absorb on the mineral surfaces (Figure 13). The absence of these competing ions resulted in a significantly higher pyroxene recovery. This is consistent with data from a study carried out on a UG2 ore sample. The study involved the evaluation of synthetic water and NaCl solutions on the grade-recovery relationship (Figure 14). The results showed that the water source which contained ~25g/L NaCl, improved the flotation recovery of the valuable minerals but at a slightly lower concentrate grade showing that selectivity between the value and gangue minerals was reduced.

The ToF-SIMS data for pyroxene also shows a significantly higher calcium surface coverage in the sodium chloride electrolyte compared to the DI and synthetic water trials irrespective of whether the ionic strength is equivalent or not (Figure 5). This implies that sodium and chloride ions modify the mineral surfaces which allow better adsorption of the abovementioned ion even though the levels were equivalent to those found in the DI water trials. What effect the sodium and chloride ions have on flotation is unclear at this stage.

There seems to be a synergistic effect between an ionic strength of ≥0.1 and ion concentration in process water. However, the ions in the process water play a more significant role in mineral surface alteration and floatability of pentlandite and pyroxene compared to ionic strength.

CONCLUSIONS

The present study has compared the floatability and surface characteristics of synthetic pentlandite and natural pyroxene mineral samples using various concentrations of key ions typically found in process water.

- In the absence of reagents the additional cation and anion concentrations compared to those found in deionised water depressed the pentlandite and pyroxene recovery.
- Once xanthate is added to the system then the recovery of pentlandite increases as the ion concentration in the water increases. This may be either due to the higher nickel concentration on the surface and/or the removal of hydrophilic compounds such as metal oxides and hydroxides on the sulphide mineral surface.
- Xanthate seems to counteract both the effects of ionic strength and cation and anion concentrations in process water.

Ionic strength was evaluated using sodium chloride as the electrolyte in order to de-couple ionic strength from the cations and anions present in solution.

- The results show that selectivity between pentlandite and pyroxene is significantly reduced in the presence of sodium chloride.
- The data for pentlandite suggests that ionic strength plays a role but only at concentrations of ≥0.1.
- The recovery of pyroxene was significantly enhanced in the 2.5 and 25g/L NaCl trials compared to the 10× synthetic water test. This suggests that the ions present in the 10× synthetic water test depress the pyroxene recovery.
- When comparing the 2.5 and 25g/L NaCl trials, the pyroxene recovery is reduced by ~20% which may be due to an increase in the concentration of ions present in solution which now begin to depress the pyroxene recovery.

The results have shown that the cations and anions present in process water have a dual effect on the recovery of pentlandite and pyroxene. There seems to be a synergistic effect between an ionic strength of ≥0.1 and ion concentration in process water, however, the ions in the process water play a more significant role compared to ionic strength in mineral surface alteration and floatability of pentlandite and pyroxene. Calcium in particular is readily adsorbed onto the pentlandite mineral surfaces and no further adsorption at calcium solution concentrations of >400 ppm indicating a possible threshold value.

ACKNOWLEDGMENT

The authors thank Anglo American Platinum management for support during the study and for permission to publish this paper.

REFERENCES

Blake, P., and Ralston, J. 1985. Particle size, surface coverage and flotation response. *Colloids and Surfaces*, 16:41–53.

Crawford, R., and Ralston, J. 1988. The influence of particle size and contact angle in mineral flotation. *International Journal of Minerals Processing*, 23:1–24.

Hewitt, D., Fornasiero, D., and Ralston, J. 1974. Bubble particle attachment efficiency. *Minerals Engineering*, 7(5-6):657–665.

Hodgson, M., and Agar, G.E. 1989. Electrochemical investigations into the flotation chemistry of pentlandite and pyrrhotite: Process water and xanthate interactions. *Canadian Metallurgical Quarterly*, 28(3):189–198.

Levay, G., Smart, R.St.C., and Skinner, W.M. 2001. The impact of water quality on flotation performance. *The Journal of the South African Institute of Mining and Metallurgy*, 101(2):69–75.

Malysiak, V. 2003. Pentlandite-pyroxene and pentlandite–feldspar interactions and their effect on separation by flotation. *Doctoral thesis*, UCT.

Smart, R.St.C. 1991. Surface layers in base metal sulphide flotation. *Minerals Engineering*, 4(7-11):891–909.

Shackleton, N.J. 2003. The role of complexing agents in the flotation of pentlandite–pyroxene mixtures. *Master's thesis*, UCT.

Wesseldijk, Q.I., Reuter, M.A., Bradshaw, D.J., and Harris, P.J. 1999. The flotation behaviour of chromite with respect to the beneficiation of UG2 ore. *Minerals Engineering*, 12(10):1177–1184.

Frothing in the Flotation of Copper Sulfide Ores in Sea Water

S. Castro, O. Ramos, and J.P. Cancino
Department of Metallurgical Engineering, University of Concepción, Chile

J.S. Laskowski
Department of Mining Engineering, University of British Columbia, Vancouver, Canada

ABSTRACT

The selection of appropriate frothing agents for flotation of porphyry copper ores in sea water is becoming an important issue. In this paper a specially modified laboratory flotation cell was designed to study both, the two-phase and three-phase foam/froth layer growth kinetic under rougher flotation conditions. This method allows a direct comparison between the foaming and frothing properties for a given frother in the environment of fresh water and sea water. First, the two-phase foam layer thickness for mixtures of fresh water/sea water and different frothing agents, such as, pine oil, X-133/pine oil; DF-1012, DF-250, MIBC, and Matfroth-355, were investigated. Most of them show increased foamability in sea water, with exception of pine oil whose foamability in fresh water and sea water is nil. Then, the three-phase frothing kinetics as a function of pH for samples of three different porphyry copper ores was studied. It was found that pine oil is a very good frothing agent in both fresh water and sea water. The DF-1012 and DF-250 are strong frothing agents in fresh water, but not in sea water. The froth layer formed in sea water is usually smaller in thickness, with lower amount of water, and is little affected by pH, (as compared with that obtained in fresh water). The froth layer thickness in fresh water is strongly affected by pH, with a peak around pH 10.0–10.5.

INTRODUCTION

The evaluation of frothing agents in flotation systems in usually carried out in a rather arbitrary way (Laskowski, 2004; Tan et al., 2005). The most common method consists in assuming that the two-phase foam height or volume produced by a given frother, is directly related to its three-phase frothing performance (Malysa et al., 1987; Xia and Peng, 2007). However, some compounds with a good foamability cannot be used as flotation frothers because their hydrodynamic, frothing or decay properties do not provide adequate operability (Cappuccitti and Finch, 2008). Hence, direct evaluation of froth stability in industrial flotation cells seems to be necessary (Tsatouhas et al., 2006; Zanin et al., 2009).

Fundamental properties of frothing agents have been extensively studied over the last years. These studies have led to the development of standardized procedures to characterize frothers in terms of their abilities to reduce bubble size and increase foam stability. The former is commonly characterized by the CCC values (critical coalescence concentration) (Cho and Laskowski, 2002a,b; Laskowski et al., 2003; Laskowski, 2004; Grau et al., 2005), while the latter by the DFI (dynamic foamability index) (Malysa et al., 1981; Czarnecki et al., 1982; Malysa et al., 1987; Castro et al., 2010).

The frothers are also assessed in terms of their ability to promote the transport of water towards the froth collection zone (Rahal et al., 2001; Schwarz and Grano, 2005; Finch et al., 2006; Melo and Laskowski, 2006; Melo and Laskowski, 2007). While all this further improves our understanding of the properties of flotation frothers, and the properties of foams generated in solutions of these frothers, the properties of flotation three-phase froths also depend on solid particles, their size, their surface properties and their concentration. Although Malysa suggested (1998) that the general relationship between the liquid content and frother concentration measured in foams should not be altered by the presence of solid particles, some results suggest that such relationships may be different for different frothers (Melo and Laskowski, 2007; Kuan and Finch, 2010). Melo and Laskowski (2007) reported for DF-1012 (the most surface active polyglycol frother among those studied in that project) that while the foams generated with the use of this frother were the most stable and carried most water, the froth studied in the presence of hydrophobic bituminous coal particles was remarkably less voluminous when this frother was utilized. The same effect was reported by Kuan and Finch (2010) who studied the effect of hydrophobic talc particles on the properties of foams in the presence of a polyglycol frother.

In the case of frothing in sea water additional factors must be considered. Surface tension at the bubble/solution interface is increased by inorganic electrolytes (Jones and Ray, 1941) and, as it is well known, decreases in the presence of surface-active compounds. Bubble size decreases in solutions of inorganic electrolytes and in solutions surface-active compounds, because both, inorganic salts as well as surface active compounds stabilize bubbles against coalescence (Laskowski et al., 2003b; Quinn et al., 2007). Dynamic foamability of common flotation frothers increases in aqueous sodium chloride solutions and sea water (Castro et al., 2010; Castro et al., 2012). Since the role of flotation frothers is to decrease bubble size and increase froth stability, in flotation systems inorganic electrolytes may exhibit capabilities similar to those exhibited by frothers. However, the three-phase froth properties, when copper ores are floated in sea water, have not been extensively studied. The aim of the present work is to investigate the frothing capability of traditional frothing agents in a sea water environment.

EXPERIMENTAL

Materials

Methyl isobutyl carbinol (MIBC) was provided by Cytec-Chile; DF-250 and DF-1012 polyglycol frothers were from Moly-Cop Chile S.A.; Matfroth-355 was provided by Mathiesen S.A.C. Other frothers like pine oil and the mixture X-133/pine oil; and collecting reagents were commercial grade products provided by the mining industry. A local sample of sea water from Concepcion city (*Bellavista-Tomé*) with salinity of 33.5% was employed. Fresh water was local tap water, and pH was adjusted with lime.

The samples of the porphyry Cu-Mo ores were provided by three large processing plants in Chile, and the respective standard laboratory flotation tests were applied. When sea water was employed, its usage includes grinding and flotation stages.

Methods

Froth and Foam Layer Thickness Measurements

The froth phase was studied by measuring the maximum froth layer thickness in a laboratory flotation cell so modified that it operates without discharge of concentrate. A LA-500 Agitair flotation cell with a volume of 2.7 L (Figure 1) was adapted for the foamability and frothability tests, as was described elsewhere (Castro et al., 2010). The froth thickness was measured during 60 sec by using a digital photographical method coupled to image analysis

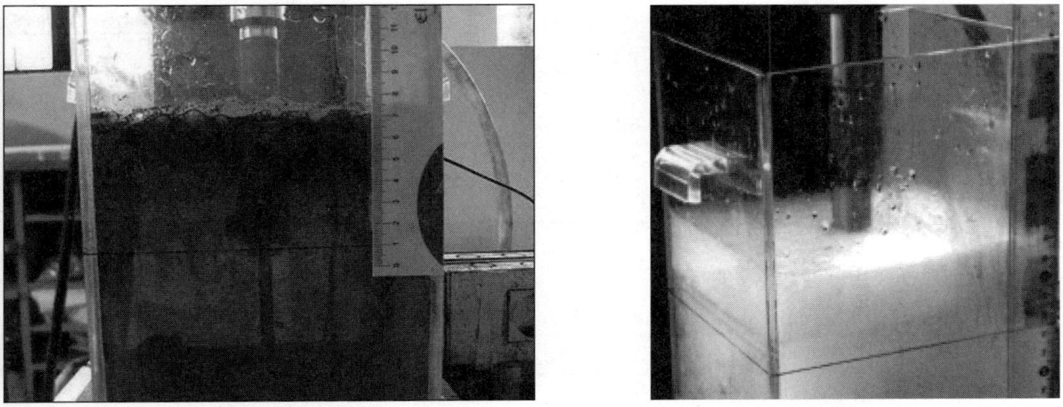

Figure 1. A 2.7-L LA-500 Agitair laboratory flotation cell adapted for the measurement of froth and foam thickness

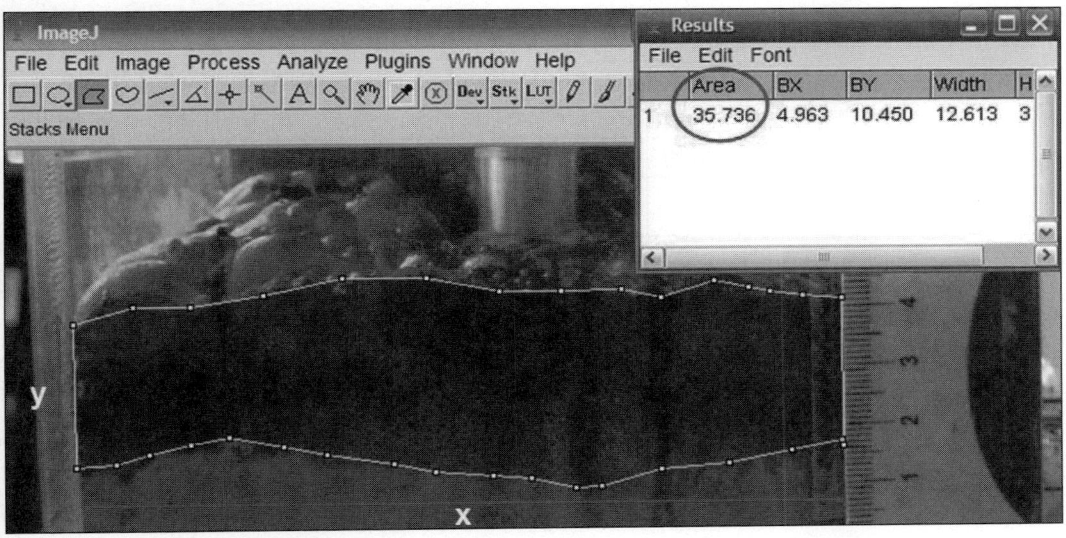

Figure 2. Determination of the area occupied by the froth by using the ImageJ program

with the ImageJ software (Figure 2). The height of the foam/froth in the cell was measured versus time (growth kinetic), and the maximum foam/froth layer thickness was determined at the equilibrium time (t_∞).

Experimental Conditions in Frothing Tests

Copper ore (sample 1). The experimental conditions used for sample 1 in Table 1 (feed assay: 0.99%Cu, 1.7%Fe and 0.007%Mo) were as follow: ore sample weight, 1,015.7 g; impeller speed, 900 rpm; air flow rate 7 L/min; 30% solids (by wt.); pH 10, particle size distribution of the flotation feed: 30% +100 Tyler mesh (P_{80} = 198.3 μm). Rougher flotation conditions: 4 min conditioning time.

Copper ore (sample 2). The experimental conditions used for sample 2 in Table 2 (feed assay: 0.56%Cu, 5.76%Fe and 0.017%Mo) were as follow: ore sample weight 1,204.4 g; impeller speed rate, 900 rpm, air flow rate 10 L/min; 35% solids (by wt.); pH as indicated

Table 1. Flotation reagents used for sample 1 during the rougher three-phase froth measurements

	Flotation Reagents, g/ton				Time, min		
Stage	Matcol D-101	IsobX	X-133/ Pine oil	Diesel	Grinding	Conditioning	Flotation
Grinding	—	—	—	20	7' 05"	—	—
Flotation	35	5	10	—		4	1

Note: As activator 100 g/ton NaSH.

Table 2. Flotation reagents used for sample 2 during the rougher three-phase froth measurements

	Flotation Reagents, g/ton				Time, min		
Stage	Sascol -95	Matcol TC-123	Matfroth-355	Diesel	Grinding	Conditioning	Flotation
Grinding	—	—	—	10	3' 30"	—	—
Flotation	11	22	10	—	—	5	1

Table 3. Flotation reagents used for sample 3 during the rougher three-phase froth measurements

	Flotation Reagents, g/ton			Time, min		
Stage	MX-7017	MX-945	MIBC	Grinding	Conditioning	Flotation
Grinding	—	—	—	7' 18"	—	—
Flotation	26	21	21	—	5	1

in figures; flotation feed particle size distribution: 29.3% + 100 Tyler mesh (P_{80} = 210 μm). Rougher flotation conditions: 5 min conditioning time.

Copper ore (sample 3). The experimental conditions used for sample 3 in Table 3 (feed assay: 0.43%Cu, 6.26%Fe and 0.08%Mo) were as follow: ore sample weight, 1,161 g; impeller speed rate, 900 rpm; air flow rate 10 L/min; 34% solids (by wt.); pH as indicated in figures; flotation feed particle size distribution: 20% +100 Tyler mesh (P_{80} = 150 μm). Rougher flotation conditions: 5 min conditioning time.

RESULTS AND DISCUSSION

The experimental technique employed in this work for characterization of foaming or frothing has the advantage of using the same procedure and equipment which is utilized in the rougher laboratory flotation tests.

The kinetics of two-phase foam or three-phase froth formation, and the maximum froth/foam thickness were obtained according a mathematic model previously published (Castro et al., 2010).

$$h_t = h_\infty(1 - e^{-K(t-\theta)}) \tag{1}$$

where

h_t = froth layer thickness after time t
h_∞ = maximum froth layer thickness at $t \rightarrow \infty$
K = froth growth rate constant
t = time corrected by the induction time θ

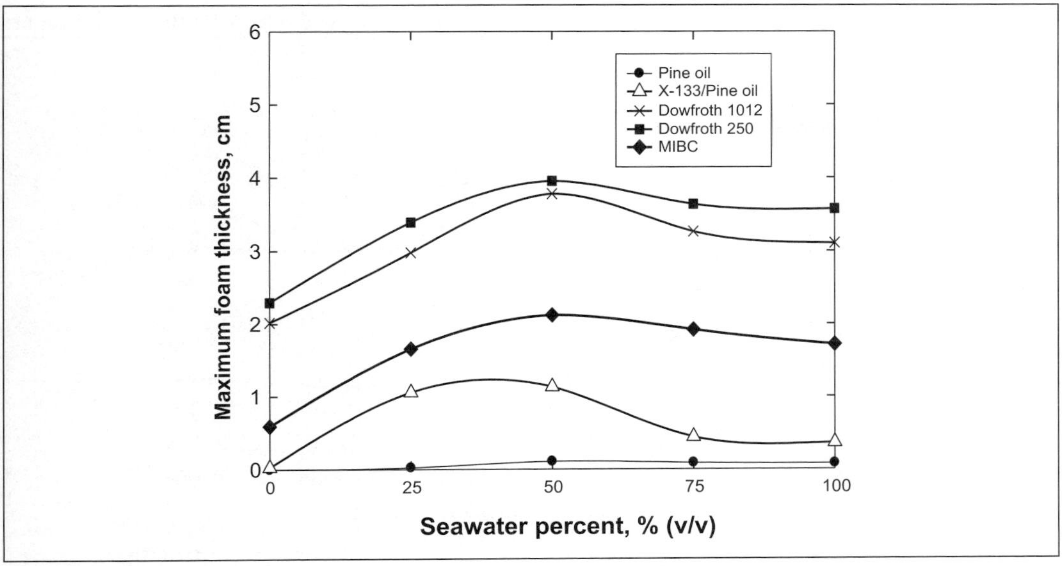

Figure 3. Foam layer thickness as a function of sea water/fresh water ratio for different frothing agents (frother dosage, 10 g/ton; pH 10.0; air flow rate, 7 L/min; impeller speed rate in the cell, 900 rpm)

Two-Phase Foaming in Seawater

In this work the foaming properties of common flotation frothers in sea water and fresh water were studied. Figure 3 shows the maximum foam layer thickness (h_∞) determined in our modified flotation cell. These tests were carried out with five different frothing agents (Pine oil, the mixture X-133/pine oil, DF-1012, DF-250, and MIBC) in blends of sea water with fresh water. Similar results are reported in Figure 4, were Matfroth-335 frother was also included.

It was found that polyglycol frothers (DF-250 and DF-1012) are better foaming agents than MIBC, which is rather well known (Tan et al., 2005). It is also clear that most tested frothers behave like stronger foaming agents in sea water, in good agreement with the previous DFI measurements (Castro et al., 2010). Figures 3 and 4 indicate that the foam layer thickness is higher in sea water than fresh water for most studied frothers (with exception of pine oil). In the case of MIBC, a week foaming agent, an increase in its foamability in diluted sea water solutions (up to about 50% sea water, v/v) was observed; with a subsequent decreasing at higher sea water concentration. It should be noted that pine oil practically does not form a layer of two-phase foam under the experimental conditions (in two different set of tests, as reported in Figures 3 and 4); and the DF-250 and DF-1012 frothers form the thicker foam layer in both fresh water and sea water.

These results indicate that there is good correlation between foamability in fresh water and in sea water, i.e., those frothers which behave like week or strong foaming agents in fresh water maintain the same character in sea water. However, for most of them, their foaming properties are augmented in sea water.

Three-Phase Froth in Copper Ore Rougher Flotation

Contrary to what can be expected from the improved two-phase foamability observed in sea water (Figures 3 and 4) the three-phase frothability measured with a copper ore (sample 1)

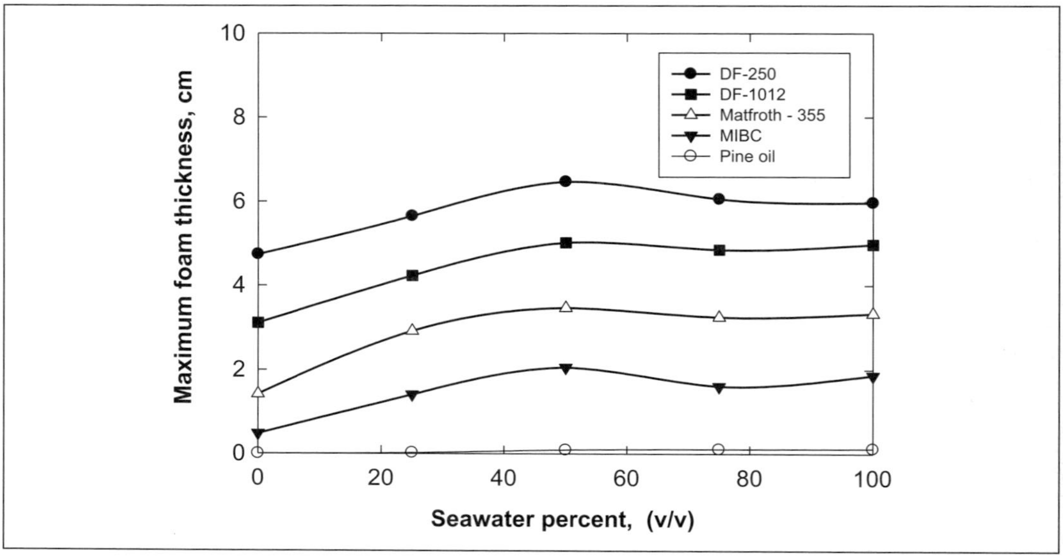

Figure 4. Foam layer thickness as a function of sea water/fresh water ratio for different frothing agents (frother dosage, 10 g/ton; pH 9.5; air flow rate, 10 L/min; impeller speed rate in the cell, 900 rpm)

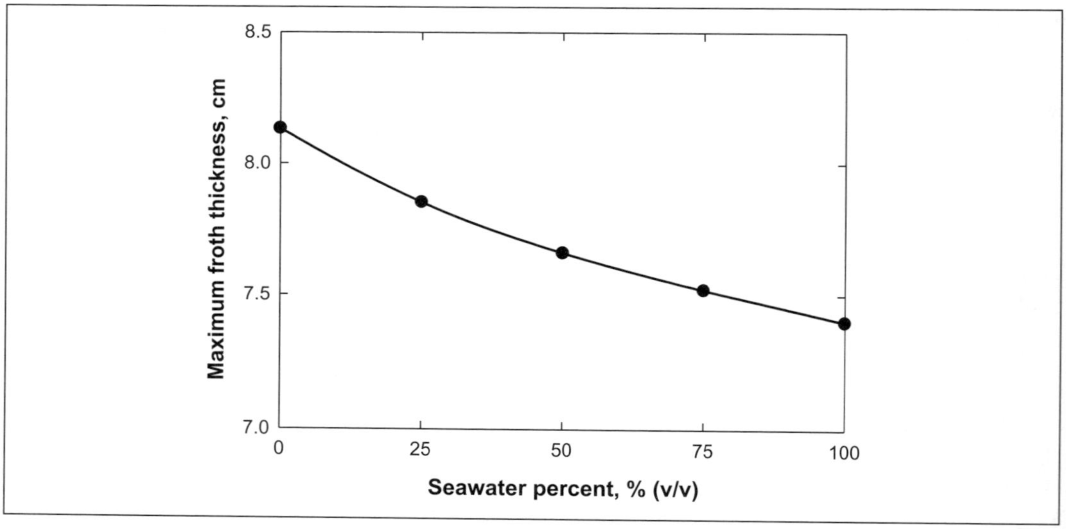

Figure 5. Froth layer thickness as a function of sea water percent (v/v) (pH 10.0; frother: X-133/ Pine oil, 10 g/ton) (ore sample 1)

and X-133/pine oil frother mixture l, was lower in sea water than in fresh water (Figure 5). This behaviour was also observed with other copper ores. Hence, these results indicate that less voluminous froths are commonly formed in sea water.

It was also found that the froth layer is a function of pH. Figure 6 shows the effect of pH on froth thickness (h_∞) in fresh water and sea water; a peak in frothability was recorded around

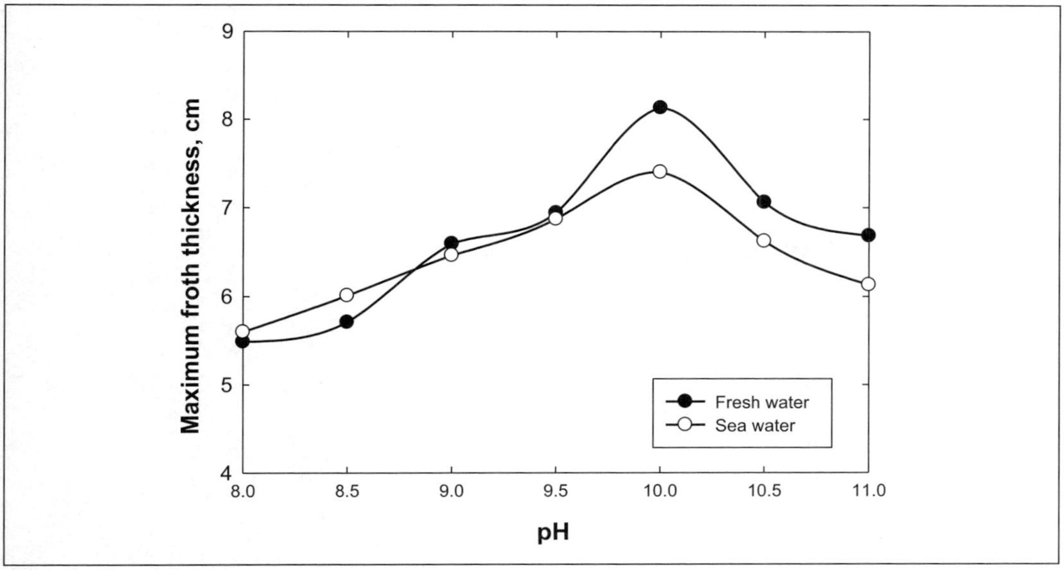

Figure 6. Effect of pH on froth layer thickness in fresh water and sea water (frother: X-133/Pine oil, 10 g/ton) (ore sample 1)

pH 10, where also the major difference in froth layer thickness between fresh water and sea water was observed.

In order to clarify the effect of pH on the frothing in the pulp of copper sulfide ores, in Figures 7 and 8 more examples obtained with other two different copper ores are shown. In all studied cases, an increase of the froth layer thickness in fresh water around pH 10.0–10.5 was observed. The thickness of froth layer in sea water was smaller than in fresh water, and it was less sensitive to pH. Therefore, the major difference in frothability between sea water and fresh water appears around pH 10.0–10.5.

These most important results can be summarized as follows: (a) The froth layer thickness in fresh water increases with pH, exhibiting a maximum value around pH 10.0–10.5; and (b) the froth layer thickness in sea water is significantly smaller in comparison with the fresh water results.

Water Recovery in Concentrate

Figure 9 shows the amount of water transferred to froth in copper rougher concentrates in the batch flotation tests with ore sample 1. It was found that the water content was lower in the froths formed in sea water. These results indicate that dryer froths are produced when sea water is employed in flotation tests.

It is well known that in froth flotation bubbles transport mineral particles and water. The presence of frothers affect the amount of water recovered in a concentrate (Rahal et al., 2001; Boylu and Laskowski, 2007). Usually water recovery increases when bubble size is reduced by frothers because the water carrying rate is proportional to frother concentration and gas holdup (Melo and Laskowski, 2006; Finch et al., 2006). However, sea water is able to prevent bubble coalescence reducing bubble size, in the same way than frothers. Thus, due to this joint

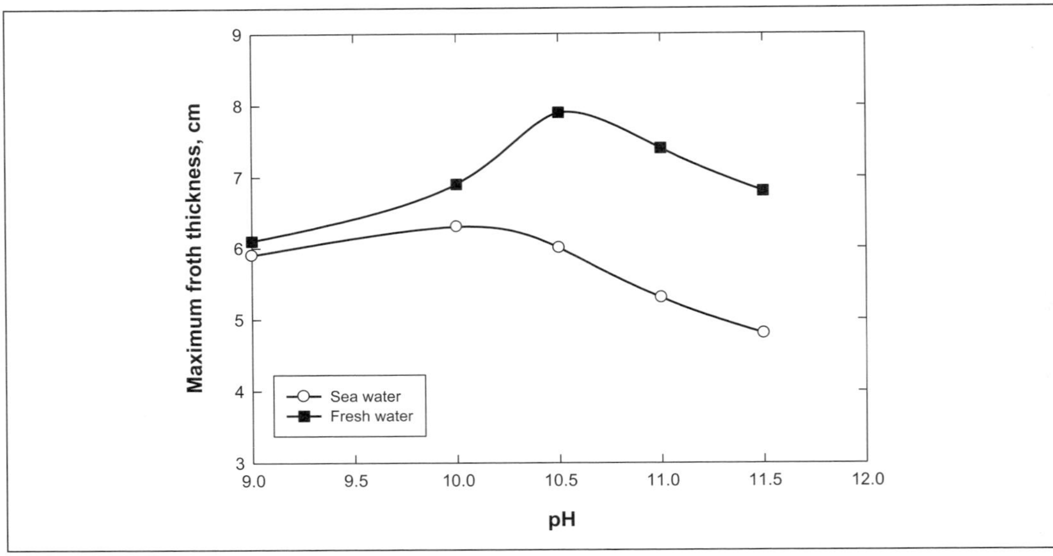

Figure 7. Effect of pH on froth layer thickness in fresh water and sea water (frother: Matfroth-355, 10 g/ton) (ore sample 2)

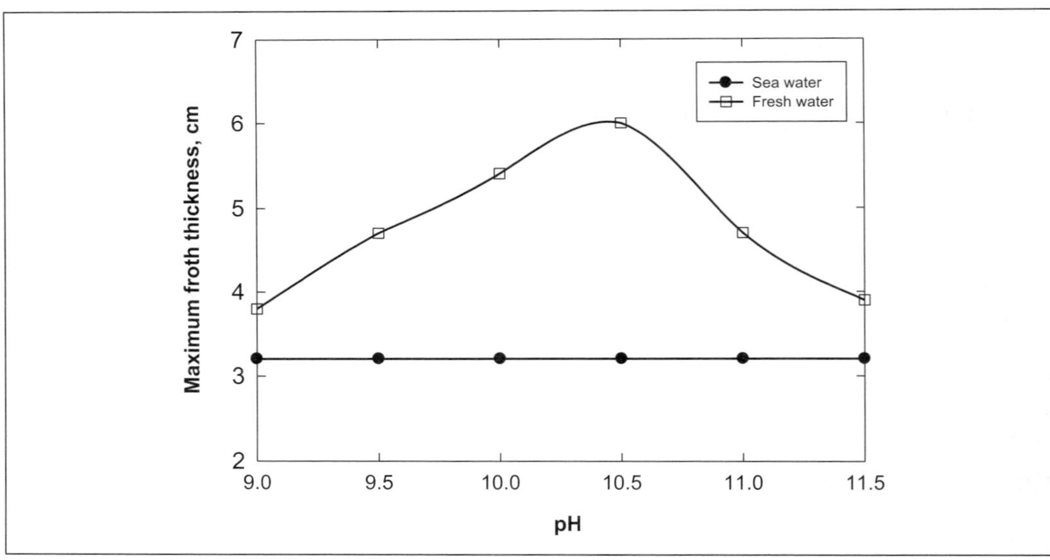

Figure 8. Effect of pH on froth layer thickness in fresh water and sea water (frother: MIBC, 21 g/ton) (ore sample 3)

effect, it could be expected that water recovery be higher in sea water. In practice this is very different. Water is transported to froth as thin liquid films surrounding the bubbles, and when the thickness of this film increases the water carrying rate increases as well (Finch et al., 2006). Our results show that in the case of sea water, the electrolytes contained in it reduce the water carrying rate, which is in good agreement with previous results (Melo and Laskowski, 2006). Therefore, the smaller frother layer thickness and the less water content in sea water froths are experimental pieces of evidence showing that inorganic electrolytes decrease the thickness of the liquid films surrounding bubbles.

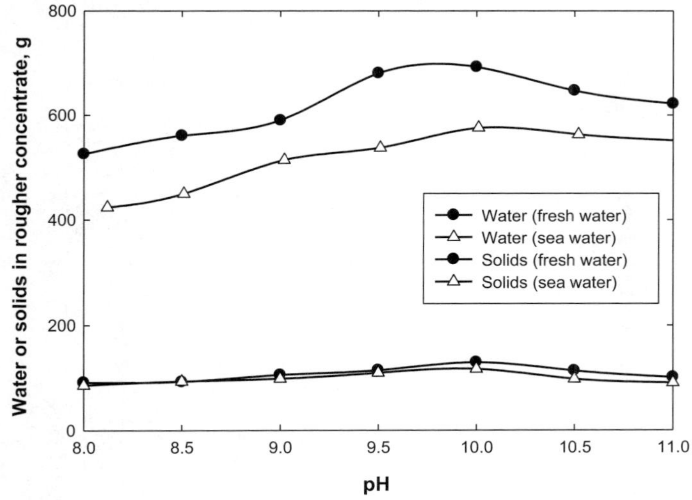

Figure 9. Effect of pH on water and solid contents transferred into the rougher concentrate in the flotation tests carried out in fresh water and sea water (frother: X-133/Pine oil, 10 g/ton) (ore sample 1)

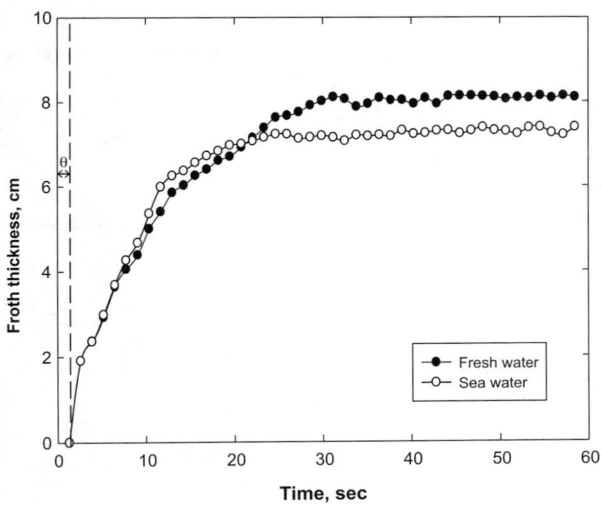

Figure 10. Froth layer growth kinetics for the X-133/Pine oil frother (10 g/ton) in fresh water and sea water at pH 10 (ore sample 1)

Selection of Flotation Frothers in Sea Water

Selection of frothing agents for flotation of copper ores in sea water emerges as an important issue. In Figures 10–14, the kinetic of froth layer growth in fresh water and sea water is compared (copper ore sample 1).

Figures 10, 11 and 14, show that pine oil, MIBC and the X-133/pine oil mixture practically maintain their frothability when fresh water is replaced by sea water. However, the frothabilities of DF-1012 and DF-250 are markedly reduced in sea water.

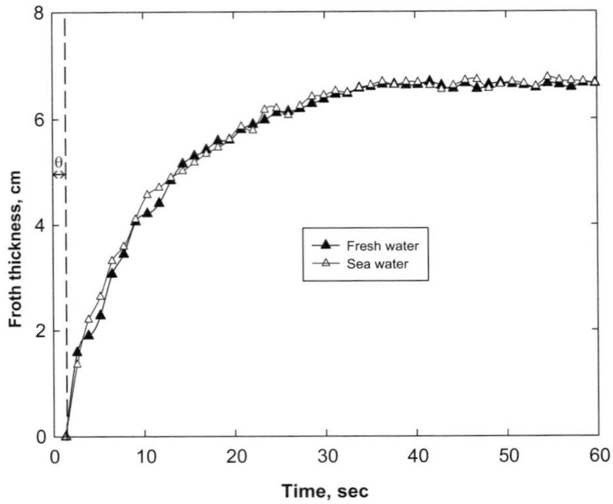

Figure 11. Froth layer growth kinetics for Pine oil frother (10 g/ton) in fresh water and sea water at pH 10 (ore sample 1)

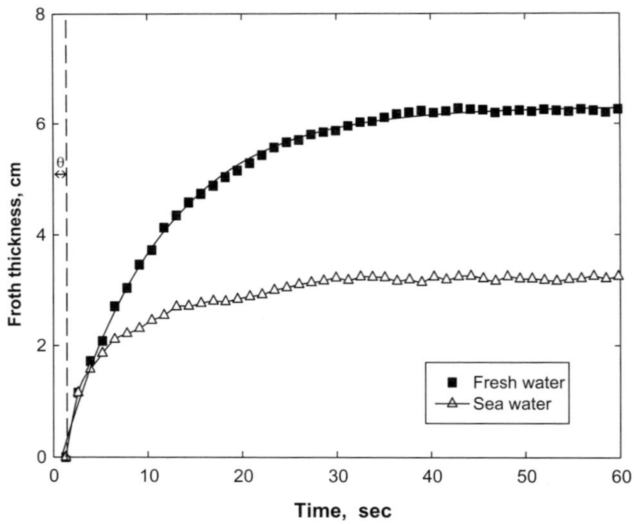

Figure 12. Froth layer growth kinetics for Dowfroth-1012 (10 g/ton) in fresh water and sea water at pH 10 (ore sample 1)

Figure 15 summarizes these results. This comparison confirms that MIBC is the weaker frothing agent of the group (in both fresh water and sea water); while the mixture X-133/Pine oil is the stronger one. In both cases, the differences in frothability between fresh water and sea water were very small. However, the major differences in frothability were observed for polyglycol frothers (DF-1012 and DF-250), which loose frothability in sea water. These results strongly suggest that polyglycol frothers are not so good frothers in sea water as they are in fresh water. In addition, pine oil, with a very low foamability in two-phase systems, turned out to be one of the best three-phase frothers for copper ores in both fresh water and sea water.

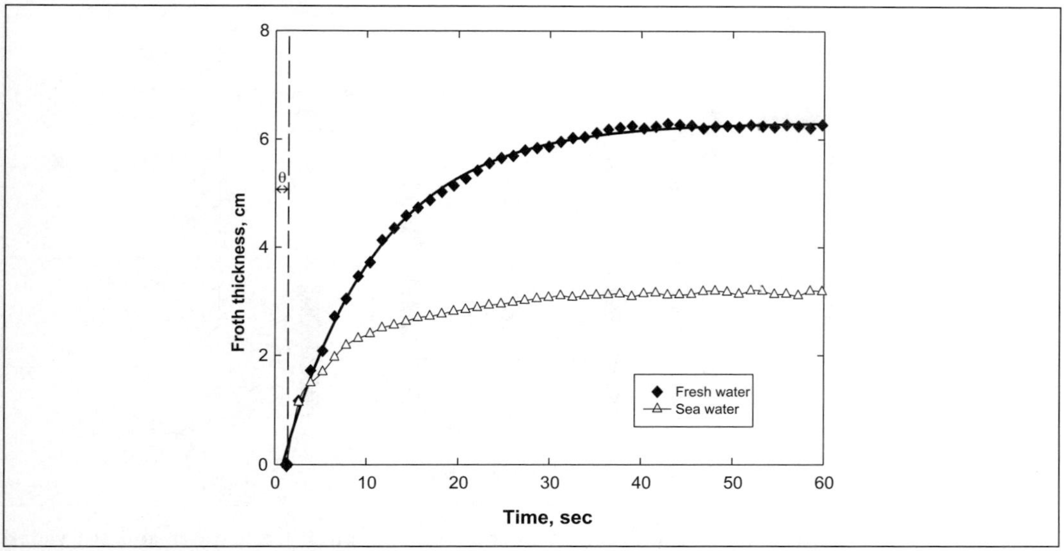

Figure 13. Froth layer growth kinetics for Dowfroth-250 (10 g/ton) in fresh water and sea water at pH 10 (ore sample 1)

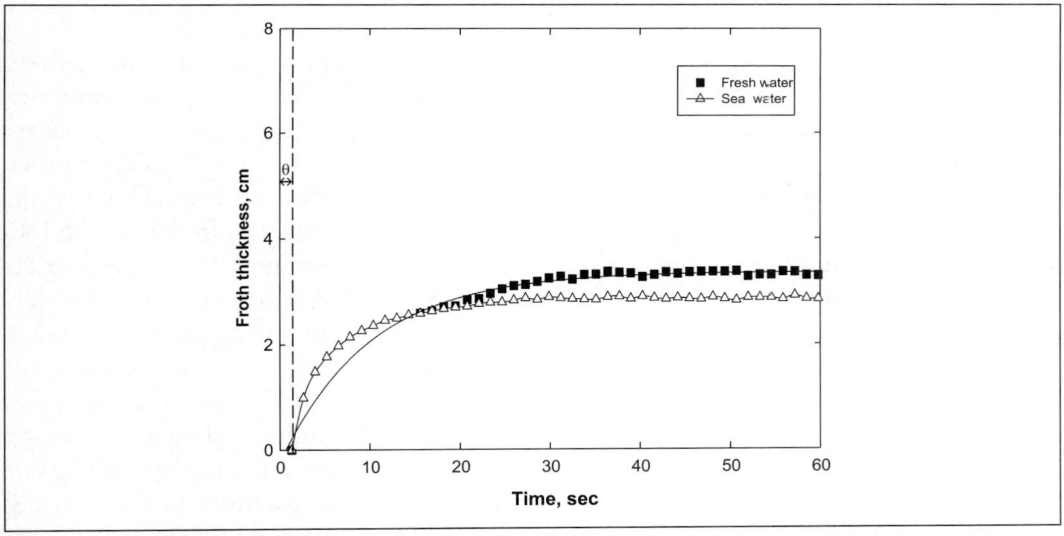

Figure 14. Froth layer growth kinetics for MIBC (10 g/ton) in fresh water and sea water at pH 10 (ore sample 1)

Therefore, the three-phase frothability of a given frother in sea water does not necessarily exhibit a good correspondence with its two-phase foamability. Pine oil and the X-133/pine oil mixture are week foaming agents in fresh water and even in sea water, but very good frothing agents (flotation frothers) in fresh water and sea water. Along the same line, DF-250 and DF-1012 are very good two-phase foaming agents in fresh water and sea water, but weak flotation frothing agents in sea water. MIBC maintains its weak foaming/frothing properties under all conditions. The differences in foaming properties described here for polyglycol frothers and MIBC are in good agreement with literature data (Tan et al., 2005).

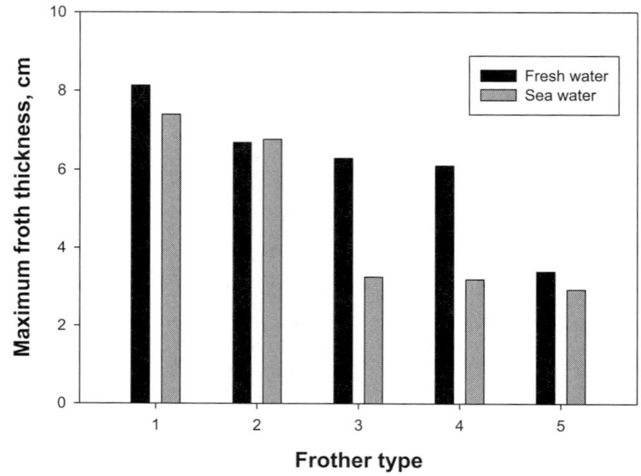

Figure 15. Froth layer thickness as a function of the frother type in fresh water and sea water: (1) X-133/Pine oil; (2) Pine oil; (3) Dowfroth-1012; (4) Dowfroth-250; (5) MIBC (10 g/ton; pH 10) (ore sample 1)

CONCLUSIONS

The modified laboratory flotation cell employed in this work was found to be useful in assessing foaming and frothing properties of flotation frothers in sea water and fresh water. It was found that most common frothers increase their foamability in sea water, i.e., behave like stronger two-phase foaming agents in sea water. However, there is no a direct relationship between the two-phase behaviour, and the subsequent copper sulfide ore three-phase frothing properties in sea water. For example, the foaming property of pine oil in fresh water and sea water is practically nil, but it is one of the best in producing voluminous froth in copper ore flotation pulps, both in fresh water and sea water environment. Along the same line, the polyglycol frothers DF-1012 and DF-250 are strong foaming and frothing agents in fresh water, but their frothabilty strongly decrease in sea water. In general, if the frothability of copper ore pulps in fresh water is compared with that in sea water, the froth layer thickness is usually lower in sea water. In addition, in fresh water the frothability is highly influenced by pH showing a peak around pH 10.0–10.5; this is much less sensitive to pH in sea water. The carrying rate of water into copper rougher concentrates is lower for sea water froths (dryer froths), which is consistent with the thinner froth layer thickness observed under such conditions.

ACKNOWLEDGMENTS

Funding for this project was provided by the CORFO-INNOVA CHILE (Project 08CM01-18), with industrial sponsorship from: Antofagasta Minerals (Minera *Esperanza*); BHP Billiton (*Minera Escondida* Ltd.); Anglo American Chile (*Mantos Blancos*); and Teck (*Carmen de Andacollo*), through AMIRA-International operating Research Grant P968.

REFERENCES

Boylu, F., and Laskowski, J.S. 2007. Rate of water transfer to flotation froth in the flotation of low-rank coal that also requires the use of oily collector. *Int. J. Miner. Process.*, 83: 125–131.

Cappuccitti, F., and Finch, J.A. 2008. Development of new frothers through hydrodynamic characterization. *Miner. Eng.*, 21: 944–948.

Castro, S., Venegas, I., Landero, A., and Laskowski, J.S. 2010. Frothing in seawater flotation systems. *Proc. XXV Int. Mineral Processing Congress*, Brisbane, pp. 4039–4047.

Castro, S., Toledo, P., and Laskowski, J.S. 2012. Foaming properties of flotation frothers at high electrolyte concentrations (*this symposium*).

Cho, Y.S., and Laskowski, J.S. 2002a. Effect of flotation frothers on bubble size and foam stability. *Int. J. Miner. Process.*, 64: 69–80.

Cho, Y.S., and Laskowski, J.S. 2002b. Bubble coalescence and its effect on dynamic foam stability. *Can. J. Chem. Eng.*, 80: 299–305.

Czarnecki, J., Malysa, K., and Pomianowski, A. 1982. Dynamic frothability index. *J. Colloid Interface Sci.*, 82(2): 570–572.

Finch, J.A., Gelinas, S., and Moyo, P. 2006. Frother-related research at McGill University. *Miner. Eng.*, 19: 726–733.

Grau, R.A., Laskowski, J.S., and Heiskanen, K. 2005. Effect of frothers on bubble size. *Int. J. Miner. Process.*, 76: 225–233.

Jones, G., and Ray, W.A. 1941. The surface tension of solutions of electrolytes as a function of the concentration. II. *J. Am. Chem. Soc.*, 63: 288–294.

Kuan, S.H., and Finch, J.A. 2010. Impact of talc on pulp and froth properties in F150 and 1-pentanol frother systems. *Miner. Eng.*, 23: 1003–1009.

Laskowski, J.S. 2003. Fundamental properties of flotation frothers. *Proc. 22nd Int. Mineral Processing Congress*, Cape Town, Vol. 2, pp. 788–798.

Laskowski, J.S. 2004. Testing flotation frothers. *Physicochem Probl. Miner. Process.*, 38: 13–22.

Laskowski, J.S., Cho, Y.S., and Ding, K. 2003b. Effect of frothers on bubble size and foam stability in potash ore flotation systems. *Can. J. Chem. Eng.*, 81: 63–69.

Małysa, K., Cohen, R., Exerowa, D., and Pomianowski, A. 1981. Steady-state foaming and the properties of thin liquid films from aqueous alcohol solutions. *J. Colloids Interface Sci.*, 80: 1–6.

Malysa, E., Malysa, K., and Czarnecki, J. 1987. A method of comparison of the frothing and collecting properties of frothers. *Colloids Surf.*, 23: 29–39.

Malysa, K. 1998. Water contents and distribution in flotation froth. *Frothing in Flotation II.* J.S. Laskowski and E.T. Woodburn, eds. Gordon and Breach, pp. 81–108.

Melo, F., and Laskowski, J.S. 2006. Fundamental properties of flotation frothers and their effect on flotation. *Miner. Eng.*, 1: 766–773.

Melo, F., and Laskowski, J.S. 2007. Effect of frothers and solid particles on the rate of water transfer to froth. *Int. J. Miner. Process.*, 84: 33–40.

Quinn, J.J., Kracht, W., Gomez, C.O., Gagnon, C., and Finch, J.A. 2007. Comparing the effect of salts and frother (MIBC) on gas dispersion and froth properties. *Miner. Eng.*, 20: 1296–1302.

Rahal, K., Manlapig, E., and Franzidis, J.P. 2001, Effect of frother type and concentration on the water recovery and entrainment recovery relationship. *Miner. Metall. Process.*, 18: 138–141.

Schwarz, S., and Grano, S. 2005. Effect of particle hydrophobicity on particle and water transport across a flotation froth. *Colloids Surf. A*, 256: 157–164.

Tan, S.N., Pugh, R.J., Fornasiero, D., Sedev, R., and Ralston, J. 2005. Foaming of polypropylene glycols and glycol/MIBC mixtures. *Miner. Eng.*, 18: 179–188.

Tsatouhas, G., Grano, S.R., and Vera, M. 2006. Case studies on the performance and characterization of the froth phase in industrial flotation circuits. *Miner. Eng.*, 19) 774–783.

Xia, Y.K., and Peng, F.F. 2007. Selection of frothers from residual organic reagents for copper-molybdenite sulfide flotation. *Int. J. Miner. Process.*, 83: 68–75.

Zanin, M., Wightman, E., Grano, S.R., and Franzidis, J.P. 2009. Quantifying contributions to froth stability in porphyry copper plants. *Int. J. Miner. Process.*, 91: 19–27.

Impact of Total Dissolved Solids in Process Water on the Surface Properties of Silica and Sphalerite Minerals

Meijiao Deng, Qingxia Liu, and Zhenghe Xu
Dept. of Chemical and Materials Engineering, University of Alberta, Edmonton, Alberta, Canada

ABSTRACT

High total dissolved solids (TDS) such as calcium sulfate in process water and recycle water are considered to have detrimental impact on selective flotation of desired minerals. In some sulphide mineral flotation operations, the recycled water contains high concentrations of Ca and SO_4 ions which exceed the solubility limit of gypsum. It is speculated that the gypsum would crystallize on mineral surfaces, resulting in reduced flotation selectivity and recovery. In this study, gypsum supersaturated solution is used to represent high TDS water. The impact of TDS on the surface properties of silica and sphalerite minerals is investigated using zeta potential distribution measurements and scanning electron microscopy (SEM). Our results demonstrate that silica and sphalerite minerals carry similar charge in the gypsum supersaturated solution, with zeta-potential of −11 mV. Both silica and sphalerite surfaces are indiscriminately coated with Ca species. The adsorbed layers are mostly Ca ions, not the nucleation of gypsum crystals. Needle shape gypsum precipitates are found on both silica and sphalerite minerals surfaces conditioned with gypsum-supersaturated solution. However, the gypsum does not grow on the mineral surfaces but form in the bulk solution. Zeta potential distribution results indicate that the hetero-aggregation between mineral and gypsum precipitates is not significant. The coating of Ca ions, resulting in the identical surfaces between silica and sphalerite, may cause problems during the flotation separation of silica and sphalerite. Further investigations on the influence of gypsum supersaturated solution on flotation separation of silica and sphalerite will be conducted.

INTRODUCTION

Water quality can have a significant effect on the mineral processing. An understanding of the role of water quality in operation is critically important for the improvement of process stability and performance (Johnson 2003; Levay and Schumann 2006; Levay et al. 2001; Liu et al. 1993). In many cases, high levels of dissolved inorganic and organic chemical species from soluble mineral phases, reagent additions, recycled water streams and make-up water can contribute to unrecognised and uncontrolled factors in mineral beneficiation processes (Schumann R. 2003). In sulphide mineral flotation operation, the recycled water usually contains a high concentration of soluble salts, due to the oxidation of sulphide minerals and the dissolution of semi-soluble minerals. For some operations, the concentration of Ca ions in recycled water can be as high as 1 g/L (Woodcock 1970), and sulphate ions concentration as high as 2.5 g/L, exceeding the solubility limit of gypsum (Grano 1995).

High total dissolved solids (TDS), for instance: calcium sulphate, are considered to have detrimental impact on selective flotation of desired minerals. The high concentrations of Ca and SO_4 ions in mineral process medium (water) have been a persistent concern in sulphide mineral flotation practice. It has been reported that while pH adjustment by lime to a value of 11 in order to depress pyrite it caused problems in the flotation of galena (Rao 2004). The flotation of chalcopyrite was also reported to be depressed in an alkaline pH when lime was used as a pH modifier or when Ca ions were present in the pulp (Liu and Zhang 2000). The presence of Ca ions may reduce the exchange rate with activating ions, usually copper and lead ions, in the case of sphalerite flotation, resulting in a decrease in recovery (Moignard 1977). However, the study carried out by Finch's group illustrated no change of copper and xanthate uptake in the presence of Ca and Mg ions (Lascelles et al. 2003). Their pure mineral flotation results indicated that the sphalerite flotation recovery only reduced at pH above 10 in the presence of Ca. The impact of Ca on the interactions between silica and sphalerite at various pHs has also been investigated (DiFeo et al. 2001). The results suggested a promoted hetero-aggregation of silica and sphalerite at high pH in the presence of Ca while they were well dispersed in the absence of Ca. The interaction forces between sphalerite and silica particles were reported to be attractive below pH 8 and decrease with increasing solution pH and become repulsive at pH 10 (Xu et al. 2000). The addition of Ca enhanced the attraction force between the two minerals at pH 6 while reduced the repulsive force at pH 10.

Calcium sulphate scaling is a well-documented phenomenon encountered in a wide spectrum of industrial processes and applications such as those involving heat exchange (Fahiminia et al. 2007; Hoang et al. 2007; Shams El Din et al. 2005), mineral processing (Grano 1995; Seewoo et al. 2004), desalination of seawater by reverse osmosis (Greenlee et al. 2009; Hasson et al. 2001) etc. Calcium sulphate may crystallize as gypsum ($CaSO_4·2H_2O$), calcium sulphate hemihydrate ($CaSO_4·½H_2O$), and anhydrite ($CaSO_4$) (Lambert and Schaffer 1926; Sudmalis and Sheikholeslami 2000). The solubility of all calcium sulphate forms decreases with increasing temperature (Agamaliev et al. 1993; Dutrizac 2002; Flint 1968; Furby et al. 1968; Gardner and Glueckauf 1970), a fact that is responsible for the formation of calcium sulphate scale on heat exchanger surfaces (Klepetsanis et al. 1999). The formation of calcium sulphate scale is affected by various factors including fluid hydrodynamics, supersaturation level, solution chemistry, surface chemistry and topography as well as temperature (Hoang et al. 2007; Hoang et al. 2006; Klepetsanis et al. 1999; Le Gouellec and Elimelech 2002; Lin et al. 2011). It has been long speculated that gypsum might nucleate on mineral surface during the mineral flotation process, which could change the hydrophobicity of the minerals and influence the response of target minerals to flotation reagents. Calcium sulphate precipitates were detected on the surface of galena sample from the Hilton concentrator (Grano et al. 1997b; Grano 1995). It has been implied that the flotation rate of galena was retarded due to the presence of an overlayer of precipitated calcium sulphate. The precipitates were speculated to deposit on mineral surface due to hetero-coagulation. Xu and Finch also studied the interactions in the silica and/or sphalerite—Ca-SO_4-CO_3 system using zeta potential measurement and surface analysis techniques (Beauchamp 2006; Rashchi et al. 1998; Sui et al. 1998). However, their results demonstrated no trace of gypsum precipitates on mineral surface. They found that the zeta potential of silica and sphalerite in Ca and $CaSO_4$ solution were similar. The adsorption of Ca in the electric double layer on mineral surface was believed to be responsible for the sulphide mineral flotation problems in the gypsum saturated process water. It has to be noted that the gypsum is considerably soluble (Gardner and Glueckauf 1970) (2.4 g/L at 25°C) and the formation of calcium sulphate precipitates depends on many factors. Therefore, the sample preparation procedure for the surface analysis is extremely important. It was noticed that in many of the previous studies, the sample rinsing with water was applied after conditioned in the $CaSO_4$ solutions. It is possible that this procedure would wash away the gypsum precipitates resulting in misleading surface analysis results. The adverse impact of calcium sulphate can be offset by the addition of sodium carbonate or bicarbonate. Grano and his coworkers

(Grano et al. 1997a) have shown that the flotation of galena has been improved significantly at carbonate concentrations greater than 6.2×10^{-3} M. However, the concentration of sodium carbonate should also be controlled (Rao 2004).

The high TDS water with high concentrations of Ca and SO_4 ions in mineral process water can have adverse effect by one or more mechanisms. It has been speculated that the gypsum would crystallize on mineral surfaces, resulting in reduced flotation selectivity and recovery. In this study, gypsum supersaturated solution is used to represent the high TDS water. Focus is placed on the impact of high Ca and SO_4 ions concentrations in mineral process water on silica and sphalerite mineral surface properties. This study is aiming to fundamentally understand the adverse effect of gypsum supersaturated process water on the flotation performance of zinc sulphide ore. Zeta potential distribution measurements and scanning electron microscopy (SEM) imaging are applied to examine the mineral surface properties in the gypsum supersaturated solution. The impact of gypsum supersaturated solution on mineral surface properties and flotation behavior is discussed.

EXPERIMENTS

Material and Reagentst

The sphalerite sample was obtained from Ward's Natural Science Establishment. The sample was crushed to –4.75 mm and any impurities present were hand sorted. The sample was then dry ground to –45 µm. Silica (Min-u-sil 5, 98.3% SiO_2, 97% –5µm) from U.S. Silica was used for the zeta potential distribution measurements.

The following reagents used were A.C.S. reagent grade: sodium hydroxide, NaOH (Fisher Scientific) and hydrochloric acid, HCl (Fisher Scientific), for pH adjustment; potassium chloride, KCl (Fisher Scientific), for preparing supporting electrolyte solution. Commercial grade calcium sulfate, hemi-hydrate ($CaSO_4\times\frac{1}{2}H_2O$) from US Gypsum Corporation was used as supplied.

Methodology

Gypsum Supersaturated Solution Preparation

In this research, laboratory supersaturated solution of gypsum was prepared with calcium sulfate, hemi-hydrate ($CaSO_4\times\frac{1}{2}H_2O$). The solution was prepared using deionized water from a Millipore Elix 5 purification system, and pH was adjusted using HCl and NaOH.

A new solution was prepared for each day of testing. 10 grams of calcium sulfate hemi-hydrate was dissolved in 1L of Milli-Q water. The pH value of the solution was maintained at 10. The mixture was then stirred under constant stirring with a combination of electric hot plate/magnetic stirrer at room temperature for 30 minutes, and was then filtered with a 1 mm filter to remove undissolved solids. Atomic adsorption analysis results indicate that Ca ions concentration in the gypsum supersaturated solution is 845 ppm, which translates to 3612 ppm gypsum (solubility in water: 2400 ppm at 25°C). The calcium concentration in this laboratory gypsum supersaturated solution is similar to that in the processing water of Teck Cominco's Red Dog operation (840 ppm).

Suspension Preparation

Suspensions of silica were analyzed in a supporting electrolyte solution (0.01 M KCl) and in gypsum supersaturated solution. A stock silica suspension was prepared by adding 1 g of Min-u-sil 5 to 50 mL Milli-Q water. The sphalerite was ground with an agate mortar and pestle until the entire sample passed a 325 mesh (45 μm) sieve. One gram of fine sphalerite was ground in the mortar with a few drops of water for 15 min. and then transferred to 50 mL Milli-Q water as a stock suspension. A fresh batch of sphalerite was prepared for each testing. The gypsum stock solution was prepared by adding 1 g of gypsum into 50 mL Milli-Q water.

Zeta Potential Distribution Measurement

Zeta potential distribution measurements have been recently used to study the interactions between two different particles by comparing the zeta potential distribution of the mix of the two different components and the single material system (Liu et al. 2004a; Liu et al. 2004b; Liu et al. 2005; Liu et al. 2002). With this method, interactions between two different particles could be probed. A homogenous suspension containing a single type of particles will have a certain zeta potential distribution centered at a mean value. Zeta potential distributions of the mixtures can be compared to those of the individual components to determine the nature of the particle interactions. In the absence of attractions, the peak for each individual species will appear in the distributions for the mixture. The distributions may be slightly shifted towards each other due to electrokinetic retardation or enhancement. Hetero-coagulation between the two types of particles will yield a zeta potential distribution centered in between the peaks obtained from the individual species, depending on the extent of surface coverage. With this method, it is possible to evaluate the surface precipitation/slime coating by measuring the zeta potential distribution of minerals in the gypsum supersaturated solution. The advantage of the zeta potential distribution measurement technique is that tests are performed in situ, avoiding complex changes that arise when the minerals are dried, as required for most ex situ characterizations. In this study, the zeta potential distributions were measured using a SEPHY/CAD Zetaphorometer III equipped with a rectangular quartz electrophoresis cell. A pair of reversible hydrogenated palladium electrodes, laser illumination, and a digital video capture system was operated through a Canon microscope. The microscope offers a very sharp 1.07 μm depth of focus for accurately viewing a stationary layer with little background interference. The system also employs sophisticated software which allows for accurate location of the stationary layers, measurement of electrophoretic mobility, and computation of the corresponding zeta potential distributions. The system outputs histograms of zeta potential distributions were computed from mobility measurements using the Smoluchulski equation (Shaw 1992).

Each suspension was prepared to an optimal concentration on the order of 0.1 to 1 g solids per liter so that between 20 and 150 particles could be tracked by the software during a mobility measurement. All suspensions were analyzed at room temperature to minimize convective effects in the electrophoresis cell. To ensure representative measurements, suspensions of between 80 and 200 mL were prepared. The cell was flushed and replenished for every second measurement. At least five mobility measurements in each direction were recorded for each sample, and all zeta potential distributions were measured at pH 10.

A certain amount of dry calcium sulfate was added into the supporting electrolyte solution (0.01 M KCl) to obtain a baseline zeta potential distribution. The zeta potential distributions of gypsum in its own supersaturated solutions were measured as a baseline.

For a single mineral zeta potential distribution measurement, stock suspension of either silica or sphalerite (0.5 mL) was added to 100 mL of supporting electrolyte solution or gypsum supersaturated solution to achieve an appropriate concentration. For the mixture of silica or sphalerite and gypsum zeta potential distribution measurements, 0.25 mL of silica or sphalerite stock suspension and 0.25 mL of gypsum stock suspension was added to 100 mL of gypsum supersaturated solution.

Surface Characterization

Silica/sphalerite samples (1 g) were conditioned with 100 mL gypsum supersaturated solution (pH = 10) at room temperature for 30 min. The solids were then collected by filtration, washed with ethanol and vacuum dried. For contrast, another silica/sphalerite samples were conditioned in Milli-Q water following the same procedure. Precipitates from the gypsum supersaturated solution were also collected for analysis. The samples were examined using SEM (JAMP-9500F).

RESULTS

Surface Properties of Silica and Sphalerite in Gypsum Supersaturated Solution

As baselines, zeta potential distributions of silica and sphalerite in supporting electrolyte solution (0.01 mol/L KCl), gypsum particles in supporting electrolyte solution and its own supersaturated solutions were measured. In the supporting electrolyte solution, the zeta potential distributions of silica and sphalerite have peaks centered at values of –51 mV and –27 mV, respectively (Figure 1a-a, 1b-a). These values are similar to the average zeta potential values reported for silica and sphalerite in similar supporting electrolyte solutions at similar pH (Beauchamp R.M. 2006; DiFeo et al. 2001; Duarte and Grano 2007; Rashchi et al. 1998; Sui et al. 1998). Zeta potential distribution of gypsum particles in supporting electrolyte solution (0.01 mol/L KCl) is similar to that in its own supersaturated solution, having a peak at the value around +5 mV. This value is slightly different from the values reported by others for gypsum under similar pH (Rashchi et al. 1998; Sui et al. 1998), which might be due to the different procedures in sample preparation.

When dispersed in the gypsum supersaturated solution, silica and sphalerite exhibit similar zeta potential distributions. The zeta potential distribution peaks of both silica and sphalerite shift to a much less negative value, –11 mV (Figure 1a-b, 1b-b). Electric double layer compression due to increased electrolyte concentration and the valence of Ca may well account for the shift of zeta-potential to less negative value.

Needle shape gypsum precipitates are found on both silica and sphalerite mineral surfaces after conditioned in the gypsum supersaturated solution for 30 min, as show in the typical SEM images (Figure 2). However, the gypsum precipitates are not seemed to be growing on the mineral surfaces since the size of the needle-shaped precipitates are quite big and silica/sphalerite surfaces stay very clean.

Effect of Ca on the Surface Properties of Silica and Sphalerite Minerals

In order to better understand the effect of gypsum supersaturated solution on silica and sphalerite mineral surfaces, the zeta potentials of these minerals in KCl solution as functions of Ca ion concentration and pH are investigated. The results are shown in Figure 3.

Ca ion has significant effect on the zeta potential of silica and sphalerite. Zeta potentials of both silica and sphalerite generally increase with Ca ion concentration under tested pHs. At pH 11.5, while specific adsorption of Ca ions on the mineral surfaces happens, the sign of both silica and sphalerite surface charge converts to positive value when the Ca ions concentration

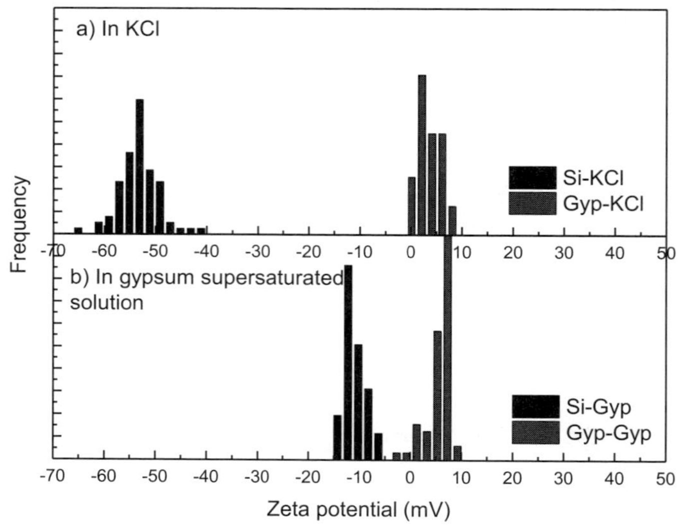

a) Silica

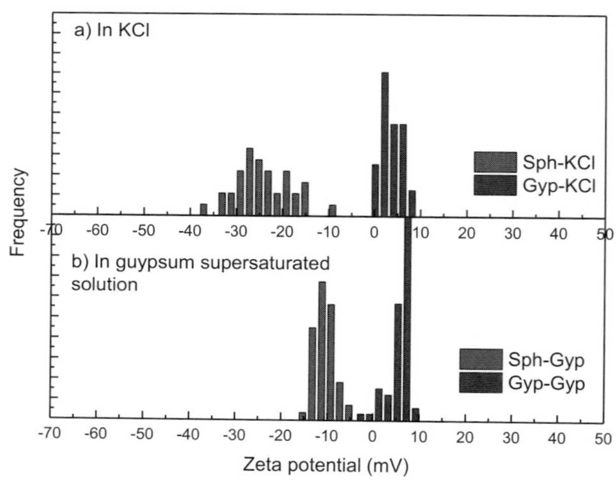

b) Sphalerite

Figure 1. Zeta potential distribution of (a) silica and (b) sphalerite in gypsum supersaturated solutions and supporting electrolyte solution (0.01 mol/L KCl) at pH 10

is sufficient high. This phenomenon doesn't exhibit at pH 10 even though the concentration of Ca ions increased up to 800 ppm.

It is interesting that at pH 10, the zeta potential of both silica and sphalerite shift to about –13 mV in the presence of 800 ppm Ca ions, which is similar to the Ca ions concentration of gypsum supersaturated solution, 845 ppm. This value is pretty similar to the zeta potential of

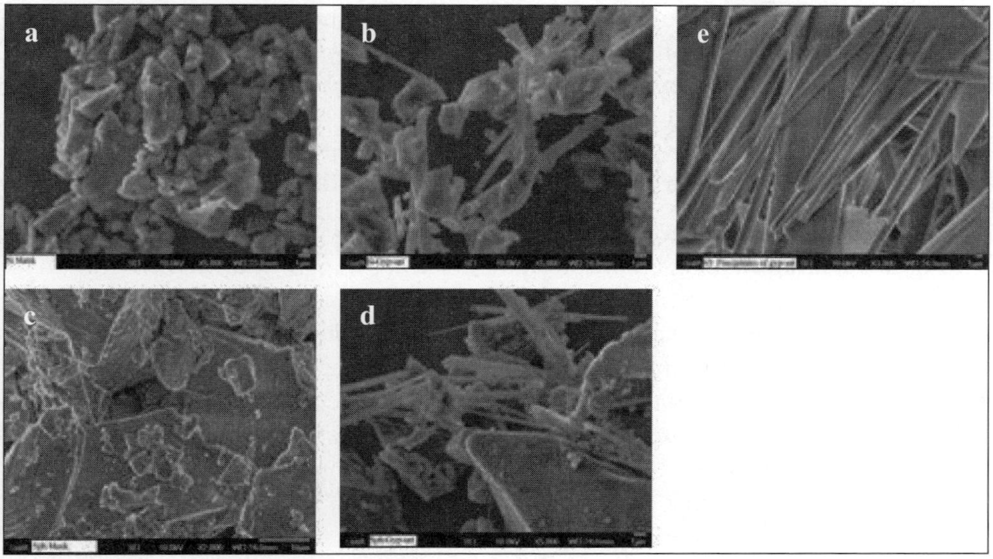

Figure 2. Typical SEM micrographs of (a) silica in water, (b) silica conditioned with gypsum supersaturated solution, (c) sphalerite in water, (d) sphalerite conditioned with gypsum supersaturated solution, and (e) gypsum precipitates

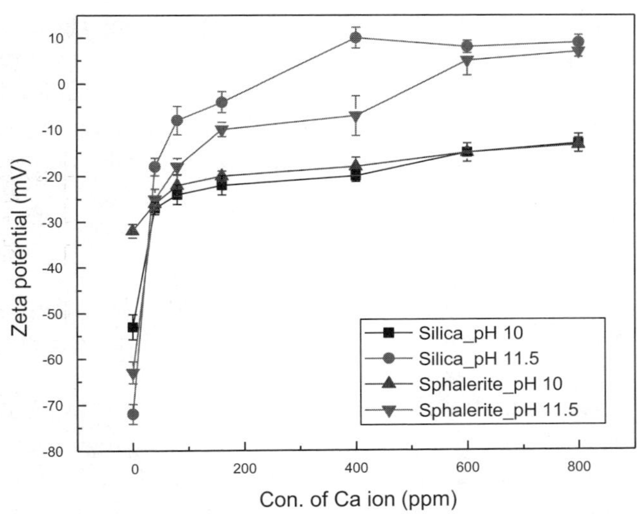

Figure 3. Effect of calcium ion concentration on the zeta potential of silica and sphalerite in 10^{-2} mol/L KCl solution at pH 10 and pH 11

silica and sphalerite in the gypsum supersaturated solution. These results suggest that the shifts of zeta potential distributions of silica and sphalerite in the gypsum supersaturated solution are mainly attributed to the compression of double layer due to the high Ca ions concentration in the supersaturated solution and the adsorption of Ca.

Effect of Gypsum Particles on the Surface Properties of Silica and Sphalerite Minerals

Our previous surface characterization results suggest that gypsum precipitates form in the supersaturated solution. Though those precipitates seemed not growing on the silica and sphalerite mineral surfaces, it is speculated that the positive charged gypsum particles might hetero-aggregate with the negative charged silica and sphalerite minerals. In order to understand the interactions between gypsum precipitates and silica and sphalerite particles, zeta potential distributions of silica and gypsum particle mixture, and sphalerite and gypsum particle mixture were investigated. The results are shown in Figure 4.

For a single mineral in the gypsum supersaturated solution, the zeta potential distribution of gypsum particles has a peak at the value around +5 mV, while silica and sphalerite have identical zeta potential at around –11 mV. The zeta potential distribution of silica and gypsum mixture has two peaks. One dominant peak at the value –5 mV (Figure 4a-a) is close to the zeta potential of silica in the gypsum supersaturated solution. The other slender peak is at around +2.5 mV, which is close to the zeta potential of gypsum particles in its own supersaturated solution. Similar phenomenon is found in the case of sphalerite and gypsum mixture in the gypsum supersaturated solution. It is very interesting that the zeta potential distributions of both silica-gypsum mixture and sphalerite-gypsum mixture exhibit two peaks. Though, in the mixture system, both the negative side peaks, representing the zeta potential of silica and sphalerite, and the positive side peak, which denotes the zeta potential of gypsum, shift toward 0 mV in compare with the single mineral system. These results imply that there is only very weak attraction between silica/sphalerite minerals and gypsum precipitates in the gypsum supersaturated solution. The hetero-aggregation between minerals and gypsum precipitates is minimal.

DISCUSSION

It has been speculated that gypsum precipitates could form in the gypsum supersaturated solution and silica and sphalerite might serve as the seeds for the precipitates. However, in many studied, SEM and XPS failed to detect any gypsum precipitates on those minerals (Rashchi et al. 1998; Sui et al. 1998). In this study, we used the gypsum supersaturated solution to study the gypsum scaling on mineral surfaces. According to our zeta potential distribution measurements results, silica and sphalerite minerals are indiscriminately coated by Ca in the gypsum supersaturated solution. SEM imaging found needle shape gypsum precipitates on silica and sphalerite minerals conditioned with gypsum supersaturated solution. One interesting discovery from this study is that gypsum precipitates indeed form in the system but those precipitates are not growing on the mineral surfaces but from bulk solution. Moreover, the zeta potential distribution results of the mineral-gypsum particle mixture system imply that the hetero-aggregation between the minerals and gypsum precipitates are not significant.

Ca ions in the system play a significant role on the minerals surface properties, the zeta potential. The adsorption of Ca on sphalerite has been reported (Moignard 1977). The changed surface properties of silica and sphalerite minerals in the gypsum supersaturated solution might contribute to the poor selectivity between those two minerals in real flotation system. Further

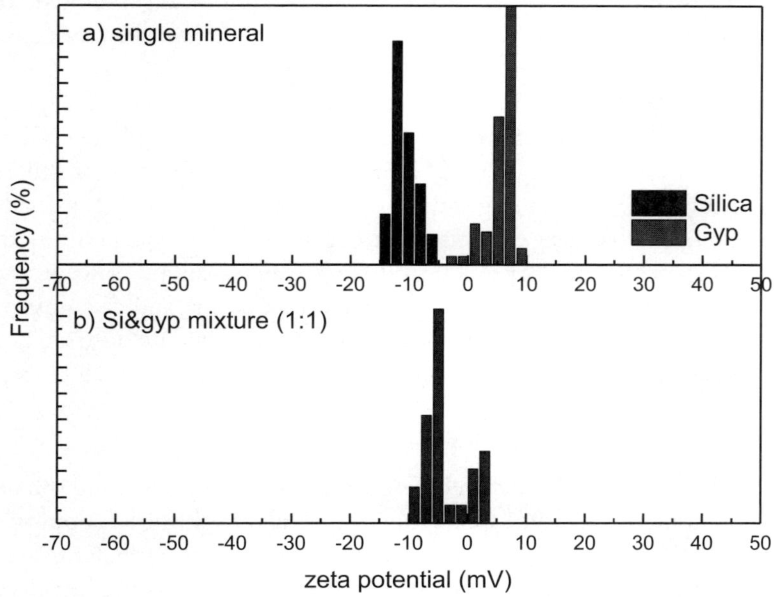

a) Silica and gypsum mixture

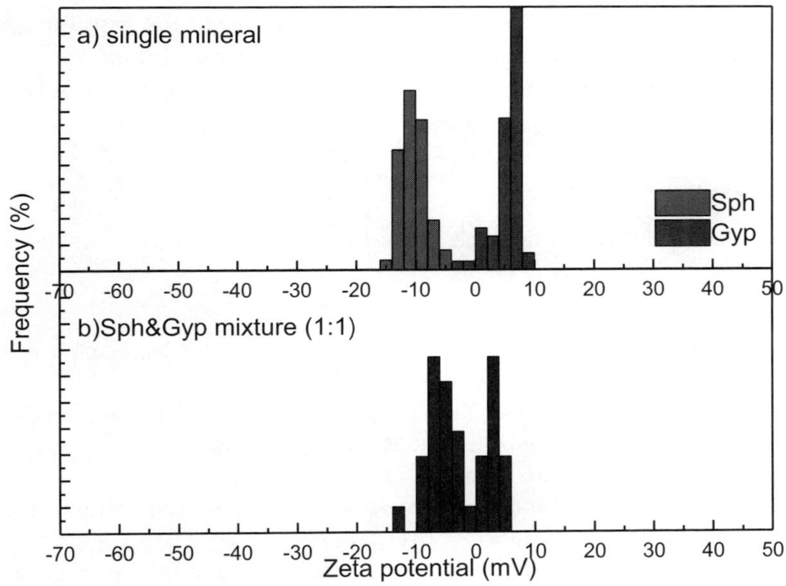

b) Sphalerite and gypsum mixture

Figure 4. Zeta potential distributions of silica and gypsum mixture, sphalerite and gypsum mixture in gypsum supersaturated solution at pH 10

investigations on the impacts of gypsum supersaturated solution on the interactions between flotation reagents and silica/sphalerite minerals and the interactions between silica and sphalerite will be conducted in the future.

An effective approach to eliminate the detrimental effect of both Ca ions and gypsum precipitates on flotation is to add carbonate (e.g., soda ash) to precipitate Ca ions as calcium carbonate or convert calcium sulfate to calcium carbonate. In the case of Hilton concentrator, the addition of soda ash to the grinding circuit significantly improved galena flotation (Grano 1995). Soda ash has also been used for control of Ca in sphalerite flotation (Nesset 1998). Surface analysis indicated a reduction in both Ca and sulfate concentration on galena with the addition of carbonate and it was speculated that galena and calcium carbonate ware mutually repelled (Grano 1995). In other studies, calcium carbonate didn't appear to disperse completely as some precipitates remained on the surfaces (Rashchi et al. 1998; Sui et al. 1998).

CONCLUSION

The gypsum supersaturated solution has significant effects on silica and sphalerite surfaces.

1. Zeta potentials of silica and sphalerite become identical. In the supporting electrolyte solution (0.01 mol/L KCl), the zeta potentials of silica and sphalerite are –51 mV and –27 mV at pH 10, respectively. However, these values become identical and shift to –11 mV in the gypsum supersaturated solution under the same pH.
2. Both silica and sphalerite are indiscriminately coated by Ca when conditioned in the gypsum supersaturated solution.
3. Needle shaped gypsum precipitates are found in silica and sphalerite samples conditioned with gypsum supersaturated solution. However, the gypsum precipitates are not growing on the mineral surface but from the bulk solution.
4. The hetero-aggregation between gypsum precipitates from the bulk solution and silica/sphalerite mineral is insignificant.

ACKNOWLEDGMENT

The authors thank NSERC and Teck Metal Resources Ltd. for their financial support.

REFERENCES

Agamaliev, M.M., Krikun, M.M., and Nasibov, A.G. 1993. Solubility of calcium sulphate in concentrates of mineralized waters. *Khimiya i Tekhnologiya Vody*, 15(1):44–50.

Beauchamp, R.M., Choung, J.W., and Xu Z. 2006. Mineral particles interactions in gypsum-supersaturated process water. *The* 6th *UBC-McGill-UA International Symposium,* 45th *Annual Conference of Metallurgists of CIM,* Montreal, Quebec, Canada.

DiFeo, A., Finch, J.A., and Xu, Z. 2001. Sphalerite-silica interactions: Effect of pH and calcium ions. *International Journal of Mineral Processing*, 61(1):57–71.

Duarte, A.C.P., and Grano, S.R. 2007. Mechanism for the recovery of silicate gangue minerals in the flotation of ultrafine sphalerite. *Minerals Engineering*, 20(8):766–775.

Dutrizac, J.E. 2002. Calcium sulphate solubilities in simulated zinc processing solutions. *Hydrometallurgy*, 65(2-3):109–135.

Fahiminia, F., Watkinson, A.P., and Epstein, N. 2007. Early events in the precipitation fouling of calcium sulpfate dihydrate under sensible heating conditions. *Canadian Journal of Chemical Engineering*, 85(5):679–691.

Flint, O. 1968. Increased solubility of calcium sulphate in sea water containing hydrochloric acid. *Desalination*, 4(3):328–335.

Furby, E., Glueckauf, E., and McDonald, L.A. 1968. The solubility of calcium sulphate in sodium chloride and sea salt solutions. *Desalination*, 4(2):264–276.

Gardner, A.W., and Glueckauf, E. 1970. Thermodynamic data of the calcium sulphate solution process between 0 and 200°C. *Transactions of the Faraday Society*, 66:1081–1086.

Grano, S.R., Johnson, N.W., and Ralston, J. 1997a. Control of the solution interaction of metabisulphite and ethyl xanthate in the flotation of the Hilton ore of Mount Isa Mines Limited, Australia. *Minerals Engineering*, 10(1):17–39.

Grano, S.R., Lauder, D.W., Johnson, N.W., and Ralston, J. 1997b. An investigation of galena recovery problems in the Hilton concentrator of Mount Isa Mines Limited, Australia. *Minerals Engineering*, 10(10):1139–1163.

Grano, S.R., Wong, P.L.M., Skinner, W., Johnson, N.W., and Ralston, J. 1995. Detection and control of calcium sulphate precipitation in the lead circuit of the Hilton concentrator of Mount Isa Mines Limited, Australia. XIX International Mineral Processing, Colorado, USA, 171–179.

Greenlee, L.F., Lawler, D.F., Freeman, B.D., Marrot, B., and Moulin, P. 2009. Reverse osmosis desalination: Water sources, technology, and today's challenges. *Water Research*, 43(9):2317–2348.

Hasson, D., Drak, A., and Semiat, R. 2001. Inception of $CaSO_4$ scaling on RO membranes at various water recovery levels. *Desalination*, 139(1-3):73–81.

Hoang, T.A., Ang, H.M., and Rohl, A.L. 2007. Effects of temperature on the scaling of calcium sulphate in pipes. *Powder Technology*, 179(1-2):31–37.

Hoang, T.A., Ang, H.M., Rohl, A.L., and Jeffrey, M.I. 2006. A study of gypsum scale formation using quartz crystal microbalance. *Developments in Chemical Engineering and Mineral Processing*, 14(1-2):313–321.

Johnson, N.W. 2003. Issues in maximisation of recycling of water in a mineral processing plant. *Proceedings Water in Mining*, Melbourne, 239–246.

Klepetsanis, P.G., Dalas, E., and Koutsoukos, P.G. 1999. Role of temperature in the spontaneous precipitation of calcium sulfate dihydrate. *Langmuir*, 15(4):1534–1540.

Lambert, B., and Schaffer, R.J. 1926. CCCXLVIII–Studies of precipitated solids. Part II. Calcium sulphate. *Journal of the Chemical Society (resumed)*, 129:2648–2655.

Lascelles, D., Finch, J.A., and Sui, C. 2003. Depressant action of Ca and Mg on flotation of Cu activated sphalerite. *Canadian Metallurgical Quarterly*, 42(2):133–140.

Le Gouellec, Y.A., and Elimelech, M. 2002. Calcium sulfate (gypsum) scaling in nanofiltration of agricultural drainage water. *Journal of Membrane Science*, 205(1-2):279–291.

Levay, G., and Schumann, R. 2006. A systematic approach to water quality management in the minerals processing industry. Brisbane, QLD, Australia, 277–287.

Levay, G., Smart, R.S.C., and Skinner, W.M. 2001. The impact of water quality on flotation performance. *Journal of the South African Institute of Mining and Metallurgy*, 101(2):69–75.

Lin, N.H., Shih, W.-Y., Lyster, E., and Cohen, Y. 2011. Crystallization of calcium sulfate on polymeric surfaces. *Journal of Colloid and Interface Science*, 356(2):790–797.

Liu, J., Xu, Z., and Masliyah, J. 2004a. Interaction between bitumen and fines in oil sands extraction system: Implication to bitumen recovery. *Canadian Journal of Chemical Engineering*, 82(4):655–666.

Liu, J., Xu, Z., and Masliyah, J. 2004b. Role of fine clays in bitumen extraction from oil sands. *AIChE Journal*, 50(8):1917–1927.

Liu, J., Xu, Z., and Masliyah, J. 2005. Interaction forces in bitumen extraction from oil sands. *Journal of Colloid and Interface Science*, 287(2):507–520.

Liu, J., Zhou, Z., Xu, Z., and Masliyah, J. 2002. Bitumen-clay interactions in aqueous media studied by zeta potential distribution measurement. *Journal of Colloid and Interface Science*, 252(2):409–418.

Liu, L., Rao, S.R., and Finch, J.A. 1993. Technical note laboratory study of effect of recycle water on flotation of a Cu/Zn sulphide ore. *Minerals Engineering*, 6(11):1183–1190.

Liu, Q., and Zhang, Y. 2000. Effect of calcium ions and citric acid on the flotation separation of chalcopyrite from galena using dextrin. *Minerals Engineering*, 13(13):1405–1416.

Moignard, M.S., James, R.O. and Healy, T.W. 1977. Adsorption of calcium at the zinc sulphide-water interface. *Australian Journal of Chemistry,* 33:733–740.

Nesset, J.E., Sui, C., Kim, J.Y., Cooper, M., Li, M., and Chryssoulis, S.L. 1998. The effect of soda ash and lime as pH modifiers in sphalerite flotation. 30th Annual CMP Meeting, Ottawa, Ontario, 460–481.

Rao, S.R. 2004. *Surface Chemistry of Froth Flotation.* New York: Kluwer Academic/Plenum.

Rashchi, F., Xu, Z., and Finch, J.A. 1998. Adsorption on silica in Pb- and Ca-SO_4-CO_3 systems. *Colloids and Surfaces A: Physicochemical and Engineering Aspects,* 132(2-3):159–171.

Schumann, R., Levay, G., Dunne, R., and Hart S. 2003. Managing process water quality in base metal sufide flotation. *Proceeding of Water in Mining 2003,* Melbourne, Australia, 251–259.

Seewoo, S., Van Hille, R., and Lewis, A. 2004. Aspects of gypsum precipitation in scaling waters. *Hydrometallurgy,* 75(1-4):135–146.

Shams El Din, A.M., El-Dahshan, M.E., and Mohammed, R.A. 2005. Scale formation in flash chambers of high-temperature MSF distillers. *Desalination,* 177(1-3):241–258.

Shaw, D.J. 1992. *Introduction to Colloid and Surface Chemistry.* Butterworth Heinemann.

Sudmalis, M., and Sheikholeslami, R. 2000. Coprecipitation of $CaCO_3$ and $CaSO_4$. *Canadian Journal of Chemical Engineering,* 78(1):21–31.

Sui, C., Rashchi, F., Xu, Z., Kim, J., Nesset, J.E., and Finch, J.A. 1998. Interactions in the sphalerite-Ca-SO_4-CO_3 systems. *Colloids and Surfaces A: Physicochemical and Engineering Aspects,* 137(1-3):69–77.

Woodcock, J.T., and Jones, M.H. 1970. Oxygen concentrations, redox potentials, xanthate residuals and other parameters in flotation plant pulps. London: Institution of Mining and Metallurgy.

Xu, Z., Chi, R., Difeo, T., and Finch, J.A. 2000. Surface forces between sphalerite and silica particles in aqueous solutions. *Journal of Adhesion Science and Technology,* 14(14):1813–1827.

Desalination of Coal Mine Water with Electrosorption

Xiaowei Sun
EST Water and Technologies Co., Ltd., China

Jiann-Yang Hwang
Michigan Technological University, Houghton, MI, USA

ABSTRACT

Desalination of coal mine water has been found necessary when the concentration of some specific elements exceeds the regulation or when the water needs to meet the salt concentrations of downstream applications. In this study, the coal mine water from a Chinese coal mine in Shangdong is desalinated to supply the cooling water needs for a nearby power plant. The re-circulated cooling water requires injecting a certain amount of fresh water to replace the amount of water evaporated during the cooling process and to lower the salt concentration increased due to the evaporation. The conductivity of the coal mine water is above 3000 μS/cm. The target is below 2000 μS/cm. The capacity requirement is 8000 m^3 per day. Electrosorption technology was selected for the installment, due to its capability to adjust the salt content of the effluent water, the low energy consumption, the tolerance to the organic content in the mine water and the economics. The results of the tests are reported in this study.

INTRODUCTION

Water shortage is a worldwide problem of the 21st century. The World Bank has reported that 80 countries have water shortages that threaten health and economies while 40 percent of the world—more than 2 billion people—have no access to clean water [1]. In the United States, the water shortage is also experienced in many regions. This issue is not limited to the western United States. States such as Alabama, Florida, Hawaii, Illinois, Louisiana, Missouri, South Carolina, Maryland and Virginia have all expressed concerns [2]. The recent droughts in Georgia and the Pacific Northwest have made the situation worse.

According to the USGS, 346 billion gallons of freshwater were withdrawn per day in the United States in the year 2000. Agricultural irrigation, the largest use, accounted for 40% of freshwater withdrawals. The second largest use, thermoelectric power generation, withdrew 39% or 136 billion gallons per day (BGD), followed by public supply, industrial uses, aquaculture, domestic use, mining, and livestock [3].

Power demand will increase with population and economic growth in the future. A Department of Energy analysis estimates that thermoelectric power capacity will increase from 700 GW in 2005 to 860 GW in 2030. The average daily national freshwater consumption resulting from U.S. thermoelectric power generation could increase from 6.1 BGD in 2005 to 8.2 BGD by 2030 [2].

The freshwater supply from the nature is limited. Growing concerns about freshwater availability must be reconciled with growing demand for power if the United States is to maintain economic growth and current standards of living. With the shortages of water, choices will have to be made regarding withdrawal and consumption of this natural resource. Agriculture and public supply will most likely be the greatest competitors due to their large water withdrawal. As with all resources, tradeoffs will occur, and concerns will increasingly be raised over which use is more important: water for drinking and personal use, growing food, or energy production.

Power generation facilities will have increasing difficulties siting new plants due to water concerns. Concurrently, existing plants will be under increasing pressure to reduce their water withdrawal and consumption. Arizona recently rejected permitting for a proposed power plant because of concerns about how much water it would withdraw from a local aquifer. In an article discussing a 1,200 MW proposed plant in Nevada, opposition to the plant stated, "There's no way Washoe County has the luxury anymore to have a fossil-fuel plant site in the county with the water issues we now have. It's too important for the county's economic health to allow water to be blown up in the air in a cooling tower." [2, 4–6].

It is necessary to develop cost-effective approaches to using non-traditional (aka impaired or alternative) sources of water to supplement or replace freshwater for cooling and other power plant needs. Examples of non-traditional waters could include mine water, coal-bed methane produced waters, produced water from geological CO_2 sequestration, and industrial and/or municipal wastewater.

The water shortage problem is much worse in China. While it is under discussion in the U.S. about using non-traditional water to supplement fresh water for power plant, the situation in China has reached the stage that this has to be implemented in some areas. It is the trend that new power plants are built near coal mine. Coal mine water is, therefore, a potential supply of the power plant cooling water. This has been investigated and the results are favorable. A commercial case study is presented here.

TECHNOLOGICAL CHALLENGES

There are many concerns from the power plants regarding cooling water quality. If the quality of water does not meet the power plant's requirements, cooling efficiency can be reduced, equipment can be damaged, and operation and maintenance costs can be increased significantly.

Scales, precipitates and other organic and inorganic solids can be present in the cooling water. Inorganic particles can precipitate out of the raw water in the form of various carbonates, phosphates, silicates and others when the saturation limit of their salts is exceeded [7]. Organic particles such as algae and bacteria or various slimes can be produced by a number of organisms found in cooling water when nutrients are available. Rust flakes can break off the walls of the piping in the system and add to the particulate load. The heat source itself can add particulates, especially if the cooling water comes in contact with the heat source as in quenching and de-scaling operations.

The scales can reduce the heat exchange efficiency and consequently require more cooling water and pumping energy since the power plant has to remove the heat at a required rate. Nozzles in counter flow cooling towers have very small orifices and can become easily plugged with organic or inorganic solids. This results in irregular distribution of water across the tower cross-section. This can lead to uneven cooling and loss of efficient transfer of heat to the atmosphere. Many towers contain packing to form a thin liquid film increasing the surface area of the water/air interface for more efficient heat transfer. Suspended solids can block passages in this packing thus decreasing the surface area for heat transfer and forming a more conducive substrate for the growth of biological materials such as algae, mosses, and even pathogens [7].

Power plants usually control the following parameters as a minimum to control the formation of scales and other organic and inorganic solids, including the total dissolved solids

(TDS), total suspended solids (TSS), chemical oxygen demand (COD), calcium, magnesium, sulfate, silica, and chloride content, and pH. Among these parameters, TDS, calcium, magnesium, sulfate and silica contents are utilized to regulate the formation of scales and precipitates such as calcium sulfate. Chloride concentration is important for the corrosion of metals in the system, such as heat exchangers and pipes. COD is an important factor to regulate the nutrients of organisms. TSS is controlled to prevent the introduction of too much suspended solids into the system. pH is an important factor to control the calcium carbonate scale formation. In addition to these parameters, power plants use continuous chlorination to maintain a residual of 0.1 to 0.2 mg/L Cl_2 in the cooling tower circulating water. A phosphonate-type inhibitor is used for scale control and an azole-type corrosion inhibitor is used to protect the copper-based admiralty brass metallurgy of the main condenser [4, 7].

Power plants use the "Cycles of Concentration" as the major means to avoid the formation of scales. It sets a target for the concentration of salts in the cooling water. For example, the threshold concentration can be set at the 70% of the saturation limit of calcium sulfate to prevent the formation of calcium sulfate scale. When the cooling water reaches this salt concentration due to evaporation, blowdown is enabled to discharge a certain amount of water and fresh water is introduced to lower the salt content. The ratio between the threshold salt concentration and the make-up water salt concentration is the "Cycles of Concentration." Conductivity is usually utilized as the means to control the blowdown since the conductivity commonly corresponds well to TDS and the relationship between TDS and scale forming ions such as calcium sulfate and others ordinarily fluctuate in a limited range.

Based on the discussion on power plant cooling water requirements, it is clear that the challenges for converting coal mine water would be to remove the suspended solids, salt, including TDS, calcium, sulfate and chloride, and the nutrients for bacteria and algae, including nitrogen (both nitrate and ammonia nitrogen) and organics (COD), from the coal mine water.

Suspended solids removal is not the technical challenge. It can be carried out with many types of commercially available mechanical filters. Organics are low in coal mine water. The major technical challenges would be the salt removal (desalination). To achieve the desired "Cycle of Concentration" of the power plant, which is usually above 7 and would preferably be higher (e.g., 10 or higher), the TDS and salts including calcium (hardness), sulfate, chloride, nitrogen (ammonium and nitrate), as a minimum on the list, need to be reduced in an economical way.

ELECTROSORB WATER TREATMENT TECHNOLOGY

Conventional desalination technologies include distillation, ion exchange, and reverse osmosis [8–16]. Compare to these technologies, Electrosorb Technology (EST) can treat coal mine water at lower cost while offers many technical advantages. Also known as the capacitive deionization technology, the EST technology is a new, unique, and revolutionary water purification technology [17–23].

Theory

EST is developed based on the phenomenon that electrodes attract and adsorb ions and charged molecules and particles in water. Purification of water from these contaminants occurs when the electrodes are (i) close to the ions and charged molecules and particles, (ii) have sufficient surface area and force to hold these ions, molecules and particles, and (iii) inert to the reaction.

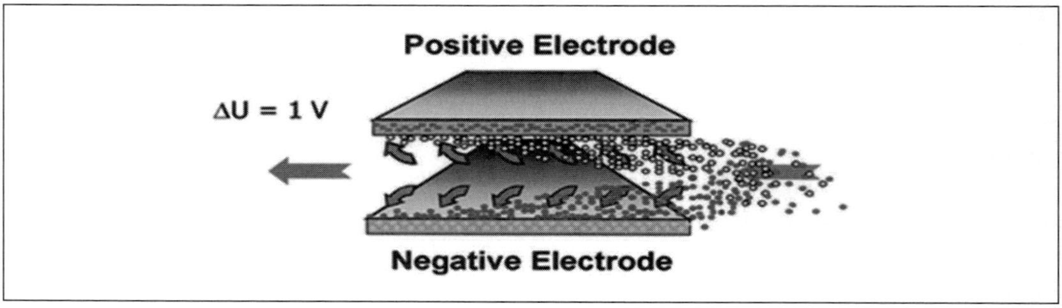

Figure 1. Electrosorb for ion adsorption in water

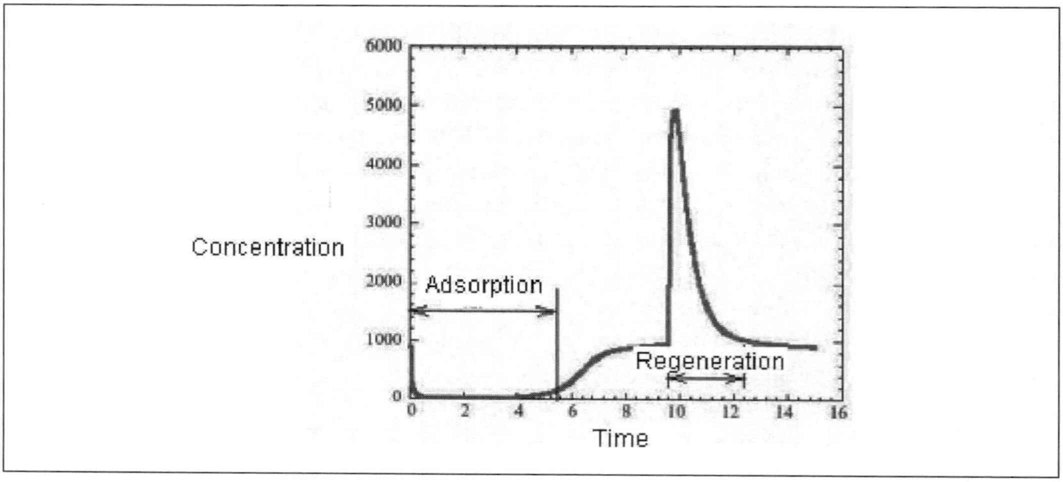

Figure 2. Concentration of salt in water in a typical cycle of EST operation

Figure 1 shows the schematics of the EST mechanism. Water flows in between the electrodes. Ions, charged molecules and particles are attracted to the electrodes of opposite charge. Once they reach the electrodes, they are adsorbed and stored there. Thus, they are separated from clean water (the adsorption stage in Figure 2). When we are finished with the purification, we short the circuit or reverse the charge of the electrodes. These ions, molecules and particles are detached from the electrodes and returned to the flush water (the regeneration stage in Figure 2). The electrodes are, then, regenerated. The amount of ions and other impurities removed can be adjusted with various operation parameters of the system, such as the voltage, flow velocity, retention time, etc.

Unlike other technologies, this technology removes salt out of water instead of removing water out of salt. It does not require high pressure to push water molecules through the membrane or high temperature to evaporate the water molecules. The energy consumption is, therefore, low. It is a physical adsorption process. It does not require the usage of chemicals. This is friendlier to the environment by not introducing more chemicals to the environment during the treatment. It contains no membrane, and hence has no membrane fouling problems. This avoids the use of fouling inhibiting chemicals, eliminating maintenance nightmares and greatly extending equipment life. The final salt concentrations of the water can be adjusted during the process. This allows the users to select ideal processing water quality at minimum cost.

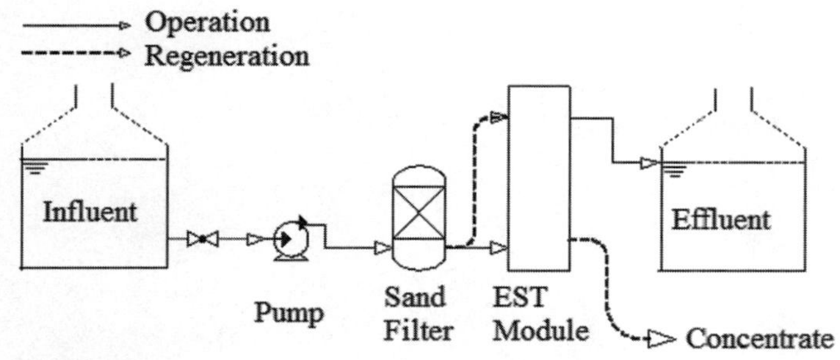

Figure 3. EST operation

Advancement of EST Technology

Although there have been many efforts to commercialize the capacitive deionization technology, EST Water and Technologies has reached the capability to offer water treatment at large industrial capacity. EST uses a module approach to manufacture the adsorption electrodes. It is able to mass produce electrode modules at 10 ton per hour (63,492 GPD), which is well beyond the capability of other developers. Depending on the plant capacity requirement, EST can quickly design and install a plant usually within months. This provides the industries with water treatment demands a new choice that can better meet their needs based on various technical, economic and environmental criteria.

APPLICATIONS TO COAL MINE WATER

A commercial plant has been designed and built to treat the coal mine water from the Jining No. 3 Mine of the Yanzhou Coal Mine Company, Yankuang Group in Shangdong, China. Shangdong is a Province located on the eastern coast of China. It has a population of about 100 million and is the most affluent province in China. Water shortage and water pollution are serious problems in this province, which strongly influence not only socio-economic development, but also ecology and quality of life. It is an outstanding example for water conflicts arising from piece meal action as well as from fast growing population, industry and agriculture. Crucial points are water shortage, the pollution of water resources, and the associated intrusion of seawater into groundwater resources.

Some power plants in Shangdong are unable to obtain fresh water for its cooling water usage and have to look at alternatives. The coal mine water is, therefore, investigated to meet this need. In order to achieve this goal, the coal mine water has to be treated to remove salt, hardness, sulfate, etc.

The design is to divert 8000 tons of mine water per day (2.1 MGD) for the cooling water supply. The average conductivity of the coal mine water is 3367 μS/cm. The target is below 2000 μS/cm at a yield of 75%. This requires a conductivity reduction of about 41%.

Figure 3 shows the schematics of the plant. For each EST module, it receives the influent from the pump during the operation mode. A sand filter is placed before the EST module for safety protection of the electrodes. The water runs through the EST electrodes at the voltage of 1.1–1.5 volts. Ions are adsorbed on the electrodes. When the electrodes are saturated with the ions, the voltages are turned off and a small amount of water is passed through the module during regeneration mode, washing the ions out as the concentrate stream. More than one set

Figure 4. Yanzhou coal mine water desalination plant

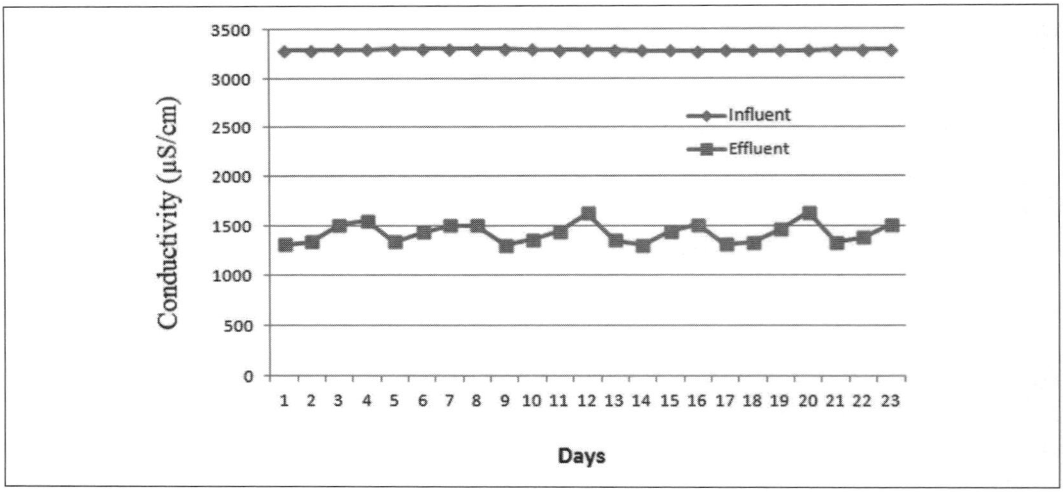

Figure 5. Average daily conductivity measurement of influent and effluent

of EST modules are installed. While one set is under the regeneration mode, the other sets are in operation modes. PLC control has been installed to automate the operations.

A picture of the Yanzhou Coal Mine water desalination plant is shown in Figure 4. In the picture, a series of EST modules can be observed. Each module is connected with several pipes. These pipes guide the influent, effluent and the reject concentrate, using valves and pumps. Conductivity of each stream is continuously measured and the results are fed to the control system. The reading is also shown on the panel in front of the module.

The plant operation has been quite smooth. Figure 5 shows the average daily conductivity measurement for the month of operation in August, 2011. It can be observed that the influent conductivity is averaged around 3300 μS/cm. The effluent conductivity is controlled around

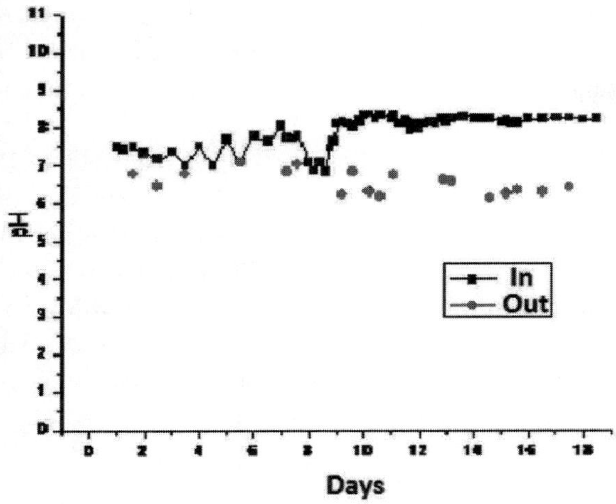

Figure 6. pH measurement of the influent and effluent

Table 1. Water analysis of the influent and effluent

No.	Item	Unit	Influent	Effluent	Removal (%)
1	Conductivity	µS/cm	2980	1450	51.4
2	TDS	mg/L	2030	640	68.5
3	Hardness	mg/L($CaCO_3$)	260	172	33.9
4	Alkalinity	mg/L($CaCO_3$)	372	208	44.1
5	Bicarbonate	mg/L	454	253	44.3
6	Sulfate	mg/L	970	223	78.0
7	Chloride	mg/L	92.5	47.9	48.2
8	Nitrogen	mg/L	0.787	0.682	13.3
9	Turbidity	NTU	2.0	0.2	90.0
10	COD_{Cr}	mg/L	<10	<10	
11	pH		7.4	7.5	

1500 µS/cm. The cooling water specification from the power plant is below 2000 µS/cm. So the specification can be well met with the EST system.

Figure 6 shows the pH measurement during the same period. The influent has pH measured between 6.5 and 8.5. The pH of the effluent is generally between 6 and 7.5, which is a little lower than the influent.

Water samples have been collected and sent to the certified lab for analysis. Table 1 shows the results for the influent and effluent samples obtained recently. The conductivity is reduced from 2980 µS/cm to 1450 µS/cm, equivalent to about 51% removal rate. The total dissolved solids (TDS) removal rate is 68.5% and the sulfate removal rate is 78%. This means that the potential for the scale built up in the cooling water system of the power plant is greatly reduced. The bicarbonate and alkalinity removal rates are about 44%. The chloride is reduced from 92.5 mg/L in the influent to 47.9 mg/L in the effluent, equivalent to about 50% removal.

This is beneficial to the prevention of the corrosion of the cooling water pipes, heat exchangers, etc. The nitrogen content in the coal mine water is low. There is a slight reduction here. High nitrogen (both nitrate and ammonium nitrogen) removal rate has been observed in other EST installations. COD is also low in this water sample. Its reduction is also common in other EST plants.

Significant economic benefit is realized by the power plant. The O&M cost for EST is about \$0.3 per thousand gallons, which is significant lower than that of the RO operation, estimating at about \$1.7 per thousand gallons. The saving is \$1.4 per thousand gallons. For 2.1 million gallons per day, the saving amounts to \$2940 per day or \$1.07 million per year.

Energy benefits are also significant. The energy consumption for EST is about 1.0 kWh per metric ton (3.78 kWh per 1000 gallons) while that for RO is 1.5 kWh per metric ton (5.67 kWh per 1000 gallons). For 2.1 million gallons per day, the energy saving would be 4,000 kWh per day or 1.46 million kWh per year.

CONCLUSIONS

A commercial plant at 8000 tons per day capacity has been built to remove ions from the coal mine water in China using the electrosorb technology. The cleaned water is utilized to supply the cooling water of a nearby power plant. The success of this project demonstrates that it is possible for the power plant to draw its cooling water supply from alternative sources. Analysis of the influent and effluent proves that conductivity, hardness, alkalinity, sulfate, chloride, bicarbonate, and nitrogen can all be reduced. O&M cost is as low as \$0.3 per thousand gallons and energy consumption is about 1 kWH per ton of the water processed.

REFERENCES

[1] Lester R. Brown. 2006. *Rescuing a Planet Under Stress and a Civilization in Trouble.* 2006 Earth Policy Institute. New York: W.W. Norton.

[2] Erik Shuster, Andrea McNemar, Gary J. Stiegel Jr., and James Murphy. 2007. *Estimating Freshwater Needs to Meet Future Thermoelectric Generation Requirements, 2007 Update.* DOE/NETL-400/2007/1304.

[3] USGS (United States Geological Survey). March 2004. *Estimated Use of Water in the United States in 2000.*; USGS Circular 1268.

[4] Kent Zammit. *EPRI Project Manager, Semi-Annual Technical Progress Report, October 1, 2005 to March 31, 2006.* DOE Award No. 41906. Use of Produced Water in Recirculating Cooling Systems at Power Generating Facilities, Deliverable Number 11.

[5] Virginia State Department of Environmental Quality. www.deq.state.va.us/wastewater/generalinfo.html.

[6] Water Environment Federation. Wastewater Treatment. www.wef.org/NR/rdonlyres/59E69C35-0E6F-4593-A4B8-D420AA9C4819/0/WastewaterTreatment912.pdf.

[7] Amiad Filtration Systems. 2006. Cooling towers: How to keep particulates at bay. Filtration and Separation, 43(6):18–22.

[8] Russell Derickson, Fred Bergsrud, and Bruce Seelig. Treatment Systems for Household Water Supplies, Distillation. www.ag.ndsu.edu/pubs/h2oqual/watsys/ae1032w.htm.

[9] Remco Engineering. Ion Exchange. www.remco.com/ix.htm.

[10] Russell Derickson, Bruce Seelig, and Fred Bergsrud. Reverse Osmosis. www.ag.ndsu.edu/pubs/h2oqual/watsys/ae1047w.htm.

[11] Buros, O.K. 1987. An introduction to desalination. In *Non-Conventional Water Resources Use in Developing Countries.* Natural Resources/Water Series No. 22. New York, United Nations, pp. 37–53.

[12] Organization by American States. www.oas.org/dsd/publications/Unit/oea59e/ch20.htm.
[13] California Coastal Commission. Seawater Desalination in California. www.coastal.ca.gov/desalrpt/dchap1.html.
[14] GE Water and Process Technologies. Reverse Osmosis Equipment (Spiral Membranes). www.gewater.com/products/equipment/spiral_membrane/index.jsp.
[15] Talbert N. Eisenberg and E. Joe Middlebrooks. 1992. A survey of problems with reverse osmosis water treatment. *Journal AWWA*, 76(8):44.
[16] J.D. Birkett. 1987. Factors influencing the economics of desalination. In *Non-Conventional Water Resources Use in Developing Countries*. Natural Resources/Water Series No. 22. New York, United Nations, pp. 89–102.
[17] Marc A. Anderson, Ana L. Cuderob, and Jesus Palmab. 2010. Capacitive deionization as an electrochemical means of saving energy and delivering clean water. Comparison to present desalination practices: Will it compete? *Electrochim. Acta*, 55:3845–3856.
[18] Q. Daoduo, Z. Linda, and H. Eric. 2007. *Res. J. Chem. Environ.*, 11:92.
[17] J. Farmer, D. Fix, G. Mack, R. Pekala, and J. Poco. 1996. *J. Appl. Electrochem.* 26:1007.
[18] J.C. Farmer, D.V. Fix, G.V. Mack, R.W. Pekala, and J.F. Poco. 1996. *J. Electrochem. Soc.*, 143:159.
[19] J.C. Farmer, S.M. Bahowick, J.E. Harrar, D.V. Fix, R.E. Martinelli, A.K. Vu, and K.L. Carroll. 1997. *Energy Fuels*, 11:337.
[20] R.W. Pekala, J.C. Farmer, C.T. Alviso, T.D. Tran, S.T. Mayer, J.M. Miller, and B. Dunn. 1998. *J. Non-Cryst. Solids*, 225:74.
[21] T.J. Welgemoed and C.F. Schutte. 2005. *Desalination*, 183:327.
[22] X. Sun. 2002. Electrosorption process and equipment. *Industrial Water Treatment*, No. 8.
[23] X. Sun. 2002. Principle and configuration of electrosorption process. *Industrial Waste and Waste Water*, No. 4.

Impact of Dissolved Gangue Species and Fine Colloidal Matter in Process Water on Flotation Performance

Mukund Vasudevan, Tarun Bhambhani, D.R. Nagaraj, and Raymond S. Farinato
Cytec Industries Inc., Stamford, CT, USA

ABSTRACT

The impact of dissolved non-sulfide gangue species and fine colloidal matter present in process water on flotation performance has been long known, but poorly understood. We present flotation data for two different porphyry Cu ore types A (characterized as 'good') and B (characterized 'bad'). Poor Cu recoveries are obtained when processing ore B and when ore B is blended with ore A. Traditional solutions investigated in the past have yielded only incremental improvements, and even then the recovery values were not commensurate with those expected from liberation analysis. The present investigations using flotation tests, spiking experiments, and water chemistry studies indicate that changes in water quality attributed to both fine colloidal matter as well as soluble mineral species resulted in poor flotation performance. Restoration of flotation performance with the use of an appropriate polymeric modifier is also discussed.

INTRODUCTION

Flotation plants very commonly process multiple ore types—over a spectrum ranging from 'sweet' to 'highly problematic.' One strategy is to process these ore types separately; however, this may not be preferable in some plants as it could result in large swings in value recovery/concentrate throughput. Alternatively, ore Blending strategies can be used to cope with ore type variability in an attempt to reduce the swings in concentrate metallurgy, often with the anticipation that there are no interactions between ore types. However, this is not always the case. For instance, a simple blending strategy would not be effective at yielding desired recoveries/grades when there are antagonistic or poisoning effects between the ore types (for instance recent studies on ultramafic Ni ores by Patra et al. 2010, 2011, Vasudevan et al. 2010, 2011). Some ores, particularly the highly problematic ones, when blended with an otherwise sweet ore, produce deleterious effects on recovery/grade far in excess of those expected from a simple weighted average of the ore types. Traditional approaches to dealing with this ubiquitous and increasingly important problem of antagonistic blending components have included both chemical and processing changes. However, virtually every flotation operation at one time or another finds that these approaches are inadequate. Over the years, we have studied many ore systems with problems discussed above and have focused on determining the underlying mechanisms and developing solutions in practical ore systems rather than merely defining the problems.

We present one case-study concerning flotation beneficiation of two types of porphyry Cu/Mo ores—A and B. Processing of ore A (characterized as 'good' ore type in this manuscript for convenience), leads to high Cu recoveries (nominally >85%) and Mo recoveries typically in the range of 60–80% in the plant (lab recoveries are generally higher). On the other hand, poor Cu/Mo recoveries are obtained when processing ore B (characterized as 'bad' ore type) despite a higher feed Cu grade relative to ore A. When ore B is blended with ore A, the Cu and

Mo recoveries of ore A are reduced. This loss in recovery is likely not entirely related to any inadequacy of liberation. Traditional solutions to the poor recovery problems in the plant, such as the use of different and higher dosages of collectors/frothers, longer flotation time, pulp dilution, addition of NaSH, high intensity conditioning, finer grinding, grinding/re-flotation of tails have provided only incremental improvements, still well below the desired level or that expected from liberation analysis.

Several hypotheses have been proposed to explain the above problems, although to date none of these have been adequately supported. One popular hypothesis has been that the poor Cu and Mo recoveries obtained when processing ore B by itself or as a blend with ore A is due to slime coating of certain deleterious gangue silicate mineral fines, such as those of clays and Mg silicates, on Cu sulfides and molybdenite. The adverse effects of silicates and slimes have been the subject of several other studies (Bulatovic et al. 1999; Chander et al. 1972; Gaudin et al. 1960; Edwards et al. 1980; Fornasiero and Ralston 2006). Other hypotheses have been concerned with certain mineralogical and physical characteristics of Cu sulfides and molybdenite, choice of collector and frother, liberation, pulp density, and kinetics of flotation. None of these hypotheses have sufficient support or confirmation, the underlying mechanism for poor Cu/Mo recoveries is still unclear, and no satisfactory solutions have yet been found. This type of problem is rather ubiquitous and has been known in the industry for a long time.

The main goals of our study were to:

- Develop a scientific understanding of the pathways by which value mineral recoveries are adversely affected by non-sulfide gangue.
- Develop robust, sustainable chemical solutions and process conditions for improving Cu/Mo recovery from difficult-to-process ores and their detrimental effects on the so-called 'good' ore types; i.e., for those cases where we cannot blend away the 'bad' ore.

Our initial hypotheses were: (a) certain gangue silicates (particularly Mg silicates: amphiboles, chlorite, serpentines and the like) and/or other non-sulfide gangue minerals in ore B have an adverse effect on Cu and Mo recovery while processing alone or as a blend, and (b) that both physical and chemical factors were responsible for poor recovery from ore B or its blends. Based on preliminary diagnostics tests, our hypotheses were refined to reflect the potential role of aqueous-species-mediated adverse effects on Cu and Mo recoveries. In order to test these hypotheses, we conducted ore (i) flotation tests, (ii) water chemistry analysis, and (iii) spiking experiments which involved investigating the effect of adding a certain fraction of the 'bad' ore (either fines or the aqueous phase from the ground pulp of the bad ore) on Cu/Mo recoveries from the 'good' ore.

This paper discusses certain key findings from this study pertaining to the effect of colloidal and aqueous solution species on flotation performance.

EXPERIMENTS AND PROCEDURES

Flotation Experiments

One kilogram of ore was ground at 60% solids in a mild steel rod mill with a 7.0 kg rod charge made of mild steel for 8'50", corresponding to a grind specification of 32% +100 mesh. Flotation was carried out at 33% solids in a 2.5-L Denver cell at 1250 rpm. Concentrates were collected for 1, 3 and 5 min. (total 9 min.). The collector (an alkyl dithiophosphate) was added either to the mill or the flotation cell. The frother was a blend of alcohol and glycol. Lime was added to the mill to obtain a pH of 9 or 10.5, as required. Frother was added during the conditioning stage. The feed grades of the ores are given in Table 1.

Table 1: Ore B and Ore A mean feed assay

Mineral	Ore B	Ore A
Cu (%)	0.7	0.5
Mo (ppm)	17	350
Fe (%)	5.8	4

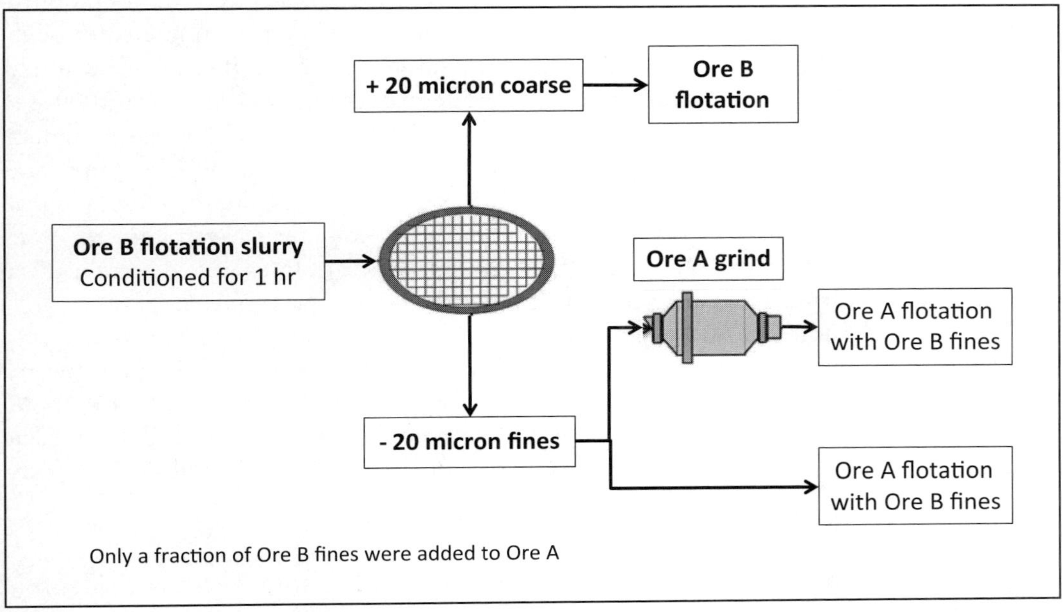

Figure 1. Schematic for the spiking experiments

Spiking Experiments

A schematic of the procedure used for spiking experiments is given in Figure 1.

One kilogram of ore B was ground at 60% solids in a mild steel rod mill with a 7.0 kg mild steel rod charge. The grind time was 8'50", corresponding to a grind specification of 32% +100 mesh. The grind discharge was size-separated using sieves, and the –31 µm (500 mesh) or –20 µm (635 mesh) fractions were collected (referred to as the slime fraction of ore B). The finer fractions (slimes) were then either added to ore A in the grind or to the ore A ground pulp in the flotation cell. The weight fraction of spiked wet fines (filter cake) was typically 4% in either case.

In order to test the hypothesis that the adverse effects could be caused by aqueous species released from certain gangue minerals, the effect of aqueous phase extracted from ground ore B pulp on Cu and Mo recovery from ore A was investigated. Thus, the filtrate (aqueous phase) was obtained from ore B grind discharge using a pressure filter wherein a packed bed of ground ore particles formed the effective filtration medium. This aqueous phase was then used to grind ore A and flotation carried out as described previously.

In this manuscript, we use 'fractionated slimes' as a generic term to describe these different fine fractions. The coarser fraction (+31µm) from the fractionated ore B was floated separately using the flotation procedure outlined previously.

Water Chemistry

Changes in aqueous solution composition arising from contact with ground ores either in the grinding or flotation stages were evaluated. The kinetics of water chemistry changes in the un-aerated ground pulp during the conditioning phase preceding flotation were also determined by taking a slurry sample at regular intervals, filtering it, and analyzing the filtrate for anions and cations using ion chromatography and ICP. Thus, 50 mL slurry was taken every 5 minutes for a total of 30 minutes from the ground slurry that was being conditioned in the flotation cell at 33% solids at 1300 rpm. The slurry sample was then filtered through a 1.5-micron Millipore™ filter, and the solids were returned to the flotation cell. The filtrate was analyzed for ionic species. Anionic species were determined using a Dionex ICS-3000 system with KOH Eluent Generator for anion separation and detection using suppressed conductivity. The DIONEX IonPac® AS19 column (250×4.0 mm), operated at 30°C at a flow rate of 1.0 mL/min, was protected by a DIONEX IonPac® AG19 guard column whose nominal packed particle diameters are 11 microns. Frits on the guard columns have ~0.5 micron pore size. The mobile phase was a KOH gradient. The cation concentrations were determined via elemental analysis carried out using a Varian Vista-Pro ICP-OES spectrometer.

The effect of addition of polymeric modifier P was determined in several 6-factor/2-level designed experiments (Nagaraj 2005; Nagaraj and Ravishankar 2007). The factors were: (1) ore type (ore B or ore B/ore A 30/70 blend); (2) grind size (32 or 38% +100#); (3) pH (9 or 10.5); (4) collector dose; (5) modifier dose; and (6) modifier type (unpublished data). Each data point in the plots that follow represents a unique set of conditions in a multivariate space.

RESULTS AND DISCUSSION

As is common in the mineral processing vernacular, we will refer to the finely divided solids that often produce deleterious effects on flotation as "slimes." This term usually carries the connotation that the adverse effects caused by these fine solids somehow involves them being attached to the surface of value mineral particles (commonly referred to as heterocoagulation). Location (coating) of these slimes on the value mineral particle surfaces then alters the value mineral wettability characteristics in a way that adversely affects recovery in a flotation process. We would like to consider in this paper a broader scope of possible mechanisms by which "slimes," and indeed other less finely divided components in the pulp, can negatively impact flotation performance through their impact on aqueous solution chemistry, microstructure of the pulp and froth phase characteristics. So we will use "slimes" to represent the fine gangue mineral species and some colloidal species generated from these and other gangue minerals that can produce a broader range of phenomena than just coating a value mineral particle surface. This is an expansion of the usual meaning of the term, and is an attempt to promote a more holistic consideration of the issues.

Spiking Experiments

Spiking experiments were initiated in order to test our hypotheses regarding the influence of slimes and aqueous species generated by gangue minerals on Cu recovery, and also to gain a scientific understanding of the slimes problem. It is important to identify the extent to, and manner in, which gangue species in ore B affect flotation performance. Our objectives for these experiments were therefore to:

- Understand the mechanism underlying poor flotation performance with ore B and the pathways by which ore B produces poisoning effects in ore B–ore A blends.

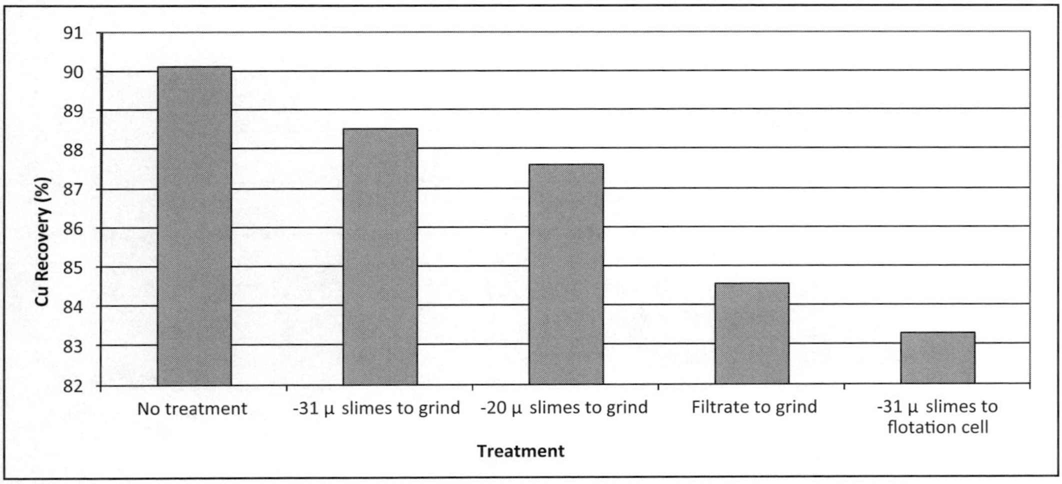

Figure 2. Reduction in Cu recovery upon spiking ore A with 4 wt% ore B "slimes" at two particle size cut-offs, and the aqueous phase (filtrate) from ore B grind

- Determine if the metallurgical performance of the ore B can be improved by the removal of slimes fractions or the soluble species derived therefrom; and what particle sizes of slimes have the maximum effect on Cu/Mo recovery.
- Ascertain if the addition of ore B slimes or its aqueous species to ore A in the grind or flotation steps affects its metallurgical performance.
- Ascertain if addition of a polymeric modifier can mitigate adverse effects of ore B addition and improve metallurgical performance in ore Blends.

We first conducted spiking experiments on ore A using fractionated ore B slimes added either in the grind or the flotation stage and measured the flotation response of this spiked ore A. Figure 2 illustrates that addition of only 4 wt% of ore B fines to either the grind or flotation stages could significantly reduce the originally acceptable Cu recovery from ore A.

The Cu recovery from ore A dropped by over 2–7% in these experiments. The reduction in Cu recovery was exacerbated with the finer fraction of the ore B slimes, and by adding them directly at the conditioning stage (rather than to the mill). When filtrate from ground ore B pulp was used in grinding ore A instead of tap water, there was an even greater loss in Cu recovery. These data support the view that aqueous solution species (generated from non-sulfide gangue) are important in, and make significant contribution to, the mechanism of Cu recovery reduction. However, it still does not rule out the possibility of fines (viz. –31 or –20 micron, or any colloidal particles in the filtrate) contributing to the recovery loss. The exact manner in which the solution species affect flotation is still not clear. One plausible scenario is that the solution species are forming colloidal precipitates which then hinder Cu/Mo flotation.

The effect of removing various fine fractions from an ore B sample on Cu recovery was studied in a separate set of flotation experiments. Figure 3 shows the modest, but positive impact of removing slimes by simple sieving and washing. This desliming (removal of mineral fines or the aqueous phase) operation increased the Cu recovery from ore B by ~2–4%. Three general pathways by which slimes might cause adverse effects may be speculated upon as follows:

1. Modification of the water chemistry in the pulp and froth phases (thereby changing surface or interfacial behavior),

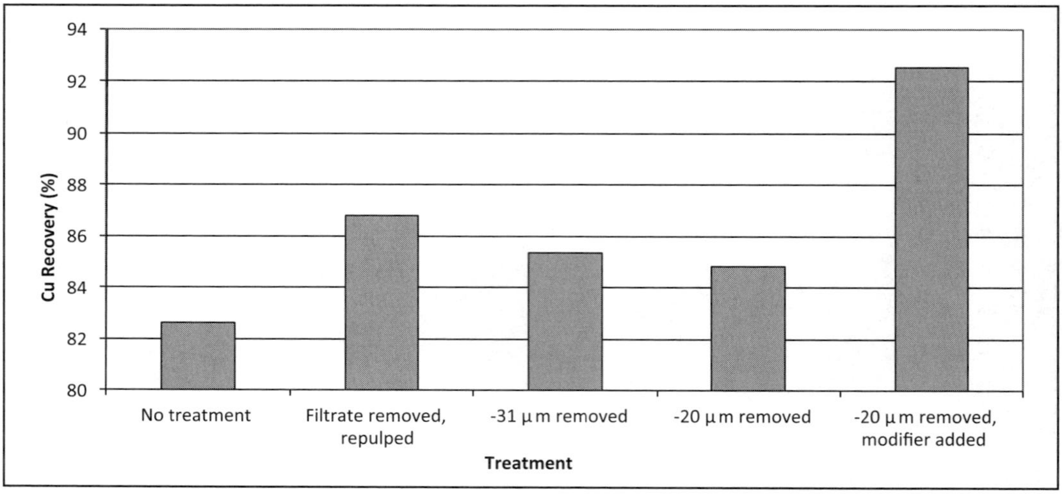

Figure 3. Effect of slime or aqueous phase removal from ore B on Cu recovery. Removing slimes increased the Cu recovery from ore B. Addition of modifier P substantially improved Cu recovery.

2. Modification of pulp and froth phase microstructures (evidenced by changes in rheology), and
3. Slime coating on Cu sulfides and MoS_2 (thereby hindering collector adsorption).

Additional experiments to shed further light on plausible pathways for adverse effects were conducted.

Water Chemistry

It is widely recognized that water chemistry plays a critical role in flotation. The term "water chemistry" in general refers to the nature and composition of soluble species in water. It should be acknowledged that water itself plays an important chemical role in grinding and flotation as a solvent for various mineral-derived species and processing chemicals, and also as a reactant with certain functional groups on mineral surfaces. Aqueous phase comprising water, soluble mineral species, and their plausible reaction products, can be considered unintended modifiers of flotation separation (as opposed to deliberately added modifiers, which are specific chemicals e.g., pH modifiers [Nagaraj and Ravishankar 2007]).

Water chemistry in the pulp, as noted previously, can change as a result of soluble ions arising from contact with the ore. The kinetics of this process might also play a role in the extent of compositional changes. Thus, since the ore grinding, conditioning and flotation stages typically require several tens of minutes to complete, there is also a chance for the soluble ion contents of the aqueous phase to change during this period. Water analysis by ion chromatography for the anions (chloride, carbonate, sulfate) and ICP for the cations (Ca^{++}, Mg^{++}, Na^+, K^+) was performed on water samples recovered from the pulp at various times throughout conditioning in the flotation cell for both ore B and a 30/70 blend of ore B/ore A; see Figure 4 and Figure 5.

One interpretation of the data is that immediately after grinding ore B had leached out less sulfate ions and less K^+ and Mg^{++} ions than did the ore Blend. An alternative interpretation is that more sulfate ions were removed from the solution phase via precipitation (e.g., $CaSO_4$ formation) which resulted in lower concentration of sulfate ions. Sulfate ion concentration in the blend pulp decreased marginally during conditioning, while it remained fairly constant in the ore B pulp. Chloride and carbonate concentrations also did not change with conditioning time. Concentrations of cations (K^+, Mg^{++}, Na^+, and Ca^{++}) remained fairly constant over the course of conditioning. The marginal decrease in sulfate concentration may be rationalized by the formation of insoluble calcium sulfate (gypsum) and possibly other precipitated hydroxy-sulfates of Ca^{++} or Ca^{++}/Mg^{++}. For example, $CaSO_4$ and $Mg(OH)_2$ have thermodynamic solubility products of 5×10^{-5} and 5.6×10^{-12}, respectively (note: the conditional solubility products, which would be more representative of the particular experimental conditions, are not available; thermodynamic data for mixed precipitates are also not available). With the removal of sulfate ions, one would expect Ca^{++} concentrations to also decrease concomitantly. However, this is not observed, probably because Ca-containing gangue minerals act as a constant source of Ca^{++} ions (due to the cation-exchange-capacity of altered silicates, and the appreciable solubility of calcite/dolomite). Importantly, in laboratory tests, it seems that changes in water chemistry due to contact with ore Are for the most part complete after the grinding operation, with only modest changes occurring in the flotation operation. The only significant difference in ion contents between slurries of ore B and the ore Blend was the sulfate concentrations. All other ion concentrations measured were essentially the same for both the ore B and its blend.

It should be noted here that water chemistry is indeed dynamic, and snap-shots of water analysis fail to capture the inherent and true dynamics of the system. There could be a steady-state between minerals and the solution, wherein species released from the minerals could precipitate in the form of hydroxides and/or sulfates, some of which may eventually be found on other mineral surfaces. It is also plausible that such precipitates would 'coat' not only Cu or Mo mineral surfaces but also other silicate and non-sulfide gangue minerals. There is insufficient information to draw definitive conclusions.

In light of our water chemistry analyses, we revisit data in Figure 2 (adverse effect of ore B). Recall that impact may be felt in the slurry, the froth, or both phases. Loss of Cu recovery from ore A became greater as the average size of the ore B fine fraction added to it decreased, all the way down to filtrate from ground ore B slurry. This filtrate contained soluble species and the fine-particle fraction below ~10 microns (although we do not know the actual solids content of the ore B slurry filtrate in these experiments). It is difficult to determine from these data the relative contributions of the colloidal and the truly soluble species to the reduction in Cu recovery, or whether some of the colloidal material formed from solubilized species that subsequently phase separated.

If a significant amount of the deleterious effects on Cu recovery were due to soluble components in the ore B acting in a negative manner in the ore B/ore A blend flotation, then one might expect to see some differences in the ion contents of the filtrate. The main differences in ion contents of the aqueous phases were lower $SO_4^=$ (~130 vs. ~190 ppm), K^+ (~8 vs. ~20 ppm) and Mg^{++} (~12 vs. ~17 ppm) with ore B compared to the ore B/ore A blend. The aqueous $SO_4^=$ concentration in the ore Blend slurry also tended to decrease with time, suggesting the potential formation of insoluble sulfates, likely with Mg^{++} and Ca^{++}, throughout the grinding and flotation stages. The surface precipitation or surface adsorption of such precipitating phases could alter Cu recovery.

Regarding colloidal particles from ore B, all we can say at this point is that the smaller particles are more egregious in reducing Cu recoveries than the larger ones. Also, placing fine ore B particles directly into the flotation cell resulted in greater reduction in Cu recovery. The Cu recovery from ore B increased by ≈ 2–4% when the fine particle fraction was removed, see Figure 3. The relative contributions of particle adsorption on minerals (i.e., slime coating of

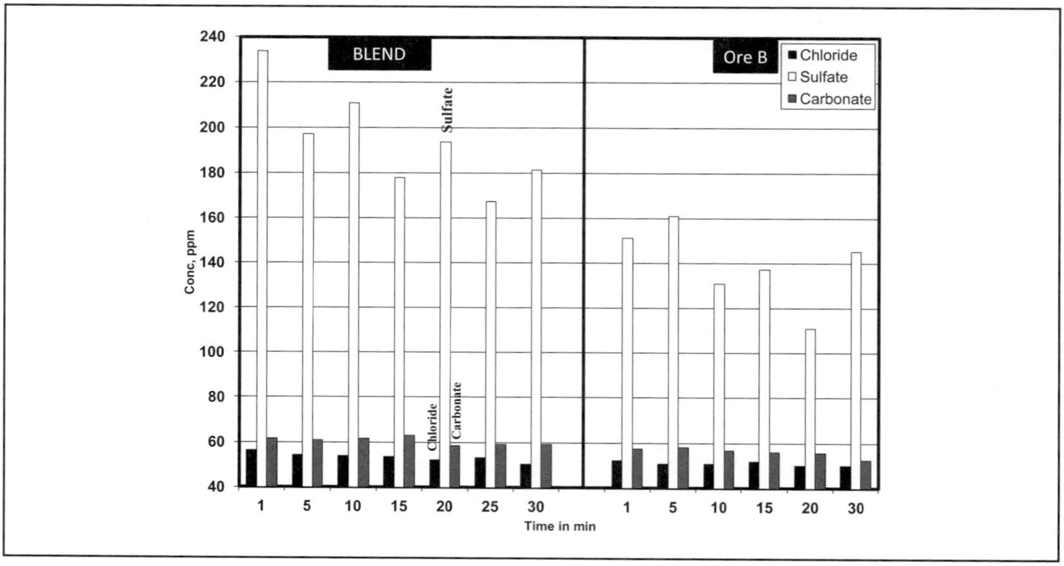

Figure 4. Anion contents of water sampled from 30/70 ore B/ore A blend (left) and ore B (right) slurry conditioned in a flotation cell without aeration

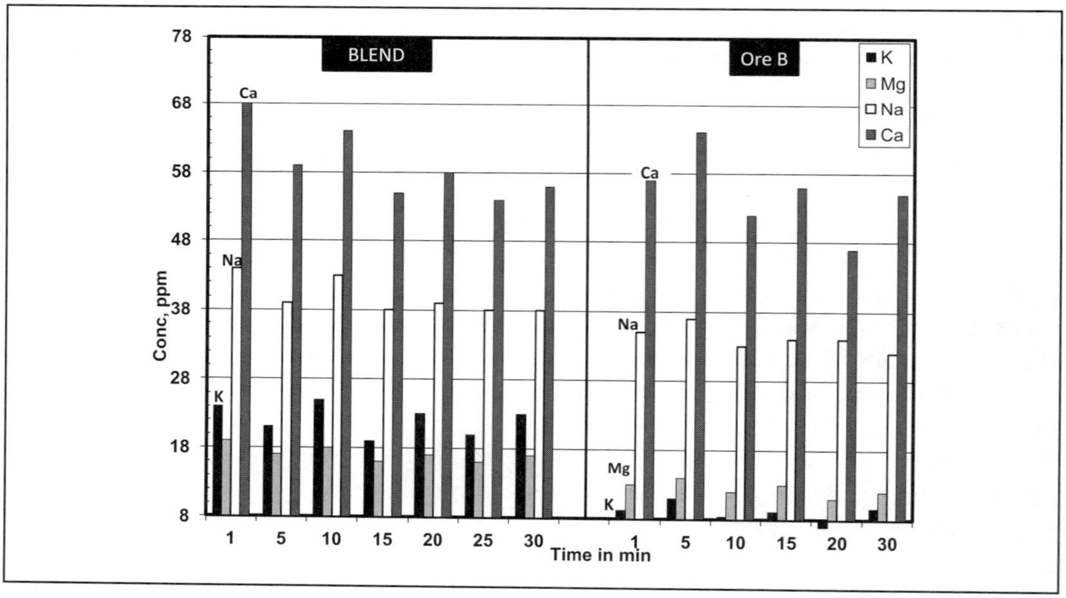

Figure 5. Cation contents of water sampled from 30/70 ore B/ore A blend (left) and ore B (right) slurry conditioned in a flotation cell without aeration

Cu-containing minerals reducing collector adsorption) vs. the small particles simply acting as a source of more readily available solution species, which may adsorb or form precipitates on minerals, remains to be explored. The data we currently have favors (at least in terms of relative contribution) an aqueous-species mediated adverse effect on Cu recovery.

Another mechanism by which colloidal species can change flotation response is via the formation of aggregates with each other or with other colloidal or larger particles. If these aggregates are transported to the froth phase through entrainment or bubble flux they may stabilize the froth by retarding water draining from the froth lamellae back into the pulp. This may cause a reduction in recovery and/or grade. Indeed we observed the appearances of the froth phases in each stage to be different for ore B and ore A.

Polymeric Modifiers

Practical solutions to improve Cu/Mo recovery in ore mixtures containing ore B are ultimately our goal. While the following discussion pertains to experiments specifically with a polymeric modifier, it is important to mention that several other solutions to the problem were explored as a part of the research program, all of which lead to, at best, only marginal change in Cu/Mo recovery. For instance, processing approaches tried were pulp dilution, finer grind, regrind of ores, and re-flotation of tailings; while examples of chemical approaches tried included testing different frothers and/or collectors, and NaSH addition. None of these led to the goal of producing a step change in Cu/Mo recovery. Since colloidal species have been at least partially implicated in the reduced recovery, we explored the addition of a polymeric modifier to mitigate this problem.

The effect of a polymeric modifier on improving flotation performance of ore A spiked with different fine particle fractions recovered from ground ore B slurries was studied. Figure 3, last entry on right, shows that the addition of polymeric modifier P in combination with the collector and frother, helped substantially improve Cu recovery from partially deslimed ore B fines (–20 micron fraction removed). Note that polymeric modifier addition led to a step improvement in Cu recovery, and was significantly different from other desliming options tried.

Figure 6 shows the significantly improved Cu recovery from ore B flotation when polymeric modifier P was added in combination with the collector and frother. Cu recovery and grade scaled with modifier dose. The improved Cu recoveries were in the range commonly associated with the good recoveries from ore A. Indeed, a marginal reduction in Cu grade was observed with the modifier addition. One could argue that this reduction was due to additional mass recovery resulting from modifier addition. However, a simple calculation assuming addition of 1% mass of the tails to the concentrate would result in <1% higher Cu recovery and approximately 5% lower Cu grade (shown as open circle in Figure 6). Clearly, the effect of modifier addition was substantially different in terms of recovery and grade, and was not related to mass recovery. In other words, the improved Cu recovery was not due to entrainment, but rather true flotation. This same polymeric modifier also improved Cu recovery in ore B/ore A (30/70) blend flotation, although there was no additional improvement between the low and high doses; see Figure 7.

These ores also contained Mo minerals, and Mo recoveries were improved in a fashion similar to Cu when this same polymeric modifier was used in combination with the prescribed collector and frother; see Figure 8 and Figure 9.

To reiterate our observations, the poor metallurgical performance in, and due to ore B likely arose from either direct adsorption of colloidal materials, solution species, and/or precipitates derived therefrom, onto Cu and Mo minerals. The addition of a specific polymeric modifier led to significantly higher Cu and Mo recoveries in both ore B and ore B/ore A blends, with only marginal reductions in grades. Such effects could not be rationalized merely by means of additional mass recovery, which would have been accompanied by significant reduction to the concentrate grade. Our explanation for this result is that the polymeric modifier interacted

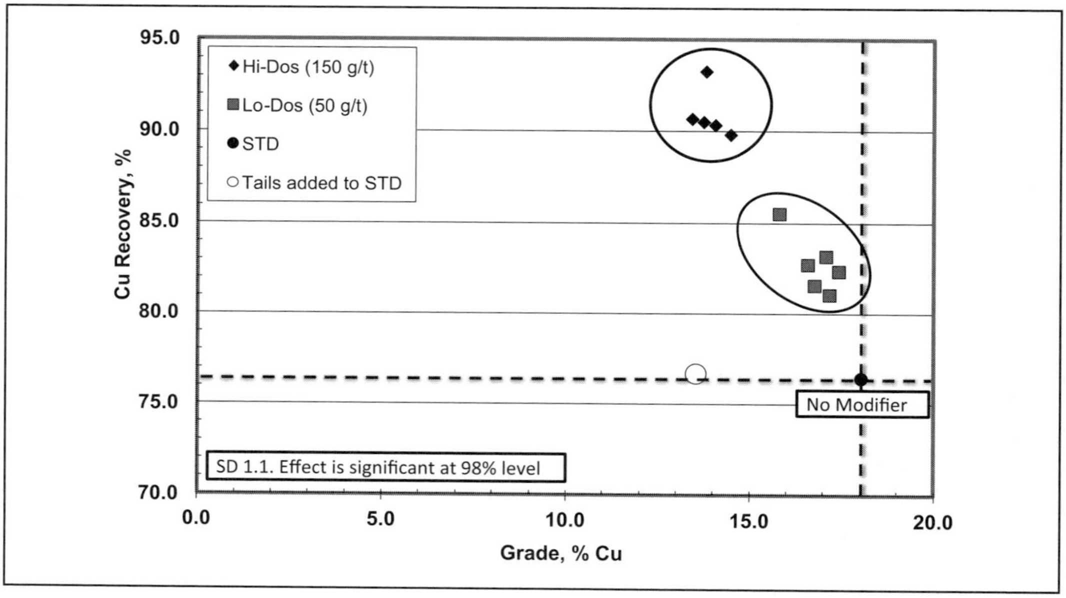

Figure 6. Effect of polymeric modifier P on Cu recovery and grade from ore B. Each data point represents an experiment in a 6-factor 2-level fractional factorial design. The open circle indicates the calculated point if 1% mass from the tails were added to the control condition (i.e., no modifier).

with colloidal materials, solution species or precipitates derived therefrom and prevented their adsorption on value minerals, thereby mitigating their effects on Cu/Mo recovery loss.

CONCLUSIONS

The effect of dissolved gangue mineral species and fine colloidal matter in process water on flotation performance from two porphyry ore types A and B (characterized as 'good' and 'bad' ores, respectively, based on Cu/Mo recoveries) was investigated. Poor Cu/Mo recoveries (at a given concentrate grade) were obtained when ore B was treated alone or in combination with ore A. In the past, many hypotheses were proposed and investigated in the plant; however, none of them could be supported, and no satisfactory solutions to the problem were found. The goals of our current study were to (a) explore the underlying mechanisms responsible for poor Cu/Mo recovery resulting from ore B, and its adverse effect on ore A, and (b) develop robust and sustainable chemical solutions.

Traditional solutions involving chemical routes (collector/frother, or addition of NaSH) and processing options (longer flotation time, pulp dilution, finer grinding, grinding/re-flotation of tails, high intensity conditioning) have only led to incremental differences in Cu/Mo recovery, and the improvements were not commensurate with expectations.

In this study, flotation tests, spiking experiments, and dynamic water chemistry studies were conducted with ore B and ore B/ore A blend. Data suggest that fine materials from ore B affected Cu recovery when added to ore A. Smaller size fractions of these fines had an increasingly severe effect. Moreover, we have provided evidence that it is not only the fines, but the aqueous species released from these minerals also affected flotation performance. In fact, these aqueous species were observed to play a significant role in reducing flotation recovery.

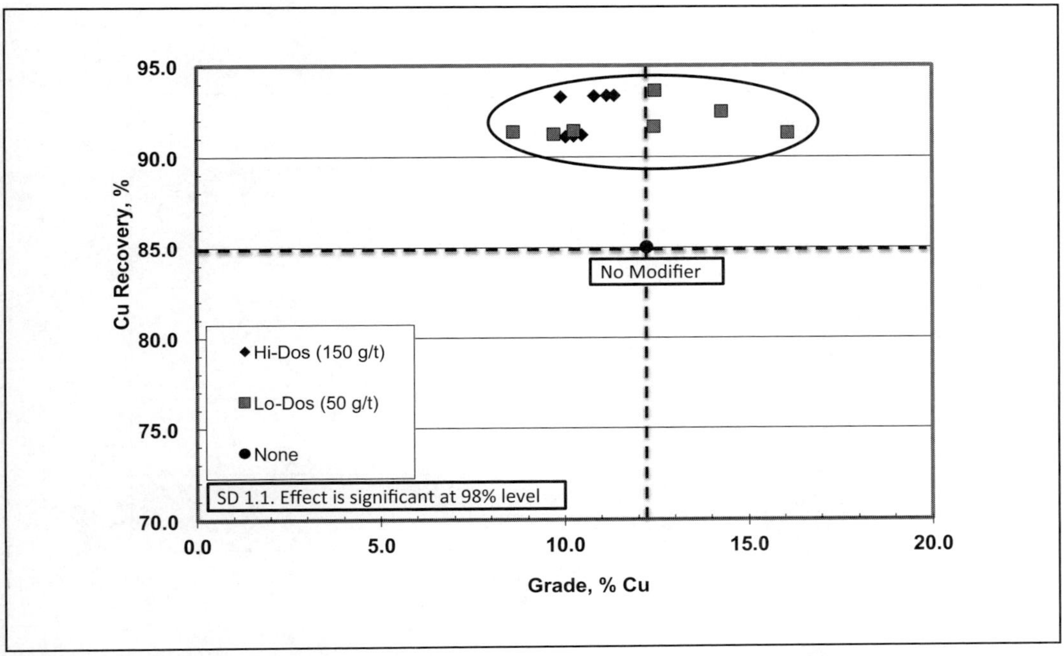

Figure 7. Effect of polymeric modifier P on Cu recovery and grade from ore B/ore A blend (30/70). Each data point represents an experiment in a 6-factor 2-level fractional factorial design.

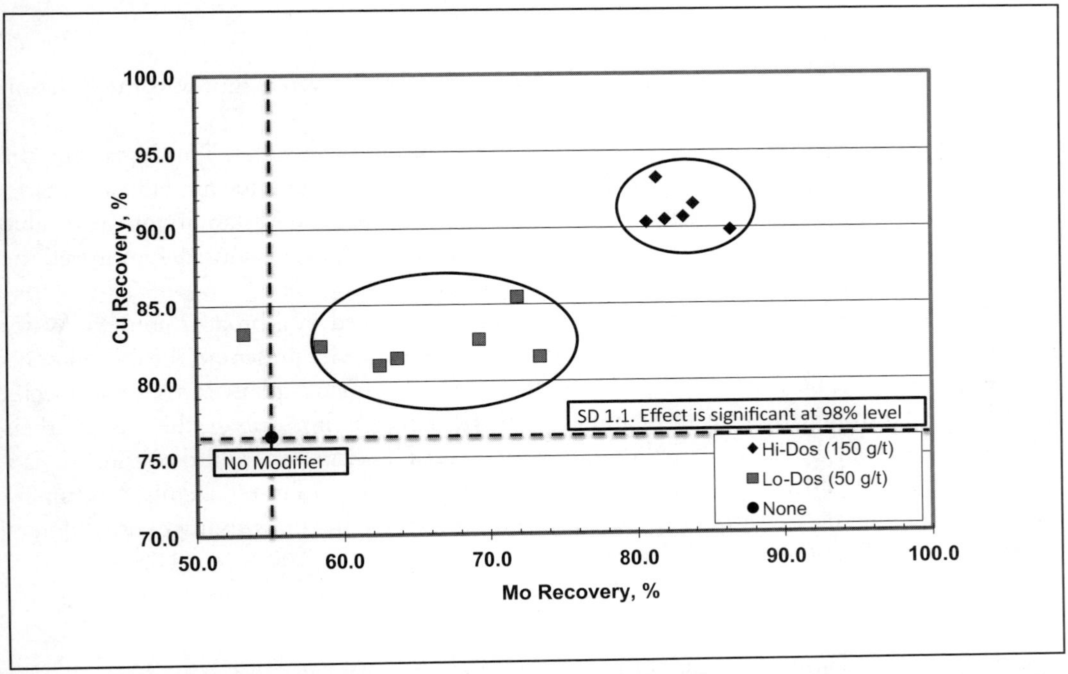

Figure 8. Effect of polymeric modifier P on Cu and Mo recovery from ore B. Each data point represents an experiment in a 6-factor 2-level fractional factorial design.

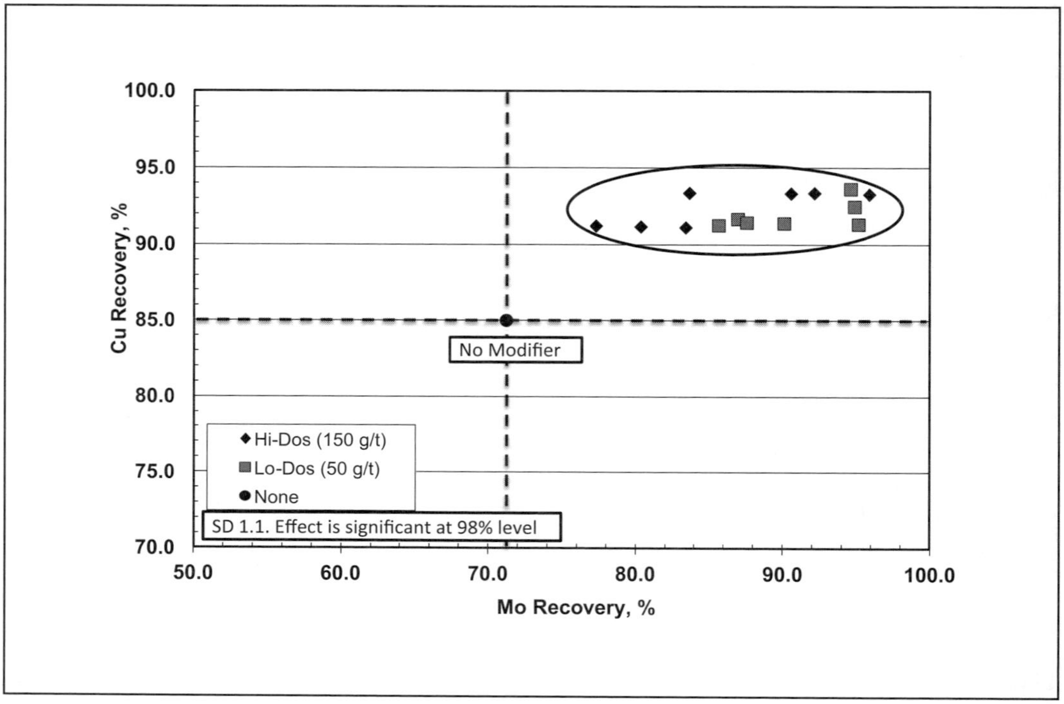

Figure 9. Effect of polymeric modifier P on Cu and Mo recovery from ore B/ore A blend (30/70). Each data point represents an experiment in a 6-factor 2-level fractional factorial design.

Moreover, replacing the aqueous phase of the slurry with fresh water improved the flotation performance of ore B.

Water chemistry studies suggested the role of Ca^{++}, Mg^{++}, and sulfate ions in causing the poor performance. We speculate that these species may precipitate on value minerals as $CaSO_4$, $Mg(OH)_2$ and mixed Ca/Mg hydroxy-sulfates, thereby altering the wettability of these value minerals. Moreover, addition of an appropriate polymeric modifier substantially improved the flotation performance of these ores.

These studies suggest that water chemistry must be viewed in a broader context. Water chemistry comprises species released from minerals, organic species present in the ore, all flotation reagents, and colloidal mineral particles. Moreover, the aqueous phase is dynamic: species are released continuously from most minerals, particularly the non-sulfides because of their relatively greater solubility compared to sulfide minerals. These species are also consumed via a variety of different pathways. Finally, understanding the dynamics of water chemistry is important in developing robust water-efficient processing, and increasing one's ability to use different sources of water, including recycled water.

REFERENCES

Bulatovic, S.M., Wyslouzil, D.M., and Kant, C. 1999. *Effect of clay slimes on copper, molybdenum flotation from porphyry ores. Proc. Copper 99,* II:95–111.

Chander, S., and Fuerstenau, D.W. 1972. *On the natural floatability of molybdenite. Trans. Amer. Inst. Min. Metall. Eng.* 2P:62–69.

Edwards, C.R., Kipkie, W.B., and Agar, G.E. 1980. *The effect of sliming coatings of the serpentine minerals, chrysotile and lizardite, on pentlandite flotation. Int. J. Miner. Process.*, 7:33–42.

Fornasiero, D., and Ralston, J. 2006. *Effect of surface oxide/hydroxide products on the collectorless flotation of copper-activated sphalerite. Int. J. Miner. Process.*, 78:231–237.

Gaudin, A.M., Fuerstenau, D.W., and Miaw, H.L. 1960. *Slime-coatings in galena. CIM Bull.*, 53:960–963.

Nagaraj, D.R. 2005. *Reagent selection and optimization: The case for a holistic approach. Min. Eng.* 18(2):151–8.

Nagaraj, D.R., and Ravishankar, S.A. 2007. *Flotation reagents: A critical overview from an industry perspective. In Froth Flotation: A Century of Innovation.* Edited by M.C. Fuerstenau, G. Jameson, and R.-H. Yoon. Littleton, CO: SME, pp. 375–424.

Nagaraj, D.R. 2008. *New approach to reagent development and applications in the processing of base and precious metal ores. Proc. International Mineral Processing Conference*, Beijing.

Patra, P., Bhambhani, T., Nagaraj, D.R., and Somasundaran, P. 2010. *Effect of morphology of altered silicate minerals on metallurgical performance: Transport of Mg silicates to the froth phase.* Rheology in Mineral Processing. In *Proc. 49th Conference of Metallurgists of CIM*, pp. 31–42.

Patra, P., Bhambhani, T., Vasudevan, M., Nagaraj, D.R., and Somasundaran, P. 2011. *Impact of aspect ratio of fibrous serpentines in the beneficiation of ultramafic Ni ores.* New Technology Implementation in Metallurgical Processes. In *Proc. 50th Conference of Metallurgists of CIM.*

Vasudevan, M., Nagaraj, D.R., Patra, P., and Somasundaran, P. 2010. *Effect of altered silicates in flotation performance: Role of changes in pulp rheology.* Rheology in Mineral Processing. In *Proc. 49th Conference of Metallurgists of CIM*, pp. 21–30

Vasudevan, M., Patra, P., Bhambhani, T., Nagaraj, D.R., and Somasundaran, P. 2011. *Processing of ultramafic ores: Atypical grade-recovery curves.* New Technology Implementation in Metallurgical Processes. In *Proc. 50th Conference of Metallurgists of CIM.*

Particle Aggregation and Sedimentation Characteristics of Kaolinite Suspensions as Explained by Surface Charge Considerations

Jan D. Miller
Department of Metallurgical Engineering, University of Utah, Salt Lake City, UT, USA

Vishal Gupta
FLSmidth Salt Lake City, Inc., Salt Lake City, UT, USA

ABSTRACT

The aggregation and sedimentation of clay mineral particles are generally controlled by the surface chemistry of the system which in turn determines the effectiveness of water recovery from tailings containing such particles. The tailings from oil sand, phosphate rock, and bauxite operations are classic examples of the technical challenge that exists for water recovery and tailings stabilization. Of particular interest are clay minerals such as kaolinite, illite, and montmorillonite. The aggregation and sedimentation of these clay minerals are complicated by particle size, shape, and surface chemistry. Results are reported for kaolinite.

Kaolinite particles naturally exist as platy shaped micron-to-nano sized particles, which are highly dispersed in some tailings streams. Consolidation of tailings containing such particles is difficult. The surface chemistry, particularly the surface charge properties of kaolinite particles, plays an important role in their dewatering characteristics and is complicated by the anisotropic, platy structure of the particles which manifests itself in edge surfaces and face surfaces. Recently, it has been established that the silica face of kaolinite is negatively charged at pH > 4, whereas the alumina face of kaolinite is positively charged at pH < 6 and negatively charged at pH > 8. The surface charge densities of the silica face surface, the alumina face surface, and the edge surface have been utilized to determine the interaction energies between different surfaces of kaolinite particles.

Results indicate that the silica face–alumina face interaction is dominant for kaolinite particle aggregation at low pH. This face–face association increases the stacking of kaolinite layers, and thereby promotes the edge–face (edge–silica face and edge–alumina face) and face–face (silica face–alumina face) associations with increasing pH to pH 5–5.5. With further increase in pH, the face–face and edge–face association decreases due to increasing surface charge density on the silica face and the edge surfaces, and decreasing surface charge density on the alumina face. At high pH, all kaolinite surfaces become negatively charged, kaolinite particles are dispersed, and the suspension is stabilized. Based on the corresponding aggregate structure, the sedimentation behavior of kaolinite suspensions is discussed, particularly the rapid settling rates at pH 5–5.5.

INTRODUCTION

The aggregation and sedimentation of clay mineral particles is generally controlled by their surface charge properties which in turn determines the effectiveness of water recovery from tailings containing such particles. Coagulants are added to neutralize the surface charge, thereby the mineral particles come closer by van der Wall attraction and form loose aggregates. Subsequently, a polymer flocculant may be added which adsorbs on particle surfaces

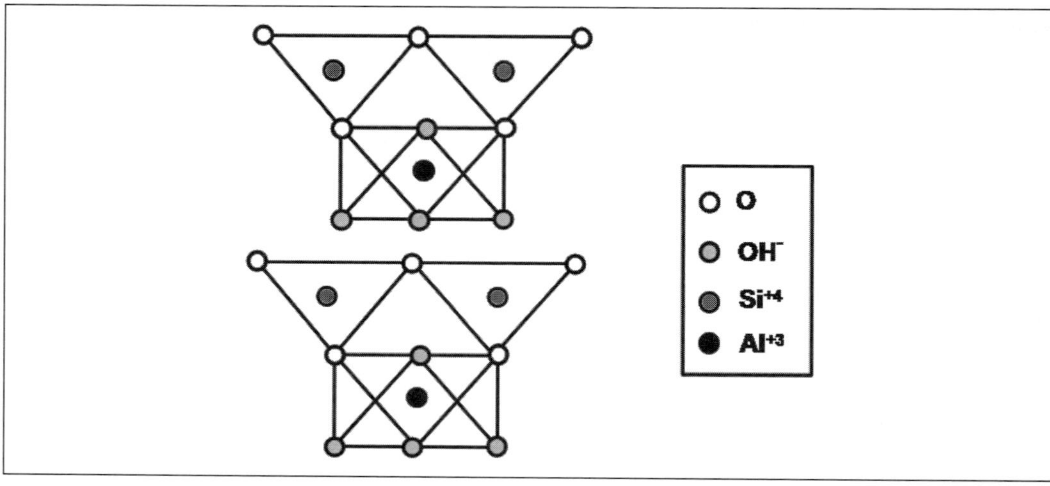

Figure 1. Structure of kaolinite along the 010 edge surface showing a silica tetrahedral layer and an alumina octahedral layer (Gupta 2011)

by attracting the oppositely charge surface of mineral particles and causes bridging, bringing several aggregates together to form inter-aggregate network, the sediment then consolidates under its own weight. The surface charge is a fundamental feature of clay mineral particles, and its understanding is desired in order to control sedimentation behavior. The surface charge of clay mineral particles in general is complicated by their size and shape, with different face surfaces and edge surfaces having different surface charge characteristics. The surface charge properties of kaolinite are discussed in this paper, and how they control the aggregate structure and sedimentation of particles.

Kaolinite particles naturally exist as pseudo-hexagonal platy shaped particles, ranging from micron to nanometer in size. The crystallographic structure suggests that kaolinite particles consist of a silica tetrahedral surface corresponding to the 001 basal plane and an aluminum hydroxide octahedral surface corresponding to the $00\underline{1}$ basal plane as shown in Figure 1. Of course, the kaolinite particles have 010 and 110 edge surfaces which are generated as a result of broken covalent bonds. The sedimentation of kaolinite particles is governed by its surface chemistry, in particular its electrokinetic features, which generally determine the particle interactions, and hence the settling rate and the degree of consolidation of the suspension.

It is traditionally believed that the 001 and $00\underline{1}$ basal plane surfaces of kaolinite particles are negatively charged due to isomorphic substitution of Al^{3+} for Si^{4+} in the silica tetrahedral layer and Mg^{2+} for Al^{3+} in the alumina octahedral layer, whereas the edge surface (010 and 110 plane surfaces) carries a positive or negative charge depending on the pH of the system (Johnson et al. 2000, Johnson, Russell, and Scales 1998, Olphen 1963, Rand and Melton 1977, Schofield and Samson 1954, Street 1956, Street and Buchanan 1956). The assumption that both basal planes carry a fixed negative charge has only recently been examined experimentally through surface force measurements using atomic force microscopy (AFM) (Gupta and Miller 2010). These surface force measurements reveal that the silica tetrahedral face of kaolinite is negatively charged at pH > 4, whereas the alumina octahedral face of kaolinite is positively charged at pH < 6, and negatively charged at pH > 8. The results suggest that the iso-electric point of the silica tetrahedral face is at pH < 4, and that the iso-electric point of the alumina octahedral face lies between pH 6 and 8 (Gupta and Miller 2010). Based on this new finding, Gupta et al. (2010) determined the interaction energies between different surfaces of kaolinite particles, which showed that the silica face–alumina face association is likely to be dominant for kaolinite particle association at low and intermediate pH conditions. However, the above conclusion of dominant silica face–alumina face association was reached by assuming that the zeta-potential of the edge was a linear combination of the zeta-potentials on silica and alumina surfaces (Williams and Williams 1978). The assumption for edge surface zeta-potential will be discussed later in this study. It seems that the edge surface zeta-potential for kaolinite is under-estimated, and that an experimental technique such as potentiometric titration should be considered in the analysis of kaolinite surface potentials.

Consideration to surface charge of the silica face, the alumina face and the edge surface of kaolinite must be given in order to determine the interaction energies when these particles come closer to each other. When there is repulsive energy between two surfaces, the kaolinite particles tend to stay away and the suspension is stable. In contrast, when there is favaorable interaction or attraction between two surfaces of kaolinite, the particles come closer to each other and form aggregate-structures. These aggregate-structure for clay minerals such as kaolinite is particularly interesting as they evolve in three-dimensions to form card-house structures (Olphen 1963). It is expected that the sedimentation of kaolinite will be maximum when there is favorable interaction between kaolinite particle surfaces, and coarser and denser aggregate-structure will be formed.

Several researchers have examined images of flocculated kaolinite suspensions using electron microscopy to determine the aggregate structure, and relate these observations to sedimentation. The earliest significant discussion of clay fabric was given by Terzaghi in 1925 (see O'Brien 1971) in which the structure of cohesive soils was discussed with regards to adhesion between adjacent minerals. Casagrande (1940) presented a theory for a honeycomb structure in soils. Idealized drawings of clay fabric forming a card-house or honeycomb structure have been proposed (Lambe 1953, Olphen 1963, Schofield and Samson 1954, Tan 1958). Rosenqvist (Rosenqvist 1959, Rosenqvist 1963) published the first electron micrograph of freeze-dried samples of undisturbed marine Oslo blue clay and supported the idea of a card-house arrangement in the undisturbed sediment. O'Brien (1971) examined the fabric of kaolinite in distilled and slightly saline water, and found that the fabric is dominated by a 3-dimensional network of twisted chains of face–face oriented flakes having the appearance of a stair-stepped cardhouse. Recently, Zbik et al. (2008) investigated the structure of kaolinite aggregates at pH 8 during sedimentation experiments using cryogenic-scanning electron microscopy (cryo-SEM). They found that the kaolinite aggregates initially show edge–face and edge–edge association, which rearrange from edge–edge chains to more compact face–face associations during settling. In another study, Zbik and Frost (2009) observed differences in the physical behavior of a number of kaolinite clays from Birdwood and Georgia. For example, SEM micrographs of the Birdwood kaolinite aggregates show the predominance of stair step edge–edge contacts forming a spacious cell structure, whereas Georgia kaolinite aggregates display edge–face contacts building a card-house structure. From these studies it is suggested that kaolinite clays are very complex and each kaolinite should be considered separately. It is evident that each kaolinite from different sources must be characterized and detailed examination may be necessary to determine the aggregate structures at low, medium and high pH values.

It is the objective of this paper to utilize the surface charge information on the face surfaces and edge surfaces of kaolinite in order to understand the aggregate structure of kaolinite suspensions and corresponding sedimentation behavior. Furthermore, the influence of particle aspect ratio and electrical double layer thickness is investigated with regard to the aggregation behavior of kaolinite particles. Finally, cryo-SEM images were used to examine the modes of particle association (face–face, edge–face, and edge–edge).

MATERIALS AND METHODS

Sample Preparation

A clean English Kaolin (Imerys Inc., UK) was obtained from St Austell area in Cornwall, UK. The sample was cleaned with water only using elutriation to achieve classification at a size of less than 2 μm. No other chemical treatment was used. Further details about the kaolinite extraction and preparation are given in the literature (Bidwell, Jepson, and Toms 1970). The kaolinite suspension was prepared in high purity Milli-Q water (Millipore Inc.). The resistivity of the water was 18.2 MΩ-cm in all experiments. Potassium chloride (1 mM solution) was used as background electrolyte for surface force measurements. The pH was adjusted to its desired value using 0.1 M HCl or 0.1 M KOH solutions. All chemicals used were of ACS reagent grade.

X-ray diffraction (Bruker AXS, Inc., Madison, WI, USA) analysis conducted on the kaolinite sample confirmed that kaolinite is the dominant mineral phase. Energy dispersive X-ray spectroscopy (FEI, Hillsboro, Oregon, USA) analysis of the kaolinite sample showed nearly 1:1 atomic distribution of aluminum (7.98%) and silicon (7.95%) with trace amounts of potassium (0.35%), calcium (0.08%) and iron (0.15%). Other details about the purity of kaolinite are provided elsewhere (Gupta and Miller 2010).

The kaolinite particles with a median diameter of 600 nm and median thickness of 11.2 nm in 1 mM KCl solution were used in this study, unless otherwise mentioned. The median diameter and thickness were determined from the imaging of at least 150 kaolinite particles using atomic force microscopy. The kaolinite particles were air-dried on a thin sheet of mica and an image of the particle was obtained using the contact mode imaging technique. Section analysis was then conducted to obtain information about the diameter and thickness of the particle. The details are provided elsewhere (Gupta et al. 2011).

Aggregate Size

The aggregate size of kaolinite particles in suspension at desired pH values was determined using photon correlation spectroscopy (PCS). In a PCS experiment, the fluctuation of scattered light from particles by their Brownian motion is collected at a scattering angle of 90° with an optical fiber, and detected by a photo-electric detector. The amplified signal from the photo-electric detector is fed into an autocorrelator for computing the intensity autocorrelation function. This autocorrelation function is used to determine the relaxation of intensity fluctuations, which in turn is related to translational diffusion coefficient of the particles. Particle size (equivalent spherical diameter) is then determined from the diffusion coefficient of the particles.

Kaolinite suspensions (4%) were prepared in 1mM KCl solution at pH 9.0, sonicated for 5 minutes and stirred for 1 hour for complete dispersion. After conditioning, 50 mL of the suspensions was taken in a volumetric flask and adjusted to the desired pH of 3, 5, 7 and 9. A small amount of kaolinite suspension (1 mL) was taken at each pH into a cuvette for PCS analysis. Each experiment was replicated three times and the average aggregate size was determined.

Table 1. Hamaker constant of different interactions between kaolinite particles

Interaction	Interaction Type	Hamaker Constant, J
1	Edge–edge	2.37×10^{-20}
2	Edge–silica face	1.63×10^{-20}
3	Edge–alumina face	3.05×10^{-20}
4	Silica face–silica face	1.11×10^{-20}
5	Alumina face–alumina face	3.90×10^{-20}
6	Silica face–alumina face	2.08×10^{-20}

Cryo-SEM

Kaolinite suspensions (1%) were prepared in 1 mM KCl solution (100 mL). About 15 mL of the sample was taken in a vial and adjusted to pH 3, 5, 7 and 9 using 0.1 M HCl or 0.1 M KOH solutions. The suspension was left overnight for conditioning. The sample was hand-shaken, and then a few microliters of the suspension were taken in a small metal rivet sealed with glue at one end. The sample was immediately plunged into a liquid N_2 slush using the Oxford LN_2 slush freezing apparatus for freezing the water without allowing crystallization, i.e., vitrifying. Vitrified samples were placed onto the liquid nitrogen-cooled specimen stage of the field emission scanning electron microscope Philips XL30 FESEM with Oxford CT 1500 Cryo stage. The sample was fractured under vacuum and a small amount of vitrified H_2O was sublimed off by raising the stage temperature to –90°C for 10 minutes to expose the aggregate structure, then lowering back to –180°C. Finally, the sample was coated with platinum before SEM imaging.

DLVO Model

Interactions between kaolinite particles were characterized by the (Deryaguin-Landau-Vervey-Overbeek) DLVO model, which considers the sum of electrostatic and van der Waals forces. An appropriate form for the van der Waals interaction energy per unit area (E^{vdW}) between two planar surfaces of thickness δ_1 and δ_2 is given as (Masliyah and Bhattacharjee 2006):

$$E^{vdW} = -\frac{A_H}{12\pi}\left[\frac{1}{h^2} + \frac{1}{(h+\delta_1+\delta_2)^2} - \frac{1}{(h+\delta_1)^2} - \frac{1}{(h+\delta_2)^2}\right] \quad (1)$$

where A_H is the combined Hamaker constant of the two surfaces interacting in the suspending medium, h is the separation distance between two interacting particles. At the closest approach, h was taken as 20 Å in all calculations for comparison with existing literature (Johnson, Russell, and Scales 1998). In the case of kaolinite, we have to consider six different surfaces for particle interaction, edge–edge, edge–silica face, edge–alumina face, silica–silica face, alumina–alumina face and silica face–alumina face interactions. Therefore, the non-retarded Hamaker constants for each of the six cases were determined from the Hamaker constant of silica and alumina as given in Table 1. Hamaker constants in Table 1 were calculated from the following equation:

$$A_{123} = (\sqrt{A_{11}} - \sqrt{A_{33}})(\sqrt{A_{22}} - \sqrt{A_{33}}) \quad (2)$$

where A_{11} is the Hamaker constant for the first interacting surface, for example the silica face, alumina face or edge surface, A_{22} is the Hamaker constant for the second interacting surface, for example the silica face, alumina face or edge surface, and A_{33} = Hamaker constant for the suspending medium which in our study is water. The Hamaker constants for silica and alumina were taken as 8.86×10^{-20} J and 1.52×10^{-19} J, respectively (Bergstrom 1997). The Hamaker constant for the edge surface was determined as 1.20×10^{-19} J from the average of the Hamaker constants for silica and alumina. The Hamaker constant for water was taken as 3.70×10^{-20} J (Bergstrom 1997).

The electrical double layer interaction energy per unit area (E^{Edl}) between two planar surfaces with surface potentials ψ_1 and ψ_2 is given by the Hogg-Healy-Fuerstenau (HHF) expression for dissimilarly charged surfaces as (Nguyen and Schulze 2004):

$$E^{Edl} = \varepsilon\varepsilon_0 \kappa \left[\frac{2\psi_1 \psi_2 \exp(\kappa h) - \psi_1^2 - \psi_2^2}{\exp(2\kappa h) - 1} \right] \tag{3}$$

where ε_o is is the permittivity of free space, ε is the dielectric constant and κ^{-1} is the Debye screening length of the electrical double layer.

For cases when the surface potential of the two planar surface are similar, i.e., $\psi_1 = \psi_2 = \psi$, the electrical double layer interaction energy reduces to

$$E^{Edl} = \frac{2\psi^2 \varepsilon\varepsilon_0 \kappa}{\exp(\kappa h) + 1} \tag{4}$$

Similarly, the electrical double layer interaction energy under the condition of constant surface charge, first established by Langmuir (1938), can be used. In this case, revising Equation 3 by replacing the two minus signs in the numerator by plus signs gives the Langmuir equation. In particular, the double layer interaction energy at constant surface potential and/or constant surface charge presents the lower and upper limits of the interaction energy. The actual double layer interaction energy with a mixed boundary condition or a surface charge regulation is between the two limits. The calculation shows that the trend of the total interaction energy for the cases is similar. Therefore, only the interaction energy described by Equation 3 is used hereafter.

The resulting DLVO interaction energy is given as:

$$E = E^{vdW} + E^{Edl} \tag{5}$$

The total interaction energy was multiplied by the measured interaction area ratio of basal plane surface to edge surface of 13.39:1 as determined by atomic force microscopy. In calculating the edge and face surface areas, the kaolinite particle was assumed to be a circular disc. The energies are scaled against the maximum predicted silica face-alumina face interaction energy, E_{max}.

RESULTS AND DISCUSSION

The particle aggregation and sedimentation of a suspension is essentially determined by the forces that control the spatial arrangement and dynamics of the suspended particles. In a suspension under the predominant influence of repulsive electrostatic energies the particles tend to take up positions as far from each other as possible. This may lead to a regular arrangement of the particles, i.e., to the development of the spatial order in the suspension. Clusters of particles, or aggregate structures, form in a suspension when the particle interactions are dominated by attractive energies. The aggregate structure or flocs will immobilize the suspending medium, and give rise to settling velocity and sedimentation of the suspension. The particle characteristics such as morphology, size, surface area, etc. will also greatly affect the suspension viscosity, and the strength of the aggregate structure. The aggregate structure plays a major part in the settling of clay suspensions. The rate and mechanism of formation of such aggregate structures and the characteristics of the aggregate structures are therefore important parameters to describe the sedimentation behavior of such suspensions. In this study, we will discuss the formation of aggregate structures of kaolinite and their validation by cryo-SEM in the following and subsequent sections.

Particle Interactions

As mentioned previously, the sedimentation of kaolinite particles is complicated greatly by the non-uniform surface charge densities on edge and face surfaces. It was realized that electrophoretic measurements of kaolinite particles do not give detailed information about surface charge characteristics, and therefore information on electrophoretic mobility is not used in this analysis (Miller et al. 2007). Instead, surface charge densities of the two faces of kaolinite particles (silica face and alumina face) were determined from surface force measurements are used in this analysis (Gupta and Miller 2010). Recently Gupta et al. (2010) used the zeta-potential for the edge surfaces as a linear combination of the zeta-potential of silica and alumina particles. It was realized that the zeta-potential does not define the surface potential of the edge surface appropriately. Instead, potentiometric titration was used to determine the surface potential and surface charge of the edge surface of kaolinite. The surface charge density for the edge surface of kaolinite was calculated from the following the charge balance as:

$$\sigma_{\text{kaolinite}} A_{\text{kaolinite}} = \sigma_{\text{silica face}} A_{\text{silica face}} + \sigma_{\text{alumina face}} A_{\text{alumina face}} + \sigma_{\text{edge face}} A_{\text{edge face}} \tag{6}$$

where $\sigma_{\text{kaolinite}}$ is the surface charge density of kaolinite particles as determined by potentiometric titration, $\sigma_{\text{silica face}}$ and $\sigma_{\text{alumina face}}$ are the surface charge densities of the silica face and the alumina face of kaolinite as determined by surface force measurements. The symbols $A_{\text{kaolinite}}$, $A_{\text{silica face}}$, $A_{\text{alumina face}}$ and $A_{\text{edge face}}$ represent the total area of kaolinite particles, area of silica face, alumina face and edge surfaces, respectively. The area of the silica face surface, the alumina face surface and the edge surface was determined from the kaolinite particle equivalent circle diameter and thickness, the details of which are provided in the Supporting Information.

With this information the surface charge densities of the edge surface of kaolinite was determined, and the results are presented in Figure 2. As shown in Figure 2, the surface charge density of the edge surface is significantly greater, over one order of magnitude greater when compared to the surface charge densities of the silica face and the alumina face of the kaolinite particles at high pH. At low pH, the surface charge density of the edge face of kaolinite is of a similar magnitude (within a factor of two) with that of the surface charge densities of the silica face and the alumina face.

Considering the surface charge data provided for the different faces of kaolinite, the net interaction energy between different surfaces of kaolinite was determined and is shown in Figure 3. The interaction energies were scaled to maximum attractive interaction energy determined for silica face–alumina face, and multiplied by the average interaction area ratio of 13.39:1 (Gupta, Miller, and Nguyen 2010).

Figure 4 shows the scaled interaction energies calculated for a separation distance, h = 20 Å, when there is no energy barrier, consistent with previous studies (Johnson, Russell, and Scales 1998). When there is an energy barrier between surface interactions, the interaction energies were calculated at the separation distance for the maximum energy barrier.

It is evident that the silica face–silica face interaction is repulsive in the pH range of 4–10 due to the strong electrostatic repulsion between negatively charged silica faces (see Figure 3

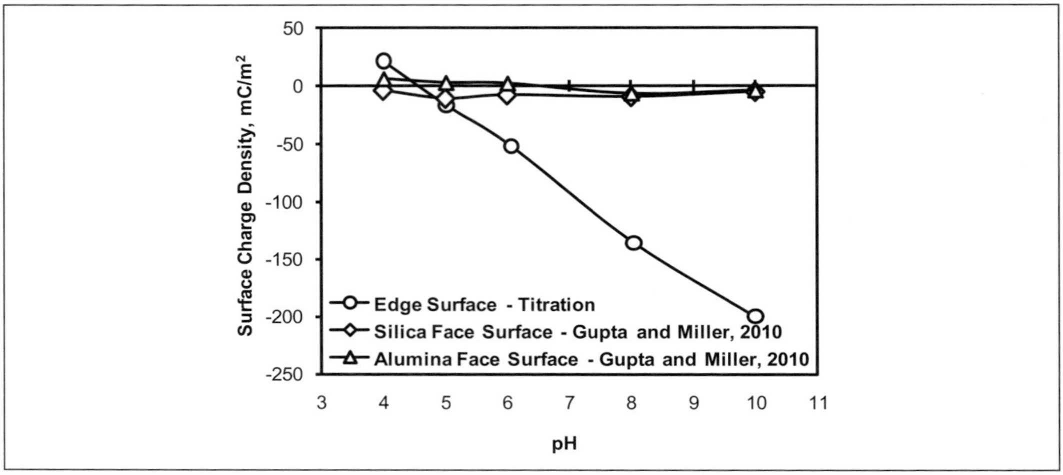

Figure 2. Surface charge densities of the kaolinite edge surface and the two face surfaces (silica and alumina face) as a function of pH in 1 mM KCl solution (Gupta et al. 2011)

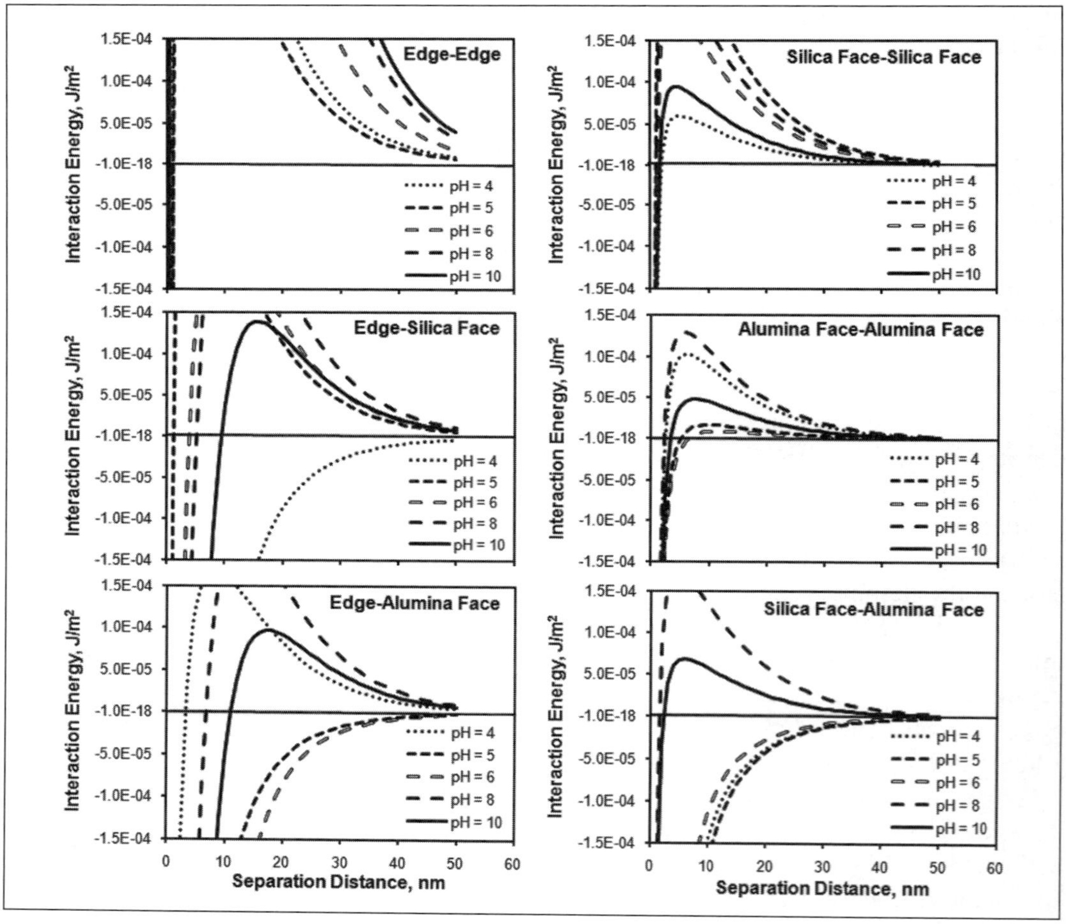

Figure 3. Interaction energy profiles for different surface interactions of kaolinite particles (Gupta et al. 2011)

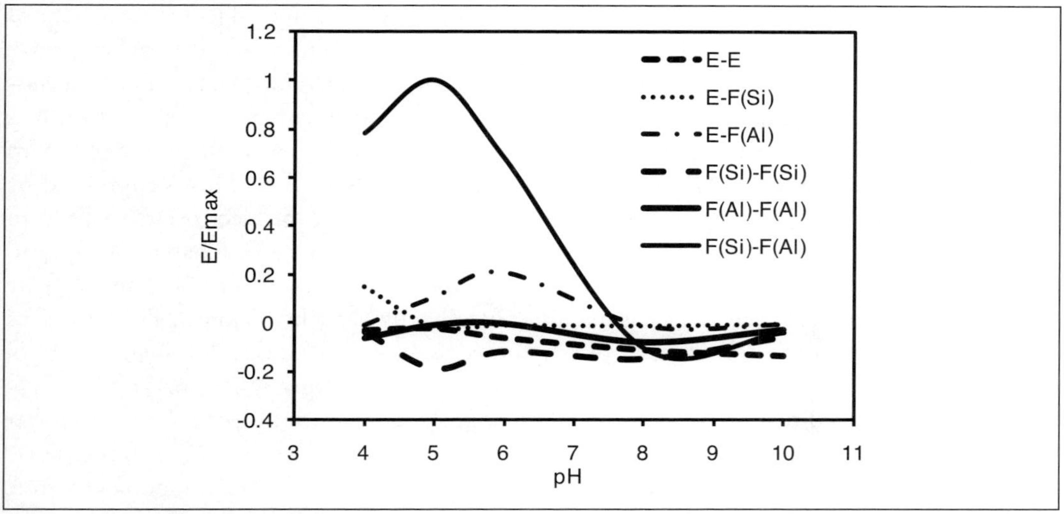

Figure 4. Predicted kaolinite edge–edge, edge–silica face (E-F(Si)), edge–alumina face (E-Al(Si)), silica face–silica face (F(Si)-F(Si)), alumina face–alumina face (F(Al)-F(Al)), and silica face–alumina face (F(Si)-F(Al)) interaction energies scaled to maximum attractive energy for the silica face–alumina face interaction. Kaolinite particles with diameter 600 nm and thickness 11.2 nm in 1 mM KCl solution (Gupta et al. 2011).

and Figure 4). The alumina face–alumina face interactions showed a slight repulsion with a small energy barrier at pH 6 due to weak electrostatic repulsion and stronger van der Waals attraction. Below pH 7.5, the silica face–alumina face interaction is attractive with a maximum at pH 5, whereas a slight repulsion was observed at higher pH (pH > 7.5). This is due to the opposite sign of the surface charge densities for the silica face and alumina face at pH ≤ 6 (silica face is negatively charged and alumina face is positively charged), causing stronger attraction both electrostatic and van der Waals attraction (see Figure 3 and Figure 4). At higher pH (pH ≥ 8), both the silica face and alumina face carry the same surface charge sign (both silica and alumina face carry negative charge), and hence a slight repulsive electrostatic interaction was found. The edge–edge surfaces showed repulsive interaction for the entire pH range, due to the similar nature of surface charge, i.e., edge surfaces carry a positive surface charge at pH 4, which becomes increasingly negative at pH values greater than 5 (see Figure 2). The edge–alumina face showed repulsive interactions at low pH (pH of 4) and at high pH (pH ≥ 8) due to strong electrostatic repulsive interactions, whereas an attractive interaction was found at an intermediate pH of 5–6. In contrast, the edge–silica face showed attractive interaction only at low pH 4 and repulsion at a higher pH (pH ≥ 5), due to similar nature of the surface charge on the edge surface and the silica face of kaolinite at higher pH. The edge–silica face and edge–alumina face interactions indicate that the edge–face interactions are favorable at pH < 8. Though the magnitude of edge–silica face and edge–alumina face interactions suggest an insignificant interaction on an interaction area basis, the interaction cannot be ignored as the edge effect could become significant in concentrated suspensions when the kaolinite particles are stacked due to the silica face–alumina face interaction.

On the basis of these different interactions shown in Figure 3 and Figure 4, it is expected that the overall interaction of kaolinite particles will be dominated by silica face–alumina face in acidic solutions (pH < 7.5). Whereas, the alumina face–alumina face interactions are unfavorable over the entire pH range of the system. Although lower in magnitude, the edge–silica face showed favorable interaction at low pH of 4, and therefore some edge–face associations will also be expected at low pH. Thus, it is anticipated that the kaolinite particles will aggregate initially in a silica face–alumina face manner forming a lamellar tactoid structure in kaolin suspensions at low pH values (Schofield and Samson 1954), but still with a relatively low settling rate. As the pH rises, the face–face association will grow, causing the formation of tactoids with thicker edge surfaces, and thereby promoting the edge–face association. This is supported by Secor and Radke (1985) who found by numerical simulation that the electrostatic field from the basal plane may "spill-over" to dominate the positive edge surface. Chang and Sposito (1994) also showed that the negative electric field from the basal plane of a disc-shaped clay mineral particle near the edge surface is mainly controlled by particle thickness. This face–face association is found to be dominant at pH 5, and edge–face associations dominant at pH 6, which will results in maximum settling rate, and corresponding larger aggregates at pH 5–5.5. As the pH is increased further, the face–face and edge–face association decreases due to lower magnitude of surface charge density on edge surfaces and face surfaces resulting in lower settling rate and aggregate size. At high pH, the edge–face and face–face interaction forces become repulsive, and the system becomes completely dispersed, and a negligible settling is expected. In this way, the maximum sedimentation behavior of kaolinite can be explained based on particle aggregation and its variation with system pH. Even if the edge surface charge density and corresponding interactions are based on zeta potential measurements as dismissed at the beginning of this section on particle interactions, the relative significance of some interactions changes but in general the same conclusion is found regarding the variation of interactions with pH.

O'Brien (1971) observed the dominant face-face and some face-edge aggregation behavior of kaolinite both in distilled water and in electrolyte solutions using scanning electron microscopy of freeze-dried kaolinite samples. However, concern was raised that the freeze-drying technique can alter the structure of these aggregates during drying. Recently, Zbik et al. (2008) observed stacks of kaolinite aggregated dominantly in face-face manner at pH 8 using cryo-vitrified technique with scanning electron microscopy. The authors could not explain the face–face type interaction based on the assumption that 001 and $00\underline{1}$ faces of kaolinite are negatively charged. Instead, it follows from our new results that the kaolinite particles are aggregating according to silica face–alumina face and alumina face–alumina face interactions.

In contrast to our results, Johnson et al. (Johnson et al. 2000, 1998) predicted that the kaolinite particles will mostly interact in an edge–face manner at lower pH. However, their studies were based on the assumption that both faces of kaolinite are negatively charged, which analysis must be reconsidered in view of the AFM surface force results reported by Gupta and Miller (2010). These different surface interactions are of importance in order to control the aggregation behavior of kaolinite particles, and the mechanical properties of such suspensions.

Influence of Aspect Ratio

The influence of aspect ratio (ratio of particle diameter to thickness) on the different surface interactions is shown in Figure 5. The aspect ratio was determined from the atomic force microscopy images of kaolinite particles. About 150 particles were imaged and analyzed using Nanoscope V7.2 software for the atomic force microscope (Veeco Instruments Inc., Santa Barbara, CA). As shown, the edge–silica face and edge–alumina face interactions are significant for particles with a low aspect ratio (9) as compared to particles with a high aspect ratio (165). Notably, the edge–alumina face interactions are dominant at pH 6, whereas silica face–alumina face interactions are dominant at pH 5 for kaolinite particles with low aspect

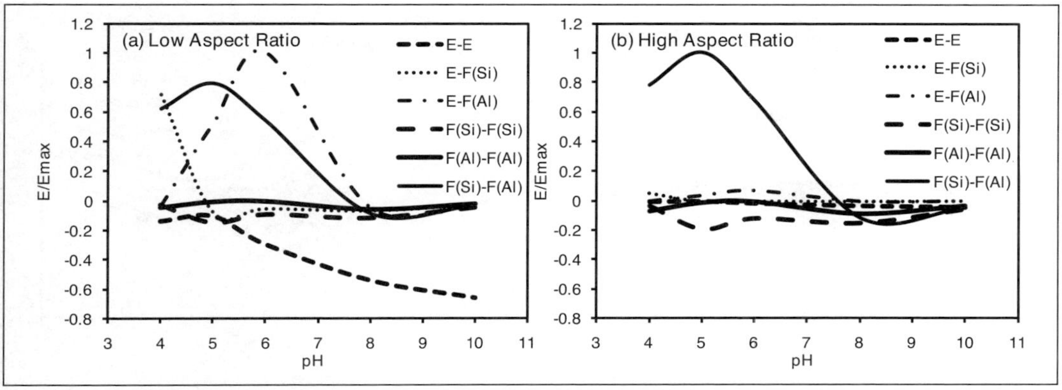

Figure 5. Effect of aspect ratio on different face type interaction of kaolinite particles (a) low aspect ratio = 9 and (b) high aspect ratio = 165 in 1 mM KCl solution. The symbols E, F(Si), and F(Al) represent edge, silica face, and alumina face, respectively (Gupta et al. 2011).

ratio. This is particularly interesting as significant edge–face interaction for particles with a low aspect ratio indicates that particles will orient themselves in both edge–face and face–face (silica face–alumina face) organizations. The particles with high aspect ratio will orient themselves dominantly in a face–face manner (silica face–alumina face). It is shown that the aggregate structures can be expected to depend on the aspect ratio of kaolinite particles, and thereby the sedimentation of the kaolinite suspensions will be affected by the aspect ratio.

Influence of Electrical Double-Layer Thickness

Figure 6 shows the effect of electric double layer thickness on the different surface interactions. Particles suspended at high ionic strength will experience a small double layer thickness (3.04 nm) (see Figure 6b). Under these circumstances there is a greater alumina face–alumina face attraction and greater repulsion of silica faces, and edge faces when compared to particles suspended in a solution with a larger double layer thickness (9.6 nm) (see Figure 6a). Also at high ionic strength, there is a greater increase in the edge–alumina face interactions which promote edge–face associations as compared to low ionic strength. It is evident that at high ionic strength, i.e., at small values of the reciprocal of the Debye constant (the thickness of the double layer), κ^{-1} = 3.04 nm, the silica face–alumina face interactions will be increased at pH 5 with improved aggregation. At the same time, the increased alumina face–alumina face interactions will expose the silica faces on the kaolinite particles, which may also further promote aggregation with alumina faces, forming a larger aggregate structure. Olphen (1963) observed that with addition of NaCl, settling rate increase, slowly at first, and rather sharply when the flocculating concentration of NaCl for the clay was approached.

Aggregate Structure

In a suspension of clay particles, three different modes of particle association or aggregate structure may occur: face–face, edge–face, and edge–edge (Olphen 1963). The DLVO interaction energies (electrostatic energy and the van der Waals interaction energy) for the three types of association are governed by six different combinations of the three surfaces interactions—the silica face, the alumina face and the edge surface, as explained previously. Consequently,

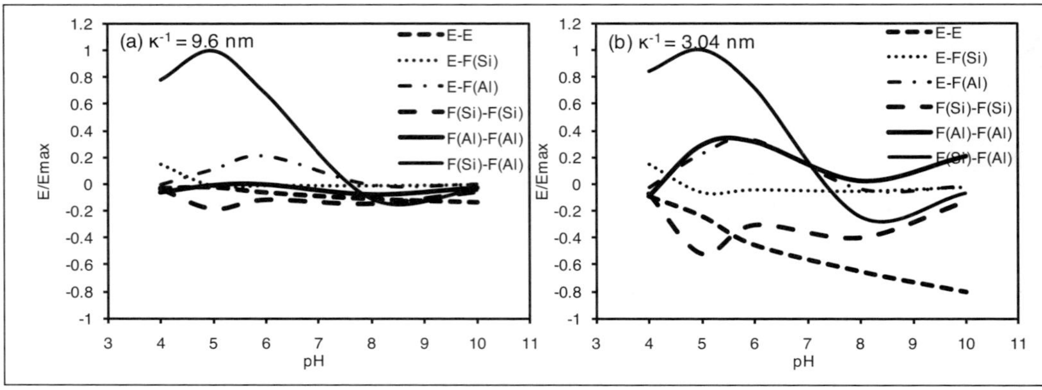

Figure 6. Effect of electric double-layer thickness (κ^{-1}) on different surface interactions for kaolinite particles with particle diameter 600 nm and thickness 11.2 nm at (a) κ^{-1} = 9.6 nm (1 mM KCl solution) and (b) κ^{-1} = 3.04 nm (10 mM KCl solution). The zeta-potentials of the silica face and the alumina face are assumed to be reduced by 30% with increasing ionic strength from 1 mM to 10 mM KCl solution. The edge surface potential data for 10 mM KCl solution were taken from literature (Williams and Williams 1978). The symbols E, F(Si), and F(Al) represent edge, silica face, and alumina face surfaces, respectively (Gupta et al. 2011).

the three types of association will not necessarily occur simultaneously or to the same extent when a kaolinite suspension is aggregated.

Face–face associations will lead to thicker and possibly larger aggregates, whereas edge–face and edge–edge associations will form three-dimensional voluminous card-house structures (Olphen 1963). The various modes of particle association are shown in Figure 7.

The SEM micrographs of kaolinite aggregates under cryogenic conditions at pH 3, 5, 7, and 9 are shown in Figure 8. It can be seen that the kaolinite particles are mostly associated in a face–face manner (silica face–alumina face and/or alumina face–alumina face), and some face–edge organization (edge–silica face) at pH 3 and pH 5 (see Figure 8). This is in good agreement with the theoretical predictions that at low pH, the attraction between the silica face and the alumina face dominates and thereby accounts for the face–face association. This face–face association also promotes edge–face aggregation of kaolinite particles with an increase in pH to pH 5.

At pH 7, the particles are mostly associated in edge–edge manner and edge–face manner (edge–silica face) (Figure 8). At higher pH (pH = 9), the particles are mostly associated in edge–edge manner creating a porous structure. Also, due to repulsion between the silica face and alumina face, face–face association was not observed at pH 9, in agreement with theoretical considerations. The edge–edge interaction at pH 9.0 should also be unfavorable due to similar surface charge density at the edge surfaces, which causes repulsion. The edge–edge associations at high pH 9 were contrary to DLVO expectation, since all the surfaces of kaolinite particles are negatively charged, and the suspension is stable. In agreement with our observation, Zbik and Horn (2003) also observed similar structures involving edge–edge association for kaolinite particles, and they proposed that hydrophobic interaction between edge surfaces could contribute to such structures. Such a proposition is unlikely, but now can be given further consideration based on the development of AFM surface force techniques for the measurement of hydrophobicity (Miller et al. 2011, Yin and Miller 2011). Alternatively, the edge–edge association observed at pH 9.0 may be due to the experimental conditions associated with freezing the sample under cryogenic conditions.

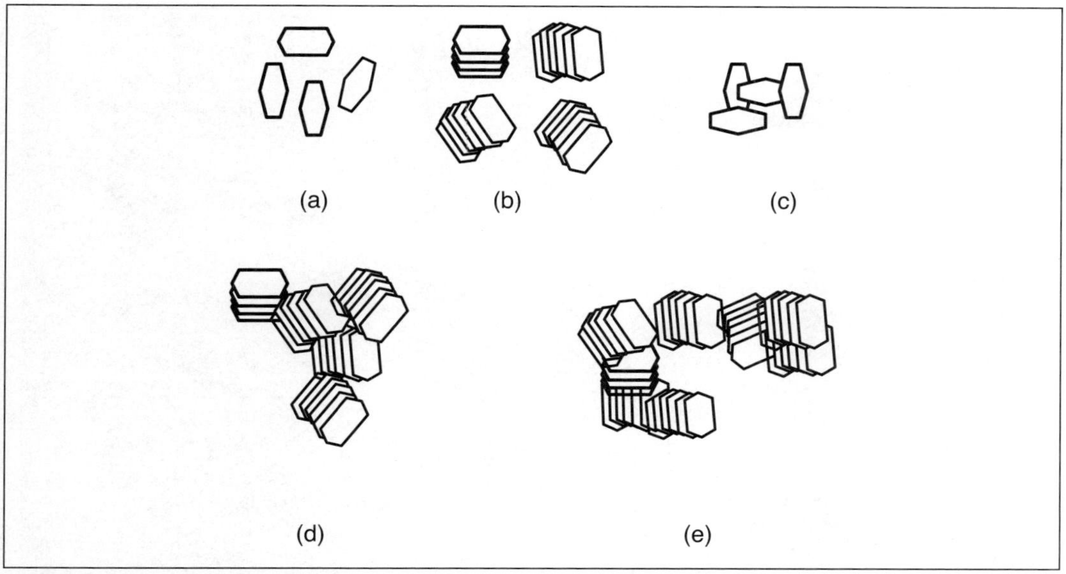

Figure 7. Formation of aggregate structures in kaolinite suspensions, as (a) dispersed, (b) face–face, (c) edge–face, (d) edge–edge, and (e) a combination of b, c, and d, depending on the solution chemistry of the suspension (Gupta et al. 2011)

The cryogenic-SEM images showed the kaolinite aggregate structure for a 4% kaolinite suspension. This concentration may well be above the gelation point (Zbik and Frost 2009), but the general trends of the aggregate structure showing dominant face–face association was also observed in dilute (0.01%) and semi-dilute (0.1%) kaolinite suspensions (SEM graphs not shown). Our results are also supported by Zbik and Frost (2009) who observed similar face–face and edge–face contacts in kaolinite from Georgia. Other modes of particle associations such as stair step edge–edge contacts were also revealed in Birdwood kaolinite aggregates (Zbik and Frost 2009).

Aggregate Size and Sedimentation

The aggregate structure of kaolinite suspensions at different pH values as revealed by cryo-SEM are also supported by further experiments conducted using photon correlation spectroscopy (PCS) to estimate the particle size of such aggregates. In our analysis, we estimated the equivalent sphere diameter of aggregate-structures for kaolinite suspensions. These aggregate structures for kaolinite suspensions are three dimensional networks of particles forming an arbitrary shape or so called "card-house" structure (Olphen 1963). The precise information for the size of aggregate structures in kaolinite suspensions could not be revealed with present instrumentation, and only qualitative trends are realized. Figure 9 shows the average particle size of kaolinite aggregates in suspension as a function of pH. The average aggregate size of kaolinite is about 16.3±0.3 nm at both pH 7 and 9. The average aggregate size of kaolinite remains unchanged when measured as a function of time at pH 9, which represents a more dispersed state for the kaolinite particles. The aggregate size grows to 38.9±0.3 nm at pH 7 in 40 minutes, which could be indicative of a loose aggregate structure formed mainly by edge–edge associations. The initial aggregate sizes for kaolinite suspensions at pH 3.5 and 5 were 266.8±12.0 nm and 127.1±7.9 nm, respectively. The aggregates of kaolinite grow over

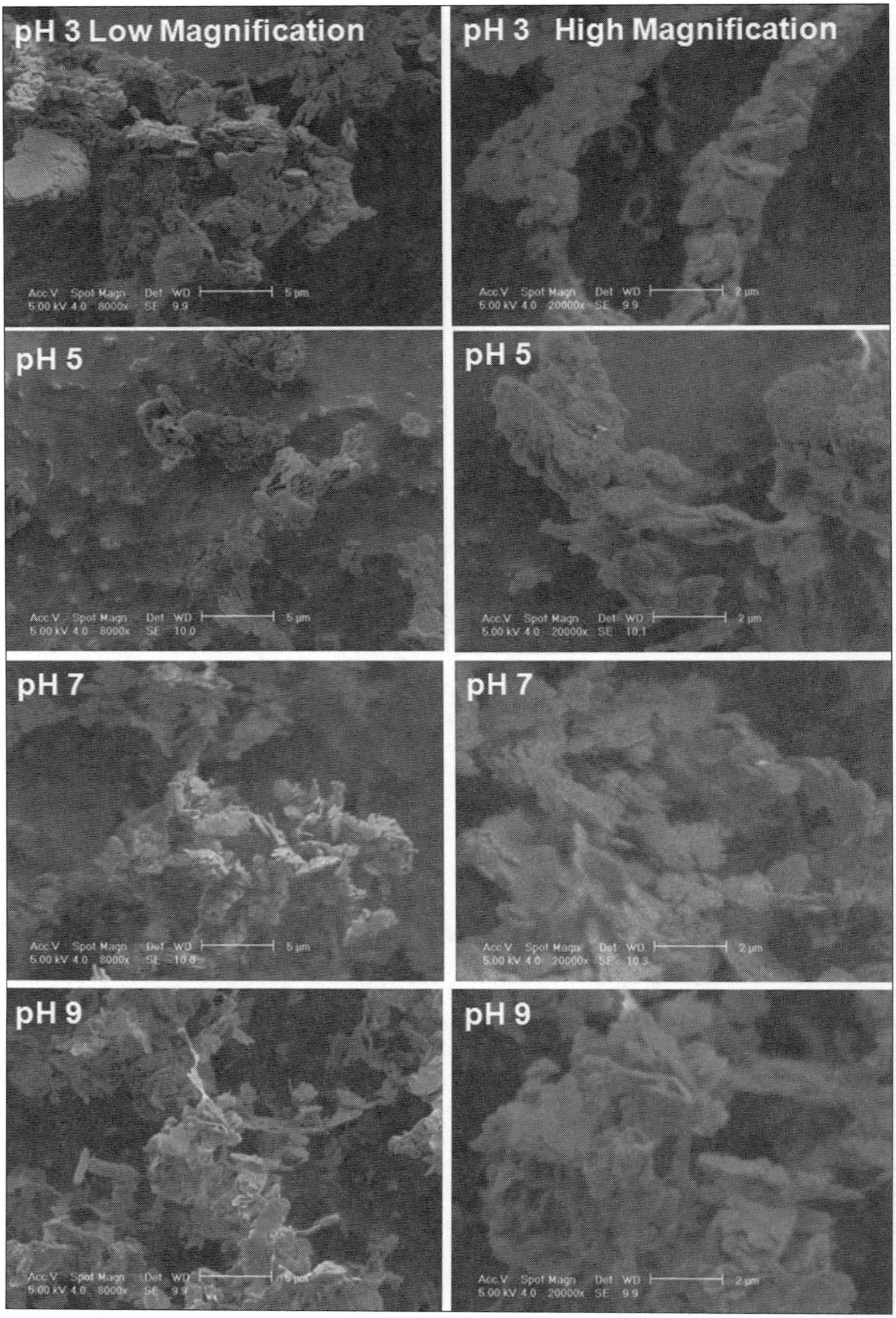

Figure 8. Cryo-SEM micrographs of kaolinite aggregates at pH 3, 5, 7, and 9 at low (left) and high (right) magnification (Gupta et al. 2011)

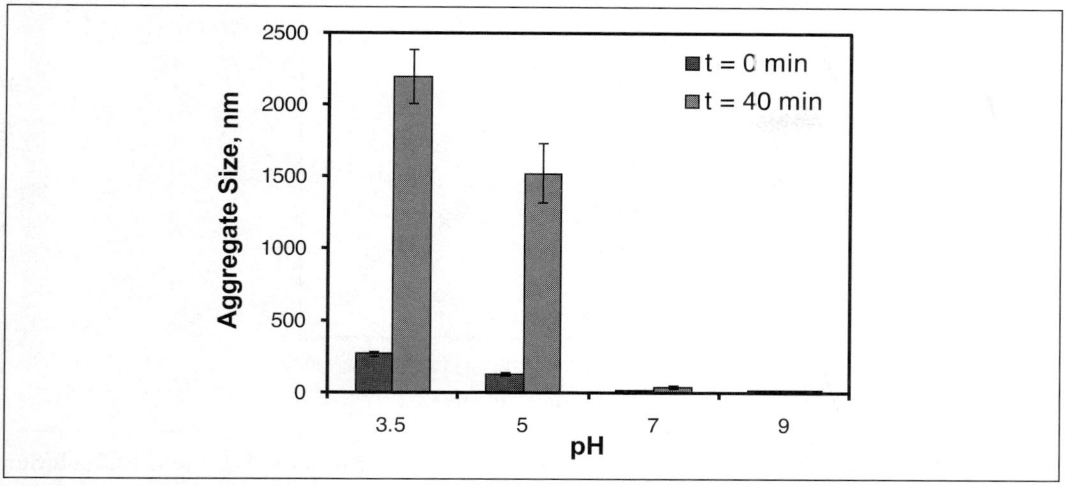

Figure 9. Average aggregate size for a suspension of kaolinite particles as a function of pH (Gupta et al. 2011)

an order of magnitude in size in just a few minutes at pH 3 and 5, and then remain constant. These results suggest that kaolinite particles are associated with all three modes of interactions—face–face, edge–face and edge-edge—forming a three-dimensional network.

Figure 10 shows the sedimentation behavior of kaolinite suspensions as a function of pH. As expected the initial settling rate of kaolinite suspension is largest at pH 5 followed by the results at pH 3.7, whereas very slow settling was observed at pH 7 and 9. Also, it was observed that kaolinite sediment was quite compacted at pH 5, whereas it remained relatively fluid at pH 3.7 (the sediment was mobile when the graduated cylinder was moved). Another interesting observation was made, when similar sedimentation experiments were conducted with the addition of 24 μm and 36 μm silica particles (0.1 wt. ratio of silica to kaolinite (see Figure 11). As can be seen, the sedimentation of kaolinite only slightly increased at pH 5 with the addition of 24 μm and 36 μm silica particles in separate experiments. The drastic effect of the addition of silica particles can be seen at pH 7 and pH 9 which show significantly increased sedimentation with the addition of 36 μm silica particles. Despite repulsive interaction between silica and kaolinite particles at pH 9, the improved sedimentation is rather surprising, and will be investigated in future research. For example, researchers from the University of Florida also conducted experiments with different amounts of sand/clay mixing and using hydrocyclones as a rapid dewatering device (El-Shall 2009). Their results indicate that up to 80% of the water could be recovered and recycled back to the plant in a few minutes (El-Shall 2009). Of course, as stated previously, the aggregate structure as revealed by cryo-SEM and PCS will influence the mechanical properties of kaolinite suspensions, and thereby affect the sedimentation of such suspensions.

CONCLUSIONS

The sedimentation of kaolinite suspensions particularly the settling rate is explained based on the aggregate structure of the kaolinite particles. A new analysis based on the surface charge densities of the silica face, alumina face and edge surfaces demonstrates the role of pair-wise interactions in controlling the sedimentation of kaolinite particulate suspensions. The maximum aggregate-size of kaolinite suspensions is indicative of the significant role played by

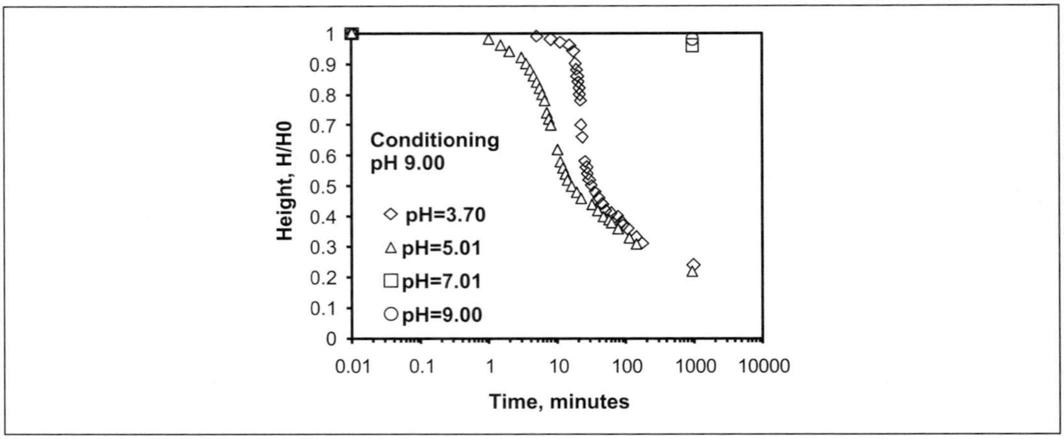

Figure 10. Sedimentation of kaolinite particles (4% w/v) initially dispersed in 1 mM KCl solution at pH 9 as a function of pH (Gupta et al. 2011)

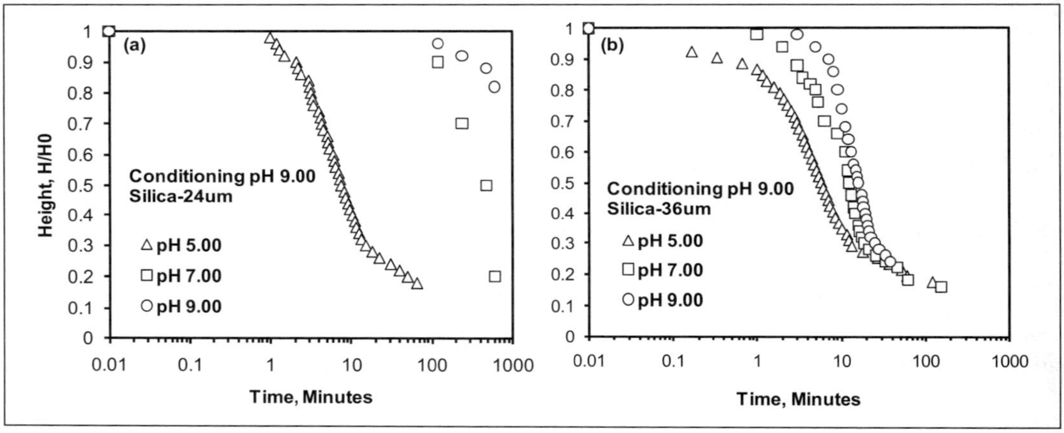

Figure 11. Sedimentation of kaolinite particles (4% w/v) initially dispersed in 1 mM KCl solution at pH 9 as a function of pH, with the addition (silica/kaolinite = 0.1 by weight) of (a) 24 μm and (b) 36 μm silica particles

particle organization in the formation of aggregates as governed by edge–edge, edge–silica face, edge–alumina face, silica face–alumina face, silica face–silica face, and alumina face–alumina face interaction energies.

The results indicate that the face–face (silica face–alumina face) is the dominant particle interaction at low pH values, which promotes edge–face (edge–silica face and edge–alumina face) and face–face (silica face–alumina face) interaction at intermediate pH values, explaining the maximum aggregate size at pH 5–5.5. This conclusion is also confirmed by cryo-SEM analysis of kaolinite aggregate structure and photon correlation spectroscopy. At a higher pH (pH = 7), the kaolinite aggregates show edge–edge and edge–face interactions, whereas a more porous structure of aggregates was observed at high pH (pH = 9).

The influences of particle aspect ratio and electrical double layer thickness were also examined to determine their effect on interaction energies. The findings show that the edge surface–silica face interactions become favorable at a low aspect ratio, whereas the silica face–alumina face and alumina face–alumina face interactions are increased at high ionic strength conditions.

These different interactions and their relative significance, provide the basis for a more detailed explanation of the aggregates of kaolinite suspensions as well as an improved foundation for the modification of kaolinite interactions and control of sedimentation characteristics.

The addition of silica particles to the kaolinite suspension can have a significant effect on the sedimentation rate and extent of consolidation depending on the size of the silica particles. Further research in this area is warranted.

ACKNOWLEDGMENTS

The authors acknowledge the financial support of grant DE-FG-03-93-ER14315 from DOE Basic Sciences and grant DP0663688 from the Australian Research Council, and the CMM at the University of Queensland for the use of the cryogenic SEM.

REFERENCES

Bergstrom, L. 1997. Hamaker constants of inorganic materials. *Adv. Colloid Interface Sci.* 70:125–169.

Bidwell, J.I., W.B. Jepson, and G.L. Toms. 1970. The interaction of kaolinite with polyphosphate and polyacrylate in aqueous solutions—Some preliminary results. *Clay Miner.* 8:445–459.

Casagrande, A. 1940. The structure of clay and its importance in foundation engineering. In *Boston Society of Civil Engineers*, pp. 72–125.

Chang, F.R.C., and G. Sposito. 1994. The electrical double layer of a disk-shaped clay mineral particle: Effect of particle size. *J. Colloid Interface Sci.* 163:19–27.

El-Shall, H. 2009. *Field Demonstration/Evaluation of a Rapid Clay Dewatering and Consolidation Process Using Other Wastes (FIPR/DIPR Process) to Minimize Clay Settling Ponds.* Florida Institute of Phosphate Research, 02-168-232.

Gupta, V. 2011. *Surface charge features of kaolinite particles and their interactions.* Ph.D. thesis, University of Utah.

Gupta, V., M.A. Hampton, J.R. Stokes, A.V. Nguyen, and J.D. Miller. 2011. Particle interactions in kaolinite suspensions and corresponding aggregate structures. *J. Colloid Interface Sci.* 359:95–103.

Gupta, V., and J.D. Miller. 2010. Surface force measurements at the basal planes of ordered kaolinite particles. *J. Colloid Interface Sci.* 344:362–371.

Gupta, V., J.D. Miller, and A.V. Nguyen. 2010. Further analysis of the rheological behavior of kaolinite suspensions. *Conference of Metallurgists, Vancouver, BC, 2010.*

Johnson, S.B., G.V. Franks, P.J. Scales, D.V. Boger, and T.W. Healy. 2000. Surface chemistry-rheology relationships in concentrated mineral suspensions. *Int. J. Miner. Process.* 58:267–304.

Johnson, S.B., A.S. Russell, and P.J. Scales. 1998. Volume fraction effects in shear rheology and electroacoustic studies of concentrated alumina and kaolin suspensions. *Colloids Surf. A* 141:119–130.

Lambe, T.W. 1953. The structure of inorganic soil. *Amer. Soc. Civil Eng. Proc.* pp. 1–49.

Langmuir, I. 1938. The role of attractive and repulsive forces in the formation of tactoids, thrixotropic gels, protein crystals and coacervates. *J. Chem. Phys.* 6:873–896.

Masliyah, J.H., and S. Bhattacharjee. 2006. *Electrokinetic and Colloid Transport Phenomena.* John Wiley and Sons.

Miller, J.D., V. Gupta, H. Du, X. Wang, X. Yin, and J. Wang. 2011. The surface chemistry of layered silicate minerals. In *Proceedings 2011 SME Annual Meeting, Denver, Colorado, 2011.*

Miller, J.D., J. Nalaskowski, B. Abdul, and H. Du. 2007. Surface characteristics of kaolinite and other selected two-layer silicate minerals. *Can. J. Chem. Eng.* 85:617–624.

Nguyen, A.V., and H.J. Schulze. 2004. *Colloidal Science of Flotation. Surfactant Science Series.* Marcel Dekker.

O'Brien, N.R. 1971. Fabric of kaolinite and illite floccules. *Clays and Clay Minerals, Proceedings of the Conference* 19:353–359.

Olphen, H.V. 1963. *An Introduction to Clay Colloid Chemistry: For Clay Technologists, Geologists and Soil Scientists*. New York: Interscience.

Rand, B., and I.E. Melton. 1977. Particle interactions in aqueous kaolinite suspensions. I. Effect of pH and electrolyte upon the mode of particle interaction in homoionic sodium kaolinite suspensions. *J. Colloid Interface Sci.* 60:308–320.

Rosenqvist, I.T. 1959. Physico-chemical properties of soils: Soil-water systems. *Amer. Soc. Civil Eng. Proc., 1959*, pp. 31–53.

Rosenqvist, I.T. 1963. The influence of physico-chemical factors upon the mechanical properties of clays. *Norwegian Geotechnical Institute 1963*, pp. 1–19.

Schofield, R.K., and H.R. Samson. 1954. Flocculation of kaolinite due to the attraction of oppositely charged crystal faces. *Discuss. Faraday Soc.* 18:135–145.

Secor, R.B., and C.J. Radke. 1985. Spillover of the diffuse double layer on montmorillonite particles. *J. Colloid Interface Sci.* 103:237–244.

Street, N. 1956. The rheology of kaolinite suspensions. *Aust. J. Chem.* 9:467–479.

Street, N., and A.S. Buchanan. 1956. The ζ potential of kaolinite particles. *Aust. J. Chem.* 9:450–466.

Tan, T.K. 1958. Discussion of soil properties and their measurement. *4th Int. Conf. Soil Mech. Found. Eng. Proc., 1958*, pp. 87–89.

Williams, D.J.A., and K.P. Williams. 1978. Electrophoresis and zeta potential of kaolinite. *J. Colloid Interface Sci.* 65:79–87.

Yin, X., and J.D. Miller. 2011. Wettability of kaolinite basal planes based on surface force measurements using atomic force microscopy. *Miner. Metall. Process.* (forthcoming, Industrial Minerals special issue).

Zbik, M., and R.G. Horn. 2003. Hydrophobic attraction may contribute to aqueous flocculation of clays. *Colloids and Surf. A* 222:323–328.

Zbik, M.S., and R.L. Frost. 2009. Microstructure differences in kaolinite suspensions. *J. Colloid Interface Sci.* 339:110–116.

Zbik, M.S., R.S.C. Smart, and G.E. Morris. 2008. Kaolinite flocculation structure. *J. Colloid Interface Sci.* 328:73–80.

Multi-Scale Investigation of Applying Secondary Effluent in Sulfide Flotation

Jinhong Zhang and Wei Zhang
Department of Mining and Geological Engineering, University of Arizona, Tucson, AZ, USA

ABSTRACT

The fact that froth flotation consumes a large amount of water, even though it is one of the most efficient methods to process sulfide minerals, brings a higher pressure from the sustainable development due to increasing population and its demand of more portable water. Pioneering work of applying secondary effluent in sulfide flotation (Fisher and Rudy, 1976) showed a 2.4% reduction in Cu recovery and 16.2% reduction in molybdenum recovery when secondary effluent was used. It was also postulated that the organic carbon, in the form of humic acid, in the effluent was the most deleterious constituent causing the losses in metal recovery. We carried out a systemic investigation in both microscopic and macroscopic aspects on the possibility of using secondary effluent in sulfide flotation. AFM images showed that collectors adsorbed on mineral surface (chalcopyrite and molybdenite) in a similar manner in both clean water and treated secondary effluent. Lab flotation tests showed that the Cu and Mo recoveries obtained with treated secondary effluent were comparable to those obtained with tap water. The findings of present multi-scale investigation will help provide a cost-efficient solution to treat low quality water and mitigate its impact, and finally succeed technically in applying secondary effluent in sulfide flotation.

INTRODUCTION

In the extractive minerals industry, valuable sulfide minerals are selectively separated from non-valuable gangue by froth flotation, which treats the highest throughput and produces the maximum economic outcome of any surface chemistry process. (Sutherland and Wark, 1955; Gaudin, 1957; Fuerstenau, 1962; Klassen and Mokrousov, 1963, Fuerstenau, 1976) Froth flotation has been widely applied in sulfide processing for more than a hundred years due to its high efficiency. However, in spite of the fact that a large portion of process water is recycled in a concentrator, the process of froth flotation involves a large consumption of fresh water, which is especially precious in arid areas, such as Arizona. For example, the water consumption at a typical copper mine is about 20,000 gal/min in total and 15,000 gal/min for the concentrator. That is, usually about 80% of the feed water to a mine is used at the concentrator for flotation.

Superficially it would seem that there is no need to use high quality fresh water for flotation because the water is mixed in a flotation cell with rock and minerals which are "dirty." This raises the logical question whether low quality water, i.e., secondary effluent (reclaimed waste water), can be used for sulfide flotation with no decrease in the efficiency of the process (Pickett, 1973; Fisher and Rudy, 1976; King et al., 1983; Amy et al., 1984; Deboer and Linstedt, 1985; Lieuwen, 1989).

It was reported in a pilot research on the use of low quality water for mineral processing that when secondary treated sewage effluent was used instead of demineralized water, copper recovery decreased by 2.4 percent and molybdenum recovery dropped by 16.2 percent. (Fisher and Rudy, 1976) It was postulated that the organic content in the effluent was the deleterious species which introduced the decrease of mineral recovery. It was also reported that by tertiary treatment with activated carbon, the detrimental effects of the effluent could be reduced.

Another prior study (King et al. 1983; Amy et al., 1984) has also been carried out to compare the mineral recovery obtained from the froth flotation using secondary effluent, tertiary effluent, groundwater, CAP (Central Arizona Project) water, and mixtures of secondary effluent with groundwater or CAP water. It was reported that the relative flotation efficiency decreased with the increase of the concentration of nonvolatile total organic carbon (NVTOC), an indicator of the organic content of the water or effluent. In addition, the reported results showed that copper recovery decreased from 92.2% to 88.7% and the molybdenum recovery decreased from 88.6% to 81.6% when secondary effluent was used. It was also suggested that the detrimental effects of the organic constituents in secondary effluent can be mitigated by dilution or by tertiary treatment.

A decrease in metal recovery is usually not acceptable. The metal that is not recovered in the flotation process becomes part of the waste stream of the tailings. Additionally, a traditional tertiary treatment makes the application economically not acceptable. These factors constitute the principal obstacles to directly using effluent in the mining industry. In practice, only small amounts of secondary effluent are used in some concentrators. Therefore, it is a major concern to increase the portion of reclaimed or low quality water in flotation without applying tertiary treatment. However, the fact that little is known about the exact mechanism of low quality water impacting flotation hinders its application.

The aim of present study is to systemically study the impact of effluent on the flotation of sulfide minerals. It will help provide solutions to mitigate the impact of low quality water and directly apply effluent in flotation without diluting it greatly or applying tertiary treatment. Due to the fact that froth flotation consumes a lot of water and the consumption of water will keep on increasing in the near future, the change of water source will greatly decrease the consumption of fresh water in a concentrator and help conserve groundwater supplies for drinking purposes to meet the increasing demands arising from the population increase. Both the community and the mines will benefit from the success of the proposed project of applying secondary effluent in froth flotation.

EXPERIMENTAL

Materials

Research grade chalcopyrite ($CuFeS_2$) and Molybdenite (MoS_2) were obtained from Wards Natural Science Establishment Inc. Chalcopyrite sample was finely polished, further cleaned by rinsing thoroughly with ethanol and water and a 1.2 cm ×1.2 cm sample was used for the surface characterization experiments. Molybdenite sample was prepared by peeling a piece of chunk along the cleavage surface. Fresh cleavage surface, with no further cleaning treatment, was used for the AFM imaging experiment. The DI water used in present work has a conductivity of 18.2 MΩ-cm at 25°C and a surface tension of 72.5 mN/m at 25°C. Ore samples A, B and C were obtained from a southern Arizona mine. Nalco 9740 and MCO are industrial flotation collectors being used in some concentrators. Nalco 9740 is a chalcopyrite collector and its MSDS shows the main components include heavy aromatic naphtha, naphthalene and alkyl mercaptan. MCO is a blend of an organosulfur hydrocarbon and heavy hydrocarbon and used as a collector for molybdenite.

AFM Surface Imaging

Surface imaging of mineral samples was carried out in contact mode using a Digital Instrument Nanoscope IIId AFM at room temperature (25±1°C). Silicon nitride NP-20 cantilevers with nominal spring constant of 0.12–0.58 N/m were obtained from Veeco, CA. Imaging was carried out immediately after DI water was injected into an AFM liquid cell as well as after a longer soaking time. No noticeable change in images was detected in present study. After the surface image in water was collected, prepared chemical solution was flushed through the liquid cell and the AFM measurement was commenced 5 minutes after the exposure of the mineral surface to solution.

Flotation Test

Collected concentrator feed was crushed to –10 mesh as lab flotation feed. The crushed ore sample was put into a ball mill filled with different water sources and ground for a specific time, which was determined beforehand through a grinding test to make a 65% (in weight) finer than 100 mesh flotation feed. Collectors were added in to the mill just before grinding. After grinding, pulp was transferred to a Denver D12 lab flotation cell and $Ca(OH)_2$ was added to bring pulp pH to 11. Preparation time is one minute with impeller string after a specific amount of frother was added into the flotation cell. Air valve was then turn on and flotation began. Flotation time in present study was set at 6 minutes. After that, concentrate and tailing were collected, filtered and dried overnight. Products were further sent out for metal analysis. Every flotation test was repeated for three times and an arithmetic average metal recovery was reported. The experiment error is usually less than ±1% in metal recovery.

The water sources are tap water, secondary effluent, the supernatant of $Ca(OH)_2$ treated secondary effluent and the mixture of tap water and secondary effluent at 1:1 volume ratio. Three different ore samples, i.e., sample A, B and C, were used for the flotation tests. Metal analysis shows that in the order of sample A, B and C, the head grade of copper is respectively 0.21%, 0.20% and 0.187% and the head grade of molybdenum is respectively 0.034%, 0.025% and 0.037%.

RESULTS AND DISCUSSION

AFM Surface Images

Figure 1 shows the AFM images of a chalcopyrite surface soaked in the solution of Nalco 9740. Figure 1A is the image of a bare chalcopyrite surface, which has been soaked in DI water in an AFM liquid cell for 10 minutes. In fact, no detectable change in surface morphology has been observed even after water was injected into the cell for one hour. The image clearly shows that chalcopyrite is relatively inert in water and no evident reaction happens at the chalcopyrite/water interface. From the image of a 10 µm × 10 µm scan area, one can see that the solid surface is quite smooth in spite of the fact that there are some scratch lines on the sample surface due to surface polishing. The smoothness also makes it applicable to study the chemicals adsorption on mineral surface using an AFM. Figure 1b is the height image of a chalcopyrite surface soaked in 0.02 g/L Nalco 9740 solution for 5 minutes with a 50 nm data scale in a 10 µm × 10 µm scan area. It can be clearly seen from the image that many small patches are uniformly distributed on the whole mineral surface. Figure 1c is the AFM deflection image, which clearly shows the shape and morphology of the patches. Figure 1d is the 3-D image of Figure 1b.

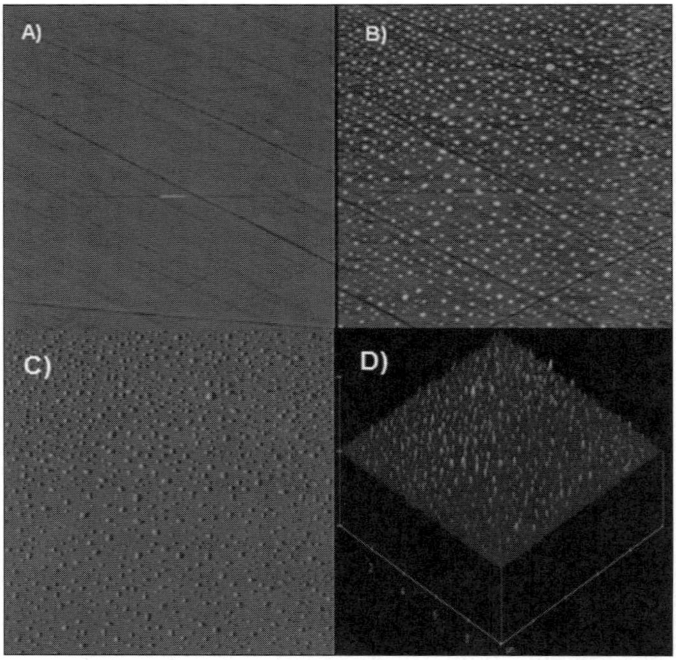

Figure 1. AFM images of a chalcopyrite surface: (a) in DI water; (b) soaked for 5 minutes in Nalco 9740 solution (0.02 g/L) prepared from DI water at pH 11 with the addition of $Ca(OH)_2$; the height image with a data scale of 50 nm; (c) the deflection image of Figure 1b with a data scale of 30 nm; and (d) the 3-D image of Figure 1b.

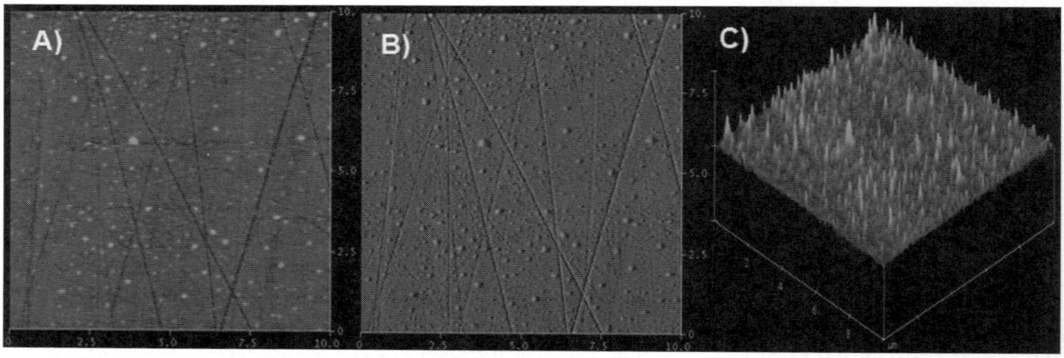

Figure 2. AFM images of a chalcopyrite surface soaked for 5 minutes in the supernatant of Nalco 9740 solution (0.02 g/L), which was prepared from the secondary effluent at pH 11 with the addition of $Ca(OH)_2$; (a) the height image with a data scale of 50 nm; (b) the deflection image of Figure 2a with a data scale of 30 nm; and (c) the 3-D image of Figure 2a.

Figure 2 shows the AFM images of a chalcopyrite surface soaked in the supernatant of Nalco 9740 solution (0.02 g/L), which was prepared from the secondary effluent at pH 11 with the addition of $Ca(OH)_2$. Figure 2a is the height image, which was taken after the sample contacted solution for 5 minutes, with a data scale of 50 nm. The images showed that the chalcopyrite surface was covered by some patches due to the adsorption of the collector. Some scratch lines due to surface polishing were also shown on the sample surface. Figure 2b is

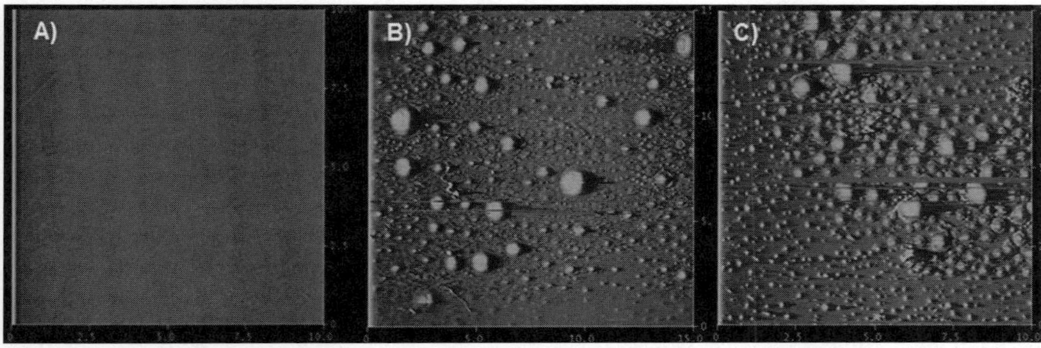

Figure 3. AFM deflection images of a molybdenite surface soaked in (a) DI water; (b) MCO solution (0.1 mg/L) prepared from DI water; (c) MCO solution (0.1 mg/L) prepared from the supernatant of $Ca(OH)_2$ treated secondary effluent

the deflection image of Figure 2a with a data scale of 30 nm to show the morphology of the patches. Figure 2c is the 3-D image of Figure 2a.

Figure 3 shows the AFM deflection images of a molybdenite surface soaked in different aqueous conditions. Figure 3a is the image with a 5 nm data scale obtained after the mineral surface contacted DI water for 10 minutes. The solid surface is very smooth without any detectable cracks or particles on the solid surface in a 10 µm × 10 µm scan area. Because the sample was prepared by peeling along the cleavage surface, no scratch as obtained with polishing chalcopyrite sample was observed on molybdenite surface. Once again, a smooth substrate is beneficial for a study of the adsorption of chemicals on mineral surface because of the minimum interference arising from the background morphology. In addition, the fact that the obtained solid surface is smooth suggested that the freshly cleaved molybdenite surface studied in present investigation was mainly the basal surface.

Figure 3b is the AFM image obtained 5 minutes after the MCO solution (0.1 mg/L) prepared from DI water flushed through the AFM liquid cell. The surface morphology changes greatly from the one obtained with Figure 3a. A lot of chemicals were observed being absorbed on mineral surface. It is also worthy while to mention that actually the surface morphology remained the same even after water was injected into the cell for more than an hour. It means molybdenite basal surface is very inert in water and some reaction such as surface oxidation is neglectable at the studied molybdenite/water interface. Therefore, the change in surface morphology is only due to the adsorption of MCO.

Figure 3c is the AFM image obtained 5 minutes after the MCO solution (0.1 mg/L) prepared from the supernatant of secondary effluent at pH 11with the addition of $Ca(OH)_2$ flushed through the AFM liquid cell. Similar to Figure 3B, molybdenite surface was full covered by chemicals. It suggests that the supernatant of $Ca(OH)_2$ treated secondary effluent has no impact on the adsorption of chemicals on molybdenite surface.

Flotation Tests

Lab flotation tests have been carried out using different water sources to study the impact of secondary effluent on metal recovery. Table 1 shows the flotation recoveries of copper and molybdenum of sample A, B and C using various water sources at pH 11 with the addition of $Ca(OH)_2$.

Table 1. Copper and molybdenum recovery of ore sample A, B, and C using various water sources

	Sample A		Sample B		Sample C	
Water	Cu Recovery (%)	Mo Recovery (%)	Cu Recovery (%)	Mo Recovery (%)	Cu Recovery (%)	Mo Recovery (%)
Tap water	87.9	86.7	86.5	89.0	89.5	88.1
Secondary effluent	86.7	84.5	86.6	87.0	—	—
Supernatant of $Ca(OH)_2$ treated secondary effluent	87.8	86.6	86.9	88.6	89.0	88.1
Mixed tap water/effluent	87.9	86.6	86.2	88.8	89.8	88.0
Tap water + (humic acid)	—	—	—	—	90.1	84.4

Table 2. Water chemistry analysis results of Alkalinity, TOC, DO, phosphate, and phosphorous in different water sources

Water	Alkalinity (mg/L)	TOC (mg/L)	DO (mg/L)	Phosphate (mg/L)	Phosphorous (mg/L)
Tap water	—	0.58	11.9	<0.1	<0.1
Secondary effluent	190	6.28	12.5	11.6	3.78
Supernatant of $Ca(OH)_2$ treated secondary effluent	151	4.82	12.8	1.15	0.375
Mixed tap water/effluent (1:1)	107	2.47	12.3	0.982	0.321

For sample A, the copper recovery obtained with tap water, the supernatant of $Ca(OH)_2$ treated secondary effluent and the mixture of tap water and secondary effluent at 1:1 volume ratio is respectively 87.9%, 87.8% and 87.9%. The recovery is 86.7% for the case of secondary effluent and the value is a little bit lower than the other three. Similarly, the molybdenum recovery obtained with tap water, the supernatant of $Ca(OH)_2$ treated secondary effluent and the mixture of tap water and secondary effluent at 1:1 volume ratio is almost constant. The value is respectively 86.7%, 86.6% and 86.6%. However, when secondary effluent is used, the recovery is 84.5%, which is lower than the other three.

For sample B, the copper recovery obtained with tap water, secondary effluent, the supernatant of $Ca(OH)_2$ treated secondary effluent and the mixture of tap water and secondary effluent at 1:1 volume ratio is almost constant. The value is respectively 86.5%, 86.6%, 86.9% and 86.2%. In the case of molybdenum, the recovery obtained with tap water, the supernatant of $Ca(OH)_2$ treated secondary effluent and the mixture of tap water and secondary effluent at 1:1 volume ratio is respectively 89.0%, 88.6% and 88.8%. However, the recovery is 87.0% in the case of secondary effluent and the value is lower than the other three.

For sample C, The copper recoveries obtained with all the four water sources are almost the same and the value lies between 89% to 90%. The molybdenum recovery obtained with tap water, the supernatant of $Ca(OH)_2$ treated secondary effluent and the mixture of tap water and secondary effluent at 1:1 volume ratio is almost constant at the level of 88%. However, when humic acid is added, the molybdenum recovery is only 84.4%.

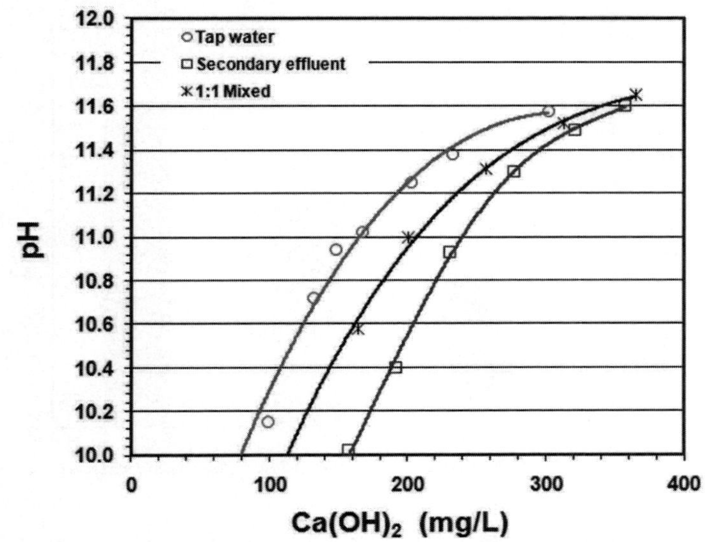

Figure 4. Relationship between the pH of solution and the amount of $Ca(OH)_2$ added for the pH adjustment

Water Chemistry Analysis

Table 2 shows the analysis results of TOC (total organic carbon), DO (dissolved oxygen), phosphate and phosphorous in different water sources, i.e., tap water, secondary effluent, the supernatant of $Ca(OH)_2$ treated secondary effluent (at pH 11) and the mixture of tap water and secondary effluent at 1:1 volume ratio (at pH 11).

For TOC, the level in tap water is 0.58 mg/L and is the lowest. The TOC amount in secondary effluent is the highest with a value of 6.28 mg/L, which can be reduced to 4.82 mg/L after the treatment with $Ca(OH)_2$. The DO level in all the four water sources is almost the same at 12 mg/L. For phosphate and phosphorous, the amount in secondary effluent is the highest and the level can be greatly reduced with dilution and the treatment with $Ca(OH)_2$. For example, the phosphate amount in secondary effluent is 11.6 mg/L; however, the amount of phosphate in the supernatant of $Ca(OH)_2$ treated secondary effluent is only 1.15 mg/L. The phosphorous amount in secondary effluent is 3.78 mg/L and the value is only 0.375 mg/L in the supernatant of $Ca(OH)_2$ treated secondary effluent.

The relationship between the pH of solution and the amount of $Ca(OH)_2$ added for the pH adjustment has been studied and the results are plotted as Figure 4, from which one can see that for secondary effluent a much larger amount $Ca(OH)_2$ has to be added to elevated the solution pH. For example, to increase the solution pH to 11, 160 mg $Ca(OH)_2$ is added in 1000 mL tap water; however, 240 mg $Ca(OH)_2$ is needed in the case of secondary effluent. The curve obtained with the mixture of tap water and secondary effluent at 1:1 volume ratio lies between the two curves of tap water and secondary effluent.

Figure 5 shows the weight of the precipitate in 1000 mL solution of different water sources at pH 11 with the addition of $Ca(OH)_2$. In tap water, the precipitate is about 160 mg/L. The value is 240 mg/L for the mixture of tap water and secondary effluent at 1:1 volume ratio. In the case of secondary effluent, the precipitate is about 320 mg/L. Actually, when secondary

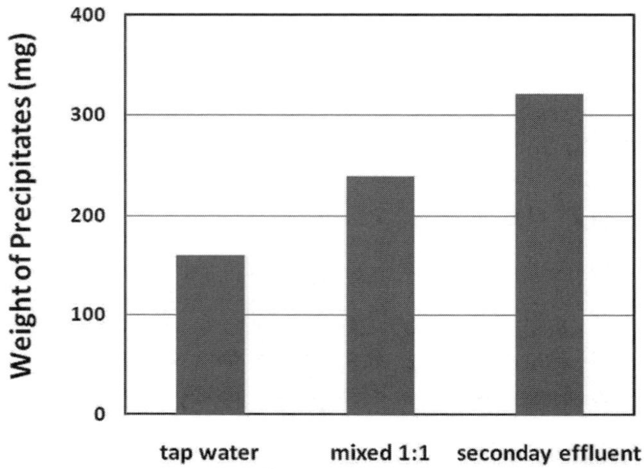

Figure 5. Weight of the precipitates in 1000 mL solution of different water sources at pH 11 with the addition of $Ca(OH)_2$

effluent was adjusted to pH 11 with the addition of $Ca(OH)_2$, a lot of precipitate was observed instantly.

Impact of Secondary Effluent on the Adsorption of Collector on Mineral Surface

In flotation, collectors are added into the pulp (mixture of water and fine-grained minerals) in flotation cells to selectively adsorb onto the target mineral and render its surface with high hydrophobicity, which is generally beneficial for a strong mineral-bubble attachment and thus a high flotation recovery. Therefore, the adsorption of collector on mineral surface is vital for a successful flotation process.

The AFM imaging technique has been used to study the adsorption of chemicals on mineral surfaces because of its advantage over other surface characterization techniques. (Zhang et al., 2005, 2006; Zhang and Zhang, 2010, 2011) For example, by employing an AFM fluid cell, one can study the in-situ absorption of chemicals on solid in solution.

In the present study, AFM image such as Figure 1 clearly show that the collector, Nalco 9740, can instantly absorb on chalcopyrite surface. After only 5 minutes, the mineral surface was fully covered by many small collector patches, which cannot be attributed to the reaction of mineral surface with water because Figure 1a remained unchanged after even one hour. The AFM images also showed that the chemical is a very powerful and efficient collector for chalcopyrite because of the short reaction time, the uniform distribution of collector on mineral surface and the small profile of the absorbed patches.

The impact of secondary effluent on the adsorption of collector on chalcopyrite surface was also studied in Nalco 9740 solution (0.02 g/L), which was prepared from the supernatant of secondary effluent at pH 11 with the addition of $Ca(OH)_2$. As shown by Figure 2, many small patches were detected on mineral surface after a 5 minute contact time. One may notice that there is some difference between the morphology of the patches as shown in Figure 1 and Figure 2 and it is likely due to the mineral's heterogeneity because the images differ slightly even when different parts of a same sample were scanned in the same solution. The mineral's heterogeneity makes an exact quantified comparison technically not applicable. However, a conclusion still can be drawn from Figure 2 that Nalco 9740 can also efficiently absorb on chalcopyrite surface in the supernatant of secondary effluent at pH 11 with the addition of $Ca(OH)_2$ because of the short reaction time, the uniform distribution of collector on mineral surface and the small profile of the absorbed patches.

It is worthwhile to mention that in present study, an attempt has also been tried to directly obtain the AFM images of chalcopyrite in Nalco 9740 solution prepared from secondary effluent without removing precipitates. Unfortunately, no good AFM image was successfully obtained because of the precipitates deposited on AFM probes, which makes it difficult for an AFM laser adjustment and therefore a potential damage to equipment.

Molybdenite is a typical mineral with natural floatability, which can be floated without the addition of collector. However, in flotation practice, hydrocarbon oil is usually added as a promoter to increase molybdenite recovery. MCO is an oily substance with hydrocarbon components and has been used as a molybdenite promoter. Figure 3b shows that MCO strongly absorb on molybdenite surface in a short time, i.e., 5 minutes. By compassion, Figure 3c shows that that MCO can also efficiently absorb in a similar manner on molybdenite surface in the supernatant of secondary effluent at pH 11 with the addition of $Ca(OH)_2$. As mentioned in above section, the attempt to directly obtain the AFM images of molybdenite in MCO solution prepared from secondary effluent without removing precipitates is not successful because of the large amount precipitates attached on AFM probes.

As a summary, the AFM images show that in the supernatant of secondary effluent at pH 11 with the addition of $Ca(OH)_2$, collectors can efficiently absorb on mineral, i.e., chalcopyrite and molybdenite, in a similar manner as obtained in the solutions prepared from DI water. It means, the supernatant of secondary effluent with $Ca(OH)_2$ treatment should have little impact on the adsorption of chemicals on mineral surface.

Impact of Secondary Effluent on Metal Flotation Recovery

A literature review shows that secondary effluent impacts metal flotation recovery by decreasing both copper and molybdenum recovery and the reason was attributed to the higher total organic content in secondary effluent. In present work, the water chemistry of different water sources has been analyzed and flotation tests have been carried out accordingly using different water sources. A lot of valuable information has been achieved and the obtained findings can be summarized as follows.

Firstly, copper recovery remains almost constant, showing no dependence on water sources. As shown by Table 1, for each ore, copper recovery did not show noticeable change when water sources changed. It is well known that for chalcopyrite flotation, the pulp potential, which is mainly determined by the dissolved oxygen content in water, is the most important factor influencing collector's adsorption on mineral surface and therefore flotation efficiency. The fact that, as shown by Table 2, the level of DO in all the four water sources is the same suggests that the adsorption of collector on chalcopyrite surface should be quite the same. This conclusion is in a line with the AFM imaging results and lab flotation study. Additionally, one has to note that the main effective component in Nalco 9740 is alkyl mercaptan, which is the strongest chalcopyrite collector with xanthate and dithiophosphate in comparison. Usually mercaptan chemical has the highest collectivity with the least requirement on pulp oxidation potential for adsorption. That is, the chalcopyrite collector, Nalco 9740, used in present investigation is not sensitive to the DO level in pulp. Another important thing worth of being mentioned is that heavy aromatic naththa is a main component in Nalco 9740. The oily substance can promote both chalcopyrite and molybdenite flotation because in flotation history oil has been used for the flotation of sulfide minerals even before the introduction of xanthate. Actually, prior results (Fisher and Rudy, 1976) also showed that when fuel oil was used in combination of PAX (potassium amyl xanthate), the copper recovery obtained with demineralized water and sewage effluent both increased a little, compared to only PAX being used, and finally showed no difference in value. All the above three reasons may account for the unchanged copper recovery in different water sources.

Secondly, molybdenum recovery obtained with the supernatant of secondary effluent at pH 11 with the addition of $Ca(OH)_2$ is comparable to the one obtained with tap water and the mixture of tap water and secondary effluent at 1:1 volume ratio. As shown by Table 1, secondary effluent impacts flotation by decreasing molybdenum recovery by 2%. However, the impact can be mitigated by either removing the precipitates after the treatment of $Ca(OH)_2$ or dilution. The impact of 2% decrease in molybdenum recovery is much smaller than the value of 16.2%, which was reported in prior study. The difference can be attributed to the following reasons. First, the 16.2% reduction of molybdenum recovery reported in prior study was achieved when no molybdenite promoter was added. This value can be reduced to 7% when fuel oil was added as molybdenite promoter. As mentioned before, hydrocarbon oil is commonly added in flotation as a molybdenite promoter in spite of the fact that molybdenite is a naturally floatable hydrophobic mineral. One reason for that is the absorption of hydrocarbon oil can increase the water contact angle of molybdenite from 85° to above 90°, which is beneficial for the formation of nanobubbles at solid/liquid interface, and therefore a high flotation rate. In present investigation, MCO and Nalco 9740 are used in the multi-component reagent scheme. The former, an oily substance, is added as the molybdenite promoter. The later has 30~60% in weight heavy aromatic naththa, which also can be beneficial for molybdenite flotation, even though it is purposefully added as the collector for chalcopyrite. It is likely that these molybdenite promoters in the flotation reagent scheme can compete with the deleterious species, for example, anionic organic constituents as suggested by Amy et al., and elevate the impact of these species on flotation. Second, the water quality of secondary effluent has improved since the 1980s. For example, the alkalinity in previous secondary effluent is 277 mg/L; however, the value is 190 mg/L as shown in Table 2. The TOC in previous secondary effluent is about 8.8 mg/L in comparison with the 6.28 mg/L in present secondary effluent. Table 2 also shows that the phosphate constituent in previous secondary effluent is 10~30 mg/L; however, the phosphate level is only 11.6 mg/L nowadays. The improvement in water quality of secondary effluent can also mitigate its impact on metal recovery.

Thirdly, treatment of secondary effluent by $Ca(OH)_2$ can improve water quality by reducing deleterious species in water. Actually, lime treatment has long been used in waste water treatment for the removal of phosphorus (De Renzo, 1978; Stumm, 1987; MWH Water Treatment, 2005). The effect is clearly shown by Table 2. For example, The TOC in secondary effluent is reduced from 6.28 mg/L to 4.82 mg/L with phosphate being decreased from 11.6 mg/L to 1.15 mg/L and phosphorus being reduced from 3.78 mg/L to 0.375 mg/L. The improvement in water quality will mitigate the potential impact of these deleterious species on flotation. It is proposed in prior work that humic acid and anionic species may account for the decrease of metal recovery when secondary effluent was used in flotation. Of course, the initial purpose of adding $Ca(OH)_2$ is not for the removal of deleterious species, but for the elevating of pulp pH. This is because in copper/molybdenum rougher flotation practice, pulp pH is usually elevated to above 11 for the depression of pyrite, which is commonly found in the porphyry copper ore. This lime treatment is different from the one applied in tertiary water. As King et al. (1983) pointed out, the former is incidental and the latter is intentional.

Fourthly, removing the precipitates of secondary effluent introduced by the addition $Ca(OH)_2$ can alleviate the impact on flotation. It is clearly shown in Figure 4 that, when compared to the pH adjustment of tap water, a much larger amount of $Ca(OH)_2$ has to be used to adjusted pulp pH to 11 because of the high alkalinity in secondary effluent. A direct result of this, as clearly shown by Figure 5, is a much larger amount of precipitates being observed in lime treated secondary effluent. Calcium carbonate was identified as the main component of the precipitates and the amount is much larger than those of TOC and phosphate. Another interesting thing worthy of being mentioned is that the result of flotation tests of tap water with the addition of humic acid is also listed in Table 1, which shows that molybdenum recovery was reduced by 3.5%. This may explain the season effect on molybdenite flotation when river water is used as process water.

CONCLUSIONS

A reinvestigation of applying secondary effluent in sulfide flotation has been carried to clarify the main reasons accounting for the reduction in metal recovery. AFM imaging study showed that collectors, i.e., Nalco 9740 and MCO, can efficiently adsorb on mineral surface in the supernatant of secondary effluent at pH 11 with the addition of $Ca(OH)_2$ because of the short reaction time, the uniform distribution of collector on mineral surface and the small profile of the absorbed patches. Flotation results suggest that copper recovery remains almost constant, showing no dependence on water sources. Applying secondary effluent decreases molybdenum recovery by 2%. However, this can be addressed by removing the precipitates due to the addition of $Ca(OH)_2$ and using the supernatant for flotation. It is also proposed that the improved water quality of secondary effluent, the addition of $Ca(OH)_2$ reducing potential deleterious species and the removal of precipitates from $Ca(OH)_2$ treated secondary effluent at high pH may account for the alleviated impact of secondary effluent on flotation efficiency.

ACKNOWLEDGMENT

J. Zhang is grateful to Freeport-McMoRan Copper & Gold, Inc. for sponsoring the Freeport McMoRan Copper and Gold Chair in Mineral Processing in the Department of Mining and Geological Engineering at the University of Arizona. The support from Science Foundation of Arizona, Freeport-McMoRan Copper & Gold, Inc. and Resolution Copper Mining for the project is also greatly appreciated. Rick Gilbert and Tomas Bolles from Freeport-McMoRan Copper & Gold Inc., Brad Ross and Darby Stracy from Resolution Copper Mining are greatly thanked for advising the project. Jeff Prevatt from the Pima Regional Wastewater Reclamation Department is acknowledged for advisory and providing secondary effluent samples.

REFERENCES

Amy, G.L., King, P.H., Chu, A., and Chandler, G. 1984. Technical and Economic Considerations in Secondary and Tertiary Effluent Reuse in Copper and Molybdenum Flotation, Vol. 2. *Proceedings of the Water Reuse Symposium III*, Denver, August 26–31.

De Renzo, D., *ed. 1978.* Nitrogen Control and Phosphorus Removal in Sewage Treatment. *Pollution Technology Review.*

Deboer, J., and Linstedt, K. 1985. Advances in Water Reuse Applications. *Water Reuse* 19(11):1455–1461.

Fisher, W.W., and Rudy, S. 1976. Utilization of Municipal Waste Water for Froth Flotation of Copper and Molybdenum Sulfides. Arizona Bureau of Mines Circular 17.

Fuerstenau, D.W. 1962. *Froth Flotation 50th Anniversary Volume.* AIME, pp. 91, 139, 382.

Fuerstenau, M.C. 1976. *Flotation, A.M. Gaudin Memorial Volume*, Vol. 1 and 2. AIME, pp. 334–357 (sulfide flotation); 275, 1307 (reclaimed water); 1145–1215.

Gaudin, A.M. 1957. *Flotation.* 2nd ed. McGraw-Hill.

King, P.H., Chu, A., and Amy, G.L. 1983. Reuse of Municipal Effluent in Mineral Flotation. *Proceed of 1983 ASCE National Conference on Environmental Engineering.*

Klassen, V.I., and Mokrousov, V.A. 1963. *Introduction to the Theory of Flotation.* Butterworths.

Lieuwen, A. 1989. PhD research report on an institutional and economic assessment of water reuse in Tucson basin.

MWH Water Treatment: Principles and Design. 2nd ed. 2005.

Pickett, D.E. 1973. Canadian Milling Practice for Water Reuse. *Canadian Milling Journal*, 94(6):29–30.

Stumm, W. 1987. *Aquatic Surface Chemistry.* Wiley.

Sutherland, K.l., and Wark, I.W. 1955. *Principles of Flotation.* AusIMM..

Zhang, J., and Zhang, W. 2010. Applying an Atomic Force Microscopy in the Study of Mineral Flotation. In *Microscopy: Science, Technology, Applications and Education.* Vol. 3. Edited by Antonio Méndez-Vilas and Jesús Álvarez. Formatex Research Center. pp. 2028–2034.

Zhang, J., and Zhang, W. 2011. The Adsorption of Collectors on Chalcopyrite Surface Studied by an AFM. Presented at R.-H. Yoon Symposium. Denver, CO, SME.

Zhang, J. 2006. Ph.D dissertation. Surface Forces Between Silica Surfaces in C_nTACl Solutions and Surface Free Energy Characterization of Talc. Virginia Tech.

Zhang, J., Yoon, R.-H., Mao, M., and Ducker, W.A. 2005. Effects of Degassing and Ionic Strength on AFM Force Measurements in $C_{18}TACl$ Solutions. *Langmuir*, 21:5831–5841.

Trace Metals Removal with Precipitated Solids from Uranium Mine Effluent at Cameco's Key Lake Operation

Kuang Lee and David Lee
FLSmidth Salt Lake City, Inc., Salt Lake City, UT, USA

Arthur Lieu
Key Lake Operation, Cameco Corporation, Saskatchewan, Canada

ABSTRACT

The removal of selenium and molybdenum from uranium mine effluent is complicated by the numerous redox states of molybdenum (Mo) and selenium (Se) as well as the increased solubility of precipitated Se and Mo in mine tailings above neutral pH. While co-precipitation with iron and the use of ferric sulfate-gypsum are accepted industrial practice for the removal of Mo and Se respectively, the molybdenum can be liberated from the $Fe_2(MoO_4)_3$ complex at pH levels greater than 6 and Se, if not in reduced forms, will remain mobilized in process solution. At the Cameco Key Lake uranium mine, innovations and extensive efforts have been applied to remove Se and Mo in the mill effluent stream to meet the regulatory permits. Additional pilot test work indicated Se and Mo levels could be reduced below permitted levels at a pH of 3.8 by using the proprietary FLSmidth MaxR™ solids recycle process. At the same time, solids concentrations in the thickener underflow can reach 35–40 weight percentages which exhibit better solid/liquid separation characteristics and a decrease in the volume of waste for tailings disposal. Long term stability of the precipitated solids at neutral pH could also be achieved due to the encapsulation of the Se and Mo containing particles with gypsum using the MaxR™ solids recycle process.

INTRODUCTION

Cameco's Key Lake Operation (KLO) has been producing uranium since 1983 and is the world's largest high grade uranium mill. Since 1999 a high-grade uranium ore from McArthur Lake has also been processed at KLO. Impurities from the high-grade uranium ore at McArthur Lake and the mineralized waste-rock from Key Lake are found primarily in the plant raffinate and include arsenic, molybdenum, nickel, radium and selenium. The majority of metal impurities (Ni, Cu, Fe, Pb, and Zn) can be removed through precipitation at high pH to form metal hydroxides followed by solid-liquid separation. Selenium and molybdenum are typically treated separately prior to this bulk impurity removal step due to their increased solubility at higher pH. In an effort to more efficiently remove Se and Mo from mill effluent and produce thermodynamically stable solids for long term tailings storage, Cameco's KLO commenced a modification project of its Bulk Neutralization (BN) circuit.

Removal of Se and Mo is a challenging task due to the complex chemical nature of the process streams. In 2003, Cameco Corp. commenced an extensive research program to develop a process to effectively remove Se and Mo from the KLO's mill effluent. This program included steps of (1) identifying the distribution of Se and Mo in the Key Lake mill; (2) determining the chemical speciation and valance states of these elements in each stream; and (3) conducting bench- and pilot-scale tests on various Se-Mo removal mechanisms simulating the operation of

BN circuit. (Lieu et al. 2010) Treatment methods were investigated including co-precipitation reaction with ettringite, selenium removal through ferrihydrite co-precipitate, co-precipitation with barium chloride, adsorption reaction with ferrous sulfate–hydroxide. These methods are discussed in the following sections.

Selenium can exist in –2, 0 +4 and +6 oxidation states in aqueous systems. The species of selenium in solution is largely dependent upon solution pH and redox potential. Selenate (SeO_4^{2-}) and selenite (SeO_3^{2-}) are the predominate species in mill process streams with protonated selenite (H_2SeO_3) being the prevailing species in the KLO raffinate stream. The complex speciation of selenium has made its removal from natural and anthropogenic sources extremely challenging. There is no "standard" method for Se removal, but known methods can be categorized in three ways: reduction to elemental Se, precipitation and co-precipitation, or adsorption. Reduction of selenate and selenite to Se^0 is done using zero valent iron particles. Co-precipitation can be achieved with the addition of ettringite (pH > 8), barium chloride (pH 6.5) and with ferric sulfate-gypsum systems (pH 3.5–4.2). (Nishimura, Hata, and Umetsu, 2000) The general reactions for the precipitation of Se are represented by the following Equations 1 through 8b.

$$\underset{\text{(ettringite)}}{Ca_6Al_2(SO_4)_3(OH)_{12}\cdot 26H_2O} + 3SeO_4^{2-} \rightarrow Ca_6Al_2(SeO_4)_3(OH)_{12}\cdot 26H_2O\downarrow + SO_4^{2-} \quad (1)$$

$$Ba^{2+} + SeO_3^{2-} \rightarrow BaSeO_3\downarrow \quad (2a)$$

$$Ba^{2+} + SeO_4^{2-} \rightarrow BaSeO_4\downarrow \quad (2b)$$

$$CaO + H_2O \rightarrow Ca(OH)_2 \quad (3)$$

$$Ca(OH)_2 + H_2SO_4 \rightarrow CaSO_4\cdot 2H_2O\downarrow \quad (4)$$

$$Fe_2(SO_4)_3 + 6OH^- \rightarrow 2Fe(OH)_3 + 3SO_4^{2-} \quad (5)$$

$$2Fe(OH)_3 + SeO_3^{2-} \rightarrow Fe_2(OH)_4SeO_3\downarrow + 2OH^- \quad (6)$$

$$2Fe(OH)_3 + SeO_4^{2-} \rightarrow Fe_2(OH)_4SeO_4\downarrow + 2OH^- \quad (7)$$

$$Ca^{2+} + SeO_3^{2-} \rightarrow CaSeO_3\downarrow \quad (8a)$$

$$Ca^{2+} + SeO_4^{2-} \rightarrow CaSeO_4\downarrow \quad (8b)$$

Using ettringgite and barium chloride alone was tested at KLO with some success but found unattractive due to high Se content in effluent or high reagent consumption. Test results indicated that co-precipitation of selenate/selenite with ferric sulfate-gypsum at low pH between 3.5 and 4.2 is more favorable to achieve Se concentration in the effluent to less than 0.02 mg/L. Bench-scale and plant-scale tests were performed in the high-pH stage simulating the operation of BN circuit using ferrous sulfate. This specific reaction involves $Ca(OH)_2$ neutralization with ferrous ion; selenite and selenate are reduced by Fe(II) hydroxide. Selenite is adsorbed more efficiently than selenate, as the reaction proceeds by reduction of selenite with ferrous hydroxide to form the mixed valent iron product Fe_3O_4. The general chemical reactions may be expressed as:

$$FeSO_4 + 2OH^- \rightarrow Fe(OH)_2\downarrow + SO_4^{2-} \quad (9)$$

$$SeO_3^{2-} + 3Fe(OH)_2 \rightarrow Se_0\downarrow + Fe_3O_4\downarrow + 4OH^- + H_2O \quad (10)$$

$$SeO_4^{2-} + 9Fe(OH)_2 \rightarrow Se_0\downarrow + 3Fe_3O_4\downarrow + 2OH^- + 8H_2O \quad (11)$$

Using ferrous ions as a reducing agent, selenite/selenate ions can be reduced to elemental Se at pH between 6.5 and 9.2. Bench-/pilot-scale tests showed the most consistent and prominent results, where concentrations of Se were further reduced to as low as 0.005 mg/L in the final effluent.

Molybdenum also exists in an array of oxidation states: –2, +2, +4 and +6. The predominant species of molybdenum in the KLO is molybdate (MoO_4^{2-}), which maintains redox stability throughout the KLO flowsheet. The primary method of molybdate removal is precipitation with ferric iron at a pH of 3.5 to form $Fe_2(MoO_4)_3$, shown in Equation 12 (Gupta, 1992). Iron molybdate will precipitate at this low pH and the concentration of Mo is expected to drop significantly below the regulatory limit.

$$3H_2MoO_4 + 2Fe(OH)_3 \rightarrow Fe_2(MoO_4)_3\downarrow + 6H_2O \quad (12)$$

This reaction occurs in the same pH regime as Se removal with ferric sulfate-gypsum precipitation; therefore, it is possible to conduct both removal reactions simultaneously in one reaction vessel. However, when this method is used for co-removal of Mo and Se, the tailings are highly sensitive to pH neutralization. When neutralized for disposal in the tailings pond, Mo and Se can be readily dissolved from solid tails and thus pose an environmental hazard. To avoid the re-dissolution of Mo and Se, the solids formed in the precipitated reaction of Mo and Se need to be encapsulated. The encapsulating species, such as gypsum or ferrihydrite, forms a protective layer and prevents the MoO_4^{2-} and $Fe(OH)_4SeO_3$ from coming in contact with the bulk solution.

Process Modifications of the Key Lake BN Circuit

Modification of the BN circuit at Key Lake Operation was commenced in 2007 (Lieu et al. 2010) with the additional process upgrades implemented in 2008, the circuit was completed and commissioned in spring 2009. The process flow diagram of the modified circuit is shown in Figure 1.

Raffinate reports to Pachuca #1 in the BN circuit, it mixes with the other contaminated water streams, such as mine water and reject water from the reverse osmosis plant, in Pachuca #2. Slaked lime is added to Pachuca #2 and the pH is controlled between 3.5 and 4.2. This process solution is then pumped to the FLSmidth High-Rate thickener (Se-Mo Removal thickener) where Se and Mo begin to co-precipitate with ferrihydrite. These chemically precipitated particles are thickened to a 30 wt% design solids and report to the thickener underflow slurry before it's pumped to a new pH adjustment tank. The slurry is adjusted to pH 6–7 with the slaked lime, then blended with other tailings streams and pumped to the tailings pond. The supernatant of the Se-Mo thickener is pumped to Pachuca 3 and overflows to Pachuca 4, where the pH of the process solution in these pachucas is maintained at pH 6.5 and 9.2, respectively. The majority of trace metal precipitates at the high pH stage are in the form of hydroxides (Ni, Cu, Pb, and Zn) or ferrihydrite (As). These chemical precipitates are captured in the Lamella Thickener. The supernatant overflows to a 3-stage pH adjustment tank series, where diluted sulfuric acid is added to reduce the pH of this process solution to between 6 and 6.5.

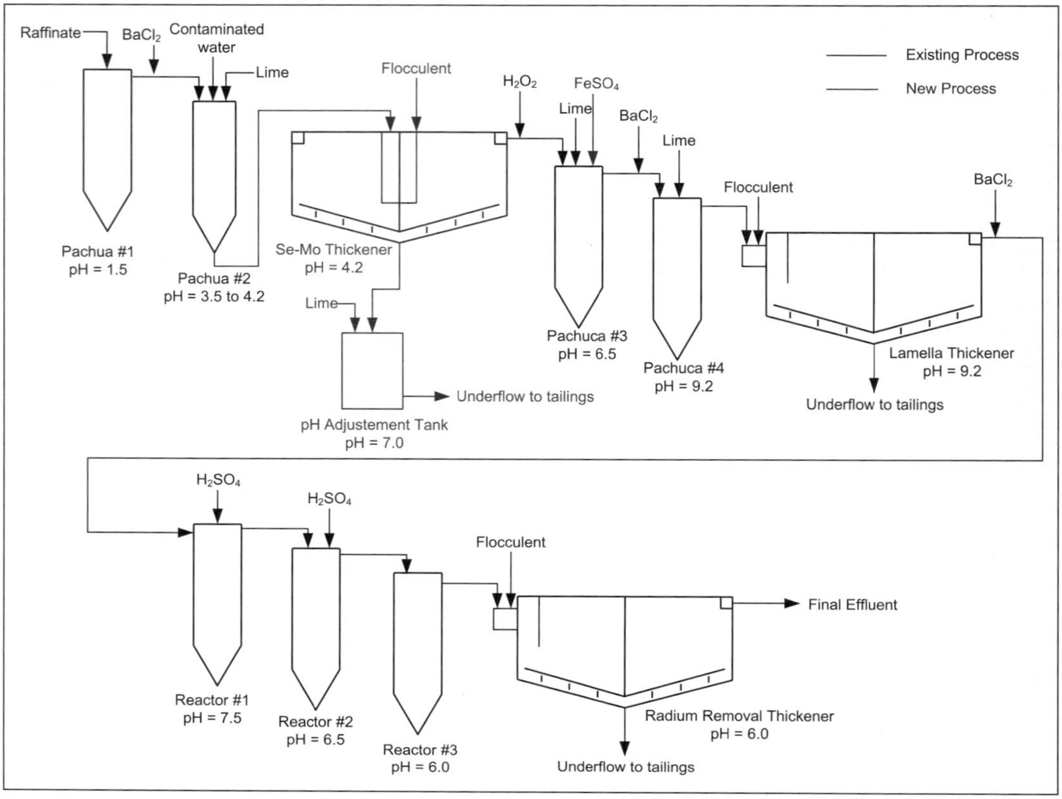

Figure 1. Process flow diagram of the modified BN circuit of Key Lake Cameco Operations

This process solution is further clarified with the Radium Removal Thickener to capture any suspended solids carried forward in the effluent. The final effluent is sent to monitoring ponds where composite samples are collected and analyzed. Effluent is released to the downstream river systems when all the regulatory limits established by CNSC are met.

Through the BN process, barium chloride is added at the discharge of Pachuca #1, #3, and Lamella Thickener to remove radium from the process solution. The entrained organic in Raffinate from the SX circuit, is treated with hydrogen peroxide at the overflow of the Se-Mo thickener. This will oxidize any organic materials to carbon dioxide and ultimately reduce the toxicity in the final effluent.

RESULTS AND DISCUSSION

Commissioning of the modified BN circuit began in March 2008 (Lieu et al. 2010). An intense sampling campaign was augmented to assess the performance of the modified BN circuit. In the first 5 months, there were process issues that affected the operation of the Se-Mo thickener, therefore sampling results were not reliable during this period. These issues were resolved in September of 2008. Sampling campaign lasted 431 days during the commissioning process except a scheduled short maintenance shutdown in May 2009. Samples were collected at the feed, Se-Mo thickener, and the final effluent streams. The feed streams consist of: (1) raffinate from the solvent extraction plant, (2) contaminated water from the DTMF, AGTMF and mill waste water.

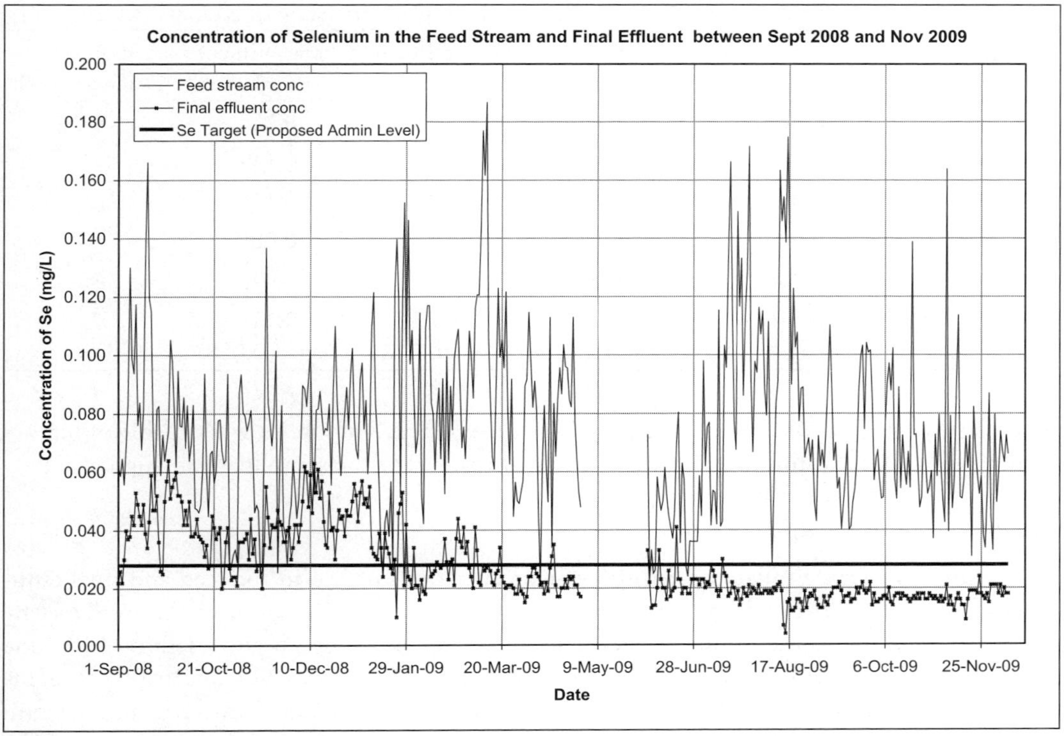

Figure 2. Concentration of selenium in the input stream and final effluent

The following sections present the results of concentration and loading of Se and Mo in the input and final effluent streams during the sampling campaign between September 2008 and November 2009.

Selenium Removal Efficiency

Figure 2 shows the comparison of concentrations of Se in the feed and final effluent streams. The horizontal line represents the proposed regulatory limit of 0.028 mg/L set by the CNSC. The input stream shows great variation of Se due to the high variability of Se in the ore body. Sustained control of the concentration of Se in the final effluent was established at the end of June 2009, where Se was consistently maintained below the regulatory limit. This decrease in concentration was due to the completion of the process improvement and establishment of target operating parameters allowing the circuit to operate at the equilibrium stage.

Table 1 summarizes the average daily concentration, loading, and removal efficiency of Se in the BN circuit. The compiled results were grouped into quarter-year format for performance comparison and simplicity purpose.

It's clear that the concentration of Se in the final effluent decreased chronologically from 0.041 to 0.017 mg/L while the concentration of Se in the feed stream varied between 0.062 to 0.093 mg/L. The concentration of Se had been sustained below the regulatory limit since the beginning of quarter 3 of 2009. The standard deviation of the Se in final effluent decreased from 0.010 mg/L to 0.002 mg/L indicating the circuit had reached steady state for Se removal at the end of quarter 4 of 2009. The final loading of Se to the environment was reduced from

Table 1. Summary of Se concentrations, loading, and removal efficiency in KLO's BN circuit

Selenium	Feed (mg/L)	Feed (kg/day)	Final Effluent (mg/L)	Final Effluent (kg/day)	Removal Efficiency (%)
Q 3, 2008	0.072	0.444	0.041	0.165	61
Q 1, 2009	0.087	0.504	0.030	0.114	75
Q 2, 2009	0.062	0.443	0.022	0.115	69
Q 3, 2009	0.093	0.624	0.018	0.088	83
Q 4, 2009	0.066	0.435	0.017	0.079	85
Overall Average	0.076 (s.d. = 0.030)	0.483 (s.d. = 0.190)	0.027 (s.d. = 0.012)	0.117 (s.d. = 0.058)	73 (s.d. = 15)

0.165 kg/day (Q3, 2008) to 0.079 kg/day (Q4, 2009). The new circuit was capable of removing 85% of selenium from the process solution at the end of the sampling campaign.

Molybdenum Removal Efficiency

Figure 3 shows a similar comparison of concentrations of Mo in the feed and final effluent streams. The horizontal line represents the proposed regulatory limit of 0.6 mg/L set by the CNSC. Similar to the profile of Se, the input stream also shows high variation of Mo due to the variability of Mo in the ore body. The removal mechanism of Mo is relatively straightforward compared to Se. The protonated molybdate is precipitated with ferric and calcium ions at the optimum pH between 3.5 and 4.2. The first tipping point when the concentration of Mo dropped below the regulatory limit was in November 2008. Due to pumping and scaling issues in the Se-Mo thickener, a lot of precipitated molybdate was carried forward to the downstream of the BN circuit, where it was re-dissolved when the pH increased to 9.2 at Pachuca #4. This effect is illustrated in the period between November 2008 and March 2009, where high variation in concentration of Mo was observed in the final effluent. The concentration of Mo became stable and reached steady state after June 2009; this is consistent with the profile of Se found in Figure 2.

The concentration of Mo in the feed stream ranged between 1.021 and 1.585 mg/L. The concentration of Mo in the final effluent decreased chronologically from 0.566 to 0.081 mg/L. The concentration of Mo has been sustained below the regulatory limit since the beginning of quarter 3 of 2009. The standard deviation of the Mo concentration in final effluent dramatically decreased from 0.256 mg/L to 0.035 mg/L illustrates the circuit had reached steady state for Mo removal at the end of quarter 4 of 2009. The final loading of Mo to the environment was reduced from 1.5 kg/day to 0.4 kg/day in the commissioning process. The maximum removal efficiency was at 96% for molybdenum at the end of the sampling campaign. See Table 2.

Potential Additional Improvement

Cameco's KLO management has interest to continue its efforts to further improve the BN effluent treatment. Besides supplying the Se-Mo thickener for the modification project, FLSmidth (formerly Dorr-Oliver EIMCO) has also been contracted to conduct test work using its proprietary MaxR™ solids recycle technology to enhance the particle growth of the precipitated solids which can improve the downstream solid-liquid separation and possibly the overall Se and Mo removal efficiency from the mill effluent.

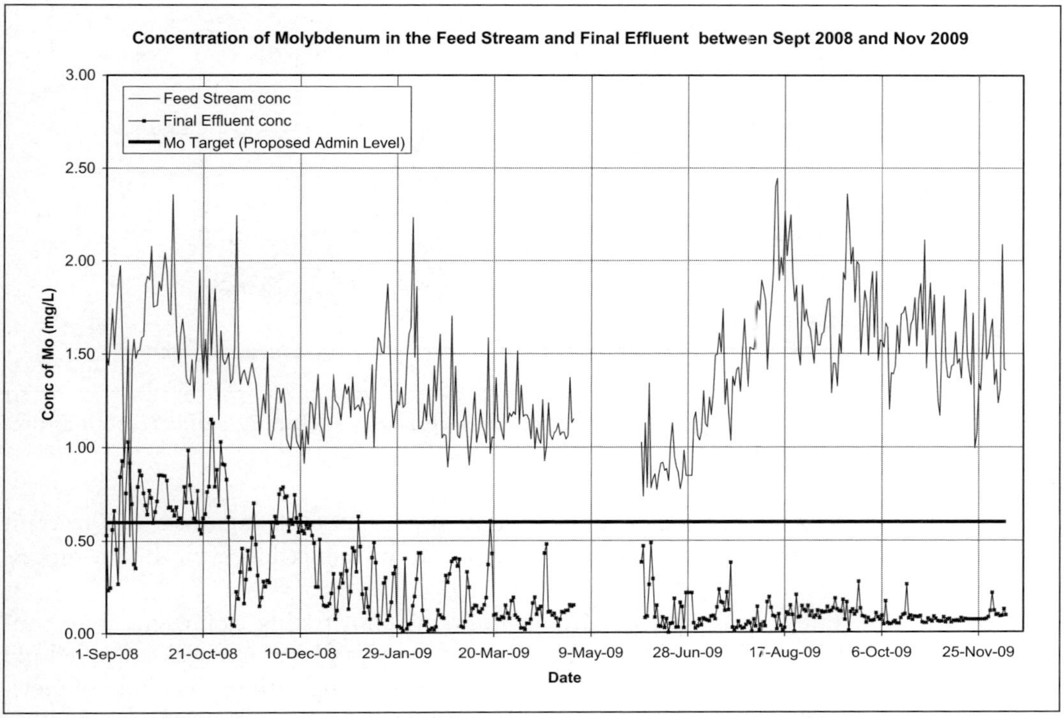

Figure 3. Concentration of molybdenum in the input stream and final effluent

Table 2. Summary of Mo concentrations, loading, and removal efficiency in KLO's BN circuit

Selenium	Feed (mg/L)	Feed (kg/day)	Final Effluent (mg/L)	Final Effluent (kg/day)	Removal Efficiency (%)
Q 3, 2008	1.429	8.6	0.566	1.5	59
Q 1, 2009	1.261	7.3	0.199	1.2	83
Q 2, 2009	1.021	7.3	0.135	0.8	89
Q 3, 2009	1.585	10.6	0.089	0.4	96
Q 4, 2009	1.550	10.3	0.081	0.4	96
Overall Average	1.402 (s.d. = 0.332)	8.9 (s.d. = 2.4)	0.214 (s.d. = 0.012)	1.4 (s.d. = 1.6)	82 (s.d. = 20)

Solids recycling has been demonstrated in commercial applications for years to improve processes where chemical precipitation, gravity separation processes, and/or filtration processes are required. The benefit of solids recycling in chemical precipitation occurs in both the reaction step and in the sludge dewatering characteristics. At the reaction step, recycled solids provides existing particles as seeds during precipitation, promoting particle growth, increasing reaction efficiency, and stabilizing the final solution with respect to super-saturation. The presence of additional solids will initiate precipitation in systems which are saturated. Particles formed are larger and more crystalline in nature. This results in improved removal of the solids in the following gravity sedimentation and filtration operations. Recycle of solids can use the reactant chemicals more efficiently, decreasing overall chemicals costs. The residual chemical becomes more associated with the solids through chemical or physical bonding at the surface

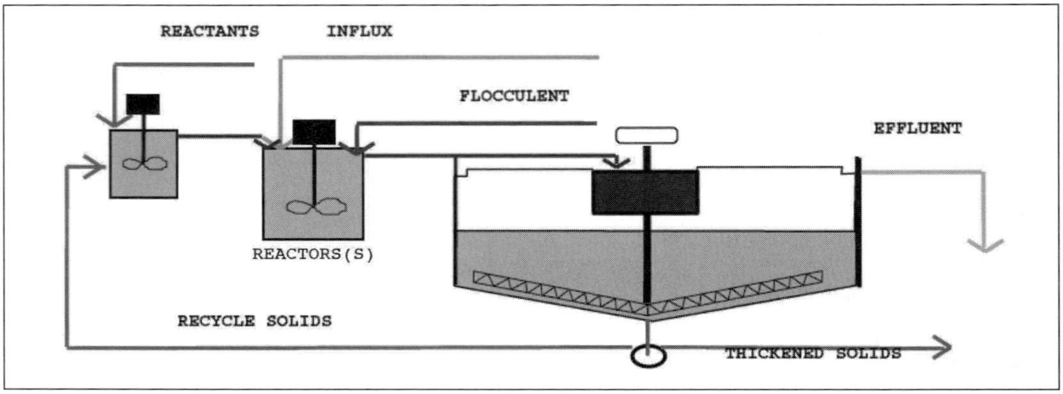

Figure 4. Flowsheet of typical MaxR™ solids recycle technology including solids recycling, pre-coating with a flash-mix tank

of the solids instead of in the bulk solution. In most cases the excess amount of the reactant chemicals required to force the precipitation reaction can be reduced. A typical pilot process flow diagram is shown in Figure 4.

For the BN effluent treatment at KLO, continuous pilot testing campaigns were conducted in 2006–2007 at KLO's mill site. The method for Se and Mo removal employed pH adjustment by lime addition to precipitate the desired metal, along with recirculation of previously precipitated solids, followed by sedimentation and/or filtration of the precipitated solids. Details of the reaction/precipitation mechanism have been described in the previous sections of this paper. The main goals of the proposed MaxR™ treatment were (1) to remove the two targeted metals—selenium (Se) and molybdenum (Mo) to low enough levels to meet environmental discharge requirements, (2) to produce solids with enhanced dewatering characteristics over conventional precipitation methods (probably through particle size growth) for tailings disposal and (3) to form solids resistant to dissolution of Mo and Se, i.e., encapsulation, in tailings storage.

MaxR™ Test Procedures and Results

The pilot plant provided by FLSmidth included the following principal components: a reactor with a volume of about 60 liters, a settling compartment with an area of ~0.1 m^2, a thickening column 1.5 m deep, a small mixing tank for blending underflow slurry and slaked lime, and associated pumps and reagent tanks (Emmett and Lee, 2006). The system was fed a combined flow of raffinate and reservoir water at a ratio of 1:2 at a rate which simulated the ~30–40 minutes available in Pachuca #2. Agitation was provided by a variable speed mixer with two impellers. Two variations were tested: (1) a "simple recycle" process combined lime, recycled underflow solids and raffinate for "flash mixing" near the impeller blades of the reactor, where the precipitation reaction proceeded to completion; and (2) a condition simulating the MaxR™ process first combined lime, recycled underflow solids and some reservoir water in a small mixing tank with about 1 minute residence time. The mixture then overflowed into the reactor and was mixed with the raffinate at the impeller blades. In both cases, the reactor discharged into a settling compartment and the underflow transported into a thickening column. Portion of the underflow solids from the thickening column was recycled back to the mixing tank. Temperature was not regulated but allowed to remain at ambient conditions,

Table 3. Summary of MaxR™ continuous testing at KLO's BN circuit

	Pilot Conv. Solids Recycle*	2-L Static*	Pilot MaxR™†	Pilot MaxR™†	2-L Static†	Design MaxR™ Condition†
Test number		4			13b	
Recycle ratio	13×	13×	4.5×	5.3×	5.3×	8–10×
Reactor pH	3.8	3.8	3.9	3.75	3.8	3.8
Reactor % solids (w/w)	5.1	4.3		3.42	6.4	>5.0
Flocculant dose, (g/t)	30	95			77	50
Underflow solids % (w/w)	44.5	35.4		20.8	41.0	35-40
Unit area (m^2/tpd)		0.26			0.12	0.3
Se conc. in feed, (mg/L)	0.031		0.026	0.05		
Se conc. in overflow, (mg/L)	0.022		0.013	0.034		<0.025
Se removal efficiency, (%)	29	21	50	32	29	
Mo conc. in feed, (mg/L)	1.38		1.24	1.39		
Mo conc. in overflow, (mg/L)	0.079		0.007	0.54		<0.1
Mo removal efficiency, (%)	94	93	99	61	94	

*Add lime and underflow separately.
†Pre-mix lime and underflow.

about 30°C. Static 2-liter settling tests were used to evaluate thickener parameters for samples taken at key times during testing.

One of the principal objectives of the test program was to determine the necessary amount of solids recycling in order to increase particle size of the precipitate and, thereby, the underflow solids concentration. This was supposed to reach at least 30% by weight in order to enhance the bulk neutralization process.

Continuous pilot test runs had achieved some promising results. Se and Mo concentrations could be reduced below the effluent limits at the low pH level of 3.8 and a recycle ratio of about 10 times of amount of precipitated solids in a single-pass condition. Underflow solids concentration as high as 44 wt% could also be achieved. Despite a large number of variables and operating conditions to test, stable conditions are necessary to obtain meaningful results. This includes careful control of pH, flows of fresh feed, lime addition, and recycle stream, and solids concentration in the reaction zone. Excursions from these conditions tend to reverse the process and yield poor settling properties. Some typical test conditions and results are listed in Table 3.

CONCLUSIONS

A comprehensive project was undertaken by Cameco Key Lake Operation to study the removal of selenium and molybdenum from its uranium mill effluent. Fundamental research work to understand the species distribution, reaction mechanisms, and proper treatment methods resulted in the engineering, implementing, and commissioning of modification in the Bulk Neutralization (BN) circuit. New methods of co-precipitating selenite/selenate with ferric sulphate-gypsum and precipitating molybdate with ferric ions have been experimented with bench- and pilot-scale testing and found to be effective. Implementation followed with careful adjustment of the circuit resulted in lowering the Se and Mo content in the final effluent to below the regulatory limits. Additional efforts have been continued with new method such as FLSmidth's MaxR™ solids recycle technology to further improve the effluent quality and minimize the volume of waste material to the tailings pond while ensuring the stability of the precipitated solids to prevent re-dissolving of Se and Mo at the final treatment steps.

ACKNOWLEDGMENTS

The authors wish to express their appreciation to the operating personnel of Cameco Corp. and FLSmidth for their valuable assistance during the course of this study. The permission from the management of both companies to publish these results is also appreciated.

REFERENCES

Emmett, B., and Lee, D. 2006. *Report of a MaxR Recycled Solids Precipitation and Clarification Test in a Pilot Plant Unit in Cameco Corp., Key Lake, Saskatchewan.* Internal report of Dorr-Oliver EIMCO, Salt Lake City, UT.

Gupta, C.K. 1992. *Extractive Metallurgy of Molybdenum.* CRC Press.

Lieu, A., et al. 2010. Selenium and Molybdenum Removal from Contaminated Mill Process Effluent: Cameco Key Lake Operation. In *Proceedings of the 3rd International Conference on Uranium.* Edited by E.K. Lam, J.W. Rowson, and E. Ozberk. Westmount, Quebec: CIM. pp. 749–759.

Nishimura, T., Hata, R., and Umetsu, Y. 2000. Removal of Selenium from Industrial Waste Water. In *Minor Elements 2000: Processing and Environmental Aspects of As, Sb, Se, Te, and Bi*, Edited by C. Young. Littleton, CO: SME. pp. 355–362.

Alternatives to Coal Mine Tailings Impoundment—Evaluation of Three Dewatering Methods at Rockspring Coal Mine

Charles Murphy and Christopher Bennett
Alpha Natural Resources, Logan County, West Virginia, USA

Greg Olinger and Bret Cousins
Ashland Water Technologies, Wilmington, Delaware, USA

ABSTRACT

In 2008, Alpha Natural Resources initiated investigations to determine the most desirable technology to facilitate the continuation of fine tailings disposal and water management at the Trace Branch tailings impoundment site during the capping stage at the Rockspring coal preparation plant. This 1800 ton per hour thermal coal facility was nearing capacity of its existing tailings impoundment and needed to determine and evaluate alternative methods to increase impoundment capacity and improve water management. This paper discusses the findings of various assessments and trade off studies that resulted in a decision to install eight three-meter belt presses capable of dewatering 180 tons per hour of tailings solids in a slurry so they could be mixed with coarse material and used as capping material.

INTRODUCTION

The Rockspring coal preparation plant has a throughput of 1800 tons of coal per hour and in the process, produces approximately 180 dry tons of fine tailings material each hour as 30% solids slurry. The slurry is pumped to the tailings impoundment area where the solids settle and then the water is recycled back to the plant. In 2008, the impoundment area was reaching capacity. If nothing changed, a new impoundment area would need to be built and there was not enough time to get a new impoundment area built and commissioned before the plant would need to be shut down. The operator of Rockspring, Foundation Coal (merged with Alpha Natural Resources in July 2009) undertook an extensive study of dewatering solutions, including geotextile tubes, deep cone paste thickening and filter belt presses, to decrease the volume sent to tailings to significantly lengthen the life span of the current impoundment area. Investigators concluded the filter belt presses provided the most practical and economical solution. Water management at the plant also became more efficient since much of the slurry water no longer left the plant, creating a reduction of water loss due to evaporation out at the impoundment site.

TAILINGS IMPOUNDMENT

Figure 1 is an aerial photo of the Trace Branch tailings impoundment. When regulators moved the area into closure, which involved the placement of capping material, the Rockspring plant, to continue operating, set up a mini impoundment area with a life span of 18 months. While regulatory processes were underway to establish a new full scale tailings impoundment area for the plant, it became apparent that 18 months would not be enough time to get a new impoundment area established and in operation.

Figure 1. Trace Branch tailings impoundment, Logan County, West Virginia

The capping process for impoundment areas includes placing coarse material over the surface of the pond to protect the fines in the area from wind, rain and other meteorological events after all the water drains away. The process creates a stable system that can then be reclaimed back to its original environment.

With the capping process underway, investigators studied how much fine material could be added to the capping material without significantly undermining the capping stability. Since the coarse material contained 5% to 6% moisture and the finished cap could contain up to 16% to 18% moisture and still maintain stability, there was opportunity for the fine material to be added. To reach the 16–18% total moisture target for the cap while handling the total 180 tons per hour of dry fines being produced by the plant (estimated to be approximately ⅓ of the capping material mixture), the fines slurry would need to be dewatered from its current 30% solids to approximately 60% solids.

If the tailings could be dewatered to 60% solids and mixed with the coarse material to cap the Trace Branch tailings impoundment, then the life span for the existing impoundment area combined with the mini impoundment area would allow ample time, approximately four and a half years, to permit and build a new, permanent impoundment area. The higher percent solids also would extend the life of the new impoundment area and significantly improve water management at the Rockspring plant.

TAILINGS SLURRY

The fines in the tailings slurry sent to the impoundment area consist of coal and rock material that has been consolidated to 30% of the slurry weight. The slurry contains fairly high clay content and approximately 80% of the solids are smaller than 40 microns in size, which creates problems when trying to increase solids content.

Figure 2. Dewatering methods that were evaluated at Rockspring

Table 1. Operational and economic drivers for dewatering Rockspring plant tailings

Method	Footprint (m^2)	Best % Solids	Capital	Operating (yr)	Over 4.5 Years
Geotubes	184000	85%+	—	\$10.5M	\$47.3M
Deep cone paste	400 (20 m diameter)	48–50%	\$5.5M–6.0M	\$2.5M–3.0M	\$16.8M–19.5M
Filter belt press	930 (Building + belt)	65–70%	\$4.5M	\$3.5M	\$20.3M

Before the new dewatering plant was commissioned, all of the plant's refuse streams were allowed to settle in conventional thickeners to approximately 30% solids and then were pumped to the tailings impoundment. The settling time was slow (30 plus days) with the tailings liquid effluent recycled back to the plant. To speed up water turn around and keep the impoundment in compliance with state and federal regulations, the slurry was treated with polymer before injection into impoundment improving turn-around to two to five days.

Regardless of settling time, approximately 100,780 gallons of water per hour were sent to the impoundment area when the tailings slurry was at 30% solids. Increasing the slurry to the 60% solids so that it could be used during the capping process would reduce the water flow to the impoundment to approximately 28,800 gallons per hour further improving the plant water balance. This increases the water re-circulated within the plant by 71,980 gallons per hour and places much more control of the water balance in the hands of operations.

Because of the difficulty in dewatering these solids, pressure would need to be applied to obtain at least a 60% solids level. Thus, as mentioned previously, investigators studied three dewatering methods: geotextile tubes, deep cone paste thickener and filter belt presses. Figure 2 illustrates each of the three methods that were evaluated at Rockspring.

DEWATERING METHODS

All three dewatering methods were evaluated and the results are shown in Table 1. Comparisons were made on performance, size of area needed for operation (footprint) and capital and operating costs extrapolated over the four and a half year extended life span of the Trace Branch impoundment area and the mini impoundment. Operational drivers were determined through lab and small pilot testing. The economic drivers considered both capital and operational costs.

Geotextile Tubes

The geotextile tube technology involves flowing slurry through a filter tube where the water passes through and the solids get compacted with the help of a flocculating chemical. Pressure is applied as water is forced through the membrane by the flow of slurry into the tube. Once the tube is filled, the geotextile tube is left in the impoundment area and a new geotextile tube is set up to continue the process. The technology is relatively inexpensive and very effective in many applications.

Testing showed that geotextile tubes effectively improved the solids level in the tailings slurry. The clay content restricted the level of compaction the geotextile tube technology achieved, limiting the maximum percent solids to approximately 85%, which was well over the required 60% solids target range. However, at 85%, the number of geotextile tubes required for the four and a half year period exceeded the amount of available land space in which to operate them. Although all the costs can be posted as operational (geotextile tubes are consumables), the cost per year of $10.5 million made the geotextile tube option the most expensive. In addition, the effluent discharged during dewatering would still have to be pumped back to the coal plant for reuse, incorporating variability in the water balance. Therefore, the investigators' conclusions were that the geotextile tubes were neither practical nor economical for treating the Rockspring tailings.

Deep Cone Paste Thickening

The technology of deep cone paste thickening is based on a steep-angled cone and a relatively deep thickener bed to enhance underflow density. The increased bed depth gives greater compaction, and thus applies pressure to push more of the water out of the slurry. This dewatering method incorporates modern thickener technology such as flocculants, feed well dilution systems for best flocculation, more effective raking mechanism designs, and high torque drives. Deep cone paste thickening has shown significant effectiveness in dealing with high clay tailings in the phosphate industry (Tao et al. 2008). These clays tend to be under 30 microns in size and colloidal in nature, much like what is found with the Rockspring tailings slurry.

Testing was designed to determine if deep cone paste thickening was capable of obtaining 60% solids for the tailings slurry at Rockspring. Investigators, testing with a series of flocculants and operational parameters, determined the best results achieved only 50% solids at an operational cost of $2.5M to $3M. Economically, over the four and a half year period, deep cone paste thickening is the best option, but since the nature of the Rockspring tailings does not allow dewatering to the 60% solids target level, the technology is not a suitable choice.

Filter Belt Presses

Water content in most slurries can be reduced to very low levels by applying mechanical pressure, depending on the solids ability to hold onto water. Belt filter presses are fed slurry that passes through a series of drum and roller systems, each series producing lower water content. As shown in Figure 3, most belt filter press operations can be divided into three general stages: initial de-watering, which makes the sludge pulp; pressing or medium pressure filtration, which conditions the sludge for high pressure filtration quality; and high pressure filtration. The process begins as the sludge enters the press. It is mixed with a dewatering chemical either in the press or in a conditioning tank prior to the press. The sludge then enters the gravity drainage zone where a large rotating drum drains a majority of the free water. Pressure is first applied in a low pressure wedge zone, which begins squeezing the remaining water out

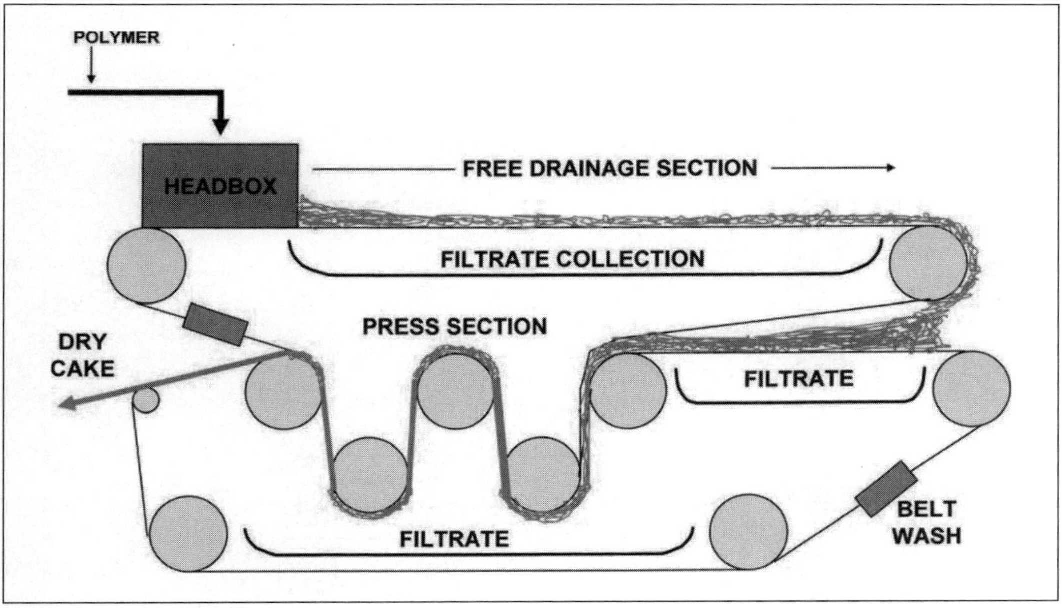

Figure 3. Filter belt press operations

of the sludge. Further de-watering occurs in the medium pressure zone, where two large, perforated drums of decreasing size apply pressure. Rollers perform the final de-watering in the high pressure zone (Jellesma 1978).

Testing the Rockspring tailings slurry on a filter belt press indicated that a higher than 60% solids target was achievable. Operational costs were higher than the deep cone paste method, but were not prohibitive.

As a result of the evaluations, the investigators determined that filter belt presses were the best option for the Rockspring coal plant to help them achieve their goal of economically lengthening the lifespan of their impoundment area and improve water balance.

PLANT DESIGN

Having chosen filter belt presses as the dewatering method, designing the entire system involved determining how many presses would be required and how they would operate. Given the current 180 dry tons per hour of tailings, six presses would be needed to handle the flow rate. The design team added two extra presses to allow for maintenance without reducing production. This also easily accommodates increased production.

Figure 4 is a flowsheet of the Rockspring dewatering plant as designed and commissioned. Design began in April of 2009 with construction initiated in April of 2010. Commissioning of the plant began in July of 2010.

The slurry line that originally flowed to the impoundment area from the existing thickener plant was tapped into for the new dewatering system and re-routed through a flow meter. Now the slurry flows into a 20,000 gallon stirred feed tank for the six operational filter belt presses (eight in total). This tank can also be fed extra water from the belt press sump for better control. The tank has an overflow system that sends slurry directly to the belt press sump as a bypass.

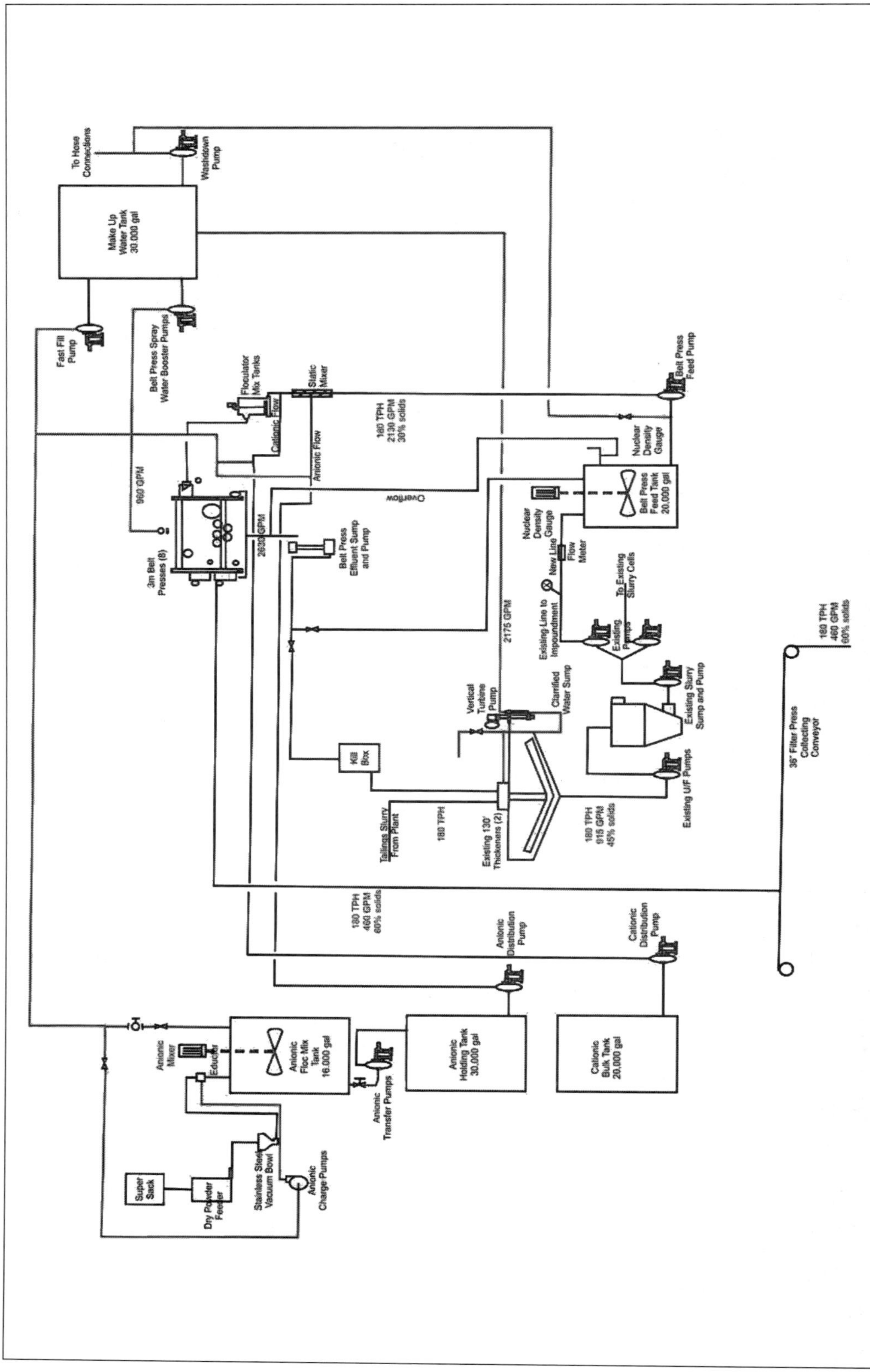

Figure 4. Flowsheet of the Rockspring plant dewatering system

From the belt press feed tank, the slurry passes through a static mixer where anionic flocculant is added to begin the dewatering process. Once through the static mixer, cationic coagulant is added to further enhance dewatering in the press. A small floculator mix tank is a pre-feed tank for the belt presses, allowing for better control of press operations. Originally, two static mixers were considered; however, due to concern that the second static mixer might cause flocculant breakdown, it was decided to utilize the floculator tank.

The slurry then enters the press, where it is dewatered to approximately 60% solids. From the press, the filter cake is discharged onto a 36 inch collecting conveyor belt that moves it to a place where it can be mixed with coarse material to make the final capping material. That material is then transferred to the Trace Branch impoundment area. The collecting belt was chosen over a radial stacker because the cake had transfer difficulties due to sticking in the radial stacker and the conveyor system required a much smaller containment area.

Flocculant System

Fully automated, the flocculant system was designed and constructed by Ashland Water Technologies. The nature of the Rockspring tailings requires a two stage flocculation process to manage the clays. The first polymer added is a very high molecular weight, anionic flocculant which adsorbs onto the surface of the particles thereby making them net negatively charged. The second polymer added is a low-medium molecular weight, cationic coagulant which interacts with both the clay particles and anionic flocculant.

Certain clay particles have both anionic faces and cationic edges. The primary anionic flocculant not only neutralizes those cationic edges, but it also provides anionic loops of polymer that extend far from the surface of the particles to make the attachment of the secondary polymeric coagulant more efficient. The second cationic polymer interacts with the anionic loops and tails of the first polymer. This process tightens the flocculated tailings and squeezes excess water from between the clay particles. The result is a much faster release of water in the free drainage section of the press and better compressibility of the sludge cake in the press section. This combination of polymers acts to increase the solids throughput of the press while, at the same time, increasing the final cake solids (alternately lowering the cake moisture).

The anionic flocculant is supplied as a dry powder in supersacs. Using four dry powder feeders, the flocculant is mixed with water to a 0.25% concentration in a 16,000 gallon mixing tank. This process is fully automated, adding the right amount of water for four supersacs to make a consistent concentration. From the mixing tank, the anionic flocculant is pumped to a 30,000 gallon holding tank for distribution to the static mixer at each belt press. The flocculant flow at each belt press is also monitored and controllable.

The cationic coagulant is in the form of a liquid polymer, ready to be added directly just before the floculator tank for each belt press. Stored in a 20,000 gallon tank, it uses the same type of pumping systems as the anionic flocculant.

Water Management

With less water being retained in the impoundment area and/or being evaporated during this process, the water management processes were more manageable. The extra water now comes off the belt presses and rather than building a catch basin for this water flow, it was found to be much more effective to let the water flow onto the plant floor and collect it with a sump system. Averaging 2630 GPM, the sump sends most of the water back to the existing

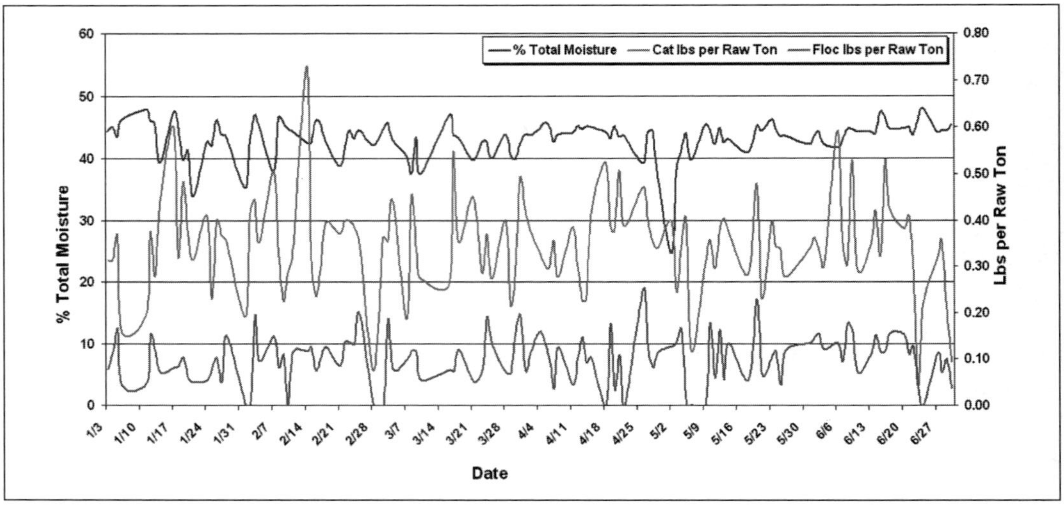

Figure 5. Cake moisture and chemical usage data

thickener. Some of the water can be re-routed to the belt press feed tank if necessary. Two nuclear density gauges are used to keep the feed to the belt presses consistent.

The existing thickener overflow is pumped to a 30,000 gallon make-up tank to feed all the other water uses in the plant. A thickener was considered rather than a make-up tank to produce a better quality of clarified water for chemical mixing and other sources, but the main consumer of the water is the existing thickener where higher quality water is not necessary. Upon determining the effects of lower quality water in other areas as being minimal, the 30,000 gallon tank and its smaller footprint was selected.

The 30,000 gallon tank is now the center of the water management system that reduces water loss. Averaging 2175 GPM of flow throughput water is distributed for anionic flocculant preparation, spray systems of the belt presses, and washdown water for the belt press feed line and hose connections throughout the plant.

PLANT COMMISSIONING

Late in July of 2010, the new Rockspring dewatering plant underwent commissioning. Slurry flow was slowly turned over to the presses to allow each press to be fully commissioned one at a time.

During the commissioning period, maximum dewatering was determined for the belt presses and produced cakes with percent solids in the 70% range, confirming pilot test investigations. The overall commissioning went smoothly with six presses converting approximately 30 tons of dry solids as a 30% solids slurry into 60% solids per hour.

Flocculant and coagulant dosages were also optimized during commissioning. Figure 5 is a graph of the press performance over the first six months of 2011. The pound per ton dosage varied as conditions in the presses and slurry characteristics varied. Cake moistures averaged in the low 40% range (high 50% solids), which is sufficient for the capping operations. Excess capacity is built into the system such that, when needed, press conditions can be modified to lower the target moisture levels.

Dosage of the anionic flocculant and the cationic coagulant are controlled through DC pumps. There is no direct automation on the polymer feed system besides specific gravity for control of slurry gravity with water. The anionic flocculant is injected into water and diluted before entering the 2-inch static mixer. After dilution, the cationic flocculant is injected into the bottom of the flocculator. To allow proper conditioning time between the two flocculant additions, the static mixer is positioned eight to ten feet before the flocculator.

CONCLUSION

The evaluation of geotextile tubes, deep cone paste thickeners and filter belt presses allowed investigators to determine which technology provided the best solution and value to the Rockspring coal plant. Each dewatering method had advantages, but the tailings' characteristics showed that filter belt presses performed the best.

The installation of filter belt presses at Rockspring extended the life of the current Trace Branch impoundment area and the mini impoundment area combined by approximately four and a half years by making the tailings slurry usable as part of capping material and by reducing the volume per day pumped to the impoundment areas. The extra time now allows for permitting and construction of a new permanent impoundment area. The new area can be designed around the lower volume that will be sent to it per day, reducing capital costs in construction and land use.

Water management is much more efficient. Most of the water now re-circulates in the plant instead of going out to impoundment and then recycling back with all of that system's impoundment retention and evaporation losses. The new water balance contains considerably less water and the plant requires less fresh make-up water.

ACKNOWLEDGMENTS

The authors thank the many people at Rockspring, Alpha Resources, and Ashland Water Technologies, who brought their expertise, hard work and dedication to the project, and in doing so, found and implemented the best solution to the Trace Branch impoundment's life span dilemma.

REFERENCES

Tao, D., Parekh, B.K., and Honaker, R. 2008. *Development and Pilot Scale Demonstration of Deep Cone Paste Thickening Process for Phosphate Clay Disposal.* Final Report to Florida Institute of Phosphate Research. FIPR Publication No. 02-162-229. Lexington, KY: University of Kentucky.

Jellesma, A. 1978. *Belt Filter.* U.S. Patent No. 4212745.

Mine Wastewater Treatment Using Novel Processes

Ronald V. Davis, Kevin O'Leary, Thomas Haynie, and Deepak Musale
Nalco Company, Naperville, IL, USA

ABSTRACT

Water used in industrial processes acquires contaminants derived from the materials and processes the water comes in contact with. In the case of mining applications, these contaminants often make the water undesirable for reuse within the mine circuit or for discharge from the site. Treatment of mine process water to make it acceptable for reuse or discharge requires methods that are resistant to a wide variety of potentially corrosive or scaling species, capable of handling a range of contaminants and function for long periods under relatively harsh conditions with minimal maintenance. This paper will discuss evaluations of mine water treatment systems that utilize several robust methods to treat mine wastewater. Concentration of wastewater contaminants using reverse osmosis followed by contaminant removal by precipitation will be discussed. The paper will focus on maximizing cost performance, lowering energy use, limiting the use of chemicals and minimizing the impact of the system on the environment.

INTRODUCTION

Growing demands on finite and variable water supplies across the world have resulted in greater government scrutiny and pressure on industry to more efficiently manage water resources. The mining industry has long been perceived as a problem area for water management due to the relatively high volumes of water used in mining applications and the wide range of contaminants that appear in wastewater derived from mining operations (Bowell 2000; Amezaga et al. 2011). As a result, managing wastewater from mining operations is often very challenging. Treatment is typically necessary to make the water suitable for reuse within the operation and is normally a requirement to meet increasingly strict discharge limits if the water is to be released back to the environment. Mine wastewater often has many contaminants exceeding use or discharge limits so treatment is accomplished by creating a purification circuit made up of multiple systems targeted to a site-specific contaminants. Identification of treatment methods that complement one another is the key to successful mine water treatment.

One of the most common mine water contaminants is sulfate ion. This ion is often the dominant contaminant in a mine operations wastewater (Bowell 2000). High sulfate water is associated with several industrial mining and mineral processing operations. One of the most common of these is Acid Mine Drainage (AMD), also known as Acid Rock Drainage (ARD), where decomposition by oxidation of pyritic solid material waste associated with mine operations yields surface water with sulfate concentrations in the 1500 to 8000 ppm sulfate range. It is worthwhile to note that AMD water also typically contains numerous other contaminants such as iron, manganese and other heavy metals. Neutral wastewaters typically contain sulfate concentrations of around 4000 ppm (Miller 2005). Since sulfate discharge limits are in the range of 250 to 2000 ppm this water must be treated before release into the environment. Much research has been directed towards sulfate removal in AMD applications due to the associated large water volumes and widespread occurrence.

Over the last 20 years several treatment schemes have emerged that are aimed at reducing the concentration of sulfate in mine wastewater to levels approaching discharge requirements. In a chemical precipitation example described by Geldenhuys an inexpensive source of metal ion is used to precipitate metal sulfate salt (Geldenhuys et al. 2003). In this method, acidic high sulfate (AMD) water is treated with limestone and then subsequently with lime. Inexpensive limestone reacts with the low pH feedstock in the first step to remove a portion of the calcium sulfate as $CaSO_4 * 2H_2O$ precipitate (Reaction 1). Subsequent treatment with lime to pH 12.26 removes additional calcium sulfate as $CaSO_4 * 2H_2O$ (Reaction 2).

$$CaCO_3 + H_2SO_4 + H_2O \rightarrow CaSO_4 {*}2H_2O + CO_2 \quad (1)$$

$$Ca(OH)_2 + H_2SO_4 \rightarrow CaSO_4 {*}2H_2O \quad (2)$$

This method is capable, with sufficient residence time, of producing an 1100 ppm sulfate water stream from a 3000 ppm sulfate feedstock. Though relatively inexpensive, this method requires a low pH feedstock for the limestone treatment portion to be effective and does not produce an effluent that will meet typical sulfate discharge requirements.

Other precipitation methods utilize barium sulfide as a barium source for barium sulfate precipitation (Maree et al. 2004) or include a source of aluminum with slaked lime to allow precipitation of ettringite or jarosite (Tait et al. 2009; Smit and Sibliski 2003). These methods, however, are dependent on regeneration of the barium ion source from the collected precipitate or require a heat source to achieve the desired precipitation efficiency.

Another example of a treatment utilized for high sulfate mine water is reverse osmosis. In this process feedstock is pretreated with scale inhibitors and filtered through a multimedia filter or an ultrafilter to limit membrane scaling as the liquid is passed across a reverse osmosis membrane under high pressure. Water migrates across the membrane, excluding most dissolved ions, producing a low sulfate permeate than can be discharged, reused or blended with other wastewater streams to meet discharge requirements (Sobana and Panda 2011). Recoveries of 60 to 70% of a 100 ppm sulfate permeate from a 5000 ppm sulfate solution may be regarded as typical. The reject stream from such an operation would contain roughly 12,300 to 16,400 ppm sulfate in 30 to 40% of the feedstock stream volume. Although this technique produces an effluent that meets typical discharge limits, reverse osmosis is more costly than chemical precipitation techniques and produces a substantial volume of highly concentrated sulfate reject water that cannot be discharged without additional treatment.

In this paper we present laboratory and pilot test results for a treatment that utilizes reverse osmosis and chemical precipitation to provide an effective means of treating high sulfate mine water for reuse or discharge. It will be demonstrated that by applying reverse osmosis to a sulfate containing mine wastewater a high sulfate reject stream is generated that is well suited for treatment with slaked lime to precipitate $CaSO_4 * 2H_2O$. Neutralization of the product water from a one hour residence time precipitation stage with carbon dioxide followed by blending with low sulfate RO permeate yielded a solution with a sulfate concentration below 1000 ppm. Results of field testing at a mine site desiring a cost-effective, robust and rapid means of meeting sulfate discharge requirements demonstrated the effectiveness of the process for meeting the locations discharge limits.

Calcium Sulfate Di-hydrate Precipitation

Calcium sulfate occurs in three forms but for most water treatment applications $CaSO_4 * 2H_2O$ (gypsum) is the predominant phase. Gypsum is a naturally occurring ionic crystalline solid that forms in many industrial processes where calcium and sulfate are present in concentrations exceeding saturation limits. Industrially, gypsum is a byproduct of the dihydrate process of phosphoric acid production, deposits on heat exchangers in cooling systems due to gypsums inverse solubility relationship with temperature and fouls reverse osmosis systems when the solubility of gypsum is exceeded near the membrane surface. As a result of its widespread occurrence, the precipitation of gypsum has been extensively studied both academically and industrially.

Gypsum precipitation occurs by a series of processes whose relative importance is dependent on the presence of inhibitors, relative supersaturation and on seed crystal surface area (Packter 1974; Gunn 1976). One process associated with gypsum precipitation is the formation of crystal nuclei from supersaturated calcium sulfate solution. In this process, prenucleation aggregates of calcium ions and sulfate ions form within the supersaturated solution. While many disassociate, some of these aggregates achieve the necessary size and structure to make dissolution energetically unfavorable and become crystal nuclei. The rate of nuclei formation is a function of the supersaturation, with several authors reporting the initial precipitation rate is proportional to the square of the relative supersaturation (low supersaturation or seeded) (Klepetsanis and Koutsoukos 1989) or to the relative supersaturation raised to the fifth power (high supersaturation unseeded) (Hamdona and Hadad 2008). In practice a period of time, known as the induction period, occurs in unseeded supersaturated calcium sulfate solutions before detectable precipitation is observed. During this induction period the concentration of nuclei grows over time until a detectable level of precipitation is reached. At high supersaturations, where homogeneous nucleation dominates (Alimi et al. 2003), the induction time is proportional to the inverse of the square of the supersaturation (Fan et al. 2010). A second process associated with the precipitation of gypsum is the growth of the existing gypsum crystals. In this process ordered growth of existing gypsum particles takes place by incorporation of calcium and sulfate ions into growth sites on the crystal surface. The rate of ordered growth is dependent on the gypsum surface area and on the supersaturation. It is worthwhile to note that when sufficient crystal surface is present no gypsum precipitation induction period is observed. Gypsum crystal growth rates are consistent with a surface controlled crystal growth mechanism (Klepetsanis et al. 1999).

Although both nucleation and crystal growth take place simultaneously within a supersaturated solution one process is often the dominant means of gypsum formation. For example, in unseeded solutions nucleation is dominant initially but becomes less important as the number of nuclei reaches a level where the gypsum surface is sufficient to allow ordered crystal growth to dominate. In seeded systems, such as continuous crystallizers, relatively few nuclei may be formed and most precipitation occurs through crystal growth. To achieve relatively rapid and predictable precipitation of crystals of a size readily removed in a continuous precipitation system it is beneficial to recirculate a portion of the precipitate back to the reaction chamber to act as seed. Use of crystallization product as seed for gypsum precipitation, rather than using synthetic gypsum seed, has no adverse impact on the rate of precipitate formation (Tait et al. 2009).

Gypsum Precipitation Using Slaked Lime

When slaked lime is used as a calcium source for gypsum precipitation the concentration of calcium present at any given time is dependent on the concentration of hydroxide ion (solution pH). As gypsum precipitation progresses the dissolved calcium that is consumed is partially replenished by dissolution of slaked lime to yield calcium ions and hydroxide ions. Calcium is removed from solution as a result of gypsum precipitation while hydroxide ion is not. As a result the concentration of hydroxide builds as dissolved sulfate is removed. Eventually the pH of the solution reaches a level that prevents sufficient calcium from dissolving and exceeding the gypsum Ksp. In practice as this limit is approached the rate of gypsum precipitation experiences a sharp decrease. The use of slaked lime to precipitate gypsum is further complicated by the tendency for gypsum to precipitate on slaked lime particles and block further dissolution of calcium derived from the slaked lime. This necessitates the use of excess slaked lime to ensure sufficient soluble calcium to exceed gypsum solubility. In addition, the solubility of gypsum increases above a pH of about 11.5, reaches a maximum at a pH of about 12.7 and then decreases at higher pH values. For example, at 35°C and pH 12.7 the solubility of gypsum is about 50% greater than at pH 7.0 (Yuan et al. 2010). This increase solubility may be due to the formation of $CaOH^+$. Thus, the overall rate of gypsum precipitation using slaked lime is a function of pH and the quantity of slaked lime present.

EXPERIMENTAL METHODS

Reagent grade chemicals were used in all laboratory testing unless otherwise noted. Concentrated reagent hydrochloric acid used in laboratory tests and concentrated industrial hydrochloric acid used in field tests were each analyzed by titration prior to use to determine acid concentration. The laboratory concentrated hydrochloric acid was 37.1% by weight and the concentrated hydrochloric acid used at the mine site was 27% by weight. Laboratory test solutions were prepared using deionized water. The 10% slurry of slaked lime used in laboratory testing was prepared using deionized water and was maintained as a uniform suspension with constant stirring using a magnetic stirrer. Solution pH was measured using an Orion Model SA720 pH meter calibrated at 25°C using pH 7.0 and 10.0 calibration standards supplied by VWR International. Immediate sulfate analyses of laboratory samples and of field samples were performed using a barium sulfate turbidity based test with aid of a Hach Model DR2400 spectrophotometer. Reverse osmosis 85% recovery concentrate was obtained using a pilot RO system operated at a mine site. Calcium analyses were performed using a Perkin Elmer Analyst 200 atomic absorbance spectrometer. Chloride concentrations were determined using a Dionex Model DX600 Ion Chromatograph.

Laboratory Gypsum Precipitation Tests

Laboratory batch studies were performed to characterize the amount of gypsum precipitate formed at varying lime and hydrochloric acid dose levels in 1 liter beakers at ambient temperature. Solutions were filtered using a 0.45 micron filter after a 1 hour residence time and the sulfate level immediately analyzed. In a typical experiment, 500 mL of RO concentrate was placed in a 1 liter beaker and stirred with a PTFE-coated magnetic stirrer. Concentrated hydrochloric acid was then injected below the surface of the solution using a syringe and then 10% slaked lime slurry was added. After 1 hour the pH of the solution was measured and the solution filtered under vacuum using a 0.45 micron filter. The sulfate content of the solution was then immediately measured and recorded. A portion of the filtered solution was then diluted and analyzed to determine the concentrations of chloride and calcium.

Analogous experiments were performed to characterize the kinetics of gypsum precipitation. These experiments were performed in a manner similar to the precipitation experiments described above with the exception that portions of the stirred test slurry were removed at desired time intervals using a 60 mL syringe. These slurry samples were then filtered using 0.45 micron syringe filters and immediately analyzed for sulfate.

Carbon Dioxide Neutralization Test

A laboratory batch carbon dioxide neutralization experiment was performed to characterize the neutralization of the slaked lime treated RO concentrate. In a 5 liter beaker a 3 liter sample of RO concentrate containing 16,500 ppm sulfate was treated with concentrated hydrochloric acid at a dose of 46.4 pounds per 1000 gallons followed by addition of 10% slaked lime slurry at a dose level of 105.3 pounds slaked lime per 1000 gallons and then stirred with a PTFE coated magnetic stirrer for 1 hour. The suspension was then filtered using a 0.45 micron filter and the filtrate placed in a 5 liter beaker. A portion of the filtrate was analyzed to determine sulfate concentration. Carbon dioxide was added to the solution at a rate of 357 milliliters per minute using a gas dispersion tube and the solution pH monitored using a pH electrode. Samples were collected at intervals of approximately 0.5 pH units using 60 mL syringes and the samples were filtered using 0.45 micron syringe filters. The filtered samples were analyzed to determine the concentrations calcium and sulfate.

Onsite Pilot Studies

A pilot study was performed over the course of 2 months at a mine site in Latin America. A skid mounted RO system with cross-wound membranes was used to treat the mine wastewater. A sand filter and 10 micron cartridge filters were operated before the membranes to protect them from soil and suspended particulate material. Sulfuric acid was added to the feed water before the membranes to adjust the pH to 7. A calcium sulfate specific RO scale inhibitor, Nalco PC504T, was fed into the feed water at 2.5 ppm in order to protect against gypsum scale formation on the membranes. Flow rate of feed water to the membranes was maintained at about 340 gallons per hour throughout the study and recovery averaged a consistent 80% with a peak of 85%. Membrane cleaning was performed approximately every 70 hours.

Concentrated hydrochloric acid containing 27% hydrochloric acid by weight was fed to the concentrate from the RO unit at a level of 60.6 pounds per 1000 gallons concentrate and the resulting solution was delivered to a precipitation vessel. This dose is roughly equivalent to a dose rate of 43.6 pounds laboratory concentrated hydrochloric acid per 1000 gallons concentrate. Slaked lime was fed to the precipitation vessel at a level of 227 pounds slaked lime per 1000 gallons concentrate. Residence time of the liquid in the precipitation vessel was one hour. The supernatant from the precipitation vessel was then treated with carbon dioxide to pH 8.1.

RESULTS AND DISCUSSION

Water Quality

The RO concentrate used for laboratory studies was obtained from a mine site in Latin America. The composition of the RO concentrate and diluted RO concentrate samples, as well as the range of composition of the RO feed water, are shown in Table 1.

High sulfate mine waste water falls within a range of concentration from 1500 to 8500 ppm sulfate. As seen in the table, the mine water RO feedstock used in this study was consistently within the typical range of high sulfate mine water. The RO concentrate collected for laboratory gypsum precipitation studies was representative the upper limit of expected sulfate concentration. Similarly, diluted RO concentrate samples prepared were representative of the middle and lower expected sulfate concentrations of the RO concentrate.

Table 1. Water composition samples used for laboratory testing

Measurement	RO Feed Water Range	RO Concentrate	Diluted Concentrate 1	Diluted Concentrate 2
pH	8.5 to 9.3	8.3	8.3	8.3
Sulfate	2320 to 3300	16500	13750	12500
Chloride	20 to 25	96	81	70

Laboratory Gypsum Precipitation Kinetics Tests

An investigation of the kinetics of gypsum precipitation in an RO concentrate sample containing 13750 ppm sulfate in the presence of slaked lime and hydrochloric acid indicates that sulfate precipitation is rapid over the first 40 to 60 minutes and proceeds very slowly thereafter as shown in Figure 1. As seen in the figure, a total of about 9750 ppm sulfate precipitates as calcium sulfate under the conditions of the test. This calcium sulfate precipitation is greater than 95% complete within 30 minutes and is 97% complete within about an hour. After about two hours the precipitation is 100% complete. No further change in sulfate level occurred after 12 hours. Under the conditions tested, a gypsum precipitation treatment system based on slaked lime addition can operate with 97% gypsum removal efficiency at a residence time of one hour.

Laboratory Gypsum Precipitation Tests

A series of gypsum precipitation tests were performed using RO concentrate generated at the Latin American mine site and diluted concentrate prepared to simulate the range of concentrate composition expected at the site described above. At this site the sulfate ion discharge limit is 1000 ppm and the chloride ion discharge limit is 500 ppm. The pH of the effluent must be no less than 6.0 and no greater than 9.0. For an average 80% recovery, this means that for the effluent to meet discharge limits the treated concentrate must contain no more than about 5000 ppm sulfate and no more than about 2500 ppm chloride prior to blending with RO permeate. Slaked lime alone fed at 82.8 pounds per 1000 gallons can only bring the sulfate concentration down to about 7500 ppm from a high of about 16,500 ppm. This does not meet the discharge limits of the mine site and is considerably higher than the 1100 ppm reported by Geldenhuys. In this case the solution pH reached a level of 12.59, preventing dissolution of sufficient slaked lime derived calcium to exceed the gypsum solubility.

Hydrochloric Acid Addition

One means of improving sulfate removal using slaked lime is to add an acid to partially neutralize the excess hydroxide formed during gypsum precipitation. A series of 1 hour residence time laboratory experiments were performed wherein hydrochloric acid was added to the RO concentrate prior to slaked lime slurry feed. The addition of acid was expected to increase the concentration of dissolved calcium derived from slaked lime and thus increase the quantity of gypsum precipitate formed. The results of experiments performed with three potential RO concentrate compositions at a constant concentrated hydrochloric acid feed rate of 46.4 pounds per 1000 gallons are summarized in Figure 2. As seen in the figure, at an initial sulfate concentration of 12,500 ppm the treated sulfate level is less than 4500 ppm over the range of slaked lime feed levels tested. At 13,750 ppm sulfate, the expected average sulfate concentration in

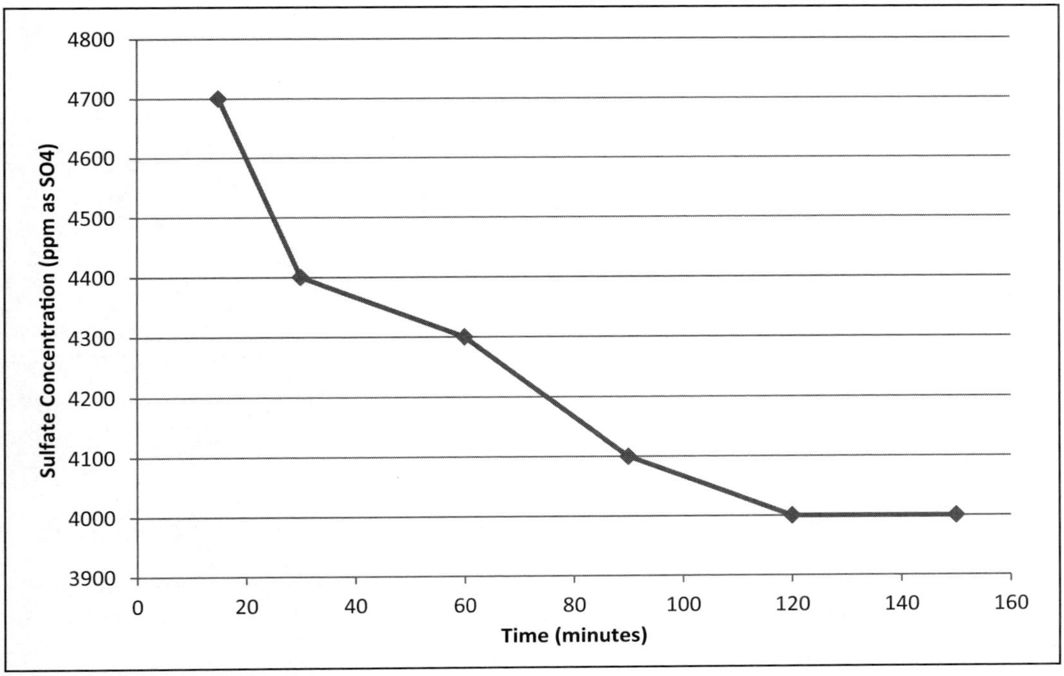

Figure 1. Gypsum precipitation kinetics at 13750 ppm total sulfate with HCl and slaked lime treatment

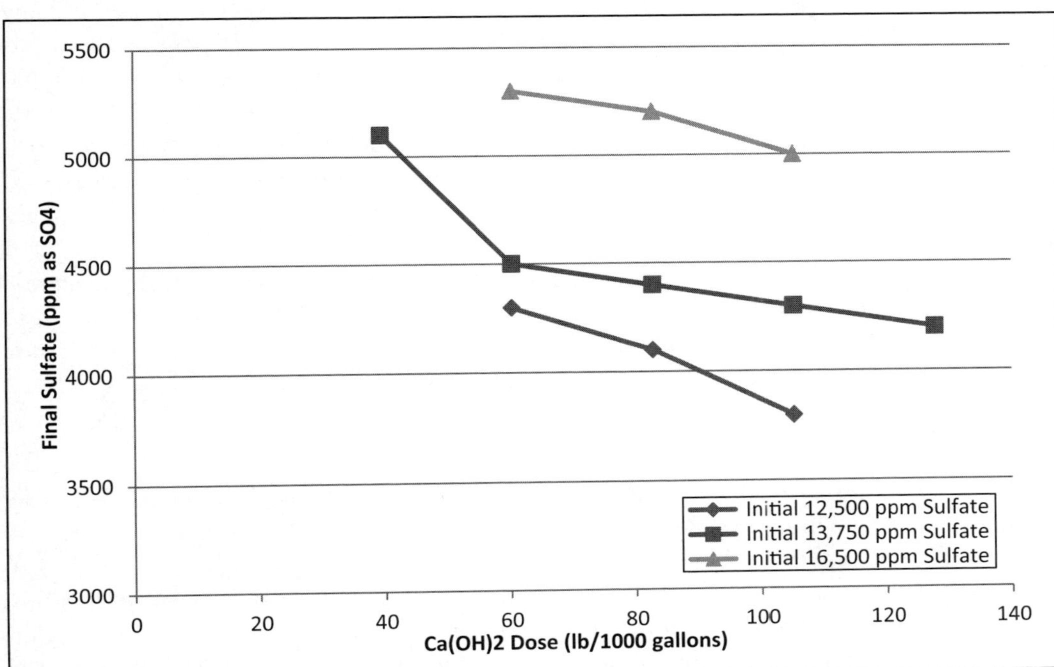

Figure 2. Sulfate removal achieved with a treatment of 46.4 pounds per 1000 gallons HCl at varying slaked lime dose

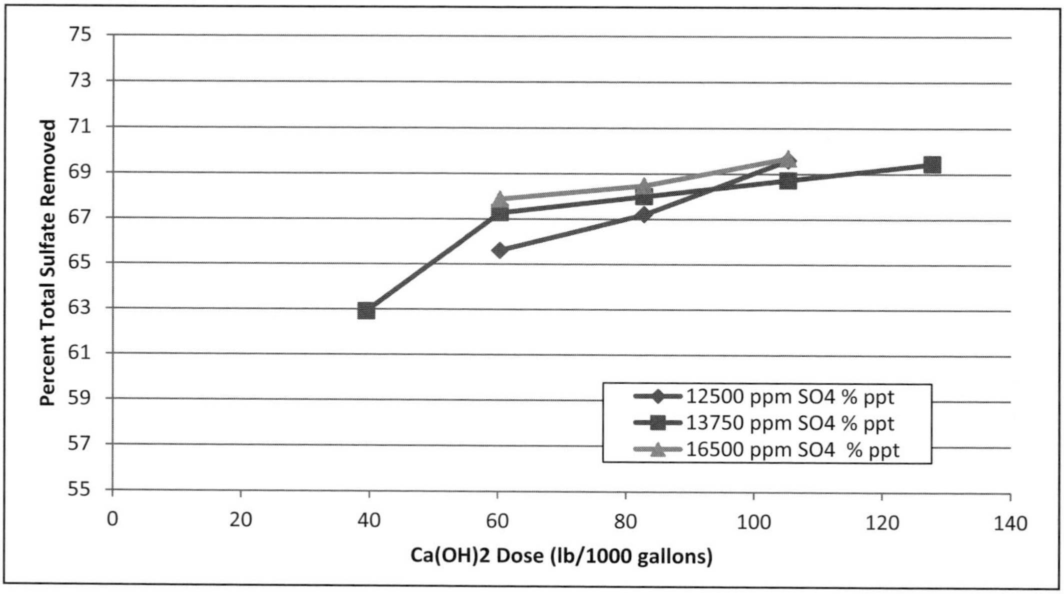

Figure 3. Percent total sulfate removed as a function of slaked lime dose at 46.4 pounds per 1000 gallons HCl treatment

the mine site RO concentrate, the treated sulfate level is less than 5000 ppm for slaked lime feed levels of 60.3 pounds per 1000 gallons and above. At the expected upper level of sulfate in the RO concentrate (16,500 ppm) the treated sulfate concentration was 5000 ppm or less only when the slaked lime feed was 105.3 pounds per 1000 gallons. Interestingly, in the case of this range of sulfate concentrations the percent total sulfate removed by this method of precipitation is about the same regardless of the initial sulfate concentration as shown in Figure 3. As seen in the figure, the percentage of total sulfate removed differs by only a few percent at each slaked lime dose level. Based on the results of these experiments the slaked lime levels needed to consistently yield a treated solution containing 5000 ppm sulfate or less at the mine site is 105.3 pounds per 1000 gallons.

The results of a series of experiments performed at varying concentrated hydrochloric acid dose at a constant slaked lime feed level of 82.8 pounds per 1000 gallons are shown in Figure 4. As seen in the figure, sulfate removal improves with addition of acid and the response is roughly linear over the range of acid levels tested.

Carbon Dioxide Neutralization Test

Neutralization of slaked lime treated RO concentrate can be readily accomplished using carbon dioxide. The lime treated concentrate contained approximately 5000 ppm sulfate. As shown in Figure 5, solution pH changes gradually between the initial value of 12.61 and 12.0. This is due to neutralization of the relatively high level of hydroxide and corresponding precipitation of calcium carbonate. Note that the dissolved calcium concentration dropped dramatically in this pH region during carbon dioxide treatment from an initial level of 1065 ppm as Ca^{2+} to 50 ppm as Ca^{2+} at pH 12.0. Approximately 98.5% of the dissolved calcium had precipitated when a pH of 11.5 was achieved. Between 12.0 and 7.5 the pH change over time is rapid. Interestingly, precipitation of calcium carbonate is associated with the removal of an additional 300 to 400 ppm sulfate from the initial slaked lime treated solution.

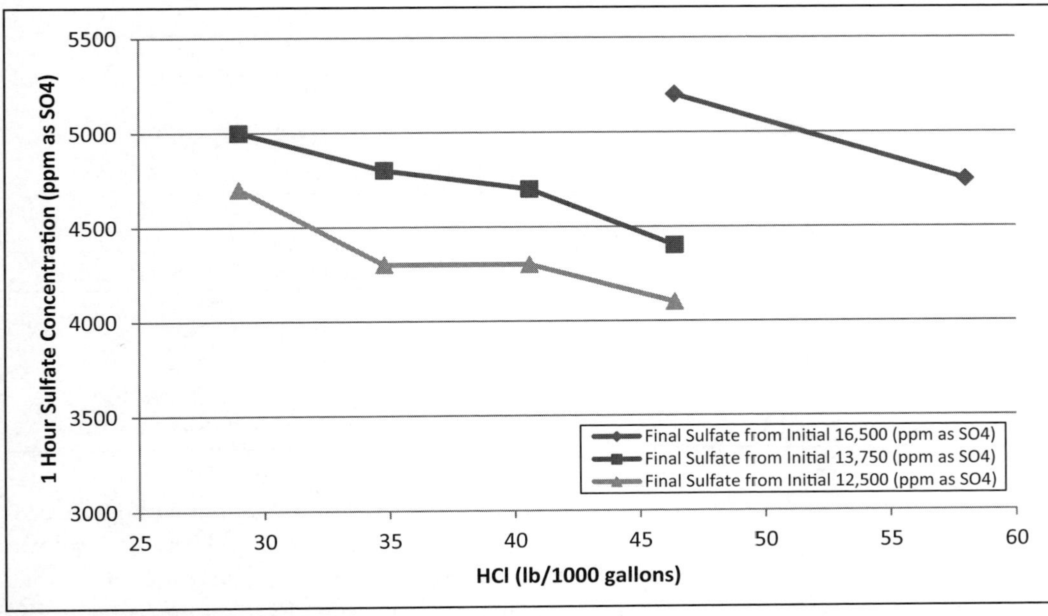

Figure 4. Sulfate removal as a function of HCl dose at 82.8 pounds per 1000 gallons slaked lime treatment

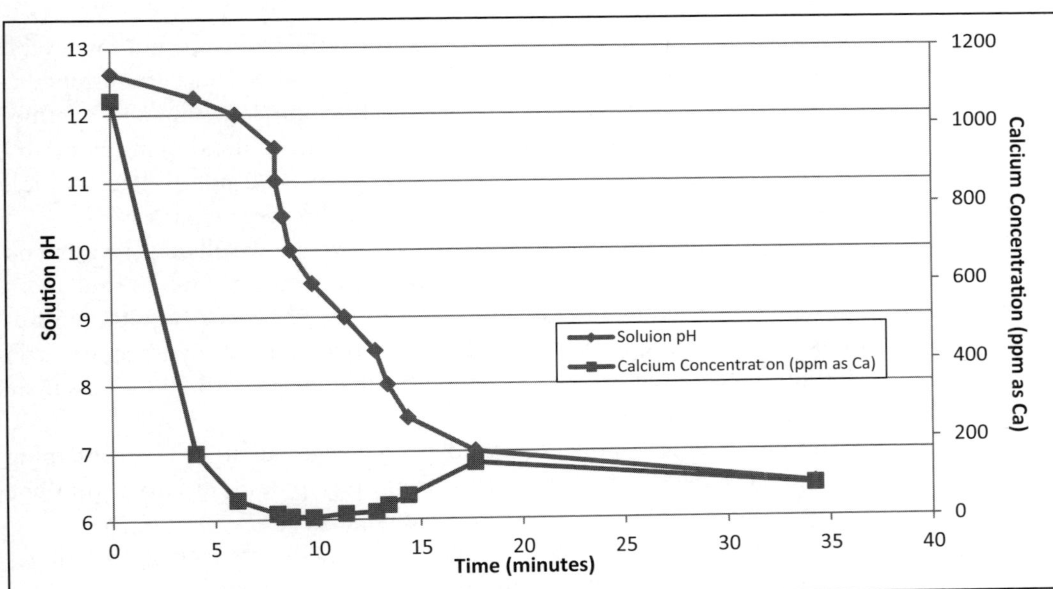

Figure 5. Carbon dioxide neutralization

Onsite Pilot Tests

The average composition of the RO influent, permeate and concentrate at the Latin American mine site are shown in Table 2. The sulfate rejection rate of the RO membrane was 99%. As seen in the table, a very high level of sulfate was produced in the RO concentrate throughout the test. Results of slaked lime treatment of the high sulfate RO concentrate followed by carbon dioxide neutralization are shown in Table 3. As seen in the table, treatment

Table 2. Average RO water parameters

Measurement	Influent	Permeate	Concentrate
pH	8.7	7	7.3
Sulfate (ppm as SO_4)	2696	27	10354
Total hardness (ppm as $CaCO_3$)	1291	79	5317

Table 3. Pilot study slaked lime treatment parameters

Measurement	Concentrate	Treated Concentrate	85/15 Effluent
pH	8.6	8.1	7.4
Sulfate (ppm as SO_4)	11909	4555	705
Chloride (ppm as Cl)	184	2269	341
Total hardness (ppm as $CaCO_3$)	6412	390	126

with slaked lime reduced the concentration of sulfate to well below 5000 ppm without exceeding the chloride limit of 2500 ppm. Prior to neutralization the solution pH was 12.7 but was readily reduced to 8.1 with carbon dioxide. Precipitation of carbonate solids lowered the total hardness to about 6% of the initial value. Blending such a treated RO concentrate with RO permeate will generate an effluent that will meet the discharge limits of the mine site.

CONCLUSIONS

- A pilot test performed at a mine site shows an RO system can operate for several months at an average recovery of 80% generating a high sulfate concentrate containing an average sulfate concentration of 10354 ppm. Blending low sulfate RO permeate with acidified slaked lime treated and carbon dioxide neutralized high sulfate RO concentrate yields a neutral solution containing less than 1000 ppm sulfate and less than 500 ppm chloride. This blend will meet mine site discharge requirements.
- Reverse Osmosis of a high sulfate mine water yields a very low sulfate permeate and a very high sulfate concentrate suitable for slaked lime treatment to remove sulfate.
- Addition of acid to neutral high sulfate water containing 12,500 to 16,500 ppm sulfate improves gypsum precipitation via slaked lime treatment. A hydrochloric acid level of 2570 ppm allows removal of 11,500 ppm sulfate from a 16,500 ppm sulfate solution in the presence of excess slaked lime.
- Reagent grade slaked lime treatment of acid treated high sulfate water containing 13,750 ppm sulfate shows 97% of the total gypsum precipitation occurs within one hour. Within two hours gypsum precipitation is complete.
- Slaked lime treatment of neutral water containing 16,500 ppm sulfate precipitates sufficient gypsum to lower the sulfate concentration to 7,500 ppm.
- Excess slaked lime improves gypsum precipitation and sulfate removal.
- Sulfate removal using slaked lime is limited by the pH achieved as gypsum precipitation occurs.

REFERENCES

Alimi, F., Elfil, H., and Gadri, A. 2003. Kinetics of the precipitation of calcium sulfate dehydrate in a desalination unit. *Desalination* 157: 9–16.

Amezaga, J.M., Rotting, T.S., Younger, P.L., Nairn, R.W., Noles, A., Oyarzun, R., and Quintanilla, J. 2011. A rich vein? Mining and the pursuit of sustainability. *Environ. Sci. Technol.* 45: 21–26.

Bowell, R.J. 2000. Sulfate and salt minerals: The problem of treating mine waste. *Min. Environ. Manage.* (May): 11–13.

Fan, C., Kan, A.T., Fu, G., Tomson, M.B., and Shen, D. 2010. Quantitative evaluation of calcium sulfate precipitation kinetics in the presence and absence of scale inhibitors. *SPE Journal* (December): 983–994.

Geldenhuys, A.J., Maree, J.P., de Beer, M., and Hlabela, P. 2003. An integrated limestone/lime process for partial sulfate removal. *J. So. Afr. Inst. Min. Metal.* (July/Aug): 345–354.

Gunn, D.J. 1976. Mechanism for the formation and growth of ionic precipitates from aqueous solution. *Faraday Discuss.* 61:133–140.

Hamdona, S.K., and Al Hadad, O.A. 2008. Influence of additives on the precipitation of gypsum in sodium chloride solutions. *Desalination* 228: 277–286.

Klepetsanis, P.G., Dalas, E., and Koutsoukos, P.G. 1999. Role of temperature in the spontaneous precipitation of calcium sulfate dihydrate. *Langmuir* 15:1534–1540.

Klepetsanis, P.G., and Koutsoukos, G. 1989. Precipitation of calcium sulfate dihydrate at constant calcium activity. *J. Cryst. Growth* 98: 480–486.

Maree, J.P., Hlabela, P., Nengovhele, R., Geldenhuys, A.J., Mbhele, N., Nevhulaudzi, T., and Waanders, F.B. 2004. Treatment of mine water for sulphate and metal removal using barium sulfate. *Mine Water Environ.* 23: 195–203.

Miller, G.C. 2005. *Reduction of sulfate concentrations in neutral mine effluent.* Final report to Nevada Department of Natural Resources and Environmental Science. Reno, NV: University of Nevada, Reno.

Packter, A. 1974. The precipitation of calcium sulfate dehydrate from aqueous solution. *J. Cryst. Growth* 21: 191–194.

Smit, J., and Sibilski, U.E. 2003. Pilot plant study to treat typical gold mine minewater using the Savmin process. *Proceedings of the Water in Mining Conference* 355–362.

Sobana, S., and Panda, R.C. 2011. Identification, modeling, and control of continuous reverse osmosis desalination system: A review. *Sep. Sci. Technol.* 46: 551–560.

Tait, S., Clarke, W.P., Keller, J., and Batstone, D.J. 2009. Removal of sulfate from high strength wastewater by crystallization. *Water Res.* 43: 762–772.

Yuan, T., Wang, J., and Li, Z. 2010. Measurement and modeling of solubility of calcium sulfate dihydrate and calcium hydroxide in NaOH/KOH solutions. *Fluid Phase Equil.* 297:129–137.

Removal of Phosphate from Mine Water Effluents—How to Meet Future Regulations for Effluent Waters

Mika Martikainen and Matias Penttinen
Kemira Oyj, Espoo, Finland

ABSTRACT

The mining industry is facing new challenges due to increasing amount of nutrients, like nitrates and phosphates, in effluent waters. Nitrogen and phosphorous compounds can originate in the soil and ore itself or from chemicals used in ore processing like explosives, reagents used in leaching (ammonia, cyanide, etc.), flotation, solvent extraction or even to pH regulation agents.

Dissolved phosphates have negative effect on nature's aqueous systems. These negative effects include for example growth of algae and depletion of dissolved oxygen.

Conventional removal methods for phosphorus removal consist of biological treatments such as the activated-sludge process, chemical treatments such as precipitation with Al-, Fe-, and Ca-reagents, or a combination of treatments. When treating to low phosphate concentrations, or treating complex waters like mine effluents, biological treatment is not necessarily an option. In very low phosphate concentrations chemical precipitation processes have some limitations due to solubility of precipitated phosphate species. Precipitation is not necessarily effective alone. Fixed-bed adsorption processes are effective removal processes for low phosphate concentrations, but higher phosphate concentrations require combination with conventional methods.

This paper is focused on iron based phosphate removal methods; both chemical precipitation methods and adsorption methods.

REMOVAL OF PHOSPHATE BY CHEMICAL PRECIPITATION

Phosphorous removal by precipitation is a physico-chemical process, where the addition of multivalent metal ion salts to the phosphorous containing solution forms insoluble precipitates of phosphate. The commonly used chemicals are aluminum (Al(III)), ferric iron (Fe(III)) added as chlorides or sulphates, and calcium (Ca(II)) typically added as lime. Chemical phosphorous removal is more complex when higher dosages of Al or Fe–salt are used, since beside simple precipitation, more complex adsorption/precipitation process also takes place.

The chemistry of phosphorus precipitation with iron is quite complex due to the formation of various metal phosphorus complexes and metal hydroxyl complexes, as well as adsorption of phosphorus onto surfaces of precipitates. Depending on the dose ratio, either only metal phosphate precipitation occurs, or both metal hydroxide and metal phosphate precipitation occurs. The lowest soluble phosphorus level that can be obtained is determined by the solubility of phosphorus, which depends on the dose ratio and pH. Residual soluble phosphorus less than the solubility limit have been observed as a result of adsorption of phosphorus onto the surfaces of precipitated ferric species.

Ferric chlorides or sulphates are all widely used for phosphorous removal. When ferric salts are added into wastewater for removal of phosphate ions, a large number of products, including complexes, polymers and precipitates are formed.

Table 1. Phosphate adsorption capacities for different iron hydroxides and oxides[1]

Material		Capacity (mg P/g)
Amorphous hydrous iron oxide (gel)	FeOOH	29.5
Akagenite	β-FeOOH	26.7
Lepidocrocite	γ-FeOOH	16.7
Goethite	α-FeOOH	6.7
Hematite	α-Fe_2O_3	5.3

The basic reaction is:

$$Fe^{3+} + H_nPO_4^{3-n} \leftrightarrow FePO_4 + nH^+$$

However in over stoichiometric dosages precipitation combined with adsorption becomes the more determinant process as hydroxides and oxyhydroxides of Fe(III) are formed.

REMOVAL OF PHOSPHATE BY ADSORPTION—PHOSPHATE ADSORPTION ON IRON HYDROXIDE AND OXIDE SURFACES

Goethite (FeOOH) and other iron (III) oxides are positively charged at neutral pH and adsorb phosphate strongly. With pure goethite as an adsorbent one has noticed that the adsorption decreases with increasing pH, but is still substantial above pH 8.0, approaching the point-of-zero charge of goethite at pH 9.4. Presence of calcium further increases phosphate adsorption at high pH.

Parfitt[1] has studied phosphate adsorption at pH 3.5 for several iron oxides and hydroxides. Amorphous hydrous iron oxide was found to have the highest capacity for phosphate adsorption. In Table 1 are shown adsorption capacities for different iron products.

Goethite's strong affinity to phosphate is allegedly attributed to ligand exchange reaction on the adsorbent surface, in which phosphate ions consistently exchange with surface structural hydroxyl groups.

The active goethite (FeOOH) surface contains oxygen bound in three ways: the A hydroxyl bound to one iron atom (Fe–OH), the B oxygen bound to three irons, and the C oxygen bond to two irons. Phosphate adsorption is thought to occur by phosphate oxygen replacing an A hydroxyl oxygen on the goethite surface, as shown below (Fe–OH represents the A hydroxyl on the goethite surface):[2]

$$\text{Fe-OH} + H_2PO_4^- = \text{Fe-OPO}_3H^- + H_2O \text{ (mononuclear)}$$

$$(\text{Fe-OH})_2 + H_2PO_4^- = (\text{Fe-O})_2PO_2^- + 2H_2O \text{ (binuclear)}$$

Antelo[3] et al. have studied phosphate adsorption on crystalline goethite surface. The combined effects of pH and ionic strength result in higher phosphate adsorption in acidic media at most ionic strengths, but result in lower phosphate adsorption in basic media and low ionic strengths.

Phosphate–goethite surface complexes are inert and desorption is slow and partly irreversible. Hence enhanced hydrolysis of organophosphates on the goethite surface does not lead directly to biologically available phosphate.[4]

Li et al. have concluded that phosphate–goethite interactions have two distinct phases; first phase described by adsorption by monolayer coverage and second phase by reaction occurring

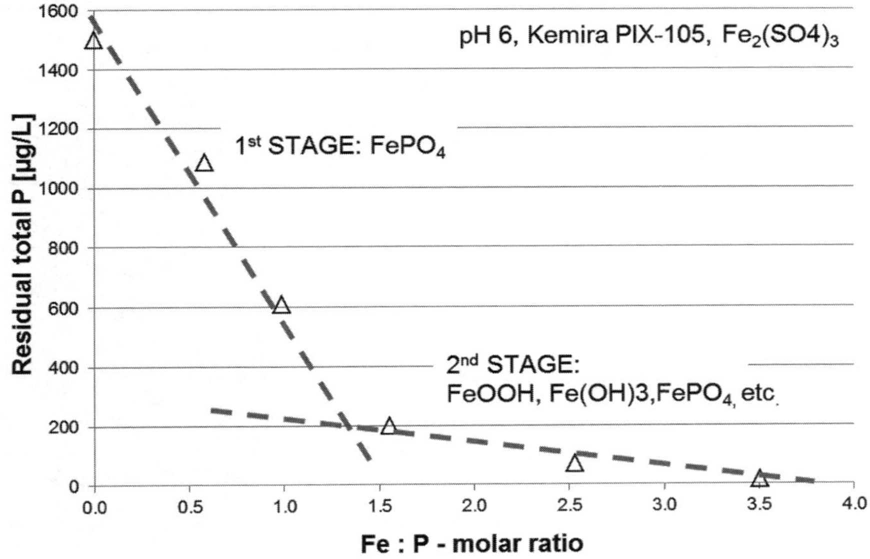

Figure 1. Residual total P as a function ferric salt dosage (indicated as Fe:P–molar ratio) when precipitation is done at pH 6. Synthetic test water.

on the surface, most likely surface precipitation. The transition between these two reactions is not apparent in the adsorption isotherms, which in general follow a Freundlich-type shape.[2]

PHOSPHOROUS REMOVAL BY PRECIPITATION WITH FERRIC SALTS

Experimental Set-Up

Three different commercial ferric salt products provided by Kemira Oyj were tested in phosphorous removal from mine water effluent. Products tested were Kemira PIX-105 ferric sulphate solution having 11.3% Fe, Kemira PIX-123 basic ferric sulphate $FeOH_x(SO_4)_{1.5-x/2}$, solution having 11.4% Fe and Kemira PIX-111 ferric chloride solution having 13.7% Fe.

Two phosphate containing test waters, synthetic water and real mine process water, as well as treated water samples were analyzed by ICP-OES. The total phosphorous content of the synthetic water was 1500 µg/L and the mine process water 7210 µm/L. Tests were carried out by taking 2 L of analyzed water, followed by changing the pH between pH 4–7 by NaOH or H_2SO_4. Ferric salt was added to effluent water with different Fe:P–molar ratio, where P is the total phosphorous in the initial water. Tested Fe:P-molar ratios were 0.5, 1.0, 1.5, 2, 2.5 where the ratio 1.0 represent the stoichiometric condition for $FePO_4$ precipitation. The pH of the effluent water was corrected, if needed, after ferric salt dosage. After 10 min mixing time sample was taken and filtered thru 0.45 µm membrane filter.

Results

Based on the results from synthetic water tests it is clear that phosphate removal goes thru two separate steps (see Figure 1). Around stoichiometric dosages phosphate is removed mainly due ferric phosphate precipitation. However this is not efficient alone but considerably higher dosages are needed at pH 6. When over stoichiometric dosages of ferric salts are used also other

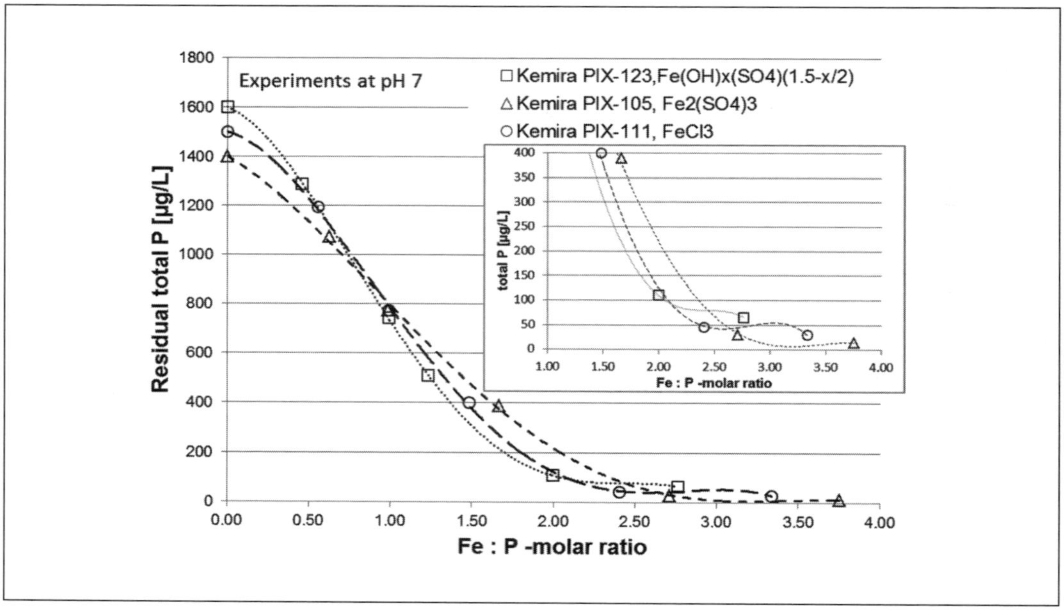

Figure 2. Residual total P as a function ferric salt dosage (indicated as Fe:P–molar ratio) when precipitation is done at pH 7. Synthetic test water.

forms of precipitates, like ferric hydroxides, are formed. In this range phosphate removal is enhanced by adsorption processes.

In Figure 2 is shown the residual phosphorous as a function of ferric salt dosage, when precipitation is done at neutral pH. It is clear that at neutral pH overdose of ferric salt is needed in order to reach low levels of residual phosphorous. What comes to ferric salt type, in principle there is not much difference between tested 3 products; both sulphates and chlorides work the same way when quite pure synthetic test water is used.

In Figure 3 is shown the residual phosphorous as a function of ferric salt dosage at pH 4. It is clear that lowering the precipitation pH greatly enhances the precipitation of $FePO_4$ and lower dosages are needed compared to experiments at pH 7. This has positive effect on residual waste, since less ferric salt is used in precipitation less solid waste is generated.

Real phosphorous containing process water, with phosphorous concentration of 7210 µg/L, was first tested as it is without pH control. All tested ferric salts worked quite well also in this high phosphorous loading water. However, more than two times the stoichiometric dosage is required in order to reach low residual phosphorous levels (see Figure 4).

When phosphate precipitation from real process water is done at pH 4, it is again seen that the performance of ferric salts is improved and significantly lower dosages are needed (see Figure 5).

It has been proven that with right dosage and pH, Kemira's ferric salt products, both sulphate and chloride based products, are efficient reagents for phosphorous precipitation.

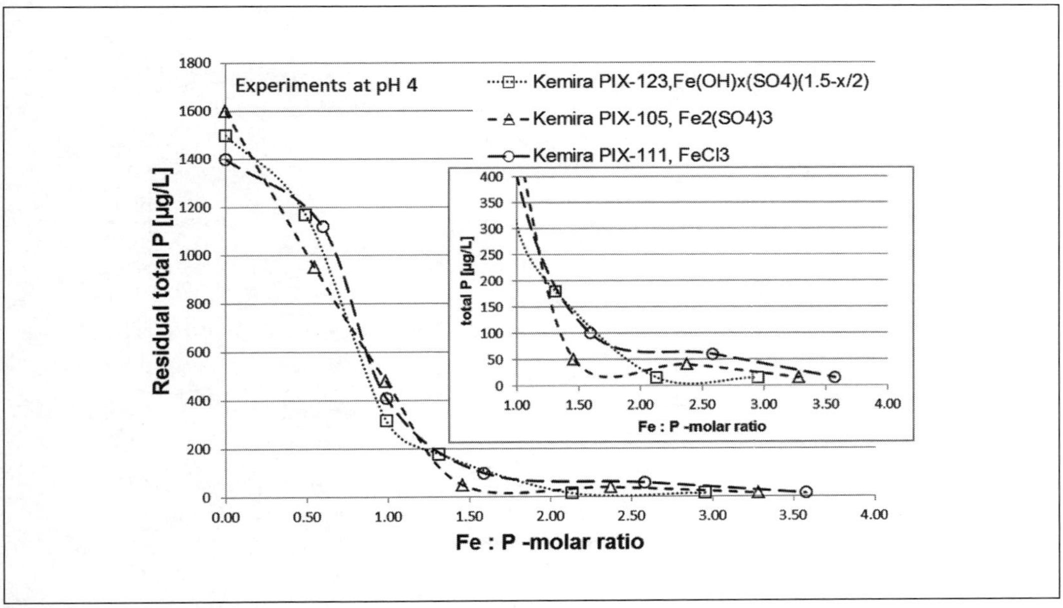

Figure 3. Residual total P as a function ferric salt dosage (indicated as Fe:P–molar ratio) when precipitation is done at pH 4. Synthetic test water.

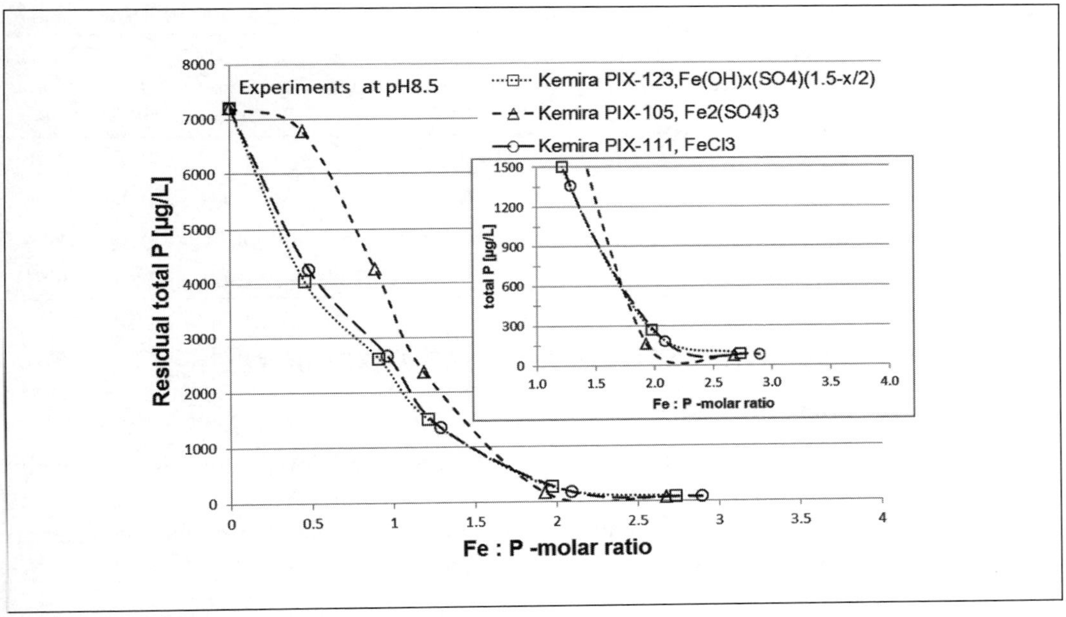

Figure 4. Residual total P as a function ferric salt dosage (indicated as Fe:P–molar ratio) when precipitation is done without pH control at pH 8.5. Real process water.

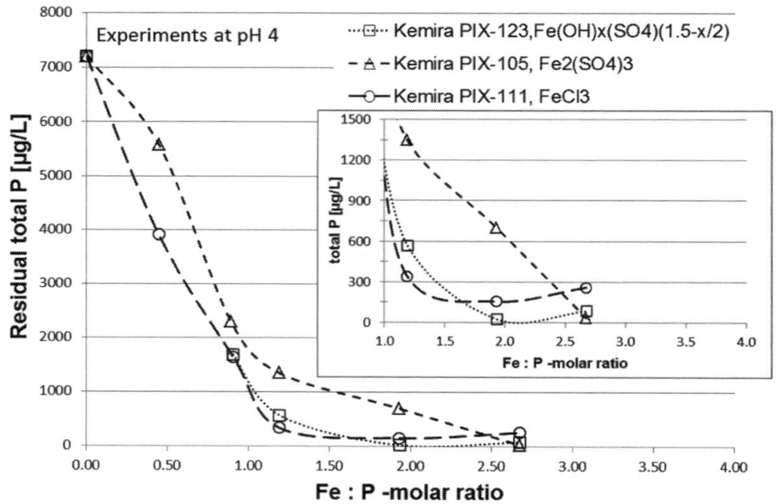

Figure 5. Residual total P as a function ferric salt dosage (indicated as Fe:P–molar ratio) when precipitation is done at pH 4.0. Real process water.

Table 2. Physical and chemical analyses of adsorbents

Sample Name	Specific Surface Area (m^2/g)	Particle Size D50 (mm)	Apparent Density (kg/L)	Moisture (%)	Fe (%)	Ca (%)	S (%)
Sample 1	30	1.5	1.23	16.3	42.4	0.60	4.46
Sample 2	81	1.5	1.26	17.5	44.7	0.28	2.05
Sample 3	110	0.9	1.20	15.7	42.1	0.26	1.59
Sample 4	23	—	0.89	13.9	11.2	18.4	9.57

PHOSPHOROUS REMOVAL WITH IRON-BASED ADSORBENTS

Analyses of Test Adsorbents

Four different iron based materials were analyzed and tested. Physical and chemical analyses of adsorbents are shown in Table 2.

XRD analyses showed great difference between products (Table 3). Some of the materials have very crystalline structure and some are very amorphous.

Equilibrium Studies

Experimental Set-Up

Isotherm tests were conducted in 200-mL bottles placed in shaker. Concentration of PO_4 was varied from test to test and the amount of adsorbent was kept constant. Each test was run for 7 days in order to reach equilibrium concentration. For the first couple of days pH was measured and adjusted if necessary to pH 6.5. Phosphorous concentration at the beginning and after one week reaction time was analyzed.

Table 3. XRD analyses of iron-based adsorbents

Sample Name	Crystallinity	XRD—Main Component
Sample 1	Amorphous	Schwertmannite $Fe_{16}O_{16}(SO_4)_3(OH)_{10}*10H_2O$
Sample 2	Amorphous	Hydronium jarosite $(H_3O)Fe_3(SO_4)_2(OH)_6$
Sample 3	Amorphous	Goethite FeOOH, hydronium jarosite $(H_3O)Fe_3(SO_4)_2(OH)_6$
Sample 4	Crystalline	Gypsum $Ca(SO_4)$, goethite FeOOH

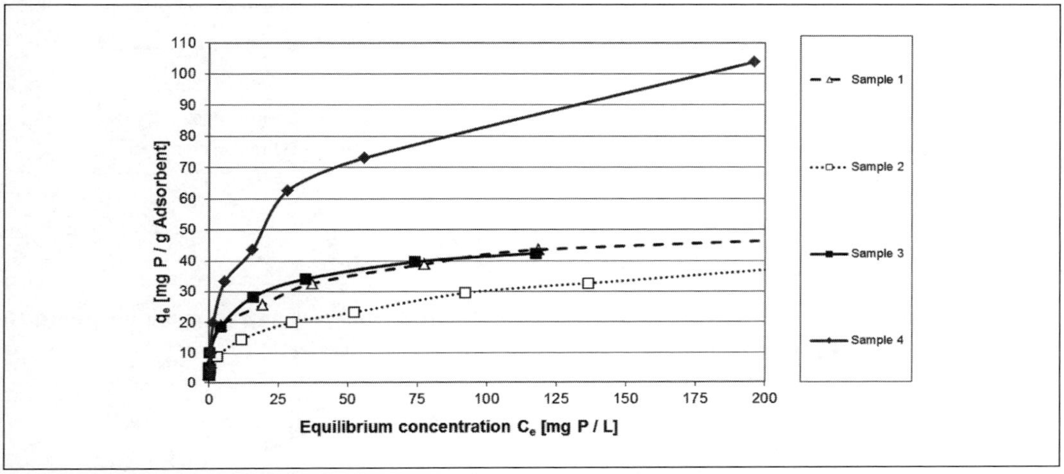

Figure 6. Phosphate adsorption capacity as a function of equilibrium concentration. Mixing time 7 days.

Isotherm Results

Adsorbed phosphorous per weight of adsorbent was calculated from isotherm results and Figures 6–9 were drawn based to these results.

Freundlich Isotherm

Freundlich equation models multilayer adsorption and the adsorption phenomena on a heterogeneous surface. Values from isotherm tests were fitted to Freundlich equation. Freundlich's adsorption coefficients n and K were solved from the linear equation.

$$q = K_f C^n$$

where

K_f and n = coefficients

q = weight adsorbed per unit weight of adsorbent

C = concentration in solution

C_{eq} = equilibrium concentration

Taking logs and rearranging:

$$\log q = \log K_f + n^* \log C_{eq}$$

By plotting log q as a function of log C_{eq} results straight line and from the intercept one can calculate K_f and from slope the value for n. In Figure 7 are the examples of measured data

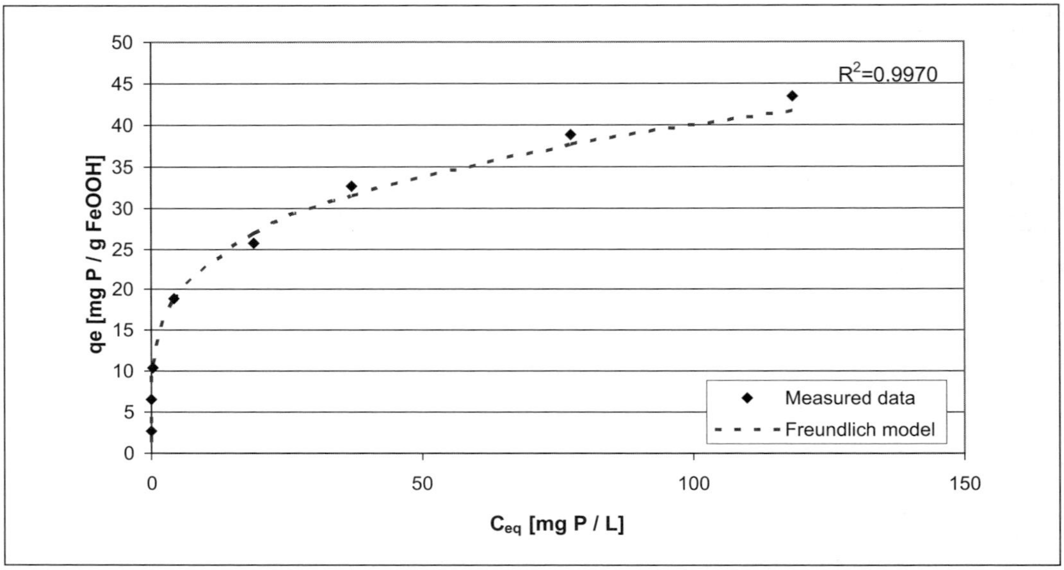

Figure 7. Sample 1's adsorption capacity as a function of equilibrium concentration of phosphorous. Dotted line is Freundlich model fitted to measured data.

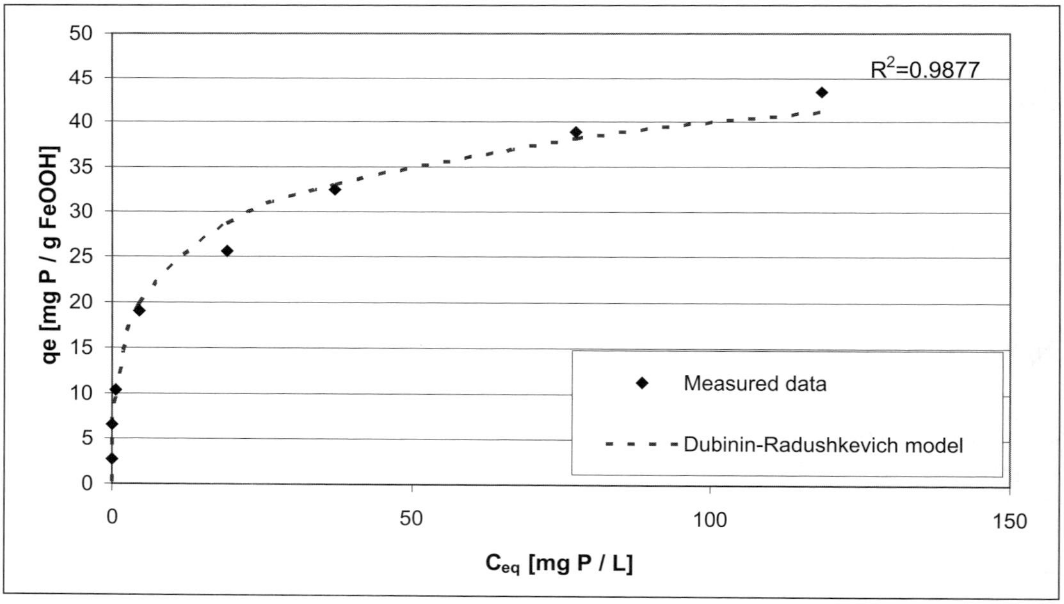

Figure 8. Sample 1's adsorption capacity as a function of equilibrium concentration of phosphorous. Dotted line is Dubinin–Radushkevich model fitted to measured data.

and fitting of the Freundlich model to data shown. The Freundlich adsorption model fits quite well to the measured data. This gives indication that studied materials have heterogeneous surface structures.

In Table 4 are the Freundlich coefficients (n and K_f) shown. In general, the measured data fits well to the Freundlich model. Based on Freundlich's equation, adsorption capacities for low equilibrium concentration (0.3 mg P/L) were calculated.

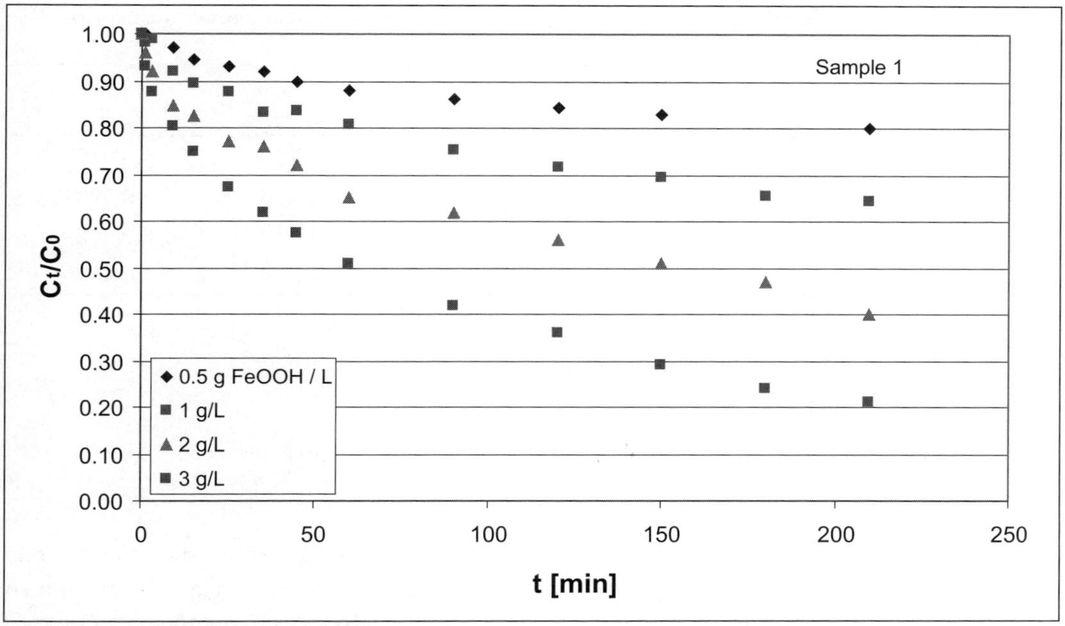

Figure 9. Phosphorous removal rate for four different adsorbents as a function of adsorption time

Table 4. Freundlich coefficients K_f and n. Capacity for 0.3 mg P/L equilibrium concentration is calculated based on Freundlich equation.

Adsorbent	1/n	K_f (mg/g)	$q_{0.3}$ (mg P/g FeOOH)	R^2
Sample 1	0.242	13.124	9.80	0.9970
Sample 2	0.347	6.033	3.97	0.9941
Sample 3	0.214	15.241	11.78	0.9941
Sample 4	0.393	15.009	9.36	0.9712

Dubinin–Radushkevich Isotherm

The Dubinin–Radushkevich (D-R) isotherm is the popular model used to estimate the characteristic porosity and the apparent free energy of adsorption.

$$q = Q_m{*}\exp(-k\varepsilon^2)$$

$$\varepsilon = R{*}T{*}\ln(1 + 1/C_{eq})$$

where

Q_m = maximum adsorption capacity
k = constant related to energy
ε = Polanyi potential
R = gas constant
T = temperature

Table 5. Dubinin–Radushkevich coefficients Q_m and k. In table is also shown adsorption free energy E.

Adsorbent	Q_m (mg P/g FeOOH)	K (mol^2/kJ^2)	E (kJ/mol)	R^2
Sample 1	66.277	–0.0026	13.90	0.9877
Sample 2	56.069	–0.0034	12.20	0.9901
Sample 3	60.837	–0.0022	15.12	0.9789
Sample 4	197.359	–0.0041	11.00	0.9880

By linearization

$$\ln q = \ln Q_m - k^*\varepsilon^2$$

Free energy of adsorption can be calculated from the k value by equation:

$$E = (-2k)^{-0.5}$$

By plotting ln q as a function of ε^2 should result straight line and from the intercept one can calculate maximum capacity Q_m and from slope the value for k. In Figure 8 are shown examples of measured data and fitting of D-R isotherm model to measurements. It can be seen that the D-R isotherm model has quite good fit to measured data.

D-R isotherm model constants are shown in Table 5. Based on results shown in the table, the sample 4 (gypsum containing FeOOH product) has the maximum theoretical adsorption capacity (Q_m).

If the free energy of adsorption (E) is between 8 and 16 kJ/mol the adsorption is considered to be ion-exchange type. If the free energy <8 kJ/mol, the adsorption mechanism is considered to be physisorption.[5] All tested adsorbents have free energy in ion-exchange/chemisorption range.

Based on the results shown in Table 5 adsorbents that are more pure and crystalline have higher free energy of adsorption indicating stronger bond between phosphate and adsorbent.

Adsorption Kinetics

Experimental Set-Up

Three different adsorbents were selected for kinetic studies: sample 1, sample 3, and sample 4. Kinetic studies were conducted with solution containing 10 mg/L of phosphorous. Volume used was 4 L and the amount of adsorbent was varied between 0.5–3.0 g/L. The pH was controlled during kinetic studies and kept on average at pH 6.5. Initial concentration was determine and after that adsorbent was added to solution. Timing was started at this point. Samples were taken at fixed intervals in order to follow adsorption of phosphate to adsorbent as a function of time. Each sample was analyzed for residual phosphate. Figure 6 shows the relation of initial concentration (C_0) and concentration at time t (C_t) for sample 1 as an example.

Elovich Equation

When adsorption to solid surface happens without desorption of products, the rate decreases with time due to an increased surface coverage. One of the most useful models for describing such 'activated' chemisorption is the Elovich equation.[6] The Elovich equation has general application to chemisorption kinetics in gas-solid systems. The explanation for this

Table 6. Constants for Elovich equation

Adsorbent dosage (g/L)	b (g/mg P)	a (mg P/g*min)	t_0 (min)	R^2
Sample 1				
0.5	0.6356	0.0945	16.7	0.9947
1	0.6502	0.0725	21.2	0.9931
2	1.0050	0.0698	14.3	0.9913
3	1.0681	0.0711	13.2	0.9990
Sample 3				
0.5	0.5364	0.1354	13.8	0.9969
1	0.5780	0.1360	12.7	0.9960
2	0.9338	0.1833	5.8	0.9996
3	1.1962	0.1623	5.2	0.9948
Sample 4				
0.5	0.5230	0.0577	33.1	0.9810
1	0.5374	0.0633	29.4	0.9960
2	0.5973	0.0507	33.0	0.9996
3	0.8250	0.0586	20.7	0.9948

form of kinetic law involves a variation of the energetics of chemisorption with the extent of surface coverage or that the active sites of adsorbing surface are heterogeneous and therefore exhibit different activation energies for chemisorption. Even though Elovich equation is well known in the chemisorption of gas molecules, equation has been applied satisfactorily also in liquid-solid systems.[7]

$$\frac{dq_t}{dt} = a \cdot e^{-bq_t}$$

After integration and linearization equation can be presented as

$$q_t = \frac{1}{b}\ln ab + \frac{1}{b}\ln(t + t_0)$$

where

q_t = adsorption capacity
t = time
a = initial adsorption rate (mmol/g min)
b = parameter related to extent of surface coverage and activation energy for chemisorption (g/mmol)
t_0 = 1/(ab)

By plotting q_t as a function of $\ln(t+ t_0)$ measured data should fall in a straight line having 1/b as slope and ln(ab)/b as intercept. Result for fitting measured data to Elovich kinetic data is shown in Table 6. One can see that for all tested materials measured data fits very well to the equation in a wide adsorption time frame. One can conclude from this that the all four adsorbents tested are heterogeneous in their surface nature what comes to adsorption of phosphate.

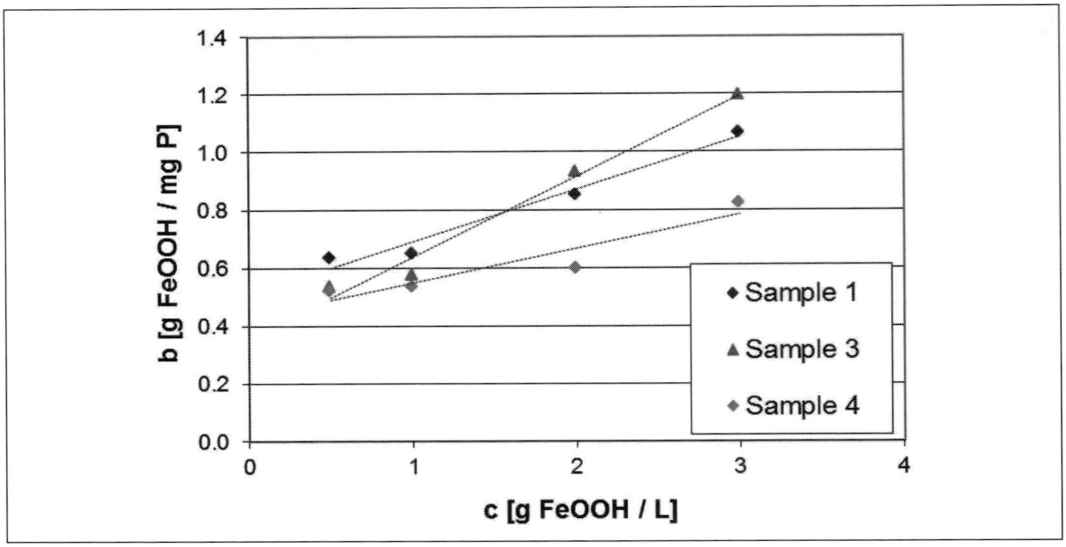

Figure 10. Elovich constant "b" as a function of adsorbent dosage

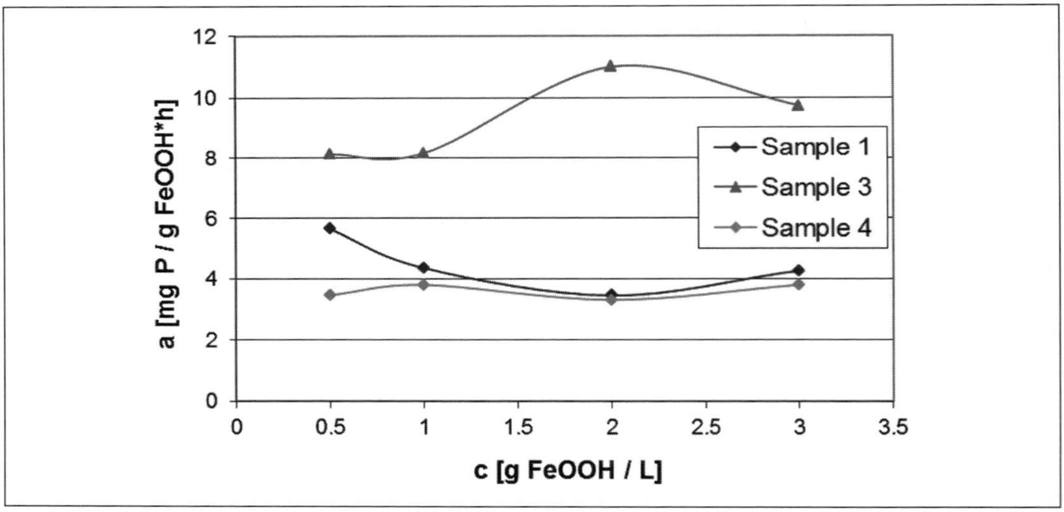

Figure 11. Elovich constant "a" as a function of adsorbent dosage

Figure 10 shows the Elovich parameter "b" as a function of dosage. As would be expected, the value of "b" increases as dosage of adsorbent increases as the constant "b" is the indication of number of active sites.

In Figure 11 Elovich parameter "a" is shown as a function of adsorbent dosage. Parameter "a" is related to initial adsorption rate. It seems that sample 3 has highest initial adsorption rate and sample 1 and sample 4 have clearly lower initial adsorption rate.

Intraparticle Diffusion Model

For intraparticle diffusivity, the adsorption rate is known to be controlled by several factors:

- Bulk diffusion–diffusion of the solute from the solution to the film surrounding the particle

Table 7. Constants for intraparticle diffusion model

Adsorbent Dosage (g/L)	a (–)	K_{id} (1/min)	R^2
Sample 1			
0.5	0.454	1.69	0.9074
1	0.481	2.78	0.9940
2	0.481	4.57	0.9916
3	0.455	7.33	0.9980
Sample 3			
0.5	0.439	2.56	0.9943
1	0.432	5.10	0.9980
2	0.413	9.42	0.9942
3	0.344	15.70	0.9944
Sample 4			
0.5	0.705	2.07	0.9592
1	0.829	1.73	0.9704
2	0.725	1.65	0.9840
3	0.691	2.64	0.9743

- External mass transfer resistance–diffusion from the film to the particle surface i.e., external diffusion
- Intraparticle resistance–diffusion from the surface to the internal sites i.e., surface diffusion or pore diffusion
- Uptake which can involve several mechanisms: physico-chemical adsorption, ion exchange, precipitation or complexation.[8]

The mechanism of adsorption is particle diffusion controlled, when intraparticle mass transfer resistance is the rate limiting step. For example Demirbas et al. have estimated intraparticle diffusivity by using intraparticle diffusion model shown in the following equation.[8,9,10]

$$R = K_{id} * (t)^a$$

where

R = percent adsorbed
K_{id} = intraparticle diffusion rate constant
t = time
a = constant that depicts adsorption mechanism

The equation can be presented in linearized form

$$\text{Log } R = \log k_{id} + a*\log(t)$$

If the adsorption is limited by intraparticle diffusion the plot log R as a function of log t will result straight line having a as a slope and log k_{id} as intercept. Results for four different adsorbents are shown in Table 7. For adsorbents sample 1 and sample 3 the intraparticle diffusion model fits quite well to the measured data. In the case of sample 4 the fit is not very good, indicating that the adsorption mechanism is not diffusion controlled.

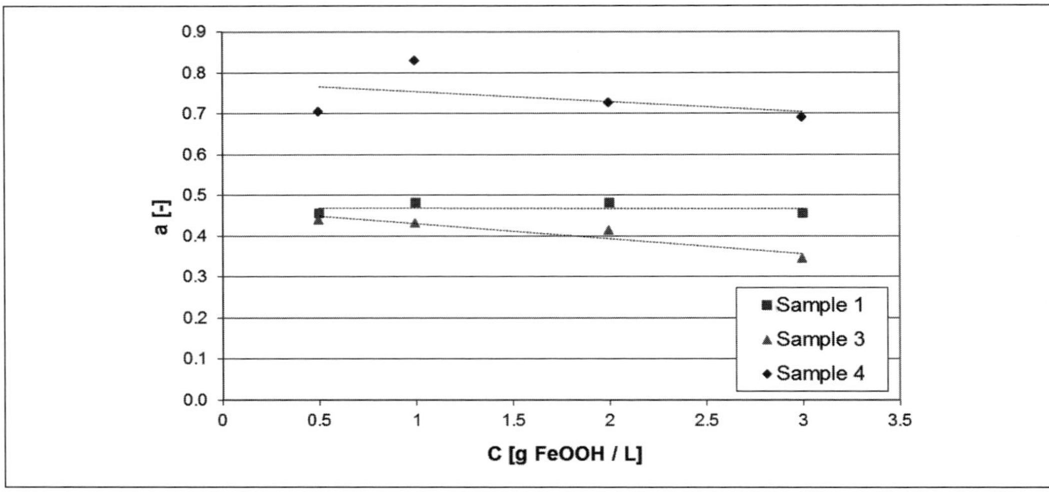

Figure 12. Intraparticle diffusion model constant "a" as a function of adsorbent dosage

Results for constant "a" as a function of adsorbent dosage is shown in Figure 12. In case of sample 1 and sample 3 one can conclude that beside the chemical adsorption/ion-exchange the intraparticle diffusion is also partly a limiting step. In the case of gypsum containing FeOOH diffusion is not the rate limiting step; instead adsorption is limited by the rate of chemisorption or ion-exchange.

In Figure 13 the rate constant is shown as a function of adsorbent dosage. Higher values of k_{id} indicate enhancement of the rate of the adsorption process. Results are logical regarding granule size; smaller the granule size higher is k_{id}.

Continuous Filtration Studies

Full scale phosphate polishing trial has been conducted with sample 1 using the treated waste water from city of Las Vegas. Results are shown in Figure 14. It is clear that sample 1 can be used for phosphate polishing in order to reach very low residual phosphate levels in effluent waters.

CONCLUSIONS

It was shown that both ferric sulphate and ferric chloride salts are efficient for treating phosphorous containing effluent waters to low residual phosphate concentrations. However, in the range of neutral pH over dosage is needed compared to the stoichiometric dosage, resulting more iron hydroxide sludge. By lowering the phosphorous treatment pH, less ferric salt is needed resulting less solid waste. Alternatively combination of phosphate precipitation with adsorption as a polishing step could result less waste.

Four different iron based adsorbent were studied in phosphate removal. It was shown that in all tested adsorbents the adsorption mechanism was thru heterogeneous surface with supporting multilayer adsorption. Also it was shown that in all cases the phosphate was bound by ion-exchange/chemisorption. The Dubinin-Radushkevich isotherm results indicate a strong bond between phosphate and adsorbents. In the case of samples 1 and samples 3 there is

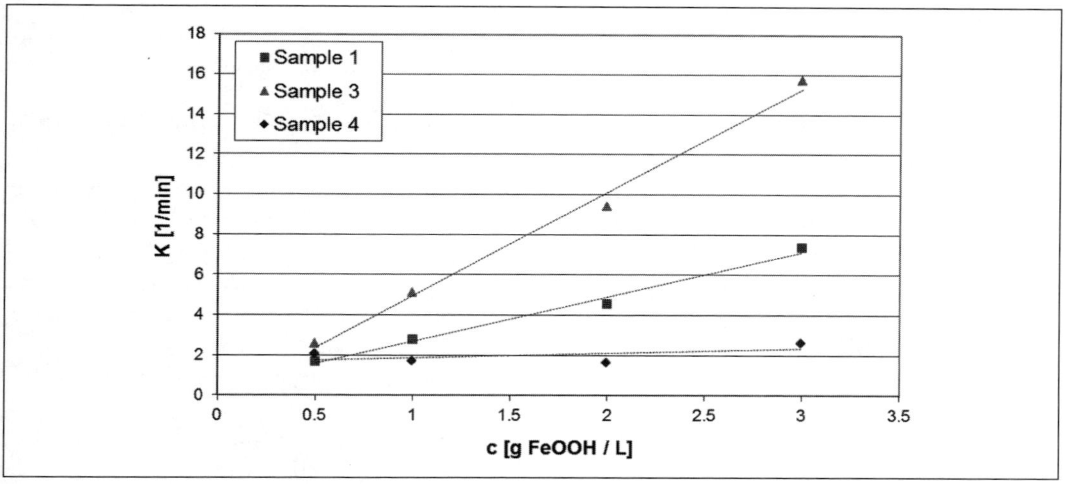

Figure 13. Intraparticle diffusion model rate constant k_{id} as a function of adsorbent dosage

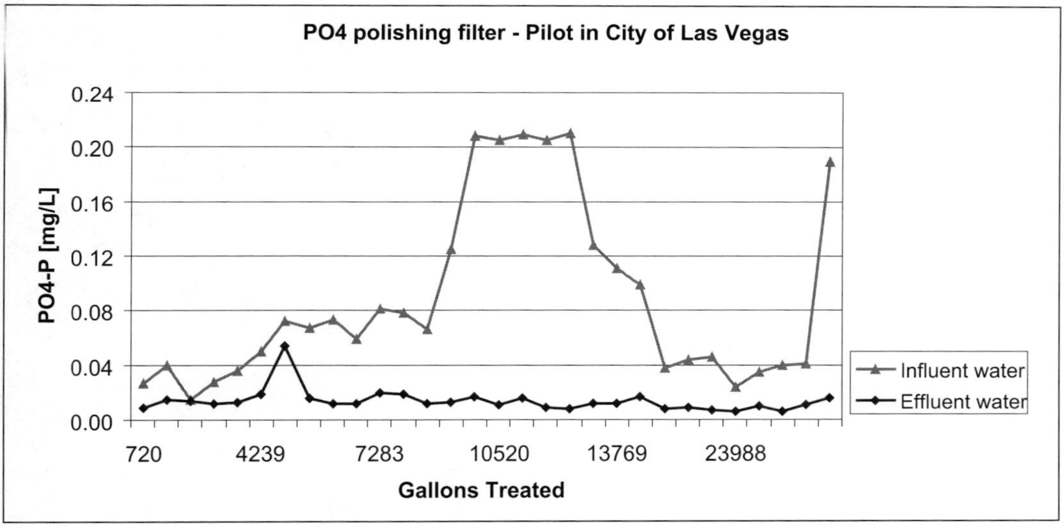

Figure 14. Effluent phosphor concentration as a function of treated gallons

evidence that beside chemisorption the intraparticle mass transfer is a rate limiting step. In the case of sample 4 it is clear that adsorption is only limited by the chemisorption step.

Continuous filtration experiments proved that the iron based adsorbents tested are an efficient treatment step for low influent phosphate concentrations. Application for adsorption process could be for example after ferric precipitation as a polishing step, when very low residual phosphate levels are needed. In some special cases iron based adsorbents could be used alone to treat waters having even higher concentrations of phosphates. Application for adsorbents in high influent phosphate concentrations could be for example an emergency backup treatment method.

REFERENCES

1. Parfitt, L., Atkinson, J., and Smart, C.St. 1975. The mechanism of phosphate fixation by iron oxides. *Soil Science Society of America Proceedings*, 39:837–841.
2. Li, L., and Stanforth, R. 2000. Distinguishing adsorption and surface precipitation of phosphate on goethite. *Journal of Colloid and Interface Science*, 230:12–21.
3. Antelo, J., Avena, M., Fiol, S., López, R., and Arce, F. 2005. Effects of pH and ionic strength on the adsorption of phosphate and arsenate at the goethite–water interface. *Journal of Colloid and Interface Science*, 285(2):476–486.
4. Turner, B.J., Frossard, E., and Baldwin, D.S. 2005. *Organic Phosphorous in the Environment.* CABI.
5. Gunay, A. 2007. Lead removal from aqueous solution by natural and pretreated clinoptilolite: Adsorption equilibrium and kinetics. *Journal of Hazardous Materials.*
6. Tsenga, R.-L., Wub, F.-C., and Juangc, R.-S. 2003. Liquid-phase adsorption of dyes and phenols using pinewood-based activated carbons. *Carbon*, 41:487–495.
7. Cheung, C.W, Porter, J.F., and Mckay, G. 2001. Sorption kinetic analysis for the removal of cadmium ions from effluents using bone char. *Water Research*, 35(3):605–612.
8. Igwe, J.C, and Abia, A.A. 2006. Sorption kinetics and intrapaticulate diffusivity of As(III) bioremediation from aqueous solution, using modified and unmodified coconut fiber. *Eclectica Quimica* 31(3):23–29.
9. Abia, A.A., and Asuquo, E.D. 2006. Lead (II) and nickel (II) adsorption kinetics from aqueous metal solutions using chemically modified and unmodified agricultural adsorbents. *African Journal of Biotechnology*, 5(16):1475–1482.
10. Demirbas, E., Kobya, M., Senturk, E., and Ozkan, T. 2004. Adsorption kinetics for the removal of Cr(IV) from aqueous solutions on the activated carbons prepared from agricultural wastes. *Water SA*, 30:533–539, 2004.

Mitigation Strategies to Reduce Diethylenetriamine (DETA) Residual in Tailings Water at Vale's Sudbury Operation

Jie Dong and Manqiu Xu
Vale Base Metals Technology Development, Mississauga, Ontario, Canada

ABSTRACT

Diethylenetriamine (DETA) has been used as an effective pyrrhotite depressant at Vale's Clarabelle Mill for more than a decade. In the fall of 2004, DETA started showing up as copper-DETA complexes in the effluent of the waste water treatment plant. Sporadic monthly copper exceedances have been observed since. DETA is a strong chelating agent and forms stable complexes with heavy metal ions (Cu^{2+} and Ni^{2+}), leading to an ineffective removal of the DETA-metal complexes by conventional lime precipitation method. In the future, more challenging ores will be processed and will require more DETA to be added in order to achieve the grade/recovery target. To avoid the exceedances of heavy metals in effluent due to DETA-metal complexation, effective mitigation strategies are being developed to reduce DETA level in the tailings water. These strategies include: (1) maximizing adsorption of DETA onto pyrrhotite and rock tailings while minimizing desorption during processing and disposal of tailings, (2) adding natural zeolite as extra sorbents to capture additional free or complexed DETA in the tailings water if necessary, and (3) finding DETA replacements as pyrrhotite depressants. In this paper, DETA adsorption and desorption properties are studied on pyrrhotite and rock tailings, natural zeolite and the mixture.

INTRODUCTION

In the processing of Cu/Ni ore at Vale's Sudbury Operation, pyrrhotite (Po) is the major sulphide gangue mineral. Po is very easy to float thus reducing concentrate grade. Po rejection is an integral part of sulfur dioxide abatement in Copper Cliff. About 15 years ago, diethylenetriamine (DETA) was introduced as an effective Po depressant at Clarabelle Mill (Marticorena et al. 1994; Xu et al. 2000). With DETA, the nickel recovery and concentrate grade are significantly improved. However, an environmental issue started showing up in the fall of 2004 at the waste water treatment plant, almost a decade after DETA was introduced into Clarabelle Mill. It is speculated that the copper exceedances in the effluent is related to DETA-Cu complexes, which are very stable and cannot be precipitated by raising the pH above 11 using lime. While the problem appeared to abate in 2005, it returned again in the winter of 2006 and 2007. To ensure no metal exceedance (i.e., Cu < 0.3 mg/L and Ni < 0.5 mg/L) in the effluent, DETA dosages at the Mill had to be reduced and this negatively impacted on the grade and recovery. In the future, it is expected that more challenging ores will be processed and that will require even more DETA. In order to avoid the heavy metals exceedances from DETA-metal complexes, effective mitigation strategies are being developed to reduce the DETA residual in the tailings water.

As shown in Figure 1, there are four streams generated at Clarabelle Mill: two concentrate streams, Cu and Ni concentrates, and two waste streams, Po and rock (Rk) tailings. The

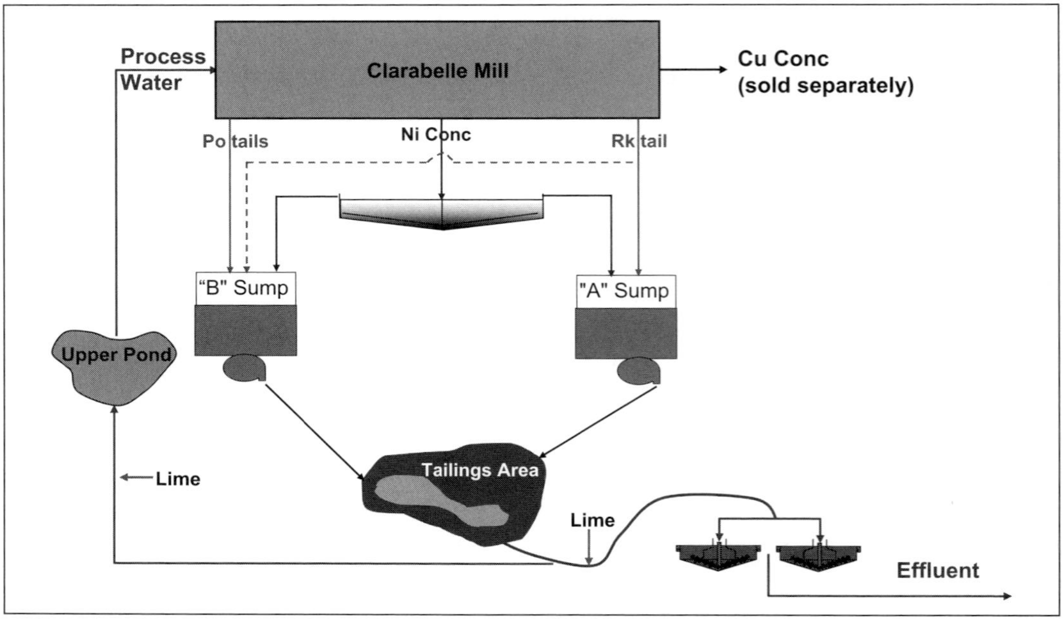

Figure 1. Tailings and process water management system at Vale's Sudbury operation

concentrations of DETA in each stream can be quite different. Rock tailings are DETA free, and DETA in Po tailings and concentrates varies according to the DETA dosage. The concentrate thickener overflow is mixed with Po and Rk tailings and the tailings are discharged into the tailings pond after pumping through a long pipeline. The process water from the tailings pond is treated at the waste water treatment plant using lime precipitation before being discharged to the environment. A portion of the process water that is recycled to the Mill is treated at the upper pond with lime addition.

It was determined that the increases of Cu level (as DETA-Cu complex) in the effluent corresponded to changes of tailings disposal in 2005. Rk tailings were used for backfill and dam construction, while Po tailings were entirely or partially beached. In 2005, when the previous tailings area reached its capacity limit, Po tailings were re-arranged to a larger tailing pond, where they were discharged into water cover and settled at the bottom of the tailings pond (Devuyst and Kerr 2008).

It was suspected that Po and Rk in tailings have some DETA adsorption capacity. During the pumping through the pipeline, DETA can be adsorbed onto the surface of the Po and Rk in tailings. However, the stability of the adsorption is not known. The best tailings disposal method is to retain DETA on the Po and Rk surfaces. Otherwise, it could be released into the tailings water upon dilution, form soluble Cu/Ni-complexes, and increase the difficulty of removal at the waste water treatment plant.

In this paper, the DETA adsorption and desorption characteristics on Po and Rk in tailings are studied. The addition of a natural zeolite to increase the capacity and stability of DETA adsorption from the tailings is investigated. To closely simulate the tailings disposal scenario, column settling tests were performed to demonstrate the differences in desorption during settling with/without zeolite.

EXPERIMENTAL

Materials

Clarabelle Mill Rk and Po tailings are used directly. Zeolite granules from the deposit at Bowie, Arizona are obtained from Zeox Corp. and used as received. DETA ($H_2N(CH_2)_2NH(CH_2)_2NH_2$), from MP Biomedicals Ltd., are used to prepare the adsorption solutions.

$CaSO_4 \cdot 2H_2O$ (Alfa Aesar), $MgCl_2 \cdot 6H_2O$ (Fisher Scientific) and $MgSO_4 \cdot 7H_2O$ (J.T. Baker) are used to prepare the synthetic process water to simulate the Clarabelle Mill flotation process water. $Ca(OH)_2$ (J.T. Baker), NaOH (Fisher Scientific) and H_2SO_4 (EMD) are used to adjust the pH to a designated value.

Characterization

The concentration of DETA is determined by ion chromatography (IC) DX120 and DX 500 from Dionex Corporation for 2~20 mg/L and 0.02~2 mg/L range, respectively. The concentrations of Cu^{2+} and Ni^{2+} are determined using a Varian atomic absorption spectrophotometer (Varian AA 200).

Procedure and Methods

Adsorption

In the adsorption kinetics studies, the substance (Po, Rk or zeolite) is added into the adsorption solution and mixed by a mechanical stirrer or on a shaking table. A sample solution is taken at different time intervals or at the end of the tests for analysis. The adsorption quantity Γ (mg/g of the substance) is calculated by the following formula:

$$\Gamma = \frac{(C_i - C_e)}{m} \times V_s \tag{1}$$

where C_i (mg/L) is the initial concentration of DETA, C_e (mg/L) is the equilibrium concentration, V_s (L) is the volume of adsorption solution, and m (g) is the mass of the substance added.

Desorption

Before starting the desorption study, a known amount of DETA is adsorbed onto the substances (Po, Rk or zeolite) (following the procedure in *"Adsorption"*). The loaded solids are filtered and subject to various levels of dilution to different solids percentage by synthetic process water at different pH values. DETA contained in the moisture of the filtered solid is subtracted in the calculation of desorption percentage using Equation 2.

$$\text{desorption\%} = \frac{(\text{DETA in solution after desorption} - \text{DETA carried from the moisture of loaded solids})}{\text{DETA adsorbed on solids before desorption}} \times 100\% \tag{2}$$

DETA Removal from Po Tailings Slurry by Zeolite

Po tailings are doped with DETA solution and mixed for one hour to obtain an approximate loading capacity of 150 g/t DETA on Po tailings and 20 mg/L of DETA in the aqueous phase. Different percents of zeolite (0, 0.5%, 1% and 1.5%) to the Po solids (dry weight based)

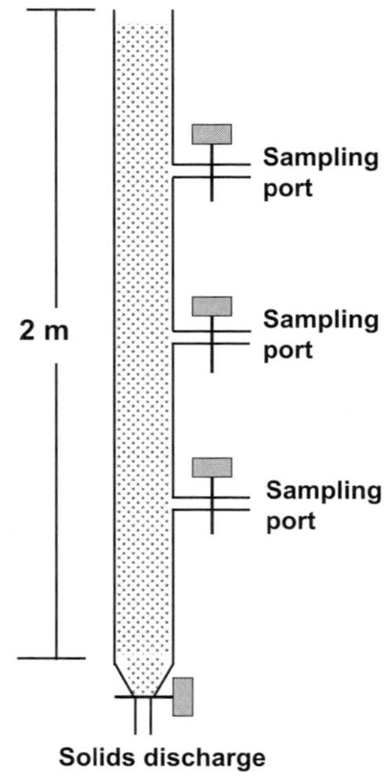

Figure 2. Experimental set-up of DETA desorption from Rk/Po tailings during settling

are added into the DETA doped Po tailings slurry. DETA concentration in the aqueous phase is measured as a function of time.

DETA Desorption from Po Tailings With/Without Zeolite

Po and zeolite are loaded with DETA as described in "*DETA removal from Po tailings slurry by zeolite*." With a known amount of DETA loaded, the mixture of Po tailings and zeolite is diluted with simulated process water at pH 8.5 to different solids percentage to determine the desorption percentage based on Equation 2.

DETA Desorption from Rk/Po Tailings With/Without Zeolite During Settling

In order to simulate the scenario of tailings disposal, a column filled with process water (as shown in Figure 2) is used to perform the desorption tests during settling. The tailings (~30% solids) loaded with DETA are pumped into the column and during this process experience dilution and settling. Liquid samples are taken at different time intervals from the three sampling ports. The desorption percentage is calculated by Equation 2. The effect of pumping rate, pH of process water in the column and zeolite addition on DETA desorption from Rk and Po tailings during settling is studied.

RESULTS AND DISCUSSION

Adsorption/Desorption of DETA on/from Po and Rk Tailings

The studies of DETA adsorption/desorption by Po and Rk tailings have been performed over a number of years by several different researchers. Due to the variation in tailings' mineralogy, particle size and the oxidation level, the data obtained are quite scattered. The equilibrium

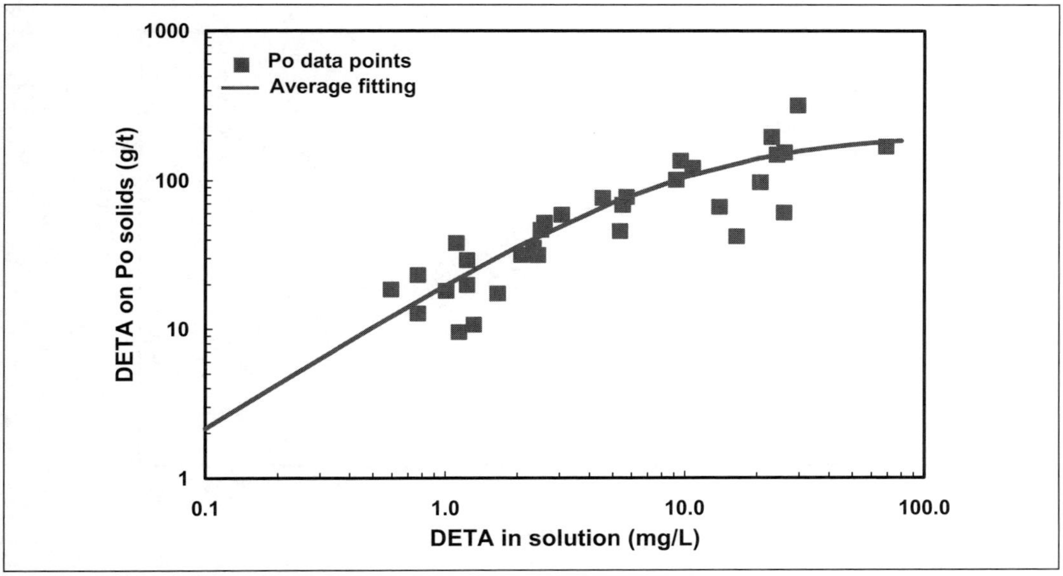

Figure 3. Adsorption isotherm of DETA on Po tailings at 20°C and natural pH

DETA concentration in solution vs. the amount of DETA adsorbed on Po and Rk solids are shown in Figures 3 and 4. An average trend line is fitted with Langmuir adsorption isotherm. The maximum adsorption capacity (Γ_{max}) is 225 g/t for Po tailings and is expected to approach ~400 g/t for Rk tailings, and the Langmuir constant (k) is 0.083 for Po tailings and 0.022 for Rk tailings. If the equilibrium DETA concentration is around 10 ppm, about 100 g of DETA can be adsorbed on one tonne of Po tailings and 70–80 g/t for Rk tailings (Santos 2009).

At Clarabelle Mill, Rk tailings (scavenger tailings) are DETA-free (DETA is only added in cleaners and scavenger cleaners). The mass of Rk tailings is about three times greater than the mass of Po tailings. Therefore, the Rk tailings can be used for DETA adsorption from Po tailings and the concentrate effluent.

Once the tailings loaded with DETA are disposed into tailing ponds, DETA desorption from the solids can be a concern. Depending on the disposal method, the tailings may experience various level of dilution. The desorption tests are performed on a shaking table and the results are shown in Figure 5. It is clear that with the decrease of solids percentage, the desorption percentage of DETA increases. However, DETA does not completely desorb, even at infinite dilution, especially for Rk tailings. This indicates that some DETA adsorption on Rk tailings is irreversible, presumably for the monolayer. In order to minimize DETA desorption from the tailings, dilution should be minimized. This could be the reason why there weren't DETA issues in the waste water treatment plant when the Po tailings were disposed on the beach with very high solids percentage.

Adsorption/Desorption of DETA on/from Zeolite

In the previous publication (Dong and Xu 2010), the adsorption properties of DETA or DETA-metal complexes on zeolite at different pH values were reported. Based on ion exchange mechanism (Breck 1974), sodium in zeolite is exchanged with DETA/Cu/Ni, and calcium and magnesium which are the major components in process water. Zeolite has a much higher DETA/Cu/Ni adsorption capacity (Γ_{max}=9.35 kg/t for free DETA and 1.91 kg/t for

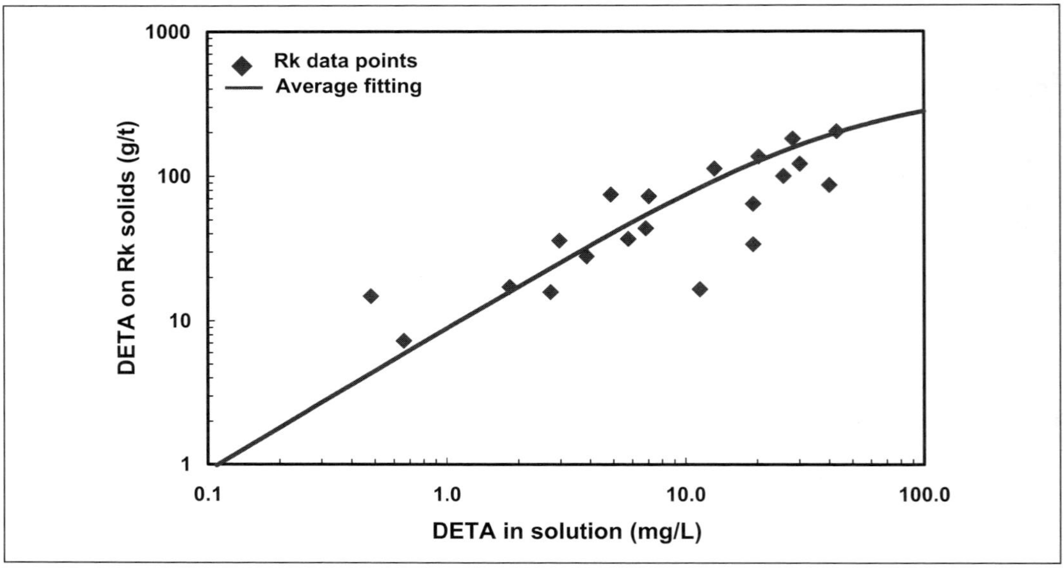

Figure 4. Adsorption isotherm of DETA on Rk tailings at 20°C and natural pH

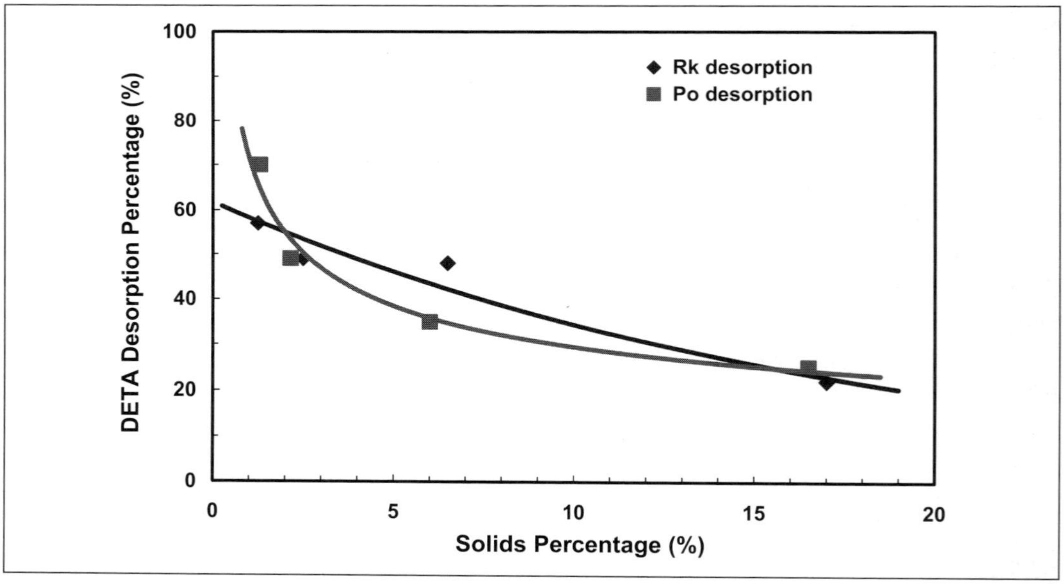

Figure 5. DETA desorption data from Po and Rk tailings

the complexed DETA) than Po and Rk tailings, leaving very low DETA concentration in solutions, as shown in Figure 6. The DETA adsorbed on zeolite is very stable over a wide range of pH (2~12). Even with high dilution there is no more than 2% of DETA desorbed from zeolite (Figure 7). Despite these advantages of zeolite, it is not economical to use zeolite to treat all of the tailings water, because of the large volume of tailings water that needs to be treated every day. One practical method is to add some zeolite into Po tailings when the DETA concentration in the Po tailings water exceeds the normal value (>15 mg/L).

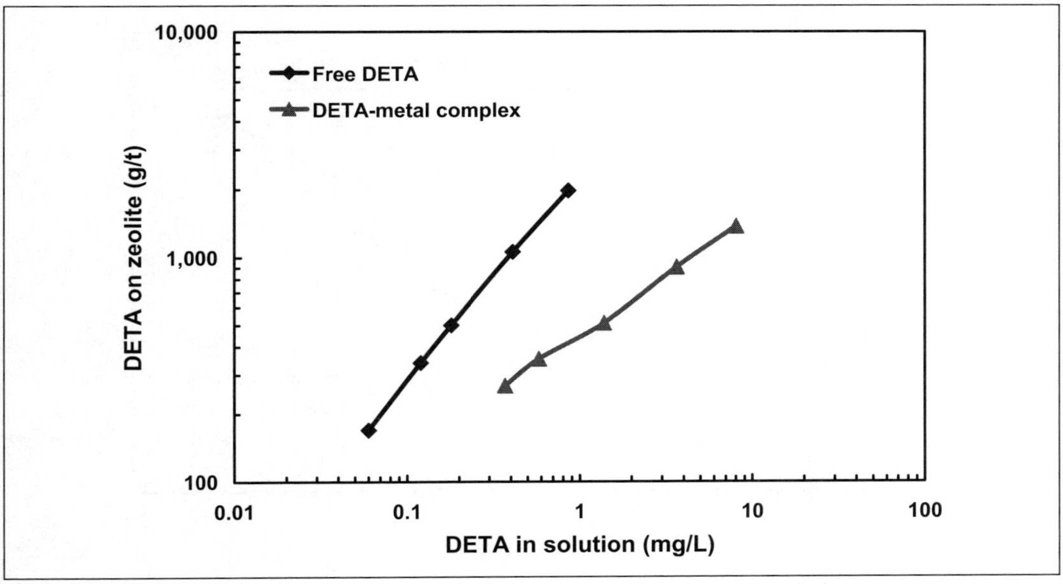

Figure 6. Adsorption isotherm of DETA and DETA-metal complex on zeolite

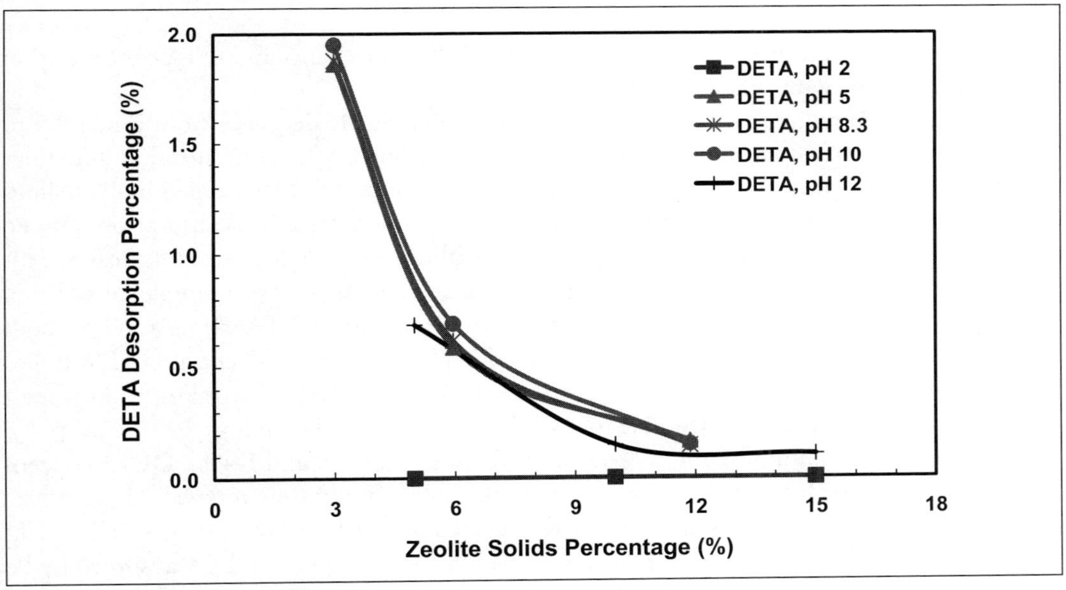

Figure 7. DETA desorption data from zeolite at room temperature

DETA Adsorption from Po Tailings Slurry with Zeolite

Currently, Po tailings are the major stream carrying DETA and experiencing dilution. Therefore, the differences in DETA adsorption/desorption from Po tailings with/without zeolite are compared.

Initially, about 150 g/t of DETA is adsorbed on Po tailings after one hour of mixing, with 16 mg/L equilibrium DETA concentration in the Po tailings water. Different percentages (0%, 0.5%, 0.9% and 1.4%) of zeolite to the dry weight of Po tailings are added into the slurry of Po

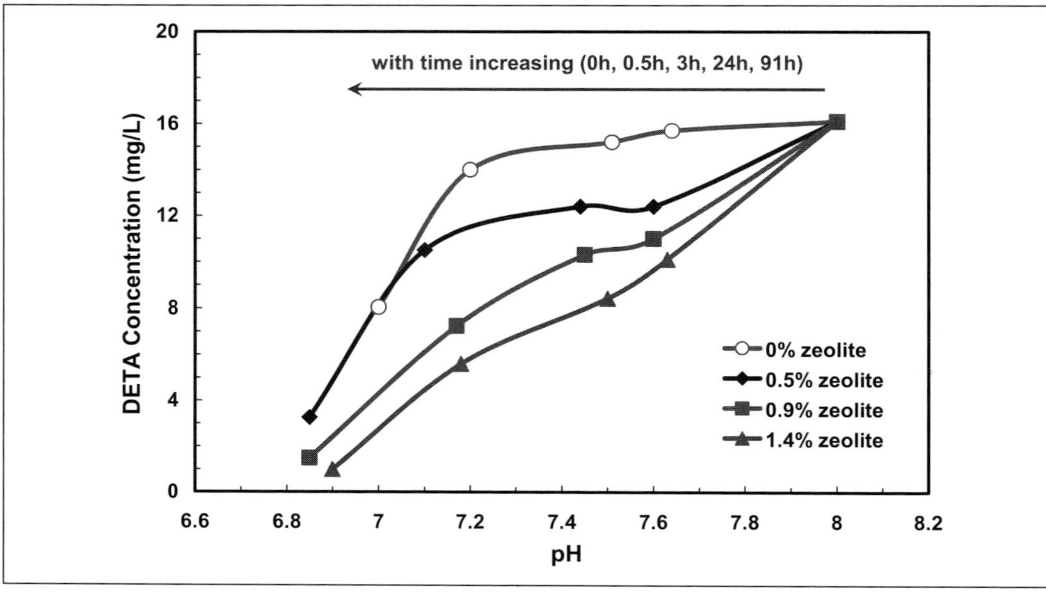

Figure 8. Changes in DETA concentration and pH values of Po tailings mixed with different percentages of zeolite as a function of time

tailings. The total solids percentage is maintained at 30%. The change in DETA concentration in Po tailings slurry with time and pH is shown in Figure 8.

The pH of Po tailings without zeolite (red circles) gradually decreases from initial 8.5 to 7.0 after 91 hours of mixing due to pyrrhotite oxidation. The DETA concentration in tailings water also decreases from 16 mg/L to 8 mg/L, indicating more DETA adsorbed by Po tailings at lower pH. Earlier research work showed that acidic pH enhances DETA adsorption onto Po tailings. DETA carries positive charges through an acidification reaction as pH decreases. This can be calculated with the equilibrium constants (Shi, et. al 2011). For example, at pH 9.5, 11% of total DETA is neutral, while 52% is DETA•H^+ and 38% DETA•H^{2+}. At pH 5, about 83% of total DETA is DETA•H^{2+} and 17% DETA•H^{3+} without any neutral DETA. It also shows that in simulated process water, the surface of Po is negatively charged over pH from 4 to 12. Therefore, lower pH is favorable for DETA adsorption on Po tailings.

With a small amount of zeolite added into Po tailings (0.5, 0.9 and 1.4%), DETA concentration decreases much faster than the case without zeolite at any time interval. With >0.9% zeolite mixed with Po tailings slurry, DETA concentration is reduced to 1.5 mg/L. The calculated DETA adsorbed by zeolite is about 830 g/t, after subtracting the DETA adsorbed by Po tailings.

DETA Desorption from Po Tailings Slurry with Zeolite

When Po tailings (with/without zeolite) loaded with DETA are diluted to lower solids percentages with simulated process water at pH 8.5, DETA desorption is expected. As shown in Figure 9a, with increasing dilution (i.e., decreasing solids percentage), the DETA desorption percentage increases as expected. Without zeolite, DETA desorption slightly decreases from 1.5 h to 91 h. This is due to the decrease in pH by the acid generated from Po tailings and consequently enhanced DETA adsorption on Po tailings. The pH change in Po tailings with time is shown in Figure 9b. As shown in Figure 9a, even at very high dilution (1% solids), DETA

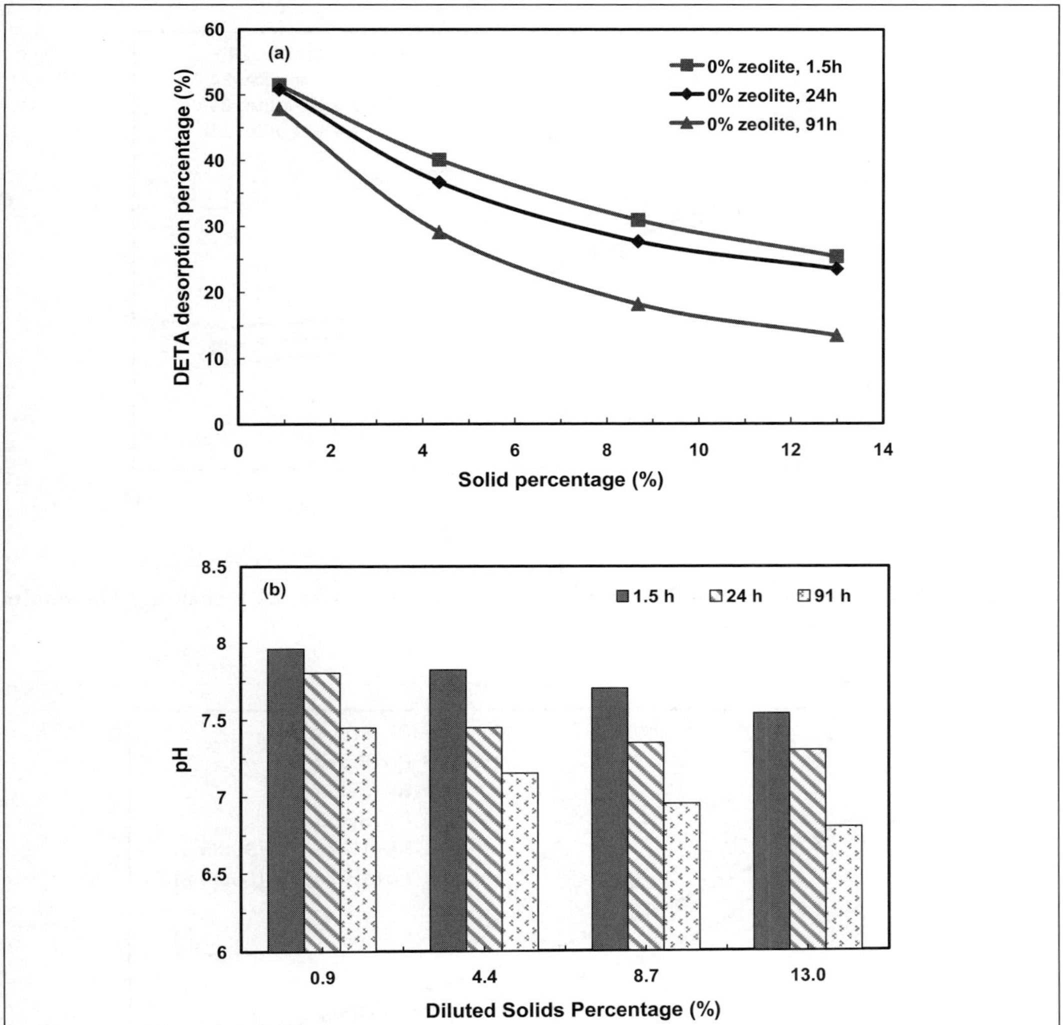

Figure 9. (a) DETA desorption from 0% zeolite in Po, and (b) the change of pH with time

desorption did not reach 100%, indicating that DETA adsorption on Po tailings is not only physisorption by electrostatic force, and some chemical bonding may also occur.

With some zeolite added into Po tailings, the overall DETA desorption over time at the same solids percentage is much lower (Figure 10). The decrease in DETA desorption with time is attributed to zeolite. It is quite possible that some of the DETA desorbed from Po tailings is re-adsorbed by zeolite.

Simulation of a Settling Disposal Scenario

Tailings disposal into a tailings pond most likely experiences high initial dilution with little or mild mixing. After settling, it will eventually reach the terminal slurry density in the range of 60–80% solids at the bottom of the pond. It is important to know if DETA desorption takes place during a short settling period with high initial dilution.

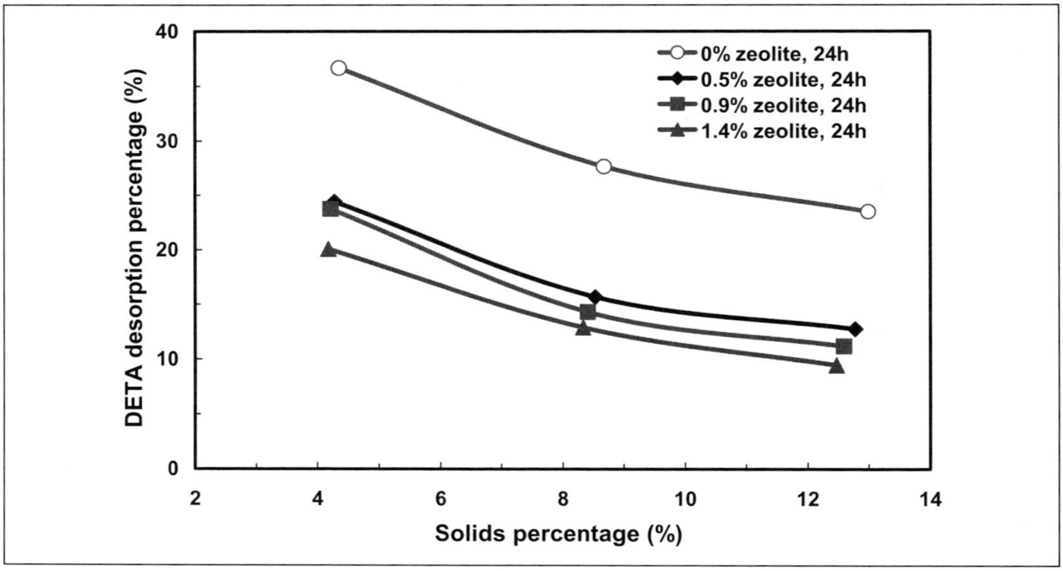

Figure 10. DETA desorption from Po tailings mixed with zeolite after 24-h shaking. The results are similar to that after 1.5-h or 72-h shaking.

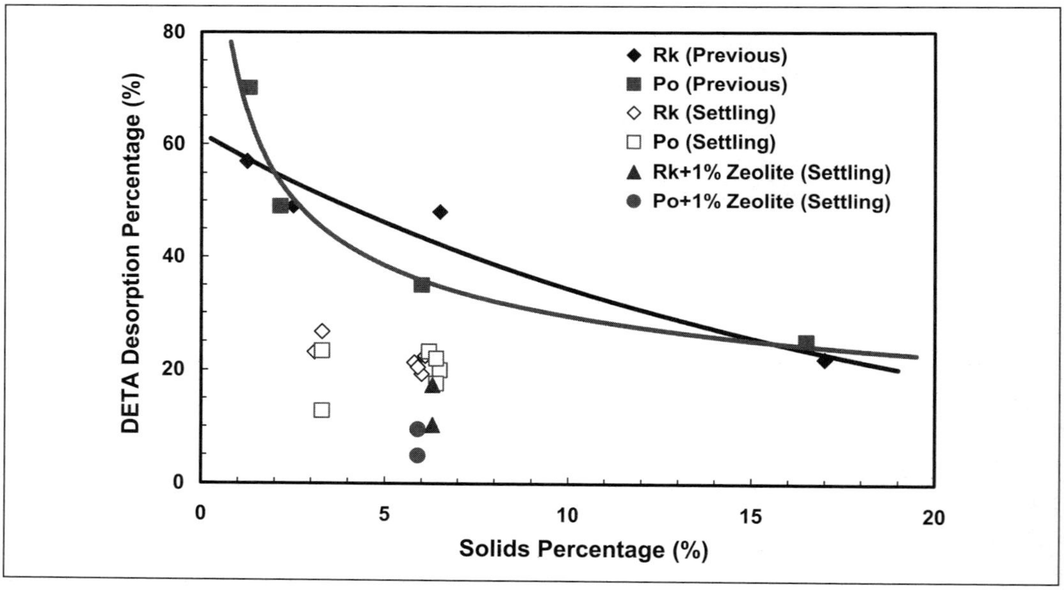

Figure 11. DETA desorption from tailings during settling, compared with the previous results from the dilution and shaking tests, and the effect of zeolite addition into tailings

The tailings (Rk and Po) loaded with DETA at 30% solids are pumped into the column (~2 m high) from the top, which is filled with simulated process water. It takes about two hours to reach terminal settling density and the supernatant is almost clear. The measured DETA desorption percentage does not change significantly after two hours. The pumping rates (50 mL/min or 200 mL/min) do not affect the DETA desorption percentage at equilibrium.

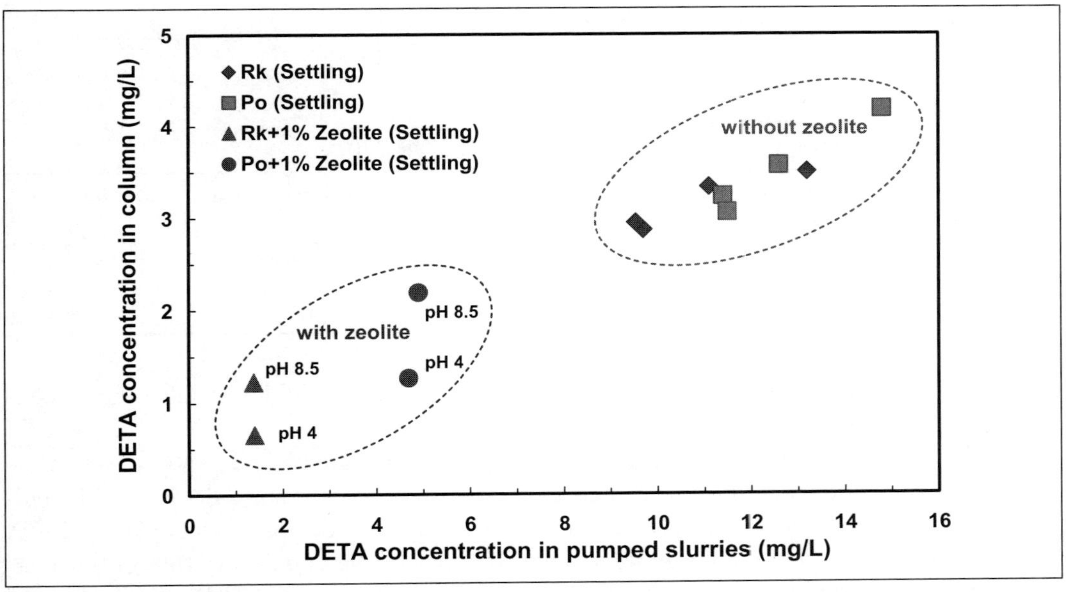

Figure 12. Equilibrium DETA concentration of the pumped tailings slurry and inside the column after settling

After 4 hours settling, the average DETA desorption percentage is plotted in Figure 11 and compared with the desorption data obtained previously from the dilution and shaking tests. The DETA desorption value obtained from the settling tests is about half of the previous value for both Rk and Po tailings at about 6% of solids. Further diluting the tailings to 3% solids does not change DETA desorption percentage much. It is speculated that under the gentle settling conditions, DETA or DETA-metal complexes remains on the solids in tailings and, consequently, higher dilution has a less impact on DETA desorption, as compared to vigorous shaking. This may more closely reflect the actual tailings disposal situation.

The effect of zeolite addition (1%) on DETA desorption during settling is also shown in Figure 11 (solid triangles and circles). The DETA desorption percentage is further reduced with zeolite addition. The reduction in DETA desorption with zeolite is more pronounced at low pH (4.0) at which DETA is much more readily adsorbed by zeolite due to the enhanced protonation reaction. Figure 12 presents the DETA concentration of the tailings slurry at the initial solids percentage (~30%) before deposition vs. the DETA concentration of the solution in the column after 4 hr settling at equilibrium. It clearly shows that there are two different regions with and without zeolite. Zeolite addition substantially decreases the DETA concentration at equilibrium, as demonstrated in earlier adsorption testwork (Figure 8).

Implementation Strategy

In the DETA mitigation study, the primary strategy is to enhance DETA adsorption on Rk and Po tailings while minimizing desorption during tailings disposal. Rock tailings are a large mass and DETA free. With a thickener and a clarifier, its effluent can be recycled directly to the Mill without having to go through tailings pond. The thickened rock tailings can act as an additional reservoir to adsorb DETA from the Po tailings and concentrate thickener overflow. Maintaining a high solid percentage during deposal is required to minimize desorption. Occasionally adding zeolite (1% of Po tailings) into Po tailings when high DETA dosage is

Table 1. Estimate of the cost of using zeolite

Zeolite Price ($/t)			300	
Po Tailings Solids (mtpd)			4000	
Total Operation Time (d)			320	
DETA Dosage (g/t)	Operating Time (%)	Zeolite Added (% of Po solids)	Zeolite Needed (t/y)	Zeolite Cost ($/y)
<60	85%	0	0	0
>60	15%	1%	1920	576,000

used (i.e., >60 g/t) would further reduce the risk of having high DETA-metal complex concentrations in the tailings water.

Table 1 shows an estimate of the cost of using zeolite. Zeolite is about $300/t. When the DETA dosage is below 60 g/t, no exceedances are expected to occur since the Rk and Po tailings have the ability to adsorb DETA-metal complexes, so no zeolite is needed. When DETA dosage is above 60 g/t for metallurgical reason, which may be the case during 15% of the operating time, zeolite can be added into the Po tailings to adsorb the extra DETA-metal complexes at the dosage of 1% of Po tailings. Approximate 1920 t of zeolite is needed per year and the cost will be $576,000/y.

CONCLUSIONS

This paper demonstrates the adsorption/desorption of DETA on/from Po and Rk tailings. Natural zeolite is an effective sorbent for DETA and DETA metal complexes. By adding a small amount of zeolite (1%) into Po tailings, it can adsorb extra DETA or DETA-metal complex by ion exchange and reduce the DETA desorption percentage during disposal dilution. Simulation of disposal by column test shows that the DETA desorption during settling (experiencing high dilution) is not as large as expected from shaking desorption tests. Zeolite mixed with Po tailings to adsorb DETA is a simple and cost-effective way to implement in practice.

ACKNOWLEDGMENT

The authors thank the management of Vale Base Metals Technology Development for permission to publish this work. Acknowledgment also goes to Erin Bobicki, Rafael Santos, Natasha Remtulla, Andy Kerr, Daryl McNamara, Virginia Lawson and Renee Barrette for their work and support.

REFERENCES

Breck, D.W. 1974. *Zeolite Molecular Sieves: Structure, Chemistry, and Uses*. New York: Wiley.

Devuyst, E., and Kerr, A. 2008. Development of DETA Mitigation Strategies for the CMERP Pre-Feasibility Study. *Vale Internal report.*

Dong, J., and Xu, M. 2010. Application of Natural Zeolite for the Removal of Diethylenetriamine (DETA) and DETA-Metal Complexes from Flotation Process Water. In *XXV International Mineral Processing Congress (IMPC) 2010 Proceedings*, Carlton South, Australia: AusIMM, pp. 4049–4056.

Marticorena, M.A., Hill, G., Kerr, A.N., Liechti, D., and Pelland, D.A. 1994. Inco Develops New Pyrrhotite Depressant. In *Innovations in Mineral Processing Proceedings of the Innovations in Mineral Processing*. Edited by T. Yalcin. Sudbury, Ontario: Laurentian University. p. 15–33.

Santos, R. 2009. Status Report: DETA Mitigation Flowsheet Modeling. *Vale Internal report.*

Shi, W., Liu, Q., and Xu, Z. 2011. *Research Progress Report II on "Fundamental Study of Diethylenetriamine (DETA) as Pyrrhotite Depressant for Selective Flotation of Pentlandite."* University of Alberta.

Xu, M., Quinn, P., Robertson, G., and Wilson, S. 2000. Development of a Two-Stage Middlings Circuit at Inco's Clarabelle Mill. In *Proceedings of 32nd Annual Meeting of the Canadian Mineral Processors.* Ottawa: Canadian Mineral Processors. pp. 349–361.

Managing Our Most Precious Resource—Quality and Quantity Issues with Water for Mineral Processing in Western Australia

Damian Connelly
Mineral Engineering Technical Services Pty Ltd (METS), Perth, Western Australia, Australia

ABSTRACT

Water for mineral processing is in short supply. Water is a precious resource in Western Australia because there are few large rivers available and groundwater has sustainability issues associated with it. Desalination is increasingly used but environmental concerns have been raised regarding the waste stream impact on the marine environment. The resource sector is growing at such a rate that new water resources will become critical and the subject of competing interests.

The history, quality and quantity, sustainability, issues and approaches, water use, balances and audits, the challenges ahead and environmental considerations are considered in relation to water in mineral processing. The presence of inorganic salts either naturally occurring or dissolved from the ore (gypsum, magnesia) has an effect on mineral processing unit operations.

Technologies such as filtering, sedimentation, clarification and microfiltration are considered.

The need to characterise the water (i.e., the alakalinity, TDS, pH, hardness, full elemental analysis and organic content of the water being used) is discussed. The impacts of poor quality process water, such as saline waters include precipitation of calcium carbonate or sulphate or iron hydroxide resulting in reduced floatability and inadvertent activation of gangue minerals resulting in loss of selectivity. Common foulants include scale, biofouling and suspended solids. Antiscalants are widely and successfully used under such circumstances.

For sustainable mineral processing water audits are necessary based on actual water balances. This will highlight balancing water quality and optimise the use of recycling. The requirements for iron ore processing are far more stringent in comparison with other commodities such as gold ore processing. A number of case studies are used as examples.

HISTORY

The history of water use in the Australian resource sector was upon small projects and low throughput. The cost of water was low and environmental aspects were not an issue. Tailings dams were crude and water recovery was unimportant and there was very little government legislation in the area. The Mundaring to Kalgoorlie water pipeline was critical to the development of the Western Australian Goldfields, improving the supply of water in Kalgoorlie and by 2007 9 million gallons were supplied daily (Ewers 2007).

WATER AND MINERAL PROCESSING

Quantity and Quality

Australia has an arid climate with few large rivers to supply the growing demand for water particularly for large projects. Water has become a scarce resource with farmers and miners competing for limited resources. Whilst groundwater is one option draw down of the aquifer limits the quantity of water that can be extracted. Seasonal climatic variations impact on the recharge rate. The quality varies from brackish to hyper saline. Quality is based on a standard laboratory water analysis referenced to World Health Organisation (WHO) potable water.

Uses and Sources

There are no large rivers in Western Australia capable of supplying water for processing ores. If sea water is used there are severe corrosion issues and the requirement to wash the concentrates with potable water. If groundwater is used there is a need to treat the water and sustainability is an issue for large projects because of the potential drawdown of the aquifer.

If sea water is desalinated the cost of the water is high because of the power requirement. At the same time the necessity to conserve water has led to flowsheets where all of the tailings have to be filtered to minimise the water requirement.

Water is used for resource projects to wet roads for dust suppression, for grinding ores, for beneficiation plants, for roasting and smelting as well as camps and cooling for power stations.

Water Conservation and Water Balances

Water recovery, recycling and conservation is determined by undertaking a water balance analysis and ranking water according to type and quality with respect to use. Seasonal variations can upset the balance. The increasing cost of water and water scarcity is driving the process in the Australian resource sector. Finding water and developing bores is expensive, the capital cost of equipping bores, pumps and gensets and operating costs continues to rise. The capital and operating costs of tailings dams have increased and water recovery is generally low. A water risk management process is now critical to resource projects. This also simplifies the environmental aspects of operating conventional tailings dams.

Environmental Considerations

Acid Mine Drainage (AMD) corrosion, scale formation and the increasing requirement for non-release systems and plastic lined tailings dams is increasing the way we value water. Contamination of groundwater with salt or heavy metals is not acceptable. A draw down of the aquifer is also not allowed and the extraction rate must be the same as the recharge rate. The annual reporting and monitoring obligations are stringent. Currently, large magnetite projects employing tailings filtration and dry stacking on a large scale all aim to reduce water consumption as a result of water scarcity.

Mineral Processing

Water quality impacts on flotation in both positive and negative terms and is project specific. The Ian Wark Research Institute in Adelaide has undertaken work in this area looking at the impact of water quality on ore processing (Ian Wark Research Institute 2009). Water quality impacts on viscosity and gold leaching particularly with respect to buffering. Thickening, filtering and stripping are all impacted by water quality.

Licenses and Regulations

No processing of minerals operates in a vacuum. An operation must be scrutinised under the environmental, land ownership and heritage acts and approved once it is clear that there is no contradiction between the acts and the future activity of the operation. In allowing an operation to go ahead, a water licence quota as well as an operational licence must be issued stipulating the amount of water that can be drawn for the operation over a period of time and from where, as well as the standards the operation should meet in regard to the environmental discharges. A series of monitoring and reporting procedures are also put in place in parallel to issuing those licences, to insure that the operation is being carried out according to them. Seepage monitoring bores adjacent to tailing dams, for example, ensure that the underground water is not being polluted by the tailing dams.

DRY PROCESSING

Currently, the price of water fluctuates according to supply and demand conditions, with the average price of water in Australia being about A$1/kL or A$1000/ML (Urban Ecology). Pressures on water supply are anticipated to result in water supplies for the minerals industry becoming increasingly compromised. Thus water availability and cost are likely to drive changes in water consumption within the minerals industry. While increased water recycling is an obvious candidate to help reduce the water consumption of mineral processing plants, there may be a more radical solution to the problem—the use of dry or near-dry processing technologies, for which the demand for water is small or zero, may offer a solution to the water consumption problem, however the introduction of dry processing will bring with it a new set of problems, including dust.

Water Treatment

The vast majority of streams in need of treatment are the result of their dissolved chemical content and not because of their appearance. Therefore, one of the first attempts in rectifying the contents of a stream is by chemical treatment. This involves pH adjustments, oxidation or reduction of ingredients, disinfection, gas/odour scrubbing and neutralisation, precipitation of ingredients, etc. On the other hand, there are streams that can be made reusable only by eliminating part of the containing salts via fine filtration like micro, ultra and nano filtration or reverse osmosis or via biological means. A list of treatments and achievements that can and have been produced through water and wastewater treatment follow.

Filtration, Sedimentation, Clarification

Process involves the use of multi-media filters which are deep bed, vertical or horizontal to gain 20–30 micron removal. This method allows for low capital and operating costs.

Clarifiers and Thickeners

Clarifiers and thickeners can also be used to encourage waste water re-use, as can ponds which have a large residence time and act as a repository for solids.

Reverse Osmosis (RO)

RO is used to separate dissolved solids and occurs in many Western Australian plants. Uses selective separation based on ionic radius and molecular charge, while also removing non-selective soluble salts. Water recovery is 50–95% vol, is limited by SiO_2 and gypsum, etc. Salt rejection is generally 95–99.8%. The capital and operating costs of this process is provided in Table 1.

Table 1. Costs of process

	Brackish	Seawater
Capital cost	$600/Ml/day product water	$1,200/Ml/day product water
Operating cost	$0.40/kl produced	$0.70/kl produced

Electrodialysis Reversal (EDR)

EDR combines membrane and electric current to achieve intermediate product water quality (1,000–2,000 ppm) and also brackish water for process or potable water. The feed water stream is drawn across the membrane and then an electric current is used to remove the anions and cations to produce a process channel and a brine channel (waste). The channels are reversed to prevent scale formation on the brine side of the membrane.

The process is particularly effective for the de nitrification of drinking water.

Electrode-Ionisation (EDI)

Unlike the EDR, EDI does not incorporate a reversal component. The feed water channel passes through the mixed bed resin in order to remove the anions and cations. The charged particles are then concentrated into the brine channel using an ion-selective membrane and rejected.

The process is very intolerant to hardness in the water.

Evaporation Ponds

The evaporation of water through the addition of heat to a water source. The major operational impediment is the formation of scale—e.g., magnesium silicate at the heat source/water contact point. They can be off line for significant time periods to deal with the removal of scale and are suffer from slow start-up time after interruptions.

Sludge Dewatering

Water can be recovered from a range of waste process streams. Based on a typical lime neutralised acid effluent, pH <2.0 the following processes can be incorporated to recover the water:

- Vacuum filter belts
  - 40–60% moisture in cake
- Belt filter presses
  - 25–45% moisture in cake
- Tube presses
  - 5–15% moisture in cake
- Centrifuges and Decanters
  - >30%

The disadvantage of using what would otherwise be a waste stream (i.e., low value) is the need for high capital and operating costs for the maintenance, wear, belt tracking, etc. This makes the use of sludge dewatering application specific.

Specific Australian Examples

1. Citic Pacific Magnetite project using desalination of seawater for process water (Citic Pacific 2010).
2. Gindalbie-Whole tailings are filtered to minimize water groundwater consumption from the Tathra bore field. Dry stacking of tailings was chosen so as to minimise water consumption.
3. The Southdown Magnetite Project comprises a greenfields magnetite mine 90 km northeast of Albany, a 120 km slurry pipeline and expansion of Albany port to export the ore. Surface runoff and groundwater are used to supply water for the project (Grange Resources 2011).
4. Olympic Dam, South Australia at the time it was commissioned in the 1980s was the largest RO plant in Australia producing around 12 ML per day (The Melbourne RO plant when fully commissioned will be the second largest in the world at 450 ML/day). The process uses cooling sand filters RO and ion exchange to produce both potable and high quality process water (BHP Billiton).

CONCLUSIONS

Water has become a critical resource for resource projects and there are competing demands from a number of sources. Groundwater is a limited resource and if drawdown of the aquifer occurs then the licence is reduced to ensure sustainability.

The history of development of the Western Australian Goldfields resulted in a 600 km pipeline being constructed from Mundaring to Kalgoorlie for potable water and mineral processing.

The increasing cost and value of water has resulted in water re use, water conservation and strategic management initiatives being put in place. A number of technologies including desalination of sea water are being included in the project design.

ACKNOWLEDGMENTS

The author thanks various companies, all colleagues, engineers at various sites, METS staff, and other consultants for their contribution and the management of METS for their permission to publish this paper and constructive criticism of various drafts.

REFERENCES

Australian Bureau of Statistics. 2006. *Water Account, Australia* [online]. Available from www.abs.gov.au/AUSSTATS/abs@.nsf/DetailsPage/4610.02004-05?OpenDocument (Accessed April 26, 2006).

Australian Government, Bureau of Meteorology. 2011. Australian Rainfall Analysis, 1 April 2010–31 March 2011—Image. *Twelve-Monthly Rainfall Totals for Australia.* Available from www.bom.gov.au/jsp/awap/rain/index.jsp?colour=colour&time=latest&step=0&map=totals&period=12month&area=nat (Accessed April 26, 2011).

BHP Billiton. *Olympic Dam Expansion EIS* [online]. Available from www.bhpbilliton.com/home/aboutus/ourcompany/regulatory/Documents/Olympic%20Dam%20Supplementary%20EIS/Information%20Sheets/Olympic%20Dam%20EIS%20Water%20Supply.pdf (Accessed July 11, 2011).

Citic Pacific Mining. 2010. *Desalination Plant a World First* [online]. Available from www.citicpacificmining.com/articles/latest-news/desalination-plant-a-world-first (Accessed July 11, 2011).

Connelly, D. 2009. *Water Management Issues with Processing Magnetite Ores*. AusIMM Water in Mining, Perth, Australia.

Ewers, J.K. 2007. Water Pipeline from Perth to Kalgoorlie: 1894–1903. *The Australian Pipeliner*, April. Available from pipeliner.com.au/news/water_pipeline_from_perth_to_kalgoorlie_1894_-1903/040096/ (Accessed July 14, 2011).

Grange Resources. 2011. *Southdown Resource Evaluation.* Available from www.grangeresources.com.au/cgi-bin/page.cgi?grangeS+21 (Accessed July 11, 2011).

Ian Wark Research Institute. 2009. *Research.* Available from www.unisa.edu.au/iwri/research/default.asp (Accessed July 11, 2011).

Newcrest Mining Limited. Cadia Hill, *Operations.* Available from www.newcrest.com.au/operations.asp?category=2 (Accessed July 11, 2011).

Urban Ecology Australia. *Water Use*. Available from <www.urbanecology.org.au/topics/wateruse.html (Accessed July 11, 2011).

Using Spreadsheets to Evaluate the Effects of Mine Water Disposal on Surface and Ground Water

Lisa M. Boettcher
Montana Department of Environmental Quality, Helena, MT, USA

ABSTRACT

Spreadsheets can accurately model effects on water quality resulting from the disposal of treated and untreated mine waters by percolation or land application disposal. User-friendly spreadsheets are a valuable tool to evaluate potential effects of a new mine, proposed revisions to an operating permit, disposal of adit water, and water disposal options when closing tailings impoundments. The spreadsheet models use basic data (hydrogeologic, wastewater quality and quantity, treatment, storage, and disposal system capacities), Darcy's equation, standard mixing zone calculations, and weighted-average equations in calculations that are consistent with NPDES permitting calculations. Formulas and input values may be adjusted to reflect life-of-mine changes in water quality/quantity, hydrogeologic data, water treatment efficiency, disposal method, volume, etc. The use of spreadsheets is an innovative approach to identify water disposal requirements and the more critical elements of a water management system so that the most effective capital investment can be made during operations and to identify operational mitigations that should be implemented to avoid costly issues during closure.

INTRODUCTION

Spreadsheet models were created for a closure water management plan draft environmental impact statement (DEIS) for two underground platinum group metal mines in Montana [Montana Department of Environmental Quality (DEQ) and US Forest Service (USFS) 2010]. The DEIS evaluated and disclosed the effects of the disposal of mine adit and tailings waters on surface and ground water. Earlier analyses did not adequately address the method(s), duration, and management of all mine water streams at closure and post-closure (Montana Department of State Lands [DSL] and USFS 1985; DSL and USFS 1989; DSL and Montana Department of Health and Environmental Services [DHES] 1992; DSL, USFS, and DHES 1992; DEQ and USFS 1998).

The mines are operated near Nye and Big Timber, Montana. Operations at the first mine began in 1985 and ore production is approximately 777,100 tons per year. Although permitted in 1993, operations at the second mine did not begin until 1998, and is approximately 407,400 tons per year. The company mills the ore at each mine by crushing, grinding, flotation, and filtration to produce a concentrate. This concentrate is shipped by truck to a base metal refinery in Columbus, MT for further processing. From the base metal refinery, the product is shipped to one of two U.S. facilities for final refining (DEQ and USFS 2010).

For both mines, every 100 tons of ore fed to the mill generates 99 tons of tailings. These tailings are pumped from the mill to underground sand and paste plants where the coarse sand fraction of tailings is separated from the slimes fraction (finest-sized particle). The sand is dewatered, cement is added, and about 58 percent of the total tailings is used to backfill underground workings. The remainder of the tailings are pumped to the respective tailings impoundments at the mine and at an off-site land application disposal (LAD) facility. The

Figure 1. The off-site land application disposal (LAD) facility center pivot system. The center pivots are used to provide a polishing treatment for nitrate and dispose of excess mine waters. These center pivots use high pressure nozzles and "drops" that are oriented upward to evaporate as much water as possible, rather than deliver it to soil. This orientation evaporates an average of about 30 percent of the end-of-pipe water volume. An additional 85 to 90 percent removal of nitrate is achieved during polishing treatment by LAD (SMC 2006). (DEQ photo)

tailings impoundments at both mines are used to balance water storage (DEQ and USFS 2010).

While mine operator and the agencies had considered various disposal scenarios for closure, the spreadsheet models helped to identify which disposal methods were optimal for the mine water constituents, where shortcomings lay in the water management plan, and how adjustments to the treatment system(s) would affect the quality of adit and tailings water at closure. The spreadsheet models provided the mine and the agencies a venue to better evaluate operational water management throughout mine life to develop mitigations that would minimize costly issues at closure.

For this DEIS, the spreadsheet models evaluated the disposal of treated and untreated mine waters using percolation and land application (specifically center pivots [Figure 1], evaporators [Figure 2], and snowmaking) during operations and at the eventual closure of two mines. The sources of water that would need to be addressed at closure were water discharging from the adits and waters removed from three tailings impoundments prior to the placement of reclamation covers (capping).

Both mines use semi-passive, fluidized bed/fixed-film biological treatment systems (BTS) for primary treatment (denitrification) of the nitrate+nitrite nitrogen that is derived from ammonia-based blasting agents (ammonium nitrate and fuel oil—ANFO). Up to 95% of the nitrate+nitrite nitrogen is removed by the BTS (SMC 2006). The second mine has a separate biological system to treat ammonia (Anox system). The first mine uses growing season LAD as a polishing treatment for nitrogen. Both mines can employ percolation for water disposal (Figure 3).

The mine water contains salts (i.e., sulfate and chloride derived from reagents added during the milling process) that are not treated in the BTS, by LAD, or percolation. Due to the nature of the ore body, the metals cadmium, chromium, copper, lead, and zinc are present in the mine water at concentrations less than human health standards or at non-detectable concentrations (the mines' water quality database). Metals are not treated in the BTS, by LAD, or percolation.

Figure 2. Evaporators/snowmakers can be used to provide a polishing treatment for nitrates and dispose of excess mine waters in the on-site LAD area. The evaporators can be used on uneven terrain to dispose of water by land application in summer and continue to provide disposal and nitrogen polishing treatment by snowmaking in winter (SMC, HKM Engineering Inc., and Knight-Piésold 2004). (DEQ photo)

Figure 3. Percolation pond used for disposal of treated mine water. Treated water is discharged to the percolation pond where it infiltrates subsurface to ground water, mixes and eventually flows to surface water. The percolation ponds at both mines are NPDES-permitted outfalls. (DEQ photo)

For the past 10 years, one of the mines has been using a 107.2 hectare (265-acre) LAD system for the operational disposal of up to 851,720 m^3 (225 million gallons) during one LAD season. The DEIS spreadsheets analyzed the existing LAD system, a new LAD system that is proposed for use at the other mine, and disposal of treated mine water by percolation to ground water at both mines. The spreadsheets also modeled the direct discharge of untreated adit water to surface water at post-closure.

Although the parameters evaluated in this DEIS were nitrogen and salts, other parameters (e.g., metals) could also be evaluated using attenuation factors. The approach used by the agencies is unique in that the DEIS analyses were not prescriptive, but considered a range of

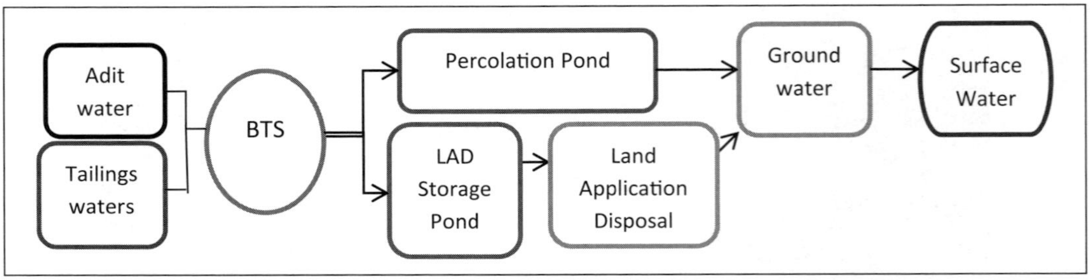

Figure 4. Typical mine water routing schematic. The mine water streams mix in the biological treatment system (BTS) during nitrate treatment, then are routed for disposal in the percolation pond and/or LAD systems, and would eventually reach ground water then surface water. This schematic does not show the routing of recycle water or assumed leakage from impoundments that were included in the agency analysis.

adit discharge flow rates using different disposal options, thereby permitting the mine to retain water management flexibility during operations and closure. These spreadsheet models are a valuable tool that can be used throughout life-of-mine to assess the elements of a water management plan, evaluate changes in water treatment methods or efficiencies, appraise potential improvements or additions to the primary treatment system(s), and include refinements in hydrologic data.

METHODS

The Microsoft Office Professional Excel spreadsheet program was used to construct the models. The spreadsheet models consisted of several sections that included the following:

- a list of the assumptions used when evaluating the alternative;
- the option analyzed (i.e., specifics of water routing);
- a list of the input parameters including citations for the data that were used in the analyses;
- an evaluation of the hydraulic capacity of the treatment system(s) and disposal method(s) considered in that option (e.g., whether the percolation pond capacity was sufficient to manage the volume of water to be disposed of);
- a calculation of surface and ground waters available for mixing;
- the projected constituent loading in surface and ground water; and
- the projected constituent concentration in surface and ground water.

Assumptions and Equations

The spreadsheets list the assumptions for each analysis that dictate how certain parameters were handled in the equations. For example, primary treatment of nitrogen (nitrate+nitrite) occurs in the fixed-film BTS in use at both mines where up to 95 percent of the influent nitrate+nitrite is removed (SMC 2006). The waste streams that were evaluated consisted of adit water (i.e., ground water inflow to the mine that is mixed with water used in the underground operations such as muck pile washdown water and excess drilling water) and treated and untreated tailings waters (i.e., the volume of supernatant plus tailings mass waters that would be removed from the impoundment so that the reclamation cover could be emplaced). The analyses considered a range of adit water flow rates from the current operational rate to the maximum anticipated rate. The BTS efficiency data were available for the current operational rate. Assumptions were made to address questions raised regarding the treatment system capacities and efficiencies if the maximum anticipated adit flow rates were reached. The assumptions also described how water would be routed for each mine and each alternative considered (Figure 4).

BTS-Treated Adit Water Nitrogen Concentration

To determine the concentration of nitrogen in adit water at closure, the agencies assumed that, regardless of adit water flow rate, the BTS would achieve nitrogen removal equivalent to at least the historic maximum nitrogen load discharged at the mines. A standard loading Equation 1 was solved for concentration (mg/L), using both the current operational and anticipated maximum adit water flow rates (volume/time) with historic maximum nitrogen load (mass/time) to find the treated adit water concentration, C_{TA} (mg/L). A conversion factor, *CF,* may be needed to obtain consistent units.

$$Load = Flow\ rate \times C_{TA} \times (CF) \tag{1}$$

BTS-Treated Tailings Waters Nitrogen Concentration

It was assumed that at least 80 percent of the nitrogen in tailings waters would be removed by the BTS based on water quality data submitted by the mines (SMC 2006). To calculate the concentration of treated tailings waters, C_{TTW} (mg/L), the median nitrate+nitrite nitrogen concentration of untreated tailings waters, C_{UTW} (mg/L), was multiplied by 20 percent (Equation 2).

$$C_{TTW} = 0.2\ C_{UTW} \tag{2}$$

Mixing Treated Adit and Tailings Waters

The adit and tailings waters would be routed to the BTS, mixed, and treated before being routed to the percolation ponds or LAD storage pond for later disposal. A standard weighted-average calculation was used to determine the resulting concentration of nitrogen or salts in the mixed waters in the LAD storage pond. Equation 3 calculates the concentration of mixed treated mine waters, C_{TMW} (mg/L), where C_{TA} is the concentration of treated adit waters in mg/L from Equation 1, Q_{TA} is the flow rate of treated adit waters (volume/time), C_{TTW} is the concentration of treated tailings waters in mg/L from Equation 2, and Q_{TTW} is the flow rate of treated tailings waters in volume/time. This weighted-average equation may be expanded by additional terms to account for other sources of water, such as water residing in the LAD storage pond over the winter, if needed.

$$C_{TMW} = \frac{C_{TA}Q_{TA} + C_{TTW}Q_{TTW}}{Q_{TA} + Q_{TTW}} \tag{3}$$

Concentration of Nitrogen in LAD-Treated Water

A polishing treatment for nitrogen occurs during growing season land application disposal where, based on a study performed by the mine, plants use at least 80 percent of the applied ammonia and nitrate+nitrite nitrogen (SMC 2006). The mine performed a study in 2002 that showed during snowmaking, 80 percent of the applied nitrogen was removed by natural processes during crystallization and sublimation of the snowpack (Cascade Earth Sciences 2008). The nitrogen not removed by plant uptake or snowmaking was assumed to enter the aquifer (Stevenson 1986), where it would mix with ground water, and eventually discharge to surface water. The concentration of nitrogen entering the aquifer from land applied water, C_{LAD} (mg/L), was calculated by multiplying the concentration of treated mine waters, C_{TMW} (mg/L) from Equation 3, by 20 percent, using Equation 4.

$$C_{LAD} = 0.2\ C_{TMW} \tag{4}$$

Percolated Waters

No nitrogen treatment occurs during percolation, so no treatment credit was given in the analyses or the mine NPDES permits. No treatment of calcium and magnesium salts occurs in the BTS or during land application. All of the salts plus the nitrogen that was not removed by plant uptake or snowmaking was assumed to enter the aquifer, mix with ground water, and eventually discharge to surface water (Stevenson 1986).

Load to Ground Water

The NPDES permits for both mines have an effluent limit for nitrogen loading to ground water. Both mines employ a BTS. The spreadsheets modeled the nitrogen load for percolated and/or land applied waters using Equations 1, 2, 3, and 4, as appropriate to the routing for the alternative. The salts load to ground water was calculated using median adit and tailings waters concentrations in Equation 1.

Precipitation

The agencies conservatively chose not to add precipitation to the spreadsheets. The dilution effect of precipitation on land applied waters may be added to a spreadsheet model using the effective precipitation rate (i.e., precipitation that reaches the soil and does not run off or evaporate but remains in the soil for plant use). The DEQ draft technical guidelines for the land application of municipal wastewater (DEQ 2005) describe the effective precipitation rate as 70 percent of the total rainfall ($f_R = 0.7$), or 50 percent (due to wind drift and sublimation) of the snowfall water equivalent ($f_S = 0.5$) where 2.5 cm of snow equals 1 cm of rain (US Department of Agriculture Natural Resources Conservation Service). The total depth of measured precipitation, P_m, then may be converted to volume, P_V, using Equation 5, where f is the appropriate correction factor for effective precipitation, and A is the area (hectares) over which LAD is applied. A conversion factor, CF, may be needed to obtain consistent units.

$$P_V = (CF)\ P_m f A \tag{5}$$

Unsaturated Zone

Nitrogen and salts are conservative parameters. That is, below the root zone no attenuation occurs. For the DEIS analyses, the water that percolates below the root zone was assumed to immediately enter the aquifer. The unsaturated zone was not modeled in these analyses.

Nitrogen or Salts Concentration in Ground Water from LAD-Treated Waters

To project the concentration of nitrogen or salts in ground water after LAD, a ground water mixing zone was first calculated. In Montana, the calculation for a standard ground water mixing zone is prescribed by regulation (Administrative Rules of Montana 17.30.517). The depth extends 4.6 m (15 ft) below the water table, and the width of the mixing zone is the width of the source plus a geometric factor that takes into account advection and dispersion along the length of the flow path. The depth times the width of the mixing zone defines the area, A (length^2), that is used in Darcy's equation (Freeze and Cherry 1979) (6), where Q_A is the flux of water in the aquifer available for mixing (volume/time), k is the hydraulic conductivity of the aquifer (length/time), and I is the hydraulic gradient (length/length).

$$Q_A = kIA \tag{6}$$

After calculating the flux of water in the aquifer available for mixing, a standard weighted-average calculation was used to determine the resulting concentration of the constituent in ground water after land application. This approach is consistent with NPDES permit calculations. Equation 7 calculates C_{MZ}, the concentration in mg/L of the constituent in the aquifer after mixing, where C_{LAD} is the maximum projected concentration in mg/L of the constituent (after plant uptake or snowmaking) in the applied water from Equation 4, Q_{LAD} is the volume of water land applied per time, C_A is the ambient concentration in mg/L of the constituent in the aquifer, and Q_A is the flux of water in the aquifer available for mixing per time from Equation 6.

$$C_{MZ} = \frac{C_{LAD}Q_{LAD} + C_A Q_A}{Q_{LAD} + Q_A} \tag{7}$$

Nitrogen or Salts Concentration in Ground Water from Percolation

To project the concentration of nitrogen or salts in ground water from percolation, C_P (mg/L), a ground water mixing zone was calculated for the aquifer beneath the percolation pond as described above, using Equation 6. Once the flux of water in the aquifer available for mixing beneath the percolation pond was known, Q_{AP} (volume/time), a standard weighted-average calculation was used to determine the resulting concentration of the constituent in ground water, C_{AP}, in mg/L. This approach is consistent with NPDES permit calculations. Equation 8 is the weighted-average calculation for disposal of both treated adit and tailings waters by percolation. Additional terms may be included, as necessary, to construct a mixing zone calculation that includes all sources of percolated water.

$$C_P = \frac{C_{TA}Q_{TA} + C_{TTW}Q_{TTW} + C_{AP}Q_{AP}}{Q_{TA} + Q_{TTW} + Q_{AP}} \tag{8}$$

Multiple Contributing Areas for Ground Water

In the analyses for the DEIS, there were several areas that contributed flow and constituents prior to discharge to the receiving stream. For example, the ground water flow path of interest might flow from beneath the percolation pond, under a tailings impoundment, beneath a LAD storage pond, under a LAD area, below a buffer area, then discharge to a receiving stream. After determining and describing the flow path for the contributing areas in the assumptions section, the ground water volumes available for mixing for each area were calculated using Equation 6. These areas may have contributed constituents (e.g., percolation, leakage from impoundments or storage ponds, LAD) or provided dilution (e.g., precipitation, buffer areas, or leaking irrigation ditches). A weighted-average calculation that included the terms for all contributing areas was used to project the final concentration of the constituents in ground water along the flow path of interest.

Such a calculation would resemble Equation 9, expanded as needed to include terms for each contribution area along the ground water flow path. Each contributing area would have terms for volume or flux, Q, and concentration, C, for both the ground water (listed below with subscript beginning with A) and constituent addition or dilution. All Q terms would be in volume/time and all C terms in mg/L.

Equation 9 would yield the ground water concentration of a constituent, C_D, as a result of water disposal from LAD (terms with subscript LAD), percolation (terms with subscript P), and leakage from tailings impoundments (terms with subscript I) and LAD storage ponds (terms with subscript SP). Parentheses can be useful to separate the terms for each contributing area.

$$C_D = \frac{(C_{ALAD}Q_{ALAD} + C_{LAD}Q_{LAD}) + (C_{AP}Q_{AP} + C_PQ_P) + (C_{AI}Q_{AI} + C_IQ_I) + (C_{ASP}Q_{ASP} + C_{SP}Q_{SP})}{(Q_{ALAD} + Q_{LAD}) + (Q_{AP} + Q_P) + (Q_{AI} + Q_I) + (Q_{ASP} + Q_{SP})} \quad (9)$$

Nitrogen or Salts Concentration in a Receiving Stream

Both mines and their LAD facilities are near surface water and a significant amount of hydrogeologic data have been collected. The spreadsheets modeled the ground water flow paths to their discharge points into the respective receiving streams. The approach used is consistent with NPDES permit calculations. The flux of ground water discharging to the stream is the sum of the volumes of the contributing areas and is equal to the denominator of Equation 9. A conversion factor, *CF,* may be necessary to obtain consistent units. Equation 10 is used to calculate Q_D in cfs, the ground water discharge to surface water.

$$Q_D = (CF) \times [(Q_{ALAD} + Q_{LAD}) + (Q_{AP} + Q_P) + (Q_{AI} + Q_I) + (Q_{ASP} + Q_{SP})] \quad (10)$$

Equation 11 calculates the concentration of the constituent in the receiving stream, C_{RS} in mg/L, after mixing with discharging ground water, where C_D is the concentration of the constituent in the ground water in mg/L discharging to the stream from Equation 9, Q_D is the volume of ground water discharging to the stream per unit time from Equation 10, C_S is the ambient concentration of the constituent in the stream in mg/L, and Q_S is the receiving streamflow (volume/time).

$$C_{RS} = \frac{C_DQ_D + C_SQ_S}{Q_D + Q_S} \quad (11)$$

The agencies used the 10-year, 7-day lowest streamflow value ($7Q_{10}$) cited in the NPDES permit when calculating receiving stream concentrations. The $7Q_{10}$ streamflow would provide the least dilution for wastewater effluent (i.e., cause the highest concentration of a constituent in the stream) and as such, is a conservative choice for most streams, except when using actual streamflow during drought conditions. Depending on the intended use of the spreadsheet models and the constituent considered, it may be more appropriate to use a larger streamflow value than the $7Q_{10}$. One suitable choice for nutrients, for example, might be based on longer annual low streamflow periods in late summer, such as the 10-year, 30-day lowest streamflow value ($30Q_{10}$). Algae grow throughout the summer months, and the discharge of nutrients during this low flow late summer period would provide the greatest potential for nuisance growth, hence the greatest potential impact to streams.

Direct Discharge of Adit Water at Post-Closure

Equation 10 may also be used to calculate the concentration of a constituent in a receiving stream, C_{RS} in mg/L, when treated or untreated adit water is discharged directly to the receiving stream. This approach is consistent with NPDES permit calculations. For this calculation, Q_S is the receiving streamflow (volume/time), C_S is the ambient concentration of the constituent in the stream in mg/L, Q_D is the volume of adit water discharging directly to the stream (volume/time), and C_D in mg/L, is the concentration of constituent in ground water discharging directly to the stream.

COMPARISON OF SPREADSHEET RESULTS TO REAL-TIME DATA

The values generated by the spreadsheets were compared to surface and ground water data collected by the mines. In general, the results were conservative with respect to real surface water data, and consistent with ground water data. That is, the spreadsheets over-predicted the concentrations of constituents in surface water and were within the observed range of constituent concentrations in ground water.

The over-prediction of surface water values is reasonable and expected because the assumptions made for surface water analyses were conservative. The spreadsheets used the $7Q_{10}$ (10-year, 7-day lowest streamflow value) for the receiving streams to account for closure during a dry or droughty year. During wetter years when the $7Q_{10}$ streamflow is likely to be met or exceeded every month of the year, more water is available for dilution of the constituent, and the in-stream concentration is over-predicted by the spreadsheet models.

Predictions for Disposal of Water Using LAD

At the LAD facility, the spreadsheets had a fairly accurate rate of prediction for constituent concentrations in ground water when compared with operational data collected between 2005 and 2009. For example, the spreadsheets predicted a ground water concentration of 1.3 mg/L nitrate+nitrite nitrogen at the compliance well, which agreed well with actual ground water nitrogen concentrations that ranged from 0.3 to 1.6 mg/L. The spreadsheets predicted a salts (total dissolved solids, TDS) concentration of 368 mg/L in ground water and a corresponding electrical conductivity of 575 µmhos/cm. Actual ground water data were about half the predicted concentration and ranged from 123 to 207 mg/L TDS with conductivity measurements from 192 to 323 µmhos/cm. The spreadsheets predicted 0.5 mg/L nitrate+nitrite in surface water, and the actual concentration of nitrogen varied between 0.1 to 0.2 mg/L. The spreadsheets predicted an in-stream salts concentration of 96 mg/L TDS, and actual concentration varied between 67 to 240 mg/L.

Predictions for Disposal of Water Using Percolation

At the second mine, where treated water is percolated, the spreadsheets were fairly accurate in their prediction when compared to operational surface and ground water data collected between 1999 and 2009. For example, the spreadsheets predicted a ground water concentration of 1.7 mg/L nitrate+nitrite nitrogen at the compliance well, which was within the actual ground water nitrogen concentration range from 0.2 to 2.7 mg/L. The spreadsheets predicted a TDS concentration of 227 mg/L in ground water and a corresponding electrical conductivity of 354 µmhos/cm. Actual ground water data were about 20 percent less than the predicted concentration and ranged from 142 to 218 mg/L TDS with conductivity measurements from

265 to 368 µmhos/cm. The spreadsheets predicted 0.5 mg/L nitrate+nitrite in surface water, and the actual concentration of nitrogen varied between 0.1 to 0.2 mg/L. The spreadsheets predicted an in-stream salts concentration of 108 mg/L TDS, and actual concentration varied between 24 to 128 mg/L.

Overall, the predicted values had reasonable agreement with actual concentrations. The difference between predicted and actual concentrations in ground water may be due to precipitation (which the spreadsheet model did not take into account), the factor used to address evaporation of LAD waters resulting in concentration of the salts, or to different volumes of water disposed of at the LAD facility than those modeled. The difference between predicted and actual concentrations of nitrogen and salt in surface water is likely due to the use of the $7Q_{10}$ streamflow value instead of the actual streamflow values.

LIMITATIONS

The spreadsheet models are best suited to less complex scenarios. Spreadsheets have limitations when aquifers are heterogeneous on a scale such that any of the following conditions are met: differences in permeability along the flow path are significant and difficult to describe geometrically or cannot be generalized; hydrogeologic parameters and ground water flow paths are unknown; a three-dimensional approach is needed; or constituents are attenuated and the attenuation cannot be addressed by retardation factors (Bair and Lahm 2006). In general, the unsaturated zone is quite complex, requiring the use of nonlinear functions (Schwartz and Zhang 2003), and is, therefore, not well-suited to spreadsheet modeling.

SUMMARY

Based on the spreadsheet modeling analyses in the DEIS, the agencies have concluded that these two mines have the water management systems available to adequately discharge water at closure to prevent violation of water quality standards. The analyses identified the critical elements of the water management systems, indicated where the systems have the potential for improvement, and spotlighted the flexibility of the water management systems to address contingencies at closure. The results of these calculations were consistent with observed concentrations of constituents from surface and ground water monitoring data.

As a result of the spreadsheet models, the DEIS agency-mitigated alternatives have added mitigations such as operational monitoring and annual reporting of changes in the volumes of tailings supernatant and water in LAD storage ponds; tracking the trend of salt and nitrate loading in adit and tailings waters; identifying efforts made by the mine to reduce salts and nitrogen concentrations each year; and monitoring soil health in the vicinity of the LAD areas. The usefulness of the spreadsheets extends throughout mine life and provides the mine and the agencies a venue to better evaluate operational water management and minimize costly issues at closure. A pdf version of the spreadsheet models is included as Appendix C of the DEIS (DEQ and USFS 2010).

While the mine operator and the agencies had considered various disposal scenarios for closure, the spreadsheet models identified some potential inaccuracies. One inaccuracy was in regard to which disposal methods were optimal for a constituent. In comparison to percolation, which is strictly a disposal option, LAD provides a polishing treatment for nitrogen, and would be a preferable disposal method to reduce the load of nitrogen to ground water. The spreadsheet analyses indicated that percolation is a preferable disposal method for salts because of the reduced potential for soil salinization and sodium-adsorption issues that can occur during LAD.

Another inaccuracy was identified with respect to the optimal disposal rate for LAD. An agronomic application rate may be preferable to maximize the LAD polishing treatment efficiency for nitrogen. A greater-than-agronomic application rate may be needed to leach accumulated salts and minimize soil salinization and sodium-adsorption.

The spreadsheets helped identify shortcomings in the water management plan, such as the hydraulic capacities of water storage and treatment systems, and increased constituent concentrations in recycled process waters. The spreadsheets can be used during mine life to evaluate how adjustments to the treatment system(s) (e.g., the addition of reverse osmosis to further reduce ammonia and its subsequent effect from disposal of brines on the salinity of mine waters) would affect the quality of adit and tailings water at closure.

ACKNOWLEDGMENTS

The author gratefully acknowledges the assistance of the environmental managers of both mines (R.W. and M.W.), who confirmed her understanding of the operation and capacities of the treatment and disposal systems, the source of nitrates and salts in the mine waters, provided data in electronic format, and verified disposal system operations and capacities. The author thankfully acknowledges invaluable comments from Patrick Plantenberg, Pat Pierson, and Peter Werner P.E. during construction of the spreadsheets, and the editorial assistance for this text by James Castro, Catherine Dreesbach P.E., Wayne Jepson, and Patrick Plantenberg.

REFERENCES

Bair, E.S., and T.D. Lahm. 2006. *Practical Problems in Groundwater Hydrology*. pp. 6-1–6-20. Upper Saddle River, NJ: Prentice Hall.

Cascade Earth Sciences. 2008. *Soils Technical Analysis: Effects on Soils from Mine Water Land Application Disposal Operations of the East Boulder Mine*. Final report to Montana Department of Environmental Quality. Spokane, WA: Cascade Earth Sciences.

DEQ. 2005. *Draft Technical Guidelines for Land Application of Municipal Wastewater.* Helena, MT: Montana Department of Environmental Quality.

DEQ and USFS. 1998. *Final Environmental Impact Statement for the Stillwater Mine Revised Waste Management Plan and Hertzler Tailings Impoundment.* Helena, MT: Montana Department of Environmental Quality and U.S. Forest Service.

DEQ and USFS. 2010. *Draft Environmental Impact Statement for the Stillwater Mining Company's Revised Water Management Plan and Boe Ranch LAD.* www.deq.mt.gov/eis.mcpx. Helena, MT: Montana Department of Environmental Quality.

DSL and USFS. 1985. *Final Environmental Impact Statement Stillwater Project, Stillwater County, Montana.* Helena, MT: Montana Department of State Lands and U.S. Forest Service.

DSL and USFS. 1989. *Decision Notice and Finding of No Significant Impact for the Stillwater Mining Company Proposed East Side Adit.* Helena, MT: Montana Department of State Lands.

DSL and USFS. 1992. *Draft Environmental Impact Statement for the Stillwater Mine Expansion 2000 Tons per Day Application to Amend Plan of Operations and Permit Number 00118.* Helena, MT: Montana Department of State Lands.

DSL, USFS, and DHES. 1992. *Final Environmental Impact Statement for the East Boulder Mining Project, Sweet Grass County, Montana.* Helena, MT: Montana Department of State Lands and U.S. Forest Service.

Freeze, R.A., and J.A. Cherry. 1979. *Groundwater.* Upper Saddle River, NJ: Prentice Hall.

Schwartz, F.W., and H. Zhang. *Fundamentals of Ground Water.* New York: John Wiley and Sons.

Stevenson, F.J. 1986. *Cycles of Soil: Carbon, Nitrogen, Phosphorus, Sulfur, Micronutrients.* New York: John Wiley and Sons.

Stillwater Mining Company, HKM Engineering, Inc., and Knight-Piésold. 2004. *Review of Snowmaker Water Treatment Efficiency: Stillwater Mining Company East Boulder Mine.* Report Revision 1 to Montana Department of Environmental Quality and U.S. Forest Service. Big Timber, MT: Stillwater Mining Company.

Stillwater Mining Company. 2006. *Biological Treatment System (BTS)—Overview of Primary and Secondary Treatment Processes.* Final report submitted to DEQ. January.

U.S. Department of Agriculture Natural Resources Conservation Service. What is the Snow Water Equivalent? www.or.nrcs.usda.gov/Snow/about/swe.html.

Wastewater Recycling Technology in Fankou Lead-Zinc Mine of China

Yuehua Hu, Wei Sun, and Runqing Liu
School of Minerals Processing and Bioengineering, Central South University, Changsha, China

Jinping Dai
Fankou Lead-Zinc Mine, Shaoguan, Guangdong, China

ABSTRACT

The effect of water quality, such as hardness, ionic strength, residual reagents and recycling water on flotation behavior of galena, zinc and pyrite was examined. To explain the flotation mechanism, electrochemical measurement and FTIR analysis were performed. Results of sedimentation and flotation experiment showed that the sedimentation time of wastewater increased with the increase of solid concentration. Dilution and addition of flocculants decreased the sedimentation time. Lead concentrate water could improve the flotation of galena while restrained the flotation of sphalerite activated by copper ion. Pyrite showed better floatability in sulfur concentrate water, zinc concentrate water and sulfur tailing water, while the flotation behavior of pyrite would be slightly depressed in galena concentrate water, zinc tailing water and sewage. Electrochemical measurement and FTIR analysis showed that between residual collector and galena, sphalertie and pyrite had different interaction mechanism and formed different surface species.

INTRODUCTION

Closely related to human life and environment, water resources have received much more concern around the world. A lot of water has been consumed and wasted in mining to obtain metal resources, which also play a very important role in the development of human society. Wastewater discharged from mines makes up 10% of the total amount of industrial wastewater discharge in China (Environment statistics communique of nation 2004). Fankou Mine is the biggest lead-zinc mine in China with a capacity of 6000 t/d. The ore of Fankou Mine is finely disseminated and complicated and the flotation technology to treat the ore is so complex that makes wastewater reuse difficult. A lot of suspended particles, heavy metal ion and residual flotation reagents have been found in it which shows a high pH value. For this reason, 7.5 million ton water has been discharged every year, resulting in waste of water and environmental pollution (Broman 1980; Forssberg, Jönsson and Püalsson 1985; Rao and Finch 1989; Basilio, Kartio and Roon 1996; Levay, Smart and Skinner 2001). Therefore, wastewater treatment and recycling have gained academic and practical support.In this paper, water treatment was conducted and the effect of water quality, such as hardness, ion strength, residual reagents and recycling water on flotation behavior of galena, zinc and pyrite was examined. The flotation mechanism was also revealed based on electrochemical measurement and FTIR analysis. A wastewater recycling technology including settlement, flocculation and usage for different circuits according to wastewater sources was developed, with which about 75% of wastewater can be reused in flotation plant. After successful pilot plant and industrial scale tests, this technology has been put into use since 2006, as a result, good economic and environmental efficiency have been obtained.

EXPERIMENTAL

Materials

Galena, pyrite and sphalerite samples used in this study were taken from Pb/Zn mine of Fankou. These mineral samples were first crushed to below 3 mm and then sent to porcelain mill where fraction of 0.147–0.045 mm was obtained by screening as flotation samples. Polyacrylamide buys form Tianjing Guangfu fine chemical reasearch institute, the molecular is about 15~80 million.

Wastewater sample was directly taken from production site, settled for some time to remove precipitated solids, and then the upper liquor was taken as flotation experimental water sample according to the experimental requirements. After treatment, wastewater could be used as sample of the site since little solid particles and suspended solid were found in it.

Flotation

Flotation was conducted in a 40 mL hitch groove flotation machine at a rotating speed of 1600 r/min. In flotation process, 2.0 g mineral sample was added into a CQ50 type ultrasonic cleaner to remove the surface oxide. The sample was then transferred into the flotation cell for further processing. Water used in the process was recycled wastewater. Reagents were added following the sequence of regulator, collector and frother. The conditioning time for collector and frother was 2 min and 1 min respectively. The flotation time was 4 minutes.

Infrared Spectrum Examination

A 1.0-g sample was immersed in 25 mL collector solution, ground with an agate pestle and mortar hand for 15 min, settled for 30 min, and then filtrated, and flushed 2~3 times using the previously recycled wastewater, after this, solid could be obtained with a vacuum drier. Infrared spectrum examination was then performed using reflection method in NEXUS-470 infrared spectrum apparatus.

Electrochemical Test

A conventional three-electrode system was used in electrochemical test. A piece of carbonaceous electrode severed as the counter electrode, Ag/AgCl electrode as the reference electrode, and a mineral electrode of 1 cm^2 was used as working electrode. Surface of the working electrode was gently polished by 600-grit silicon carbide papers. After that, the electrode was rinsed with distilled water and transferred quickly to the cell. The electrochemical measurements were performed at a scan rate of 50 mV/s with a Potentiostat/ Galvanostat Model 283A. All potentials reported in the paper were converted to the standard hydrogen electrode (SHE).

RESULTS AND DISCUSSION

Effect of Sediment Particles Concentration and Adding Flocculants on the Setting Velocity

Figure 1 was natural sedimentation velocity of beneficiation wastewater as a function of sediment concentration, which showed that when sediment concentration was 15.56%, the

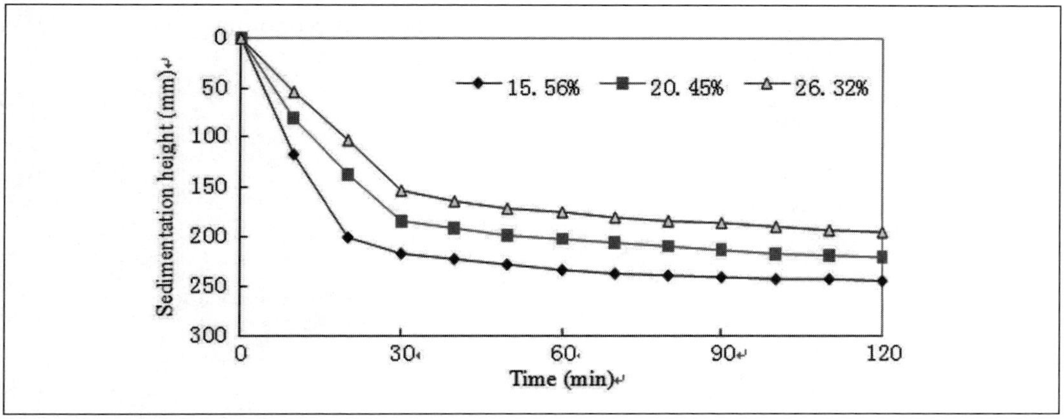

Figure 1. Sedimentation velocity of beneficiation wastewater as a function of sediment concentration

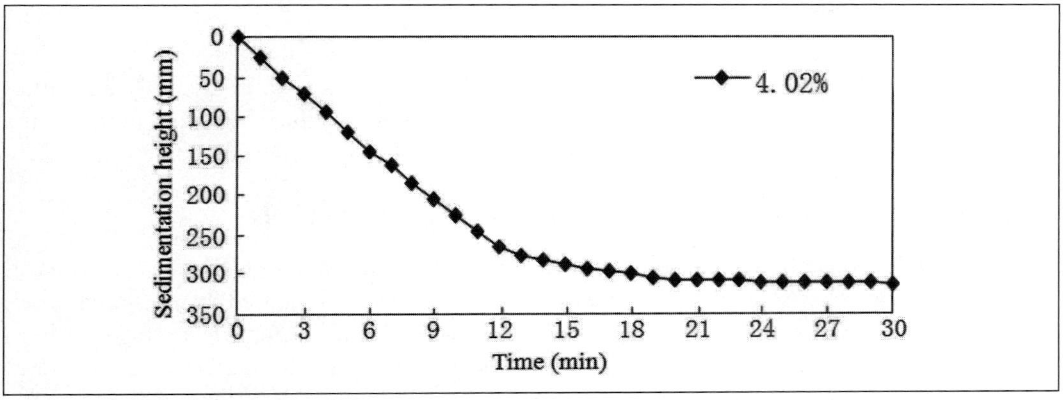

Figure 2. Sedimentation velocity of beneficiation wastewater diluted to 4.02%

settlement would be divided into two stages. In the first twenty minutes, a linear relationship between the increase of sedimentation height and time was found, then the settlement process became stable and the sedimentation height rose slowly until it remained stable after 90 minutes. When the sediment concentration increased to 20.45% and 26.32%, the settlement process was different. The increase of the sediment concentration decreased the sedimentation height, so the whole process did not become stable in the first 20 minutes. As the sedimentation process went on, wastewater of 20.45% became stable up to 2 hours, and wastewater of 26.32% became stable up to 24 hours. The results indicated that wastewater contained amounts of fine particles so the sediment concentration was high, which would directly influence the efficiency of wastewater treatment. So, the dilution tests were carried out.

Figure 2 was the settling curve of wastewater diluted to 4.02%. The settling curve showed that compared with the settling time of wastewater of 15.56% being 90 minutes, the settling time of wastewater of 4.02% decreased to 12 minutes. The results indicated that with the decrease of the solid concentration of wastewater, the sedimentation time dropped sharply till it was stable.

Figure 3 showed that 30 minutes was needed for zinc wastewater to become stable without adding polyacrylamide. While 5 g/t polyacrylamide was added to the water, only 9 minutes were needed. When the dosage of polyacrylamide kept increasing, the sedimentation time

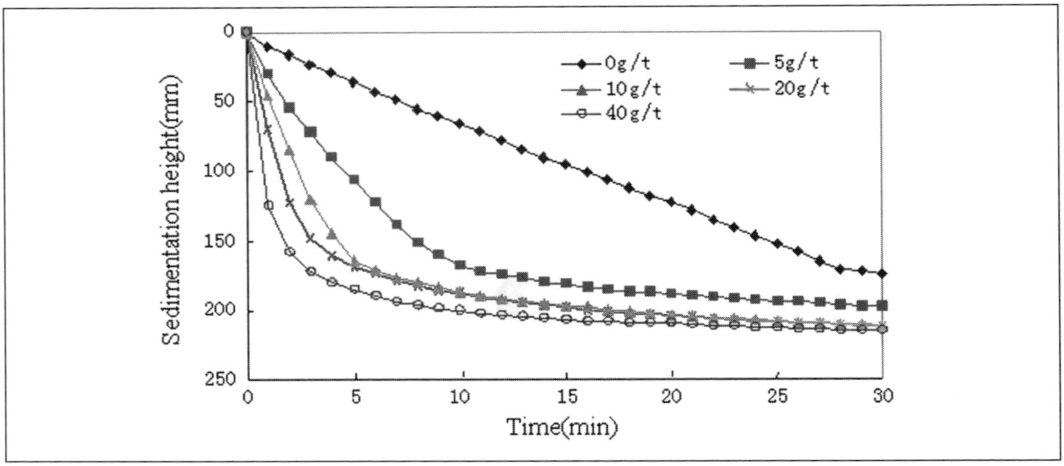

Figure 3. Sedimentation velocity of zinc wastewater as a function of polyacrylamide dosage, solid concentration of 11.03%

would continue decreasing. When the polyacrylamide dosage reached 40 g/t, only 2 minutes was needed for wastewater to become stable. The results indicated that the sedimentation time of zinc wastewater became shorter following the increase of polyacrylamide dosage.

Effect of Water Hardness on Flotation Behavior of Sulfide Minerals

As we all know, water can be divided into hard water and soft water. Water with its hardness below 8 degrees is soft water, between 8 and 16 degrees is reclaimed water. Water above 16 degrees is hard water. Water that exceeds 30 degrees is very hard water. The average hardness of production water supply in Fankou is 40.7. It belongs to very hard water. Because a lot of lime has been added in flotation and ores of Fankou contain considerable part of carbonate, this part of the minerals will dissolve in the solution due to the low pH value in pyrite flotation. So the total hardness of the final backwater is very high. The hardness of zinc tailing water is 349.1 which is far beyond the scope of normal water, thus affects the flotation of sulfide minerals. Figure 4 showed flotation recoveries of galena and copper ion activated sphalerite as a function of water hardness under high alkali conditions. Figure 5 showed pyrite flotation recovery as a function of water hardness under weak acid conditions. These two figures showed that water hardness had little effect on floatability of galena and copper ion activated sphalerite, while had great influence on floatability of pyrite. The recoveries of galena and sphalerite were about 82% and 85% respectively in the tests. When water hardness was more than 100, pyrite was suppressed obviously and the recovery decreased from 86% to 60%. This might be related to pyrite surface electrochemical reaction. The results showed that the higher the water hardness, the easier to separate lead and zinc from pyrite, but the worse for pyrite recovery, namely, backwater with high basicity was favorable for separation of lead-zinc and pyrite but unfavorable for pyrite flotation.

Effect of Ion Strength on Flotation Behavior of Sulfide Minerals

The influence of ionic strength on the flotation behavior of galena, pyrite and sphalerite activated by Cu^{2+} under high alkali conditions was shown in Figure 6. The ionic strength of the solution was prepared according to the molar ratio of sodium chloride to sodium sulfate,

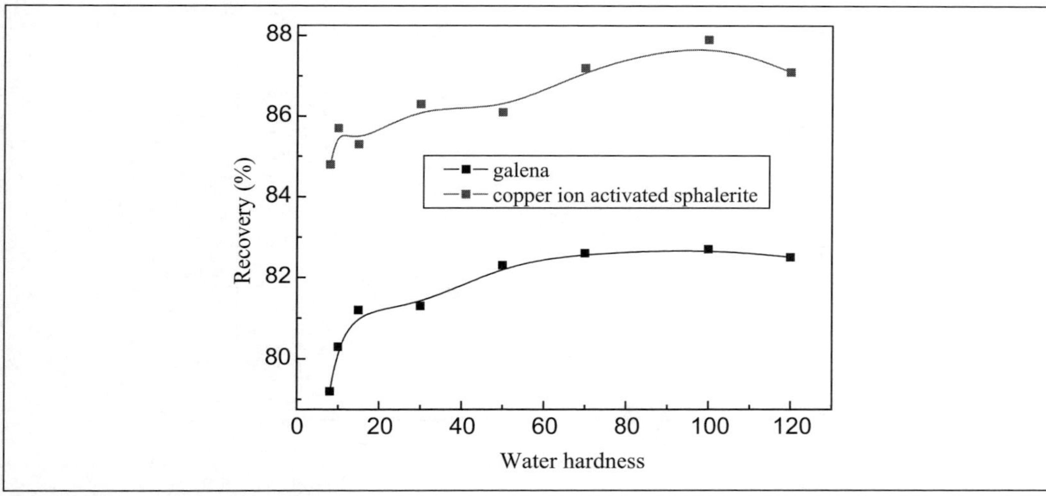

Figure 4. Flotation recoveries of galena and copper ion activated sphalerite as a function of water hardness under high alkali conditions, C(KBX)=1×10^{-4} mol/L

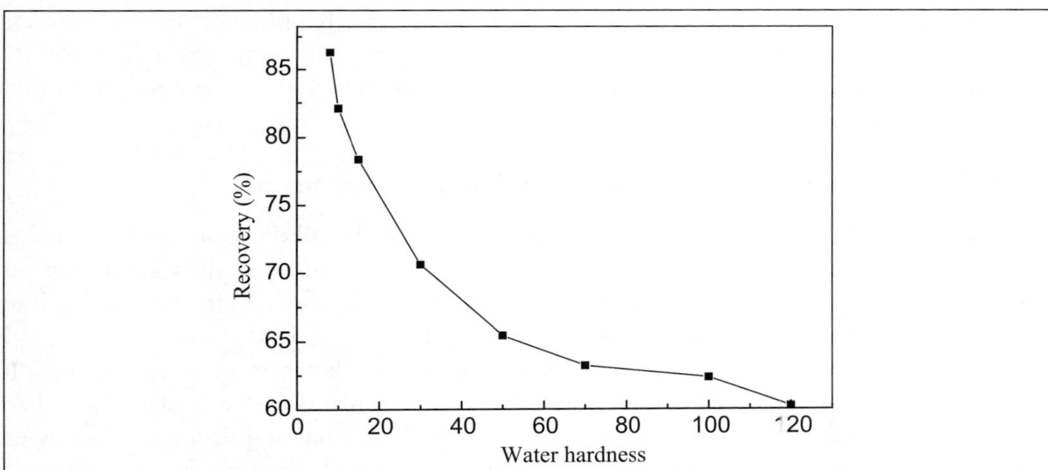

Figure 5. Flotation recovery of pyrite as a function of water hardness under weak acid conditions, C($CuSO_4$)= 1×10^{-4} mol/L

which was designed by reference to the ion weight ratio of Cl^- and SO_4^{2-}, because the concentration of Cl^- and SO_4^{2-} in zinc tailing water reached 303.7 mg/L and 1258 mg/L respectively. This ionic strength didn't consider the contribution of Ca^{2+} and Mg^{2+} influence of which had be investigated in the influence of hardness factors. Figure 6 showed that the recovery of galena was increasing gradually with the increase of the ionic strength. The flotation recovery reached the highest 85% when the concentration of ionic was 10^{-3} mol/L, then remained gentle. Promotion of the floatability of galena might be due to the agglomeration of hydrophobic particles caused by the neutralization of surface electrical property. For sphalerite, the flotation rate decreased with the increase of ionic strength, because the electric double layer was compressed, and Na^+ of high concentration could resist Cu^{2+} adsorption on the surface of sphalerite, which influenced the activation effect and led to poor flotation behavior. For

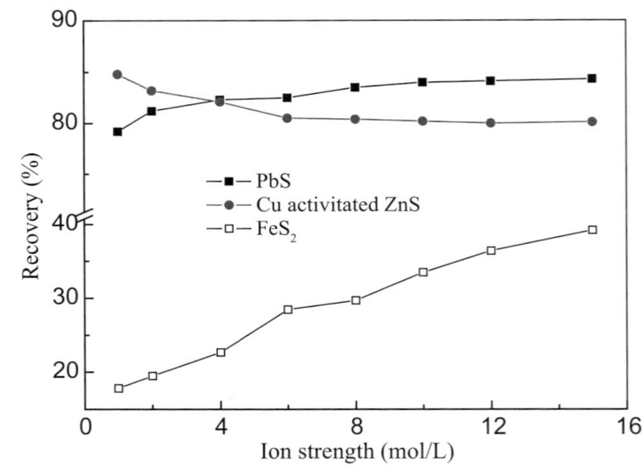

Figure 6. Flotation recovery of three kinds of sulfide minerals as a function of ionic strength under high alkali conditions, C(KBX)=1×10^{-4} mol/L

pyrite, it showed a slight activation with the increase of ionic strength. The recovery was less than 20% in distilled water under high alkali conditions, but with ionic strength increasing gradually, the recovery increased to about 40%, which was related to the variation of transfer properties of interface.

Effect of Residual Reagent on Flotation Behavior of Sulfide Minerals

The reagent in the wastewater had two sources. One was the collector and inhibitor added in the process of flotation. The other was water treatment agent added in the water treatment process. In order to eliminate the suspended and residual solid, flocculants were added into wastewater, so there was residual flocculants in wastewater.

Figure 7 showed the influence of polyacrylamide on the flotation of three minerals. It could be seen that when polyacrylamide was used as the flocculants with a dosage of less than 5 g/t, the effect on the floatability of three minerals was not obvious and the recovery changed little. When the dosage was more than 5 g/t, galena was restrained. These results indicated that low dosage of polyacrylamide was ideal for water treatment agent, which not only guaranteed better backwater, but also had no effect on separation of PbS, ZnS and FeS_2.

Effect of Recycling Water on Flotation Behavior of Sulfide Minerals

Figure 8 showed the result of galena flotation test carried out in different concentrate water. The dosage of collectors used in the test was 20% less than that in distilled water. Figure 8 showed that, following the increase of pulp pH, the recovery of galena decreased. More than 90% of galena recovery would be realized before the pulp pH reached 8. The recovery decreased from 90% to 80% when pulp pH rose from 8 to 12. The changing tendency of galena flotation behavior was similar in the three concentration water. And the recovery of galena decreased with the increase of pulp pH. The three concentrate water would influence the flotation behavior of galena differently. The flotation of galena in lead concentrate water could get higher recovery than in distilled water. On the contrary, the addition of other two concentrate water would slightly depress the flotation of galena. The floatability of galena in

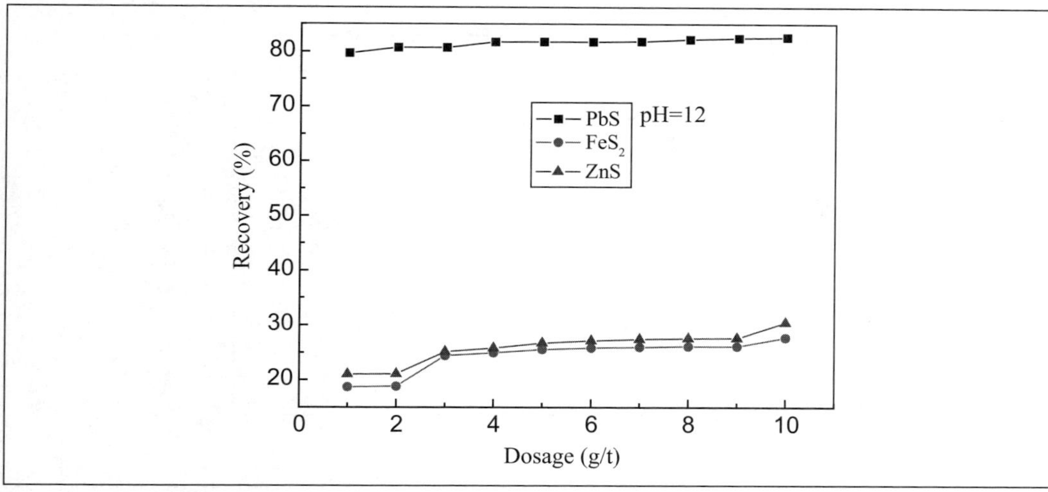

Figure 7. Flotation recovery of three kinds of sulfide minerals as a function of the dosage of polyacrylamide under high alkali conditions, $C_{(KBX)}=0.25\times10^{-4}$ mol/L

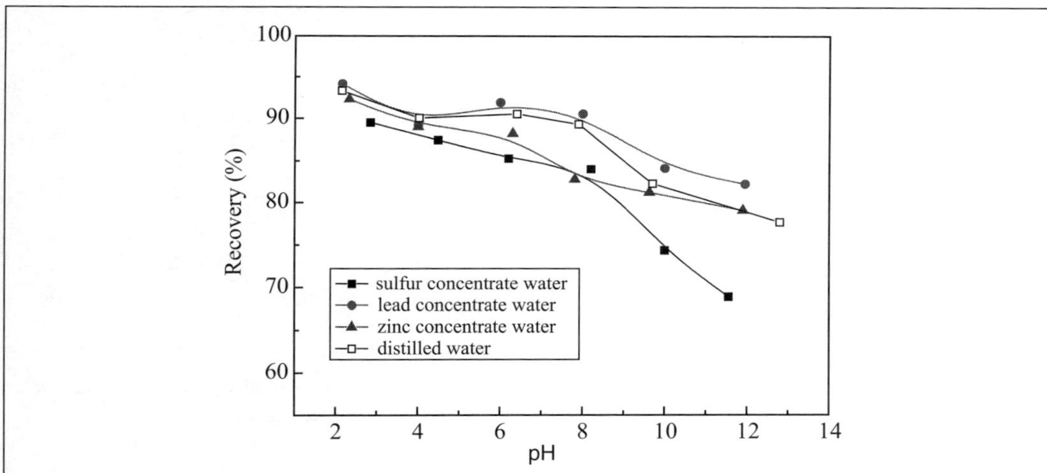

Figure 8. Flotation recovery of galena as a function of pH in different concentrate water, C(KBX)=1×10^{-4} mol/L

these four water systems was: lead concentrate water > distilled water > zinc concentrate water > sulfur concentrate water.

Figure 9 showed the result of galena flotation test in tailing water and sewage. From the result we could see that the flotability of galena would grow worse by adding sulfur tailing water, zinc tailing water and sewage in the test. Therefore, the flotation behavior of galena would be depressed by adding tailing water.

Figure 10 showed the recoveries of pyrite in concentration water as a function of pH, from which we could see that the floatability of pyrite was good, with a recovery of more than 83% under acid and alkalescent conditions in distilled water. After the pulp pH reached 8, the recovery of pyrite decreased sharply with continuous increase of pH. When the pulp pH reached 12, recovery of pyrite was only about 20%. The flotation behavior of pyrite in

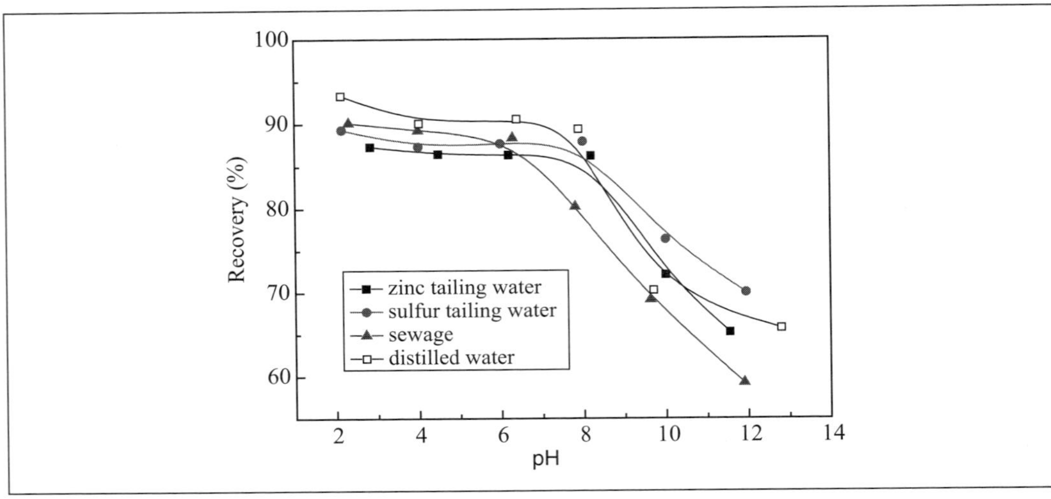

Figure 9. Flotation recovery of galena as a function of pH in tailing water and sewage, C(KBX)=1×10^{-4} mol/L

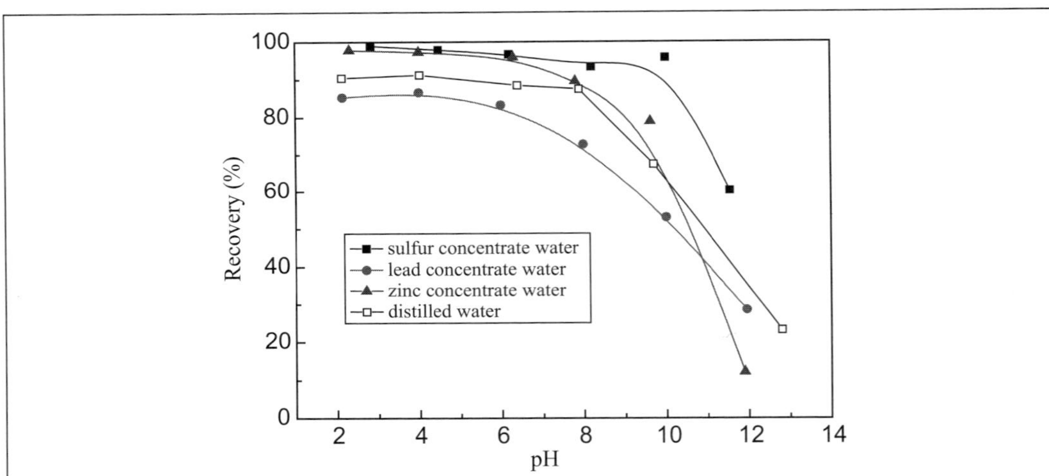

Figure 10. Flotation recovery of pyrite as a function of pH in different concentrate water, C(KBX)=1×10^{-4} mol/L

different concentrate water was similar to that in distilled water. While in zinc and sulfur concentrate water, the floatability of pyrite was better than that in distilled water. On the contrary, when the test was carried out in galena concentrate water, the floatability of pyrite would get worse, opposite to the floatability of galena affected by adding different concentrate water. Therefore, it would be good for the separation of PbS and FeS_2 by adding galena concentrate water, while adding sulfur concentrate water, adverse effect on the separation of PbS and FeS_2 could be seen. This should be considered in the mixing of wastewater test. Figure 11 showed the floatability of pyrite in different tailing water and sewage. The flotation behavior of pyrite in the four kind of water had similar change tendency. Pyrite showed better floatability in sulfur tailing water, while flotation behavior of pyrite would be slightly depressed in zinc tailing water and sewage.

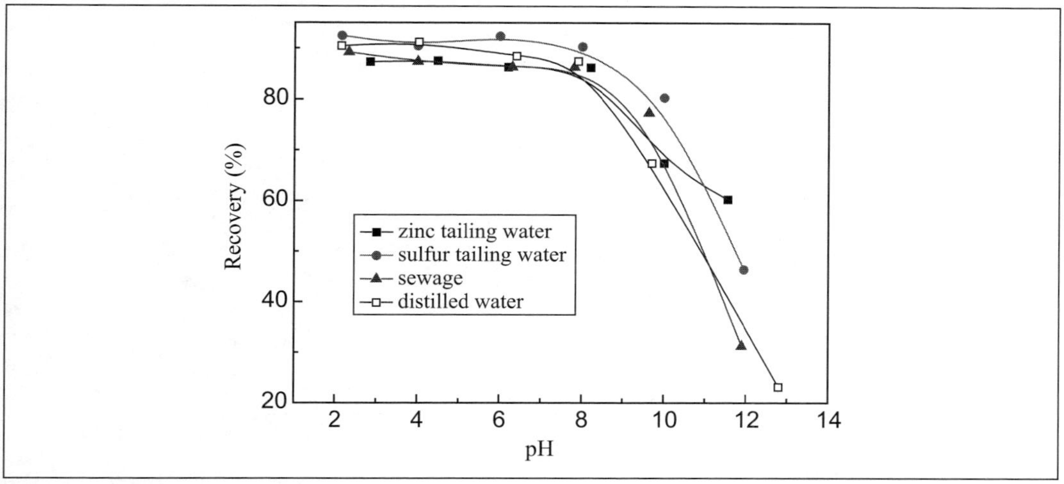

Figure 11. Flotation recovery of pyrite as a function of pH in different tailing water and sewage, C(KBX)=1×10^{-4} mol/L

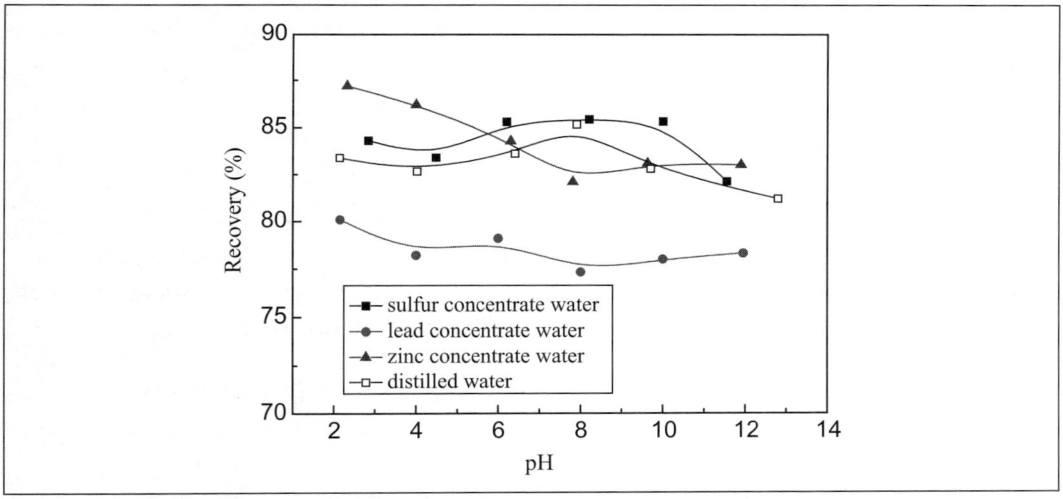

Figure 12. Flotation recovery of sphalerite activated by copper ion as a function of pH in different concentrate water, C(KBX)=1×10^{-4} mol/L

The recovery of sphalerite activated by copper ion as a function of pH in different concentrate water systems was shown in Figure 12. In the whole pH range the activated sphalerite showed good floatability in distilled water system, with a recovery of more than 80%. In zinc and sulfur concentrate water the flotation of activated sphalerite was slightly promoted, while in lead concentrate water it would be depressed to some extent, with a recovery of 75%~80%. The floatability of activated sphalerite in these four water systems was: sulfur concentrate water > zinc concentrate water > distilled water > lead concentrate water. The flotation behavior of sphalerite activated by copper ion in different tailing water and sewage was shown in Figure 13. The flotation behavior of sphalerite would be activated under acid and alkalescent conditions in zinc tailing water, lead tailing water and sewage. After the pulp pH reached 8,

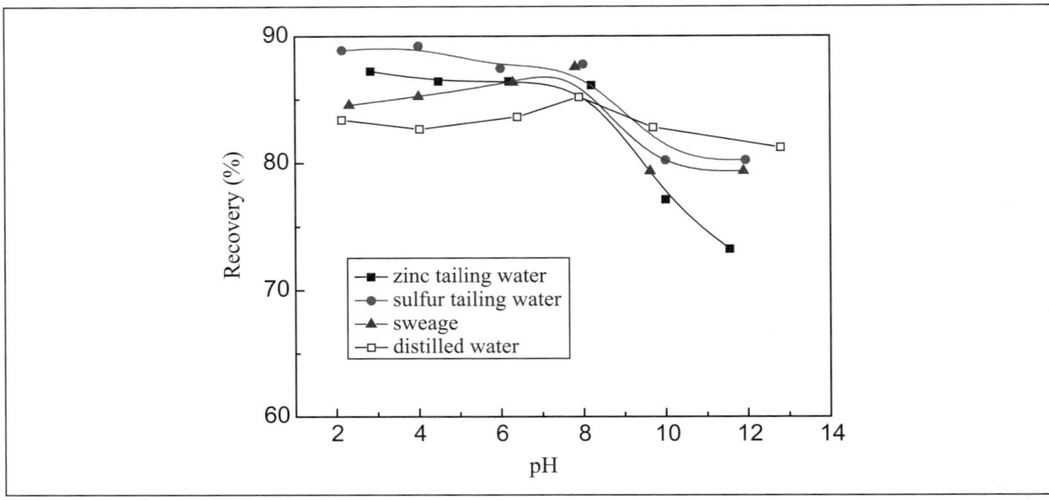

Figure 13. Flotation recovery of sphalerite activated by copper ion as a function of pH in different tailing water, C($CuSO_4$)= 1×10^{-4} mol/L

the flotation of sphalerite would be slightly depressed with a recovery of 5%, lower than that in distilled water.

MECHANISM ANALYSIS

Infrared Spectra of Sulfide Minerals in Beneficiation Wastewater

The infrared spectra of galena in xanthate solution and in wastewater-xanthate solution were shown in Figure 14. The results indicated that the surface product of galena in five water systems was basically the same, with similar absorption peaks. However, the absorption intensity was somewhat different. Absorption peaks of 2955 cm^{-1} and 2870 cm^{-1} were attributed to the C—H asymmetry and symmetry stretching vibration. 1466 cm^{-1} was derived from symmetry bend vibration of —CH_3 or shear swing vibration of —CH_2, while 1405 cm^{-1}, from CH_3 bend vibration. 1206 cm^{-1} and 1029 cm^{-1} were stretching vibration absorption peaks of C—O—C and —C=S, respectively. It could be seen from Figure14 that surface absorption characteristic peak of galena was broader and stronger in mineral processing wastewater. The order of infrared absorption intensity of galena in five water systems was: zinc tailing water > zinc concentrate water >= lead concentrate water > distilled water > sulfur concentrate water. This phenomenon indicated that collector was connected with physical and chemical reaction of galena surface. This result was in good agreement with flotation result.

Figure 15 and Figure 16 were the infrared spectra of pyrite in zinc concentrate water and zinc tailing water, showing that there was a wideband absorption at 3250 cm^{-1}, which was induced by surface hydroxyl compound because of the oxidation of mineral surface. The absorption peak at 1102 cm^{-1} was induced by the absorption of sulfate radical. The results indicated that the interaction between pyrite and collector was inhibited, and surface reaction of pyrite was dominated by the oxidation of pyrite.

Infrared spectra of pyrite interaction with three kinds of water were presented in Figure 17. It could be seen that there were characteristic absorption peak of methyl and methylene at 2963 cm^{-1} and 2880 cm^{-1}. In addition, there were also characteristic absorption peak at 1290 cm^{-1} and 1033 cm^{-1} caused by C=S and C—O—C stretching vibration of dixanthogen.

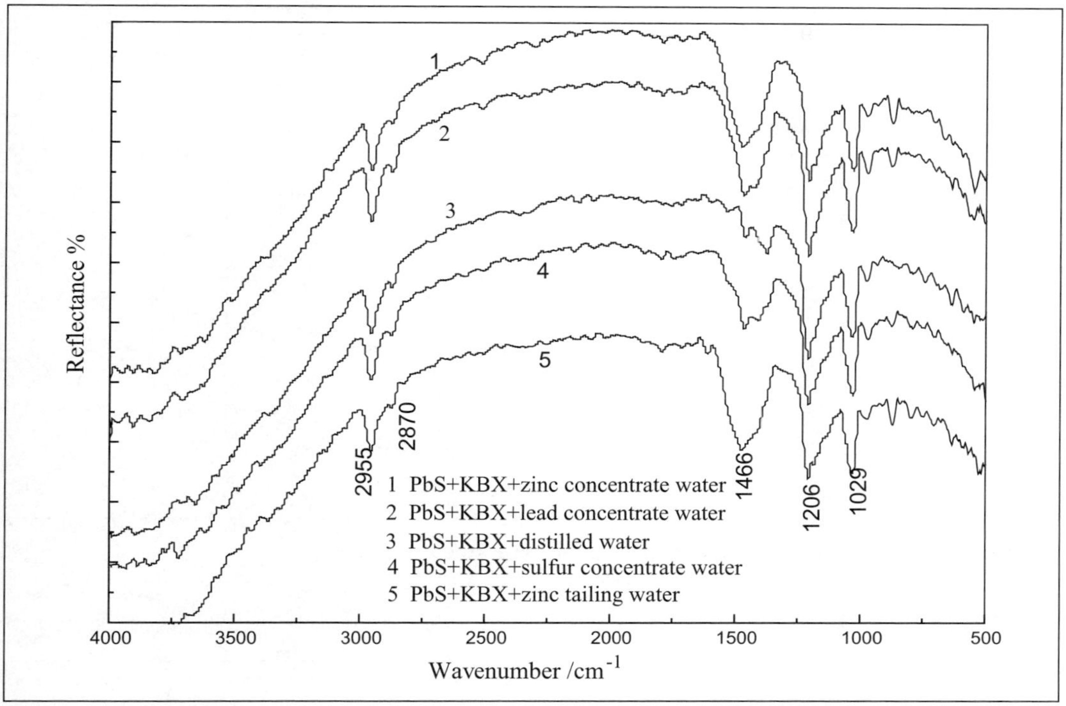

Figure 14. Infrared spectra of galena in five water systems: 1-PbS+KBX+zinc concentrate water, 2-PbS+KBX+lead concentrate water, 3-PbS+KBX+distilled water, 4-PbS+KBX+sulfur concentrate water, 5-PbS+KBX+zinc tailing water

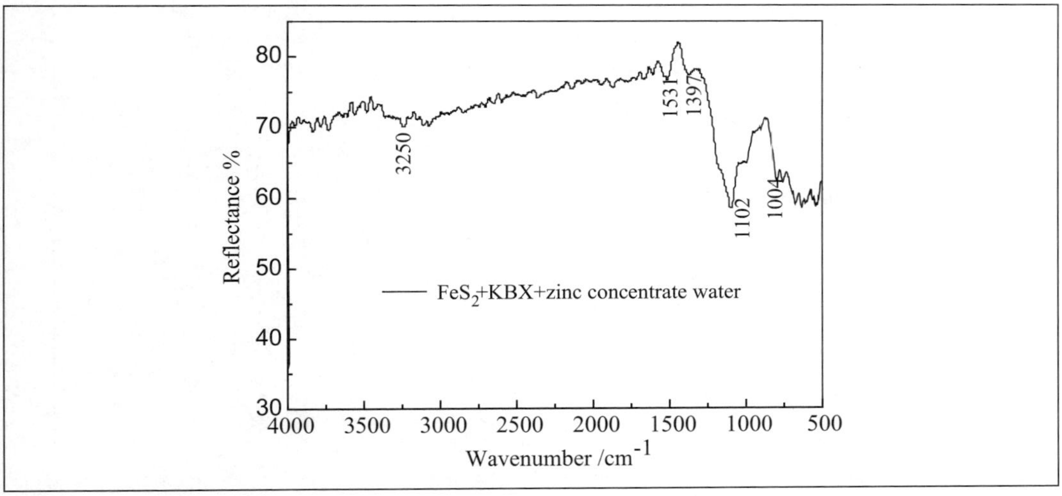

Figure 15. Infrared spectra of pyrite in zinc concentrate water

According to ZHANG[16], C=S stretching vibration peak decreased from 1049 cm^{-1} to 1019 cm^{-1} and C—O—C stretching vibration increased from 1240 cm^{-1} to 1290 cm^{-1} when xanthate was oxidated to dixanthogen. This result conformed to the literature report and the above flotation results. It could be inferred that absorption intensity of pyrite in three kinds of water was: sulfur concentrate water > distilled water > lead concentrate water.

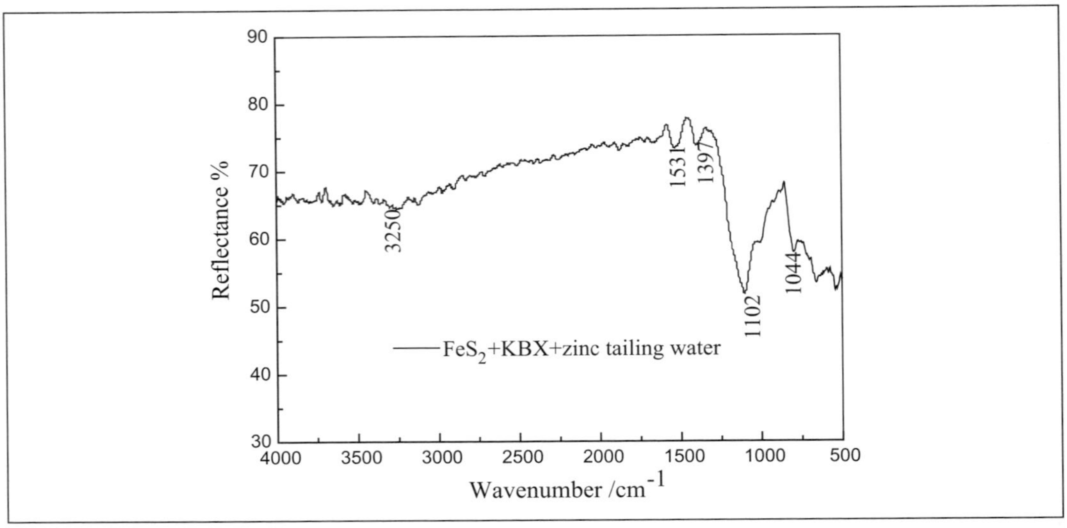

Figure 16. Infrared spectra of pyrite in zinc tailing water

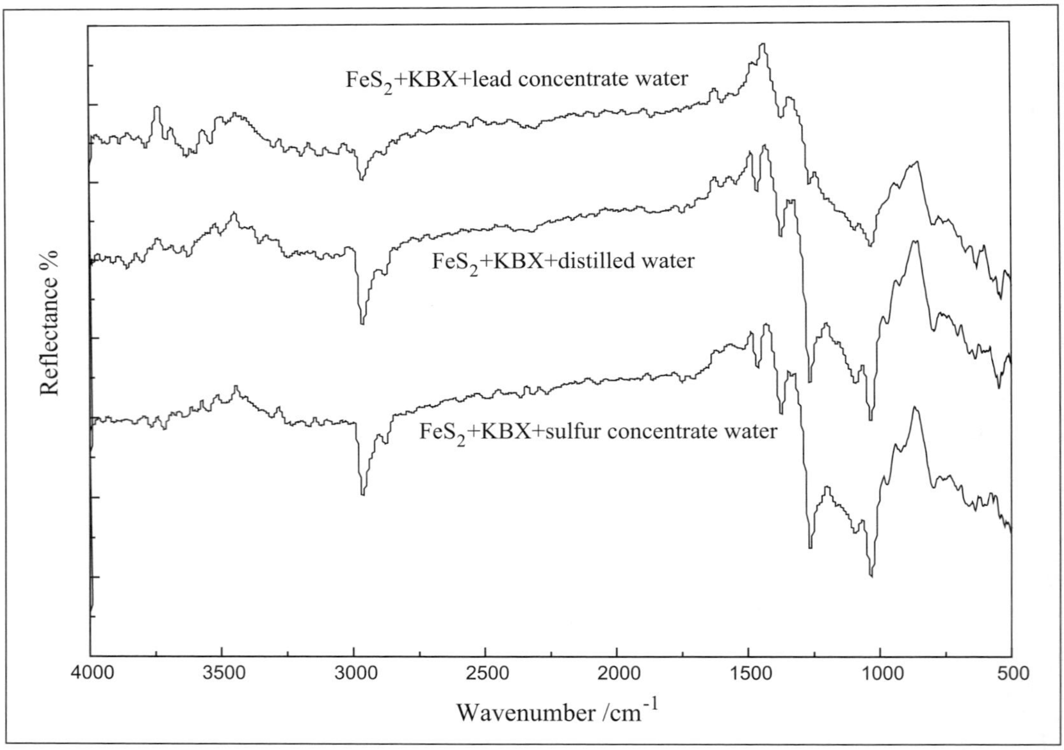

Figure 17. Infrared spectra of pyrite in three kinds of water systems: 1-FeS_2+KBX+lead concentrate water, 2-FeS_2+KBX+distilled water, 3- FeS_2+KBX+sulfur concentrate water

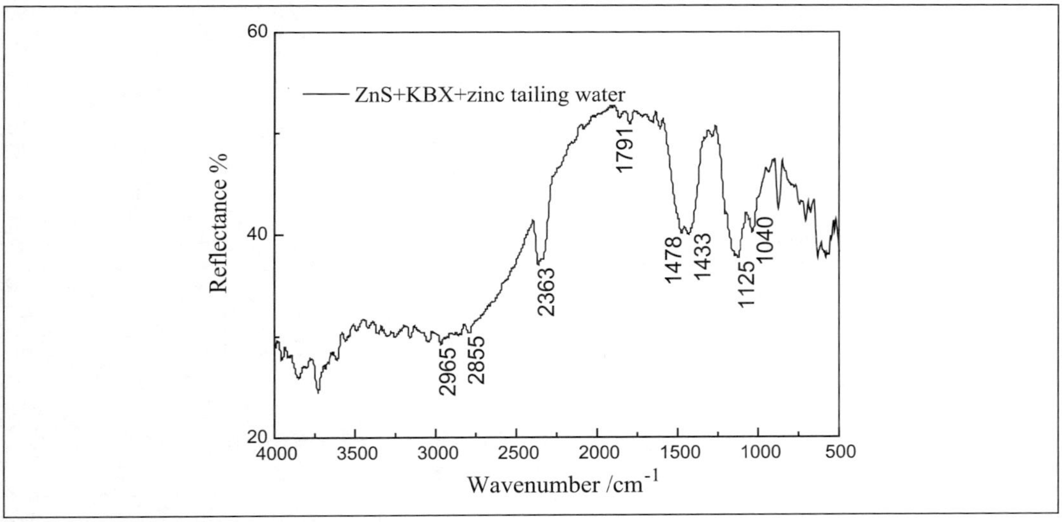

Figure 18. Infrared spectra of sphalerite in zinc tailing water

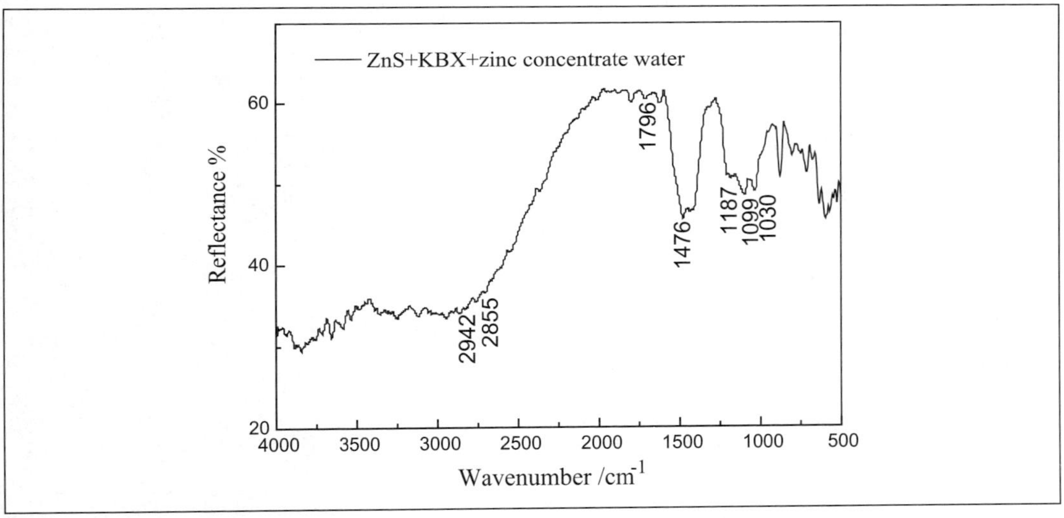

Figure 19. Infrared spectra of sphalerite in zinc concentrate water

Infrared spectra of sphalerite interaction with zinc tailing water and zinc concentrate water were presented in Figure 18 and Figure 19. The results showed that there were adsorption peaks at 2942 cm^{-1} and 2855 cm^{-1}, which were attributed to the methyl and methylene stretching vibration. Peaks at 1187 cm^{-1},1125 cm^{-1}, 1040 cm^{-1} were the stretching vibration of C-O-C group because the interaction between copper ion and xanthate salt led the peak to move to high wave numbers. So it could be inferred that surface product of sphalerite was $Cu(EX)_2$.

Electrochemical Reaction of Galena and Sphalerite in Different Water Systems

In order to study the effect of recycled wastewater on galena floatability, electrochemical technology was chosen to study the surface reaction mechanism of galena in different water systems. The test was carried out in solution with or without collector as follows:

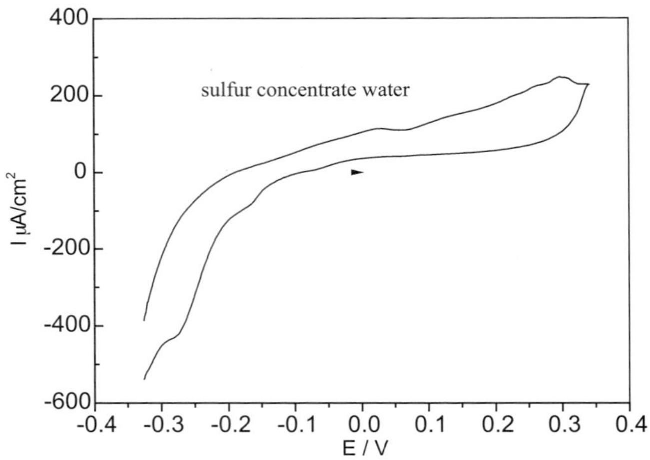

Figure 20. Voltammetry curves of galena electrode in sulfur concentrate water scan rate: 50 mV/s KNO_3:0.1 mol/L

Collectorless state:

$PbS=Pb^{2+}+S^0+2e$ (E_0=0.054V) (1)

$PbS+H_2O=PbO+2H^++S+2e$ (E_0=0.750V) (2)

$PbS+4H_2O \rightarrow Pb^{2+}+SO_4^{2-}+8H^++8e$ (E_0=0.450V) (3)

$2PbS+3H_2O=2Pb^{2+}+S_2O_3^{2-}+6H^++8e$ (E_0=0.326V) (4)

Galena interaction with butyl xanthate:

$PbS+2X^-=PbX_2+S+2e$ (E_0=–0.124V) (5)

$PbS+2X^-+4H_2O= PbX_2+SO_4^{2-}+8H^++8e$ (E_0=0.232V) (6)

$PbS+4X^-+3H_2O \rightarrow 2PbX_2+S_2O_3^{2-}+6H^++8e$ (E_0=0.101V) (7)

Galena interaction with DDTC:

$2PbS+2D^- \rightarrow PbD_2+S+2e$ (E_0=–0.121V) (8)

$2PbS+4D^-+3H_2O \rightarrow 2PbD_2+S_2O_3^{2-}+6H^++8e$ (E_0=0.171V) (9)

$PbS+2D^-+4H_2O=PbD_2+SO_4^{2-}+8H^++8e$ (E_0=0.181V) (10)

Figure 20, Figure 21 and Figure 22 showed the voltammetry curves of galena in sulfur concentrate water, lead concentrate water and zinc concentrate water. From Figure 20 we could see that the initial oxidation potential of galena was about –0.017 V in sulfur concentrate water, which was lower than the potential of collector reaction on galena surface because of the oxidation of other reduction substance and the adsorption of active substance on the surface of galena. At the potential of 0 V, oxidation of galena occurred and an obvious oxidation peak appeared. The result indicated that the oxidation of galena was attributed to Reaction 5,

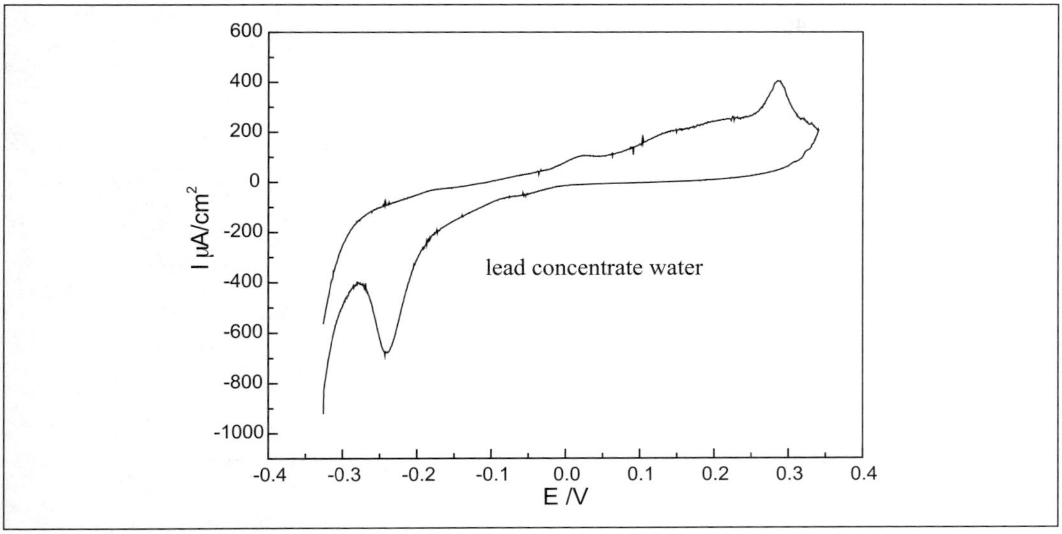

Figure 21. Voltammetry curve of galena electrode in lead concentrate water scan rate: 50 mV/s KNO_3:0.1 mol/L

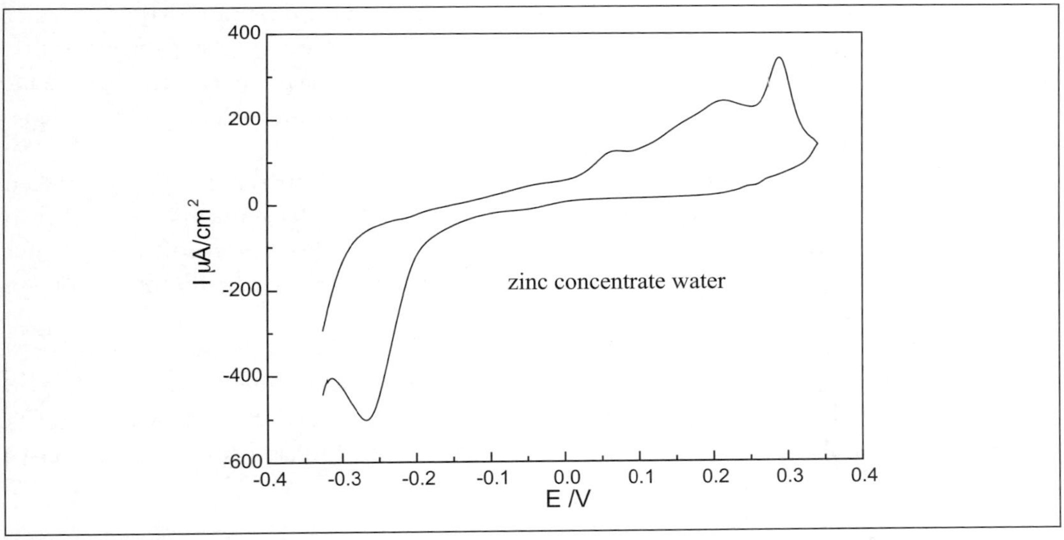

Figure 22. Voltammetry curve of galena electrode in zinc concentrate water, scan rate: 50 mV/s, KNO_3:0.1 mol/L

namely the formation of xanthate lead due to the existence of residual collector in sulfur concentrate water and the potential was lower than that of Reaction 1.

From Figure 21 we could see that the initial oxidation potential was about –0.08 V. Following the increase of the potential, two obvious oxidation peaks occurred at potentials 0.15 V and 0.3 V, respectively, corresponding to Reactions 9 and 10. It indicated that there was DDTC salt formation on galena surface, and residual DDTC of lead concentrate water played a dominant role on galena oxidation.

Figure 22 showed that the initial oxidation potential of galena was –0.1 V which was in good agreement with Reaction 5. It could be seen that the second oxidation peak corresponded

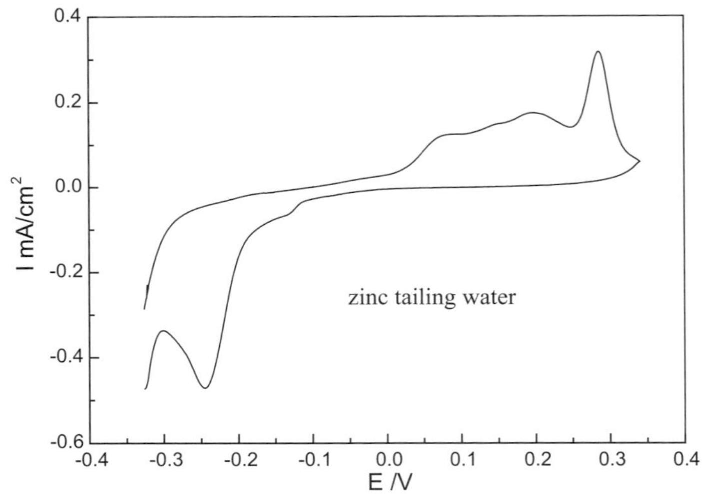

Figure 23. Voltammetry curve of galena electrode in zinc tailing water, scan rate: 50 mV/s, KNO_3:0.1 mol/L

to Reaction 7 at the potential of 0.1 V. When the potential ranged from 0.2 V to 0.3 V, further oxidation of galena would take place, as shown in Reactions 3 and 6. The result suggested that oxidation products of galena were mainly xanthate lead and sulphate, and the residual xanthate in zinc concentrate water was the dominant factor on electrochemical reaction of galena, which was consistent with the flotation results.

Figure 23 showed the voltammetry curve of galena in zinc tailing water. As could be seen from Figure 23, the voltammetry curve of galena in zinc tailing water was similar with that in zinc concentrate water. It could be concluded that chemical composition in zinc tailing water was the same as that in zinc concentrate water. Residual xanthate played a dominant role on electrochemical reaction of galena.

Voltammetry curve of galena in sewage (Figure 24) demonstrated that the oxidation potential of galena was about –0.09 V. It suggested that this reaction might be caused by two reasons: the deposition of xanthate or the adsorption of other reduction substances on galena surface. We could also see that the initial reaction products hindered further oxidation of galena as shown from the weak oxidation peak.

CONCLUSIONS

On the basis of the test results, the following conclusions could be drawn:

1. The sedimentation time of wastewater increased with the increase of solid concentration. Dilution and addition of flocculants decreased the sedimentation time. The dosage of flocculants required should be less than 5 g/t in order not to affect flotation process.
2. The physicochemical factors, such as water hardness, ion strength, residual flotation reagent etc., had effects on the flotation behavior of sulfide minerals, on which water recycling technology based.

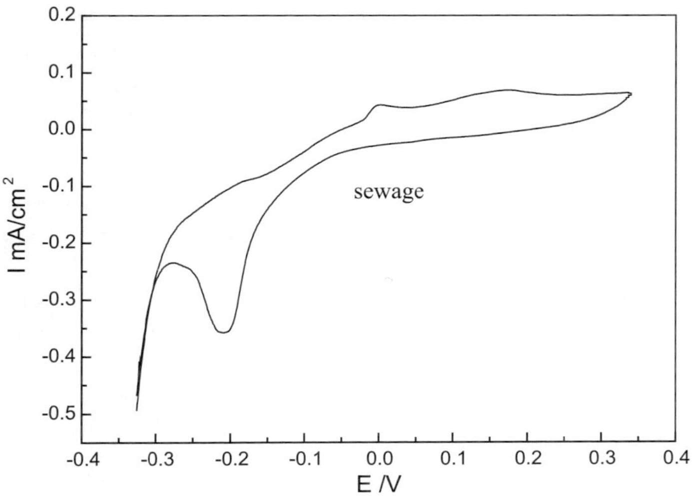

Figure 24. Voltammetry curve of galena electrode in sewage, scan rate: 50 mV/s KNO_3:0.1 mol/L

3. The floatability of galena in different water systems was: lead concentrate water > distilled water > zinc concentrate water > sulfur concentrate water > tailing water > sewage. For sphalerite activated by copper ion, the other wastewater would slightly accelerate the flotation except lead concentrate water. Pyrite showed better floatability in sulfur concentrate water, zinc concentrate water and sulfur tailing water, while the flotation behavior of pyrite would be slightly depressed in galena concentrate water, zinc tailing water and sewage.
4. Infrared spectrum indicated that surface absorption characteristic peak of galena was broader and stronger in mineral processing wastewater. The infrared absorption intensity of galena in three kinds of water was: lead concentrate water > distilled water > sulfur concentrate water. The residual collector in lead concentrate water was beneficial to the formation of lead xanthate on the surface of galena. For pyrite, the infrared absorption intensity was as follows: sulfur concentrate water > distilled water > lead concentrate water.
5. Electrochemical results indicated that the residual collector in backwater dominated the electrochemical and chemical reactions of galena surface, as a result, DDTC salt and xanthate salt would be formed in lead and zinc concentrate water. Besides, interaction between collector and galena in wastewater was stronger than that in distilled water. The reactions on the surface of galena were changed by some specific chemical composition in zinc tailing water and sewage which was quite different from that in distilled water. Reuse of the backwater might have some influence on galena floatability.

ACKNOWLEDGMENT

The authors thank the National Natural Science Foundation of China (50874117) and the freedom explores Program of Central South University (2011QNZT081) for financial support.

REFERENCES

Environment Protect Department of National. 2004. Environment Statistics Communique of Nation. [EB/OL]. www.zhb.gov.cn/plan/hitj/qghjtjgb/t20050610-67563.htm.

Broman, P.G. 1980. *Water reuse at sulfide ore concentrators in Sweden: Practice, experience and current development.* In *Complex Sulphide Ores.* Edited by M.J. Jones. London: Institution of Mining and Metallurgy, 28–39.

Forssberg, K.S.E., Jönsson, H.R., and Püalsson, B.I. 1985. *Full scale test of process water reuse in a complex sulphide ore circuit.* In *Flotation of Sulphide Minerals.* Edited by K.S.E. Forssberg. Amsterdam, Netherlands: Elsevier, 197–217.

Finch, J.A., and Rao, S.R. 1989. A review of water reuse in flotation. *Minerals Engineering,* 2:65–85.

Basilio, C.I., Kartio, I.J., and Roon, R.-H. 1996. Lead activation of sphalerite during galena flotation. *Minerals Engineering,* 9:869–879.

Levay, G., Smart R.St.C., and Skinner, W.M. 2001. The impact of water quality on flotation performance. *Journal of the South African Institute of Mining and Metallurgy,* 101:69–75.

Sulfidized Red Mud—A New Sorbent for Toxic Substances

Joseph Iannicelli
Aquafine Corporation, Brunswick, GA, USA

ABSTRACT

A powerful improved sorbent is produced by sulfidizing red mud, a noxious by-product from the Bayer extraction of alumina from bauxite. Sulfidized red mud (SRM) sorbed 90 to 100% of the following metals from laboratory solutions of Cr, Co, Ni, Cu, Zn, Se, Ag, Cd, Hg, Pb, Th, U. Discolored organic compounds (DOC) are also sorbed (90%). Sulfidization of red mud is accomplished under ambient or relatively mild conditions using exemplary compounds such as H_2S, Na_2S, K_2S, $(NH_4)_2S$, and CaS_x. Sulfur content ranges from 0.2% to 10% above the residual sulfur in red mud. The sulfidization reaction blocks leaching of metals naturally present in red mud. In some cases, (As, Mn, Sr), mixtures of sulfidized red mud plus red mud are more effective than sulfidized red mud alone. Sulfidized red mud has applications for cleaning raw industrial process water as well as effluent wastewater (and gases) for the entire range of industrial processes.

BACKGROUND

Red mud is a noxious by-product and pollutant of the production of alumina from bauxite by the process invented by Karl Bayer in 1887. This process relies on the selective solubility of aluminous minerals in hot (125–250°C) sodium hydroxide solution and the insolubility of the remaining minerals (iron, titanium, and silica) which are either insoluble or react and re-precipitate. The insoluble, iron rich residue can contain 17.4 to 37.5% (Fe). Red mud is a complex mixture of finely divided hydrated iron oxides and a wide range of lesser minerals containing Al, Na, Ti, Si, Ca, Mg plus traces of over a score of other elements including Cr, Ni, Cu, Pb, Se, Hg, As, Th, etc.

The resulting red mud has strong sorptive and complexing properties and is the subject of scores of publications. Because of its preparation, red mud is intensely alkaline, with pH values of 13 and above, but also may contain and leach toxic metals. This creates serious problems with its storage in tailings impounds which poses a toxic hazard for wildlife and personnel, and creates widespread contamination of ground water. Reduction of pH below 10 is necessary for safe storage and many sorptive applications.

It is estimated that 150 million tonnes of red mud is produced and impounded per year and that about 2.5 billion tonnes is currently stored worldwide.

Hazards of storing highly caustic and toxic red mud has been brought into focus by the bursting of a red mud impound at Ajka, Hungary on October 4th, 2010 which released 700,000 tonnes of red mud over 40 square kilometers, killing ten people and hospitalizing 120 others.

Neutralization of red mud can be accomplished with waste acid, or by washing red mud with large amounts of sea water (typically 12 to 18 times the volume of red mud). This requires seaside location, large settling basins, and of course the ability to discharge waste water back to the sea.

Red mud has been proposed as a sorbent for heavy metals, cyanides, phosphates, and the like. However, the sorptive and release properties of red mud are not always compatible. Depending on the source of a particular red mud, it can leach out significant amounts of toxic pollutants such as radioactive thorium, uranium, chromium, barium, arsenic, copper, zinc, cobalt, as well as lead, cadmium, beryllium, and fluorides.

Red mud is a very hydrophilic high pH slime which is difficult to dewater by filtration or sedimentation means. This also complicates and limits its utility as a sorbent in aqueous systems.

The potential problems involved with use of red mud to control pollution are highlighted in an e-newsletter article entitled "The Great Red Mud Experiment that Went Radioactive." This experiment conducted by the Western Australian Agricultural Department involved placing 20 tonnes of Alcoa red mud per hectare on pastureland in order to stop unwanted phosphorous from entering waterways. An unintended result of this experiment was that runoff waters showed excessive quantities of copper, lead, mercury, arsenic, and selenium. Emaciated cattle grazing on treated land exhibited high chromium, cadmium, and fluoride levels. Furthermore, each hectare contained up to 30 kilograms of radioactive thorium. The disastrous red mud application test was abruptly terminated after five years.

It is evident that extreme caution must be exercised in selecting, treating, and testing red mud before attempting to use it to sorb toxic compounds.

Furthermore, the capacity of red mud to capture and hold toxic substances such as mercury and related metals often is not adequate to eliminate traces of these metals in leachate. The possibility also exists that sorption of one toxic pollutant may release other pollutants. Therefore, use of red mud as a sorbent to purify water is problematic.

As a result of intensive investigations on methods for neutralizing and using red mud, an Australian based company, Virotec, has developed a line of red mud based products covering a wide range of pollution control applications. Virotec uses a variety of methods to neutralize red mud. These involve use of natural sea water (up to 13 washings), evaporatively concentrated sea water, saline or hard groundwater brines, salt lake brines, industrial waste brines and even solid salts.

APPLICATIONS FOR SULFIDIZED RED MUD

Heavy metal contaminated liquids and flue gases from various sources (ground, stream, runoff, mines, petroleum, industrial waste) are among the most dangerous and difficult environmental problems facing the world today. Among these metals are mercury, chromium, cobalt, nickel, copper, zinc, silver, gold, cadmium, lead, selenium, and transuranic elements.

Mercury contamination of the environment is the subject of increasing attention because it eventually accumulates at high levels in bodies of large predatory fish such as tuna, swordfish, and shark. A major concern is the atmospheric release of mercury from coal fired power plants, currently estimated at 46 tons per year in the United States. The Environmental Protection Agency (EPA) has identified women of childbearing age as especially threatened because of possible neurological damage to unborn children. It is estimated that 8% of women in this category have a methyl mercury blood level above 5.8 ppb.

On Dec. 14, 2000, the EPA issued a determination that their agency must propose new regulations under the Clean Air Act to control mercury emissions from coal and oil fired power plants by Dec. 15, 2003. One proposal was to reduce mercury emissions from power plants 90% by 2007. According to an article in Forbes, such regulation "could cost the power industry at least 8.58 billion dollars per year." More recent proposals such as the Clear Skies Act call for a 70% reduction in mercury emissions over 15 years.

Sulfidized red mud is a powerful sorbent for remediating polluted sources such as groundwater, wastewater, mine runoff, petroleum streams, and industrial waste. Of particular interest is sorbing heavy metals such as mercury (Hg), chromium (Cr), lead (Pb), copper (Cu), zinc (Zn), silver (Ag), cadmium (Cd), selenium (Se), thorium (Th), and uranium (U) from such sources. The metals may be present as free elements, ions, or in compounds with other elements.

Of special interest is remediation of over 30,000 mine drainage streams where the alkalinity of sulfidized red mud would be useful.

PREPARATION OF SULFIDIZED RED MUD

The sorbent is prepared by the sulfidation of red mud, which contains hydrated ferro ferric oxides derived from the Bayer processing of bauxite ores. Sulfidation can be achieved by reacting red mud with one or more sulfidizing compounds such as H_2S, Na_2S, K_2S, $(NH_4)_2S$, and CaS_x. Unlike red mud, which is very hydrophilic, sulfidized red mud is lyophobic. As a result, sulfidized red mud has much faster dewatering rates than red mud.

The relative amount of sulfidizing agent is selected so that the sulfur content of the reaction product is from about 0.2 to about 10% above the residual sulfur content of the red mud. The weight ratio of sulfidizing compound to red mud will vary with the type of sulfidizing compound used and the desired level of sulfidation for a particular end use. Most often, the sulfidizing compound and red mud are combined at a weight ratio usually from about 1:25 to about 1:6.

Conditions under which red mud can be sulfidized depend on such factors as the type of sulfidizing compound(s) and the intended use of the resulting sorbent. In some cases, sulfidation can be accomplished by mixing red mud and the sulfidizing compound at ambient temperature and atmospheric pressure. In general, higher sulfur contents can be obtained when the reaction is carried out at slightly elevated temperatures and/or elevated pressures. Sulfur content in the reaction product is affected by sulfur content of the sulfidizing agent. For example, compounds such as calcium polysulfide, usually yield products having higher sulfur contents.

When using gaseous sulfidizing compounds, such as hydrogen sulfide (H_2S), it is often preferred to conduct the reaction at slightly elevated temperature and/or elevated pressure to increase the rate of reaction and the sulfur content of the resulting sorbent. Suitable reaction temperatures range from about 40 to 200°C, often from about 80 to 120°C. The reaction pressure typically ranges from about 30 to about 70 psi (absolute).

USE OF SULFIDIZED RED MUD

In a typical application, the sorbent is slurried with a medium containing the contaminant(s) to be extracted. The sorbent, which forms a complex with the contaminant(s), can then be separated from the slurry using one or more conventional techniques such as filtration, sedimentation, or centrifugation.

In an alternative application, sulfidized red mud sorbent is processed into pellets using conventional pelletizing or extrusion equipment. The pellets can be used in filters of conventional construction in a variety of industrial or consumer filtration applications, including filters for preparing potable water.

Table 1. Sulfur content of RM-1 and SRM (2–6)

Code	Description	Example	S (wt%)
RM-1	Red Mud	1	0.19
SRM-2	Sulfidized Red Mud H_2S	2	0.48
SRM-3	Sulfidized Red Mud H_2S with Pressure	3	0.90
SRM-4	Sulfidized Red Mud $(NH_4)_2S$	4	0.46
SRM-5	Sulfidized Red Mud Na_2S	5	0.62
SRM-6	Sulfidized Red Mud CaS_x	6	1.19

It has been found that sulfidized red mud sorbent is effective for sorbing various contaminants, such as mercury, which are not effectively sorbed by red mud. On the other hand, red mud itself is effective for sorbing other contaminants, such as arsenic, which are not efficiently sorbed by sulfidized red mud. For treatment of media having contaminants in both categories, use of red mud and sulfidized red mud in tandem, either in the same sorbent composition or in sequential treatment stages (e.g., red mud followed by sulfidized red mud) can be more advantageous than using either sorbent alone.

RM 1. Preparation of Red Mud. A 1 kg sample of red mud received from Sherwin Alumina Company of Corpus Christi, TX was slurried at 15% solids in demineralized water and filtered on a Buchner funnel. The resulting filter cake was re-slurried with demineralized water, re-filtered, and used as the starting material in Example 2. See Table 1.

SRM 2. Preparation of Sulfidized Red Mud Using Hydrogen Sulfide (H_2S). Washed red mud (100 g) from Example 1 was slurried in demineralized water at 15% solids and the stirred slurry was saturated with hydrogen sulfide for 30 minutes at ambient temperature. The sample was dried overnight at 100°C and the resulting cake was pulverized.

SRM 3. Preparation of Sulfidized Red Mud Using H_2S Under Pressure in a Parr Bomb. The sulfidation procedure of Example 2 was repeated using a Laboratory Parr Bomb. After saturation of the slurry with hydrogen sulfide gas, the bomb was sealed and heated four hours at 100°C while stirred. The bomb was then cooled, depressurized and the contents filtered, dried, and pulverized.

SRM 4. Preparation of Sulfidized Red Mud Using Ammonium Sulfide $(NH_4)_2S$. Red mud (200 g) was dispersed in 600 grams of deionized (DI) water in a Waring Blender for 5 minutes. Ammonium sulfide (10 g) was added and the slurry was heated with stirring on a hot plate for 1 hour at 60°C. It was then filtered and dried at 90°C.

SRM 5. Preparation of Sulfidized Red Mud Using Sodium Sulfide (Na_2S). The procedure of Example 2 was repeated using sodium sulfide instead of ammonium sulfide.

SRM 6. Preparation of Sulfidized Red Mud Using Calcium Polysulfide (CaS_x). The procedure of Example 2 was repeated using 33.5 g of 30% solution of Cascade, calcium polysulfide.

A more complete analysis of RM-1, SRM (3–6) is given in Table 2. The analysis reveals that filtration and washing during preparation of sulfidized red mud extracts sodium chloride (except for SRM-5) and increases concentration of Fe_2O_3 in red mud. It is significant that very small amounts of reacted sulfur have such a strong effect on the chemical and physical properties of red mud.

Table 2. Analysis of RM-1, SRM (3–6)

		Weight %						PPM		
Code	Description	Na_2O	MgO	Al_2O_3	SiO_2	P_2O_5	Fe_2O_3	Cr	Pb	Cu
RM-1	Control	4.73	0.12	17.1	8.23	1.14	39.9	1258	144	119
SRM-3	H_2S (b)	3.94	0.14	14.6	9.14	1.38	46.2	1506	180	138
SRM-4	$(NH_4)_2S$	4.39	0.13	17.9	9.24	1.26	42.3	1379	176	146
SRM-5	Na_2S	5.20	0.11	17.2	8.56	1.15	41.5	1272	159	130
SRM-6	CaS_x	4.44	0.09	16.2	8.41	1.29	41.2	1364	165	138

Table 3. Metal concentration in leachate (ppm)

	Hg	As	Cd	Cr	Pb	Se
SRM-2	0.0026	ND*	0.0013	0.0044	ND	ND
RM-1	0.0032	0.096	ND	0.0510	0.0064	0.017

*ND—Not detectable, below limits.

Leaching of RM-1 vs. SRM-2. In part (a), a slurry of red mud (50 g) and demineralized water (450 mL) was prepared, mixed for 30 minutes, and filtered. The filtrate was acidified with 2 mL concentrated nitric acid and analyzed by ICP using EPA3050 and EPA6010 methods.

In part (b), the procedure of part (a) was repeated using sulfidized red mud (SRM-2).

Results are given in Table 3 and show that leachate from sulfidized red mud (SRM-2) gave a much reduced content of heavy metals (low parts per billion) than leachate from the red mud (RM-1) in every case, except Cd, where the difference was insignificant.

Mercuric Solution (3.5 ppm) Sorption by SRM-3. Ten grams of sulfidized red mud SRM-3 was slurried 30 minutes with 1 kg demineralized water containing 3.5 ppm mercury (5.66 ppm mercuric nitrate). The slurry was filtered and analyzed for mercury (Hg^{++}) by ICP (Method EOA 245.1).

The procedure was repeated using 22.0 ppm and 41.0 ppm mercury solutions (11–12), (13–14).

Results of tests 9–14 are summarized in Table 4 and demonstrate the superior performance of sulfidized red mud compared to red mud for sorption of mercuric ion from aqueous solutions.

Example 15 Mercury (metal) Sorption from Vapor Phase by SRM-3 and RM-1 (Spray Absorbed). In part (a), one gram of mercury metal was placed in a two necked round bottom (RB) flask on a supported heating mantle. One neck of the flask was open and the second neck was connected with a Teflon® tube to an aperture in the inlet duct of a spray dryer. The mercury was heated to 300°C. A slurry of 580 g SRM-3 in 450 mL demineralized water was sprayed by a rotary atomizer operating at 30,000 rpm. The feed rate of SRM-3 was regulated to produce an outlet temperature of 100°C from the dryer.

In part (b), the procedure of part (a) was repeated using RM-1 instead of SRM-3.

The mercury content of the spray dried SRM from part (a) and the RM from part (b) are tabulated in Table 5 and show that the SRM had a significantly improved sorption of mercury.

SRM-3 absorbed 7.5 times more mercury as RM-1 when spray dried at 300°C inlet and 100°C outlet in the presence of an air stream containing mercury heated to 250°C. Sulfidized red mud is significantly superior to red mud as a sorbent for elemental mercury metal vapor.

Table 4. SRM-3 vs. RM mercuric ion sorption from aqueous solutions

Example	Mercuric Concentration in Filtrate (ppm)	% Sorbed
Control solution	3.5	—
9 RM-1	0.56	84
10 SRM-3	0.2	94.3
Control solution	22.0	—
11 RM-1	8.0	64
12 SRM-3	0.22	99
Control solution	41.0	—
13 RM-1	23.4	43
14 SRM-3	0.04	99.9

Table 5. Mercury sorption by spray dried SRM-3 and RM-1

	Sorbed Hg Concentration (ppm)
15(a) SRM-3	61.0
15(b) RM-1	8.1

Table 6. Mercury sorption from vapor phase

		Sorbed Hg Concentration (ppm)
16(a) SRM-3	1st pass	95
16(b) SRM-3	2nd pass	340
16(c) RM-1	1st pass	43
16(d) RM-1	2nd pass	48

Example 16 Mercury (metal) Sorption from Vapor Phase by SRM-3 and RM-1 (Spray Absorbed). Example 15 was repeated except that a slurry of 100 g SRM-3(a) and also 100 g of RM-1 in 900 mL demineralized water were spray dried (b). Samples 16a and 16b were analyzed for mercury.

This experiment was then repeated using 100 g RM-1 and also 100 g SRM-3 to furnish samples 16c and 16d, which were analyzed. The results of tests 16(a)–(d) are shown in Table 6.

As evident from Table 6, SRM-3 is about twice as efficient as RM-1 on the 1st pass and about seven times as efficient as RM-1 on the second pass. The results show that the affinity of SRM-3 for mercury vapor improves with increased exposure to mercury, indicating an induction effect.

Sorption of mercury by scrubbing gases with sulfidized red mud has important potential for reducing mercury contamination of both freshwater and saltwater bodies.

Table 7 summarizes the results of Examples 19–28 using the general procedure of Example 9. The last column indicates the amount (in wt %) of the target ion that was removed by SRM. The results with thorium are especially significant.

Example 29 Comparison of SRM and RM for Sorption of As, Co, Mn, and Sr. The procedure of Example 9 was repeated using solutions of arsenic (III), arsenic (V), cobalt II, manganese (II), and strontium (I), with results summarized in Table 8.

Table 7. Summary of examples 19–28 SRM-3 vs. RM-1

Example	Element	Control Solution (ppm)	RM-1 Filtrate (ppm)	SRM-3 Filtrate (ppm)	% Removed by SRM-3
19	Chromium III	2.240	0.018	0.005	99.8
20	Copper II	1.550	0.028	<0.004	99.99
	Copper II	6.250	0.054	0.038	99.4
	Copper II	30.50	0.073	0.040	99.9
21	Zinc II	1.850	0.035	0.009	99.5
	Zinc II	2.380	0.103	0.022	99.1
22	Silver I	3.15	ND*	ND†	99.99
23	Gold I	0.703	ND	0.227	67.7
24	Cadmium II	1.850	0.035	0.009	99.5
25	Lead II	2.0	0.058	0.007	99.7
26	Selenium	2.5	2.1	0.24	90.4
27	Thorium IV	0.956	0.051	ND	99.99
	Thorium IV	4.93	0.260	ND	99.99
	Thorium IV	10.50	0.564	ND	99.99
	Thorium IV	19.40	0.921	ND	99.99
28	Uranium II	1.13	0.074	0.04	96.5
	Uranium II	10.1	2.45	0.494	95.1
	Uranium II	38.0	6.90	3.95	89.6

*ND: Not detectable.
†ND: Essentially quantitative removal of Thorium was obtained by SRM-4.

Table 8. Comparison of SRM-3 and RM-1 sorption

Element	Control Solution (ppm)	RM-1 Filtrate (ppm)	% Removed	SRM-3 Filtrate (ppm)	% Removed
Arsenic III	0.60	0.11	81.7	0.36	40
Arsenic V	1.60	0.21	86.9	1.15	28
Cobalt II	2.75	0.013	99.5	0.046	98.3
Manganese II	1.63	0.135	91.7	0.548	66.4
	2.10	0.72	65.7	0.792	62.3
Strontium II	1.90	0.10	94.7	1.10	42.1
	9.0	0.08	99.1	4.60	48.9
	27.0	0.19	99.3	11.0	59.3

These experiments reveal that sorption of red mud (RM-1) is significantly better than SRM-3 in the case of As (III), AS (V), Mn (II), and Sr (II). However, the use of red mud as a sorbent is restricted by leaching of undesirable elements which can cause serious problems. Use of sulfidized red mud in combination with red mud is useful because sulfidized red mud prevents undesirable leaching of toxic metals from red mud itself.

Table 9. Sorption of Hg(II) by SRM (3–6)

	Concentration of Hg(II) in Original Solution (ppm)	Concentration After Treatment with SRM-6 (ppm)	% Removed
SRM 4	4.5	0.001	100
5% $(NH_4)_sS$	19.6	0.0229	99.9
SRM 5	4.5	0.449	90.0
5% Na_2S	19.6	3.68	81.2
SRM 6	4.5	0.005	99.9
5% CaS_x	19.6	3.16	83.8
SRM 3	4.5	0.004	99.9
H_2S pressure	19.6	0.02	99.9

Example 30 Sorption of Hg (II) by Various SRMs. The results are summarized in Table 9. SRM-3, 4, and 6 gave excellent sorption results from solutions of Hg(II) at two concentrations (4.5 ppm and 19.6 ppm). It is significant that SRM-4 reduced Hg to 1 ppb, thus meeting current drinking water standards (3 ppb maximum).

Ammonium sulfide treatment red mud (SRM-4) was the most effective sorbent despite the fact it had the lowest S content. SRM-5 prepared by treatment of red mud with Na_2S was much less effective than SRM-4.

Example 35 Sedimentation Rates of SRM-4 and RM-1. In the course of tests on metal sorption from aqueous solutions by sulfidized red mud and red mud, it was found that in all cases, sulfidized red mud exhibited significantly faster filtration rates than red mud. Red mud is very hydrophilic but conversion of red mud to sulfidized red mud transforms it to a lyophobic sorbent which is more readily dewatered. The unexpected improvement of dewatering behavior is shown in the following experiment.

A dispersion of 50 grams of RM-1 in 500 mL demineralized water was prepared by rapid mixing in a Waring Blender for 10 minutes. The experiment was repeated using 50 grams of SRM-3 in 500 mL demineralized water. See Figure 1.

Both freshly prepared slurries were allowed to settle undisturbed at ambient temperature (25°C) for a period of 23 hours. After 23 hours, the RM-1 dispersions had settled to give a clear supernatant layer of only 1 cm. The remaining slurry consisted of dispersed RM-1 with no visible sediment.

During a 23 hour period, the SRM-3 slurry settled to furnish a sedimentary layer about 3 cm deep and a clear supernatant layer 11.5 cm above the sediment.

These results clearly show the significant alteration of surface chemistry and dewatering characteristics of red mud by relatively small degrees of sulfidation.

Example 36 Clarification of Okefenokee Swamp Water with SRM-4. 500 mL of Okefenokee Swamp water (Sample I) was adjusted to pH 7 with dilute NaOH and mixed with 10 grams of SRM-410 (made with 10% ammonium sulfide) in a Waring Blender at high speed for 5 minutes. The mixture was transferred to a beaker and allowed to stir an additional hour using a magnetic stirrer.

The suspension was filtered and the color value of the filtrate was determined with a LaMotte TC-3000e colorimeter. Another 10 grams of SRM-410 was then added and the procedure was repeated a second time (2nd Pass). The filtrate was again evaluated for color. Results

Table 10. Absorbance testing of Okefenokee "black" water (Sample I)

Sample Designation	Color Value (CV) (375nm)
Control untreated	347
1st Pass SRM-410	38.9
2nd Pass SRM-410	18.8

Table 11. Okefenokee "Black" water (Sample II)

Sample Designation	Absorbance*
Control untreated	0.063
Sample II	0.0063

*Fisher Genesys5 Spectrophotometer 500nm

Figure 1. Sedimentation

Figure 2. Okefenokee "Black" water DOC removal

are given in Table 10 and showed that the treated sample was nearly colorless (over 90% reduction in absorbance).

Another sample of Okefenokee "Black" Water (Sample II) was treated with sulfidized red mud according to the above procedure. The absorbance was reduced 90% to nearly colorless as shown in Table 11 (2 passes) and Figure 2.

REFERENCES

Forbes Magazine. 2003. April ed., p. 104.

Iannicelli. 2010. *Method and Composition for Sorbing Toxic Substances/Flue Gas Scrubbing.* USPTO Publication 2010/0218676 A1.

Iannicelli. 2010. *Method and Composition for Sorbing Toxic Substances/Combination of Red Mud and Sulfidized Red Mud.* USPTO Publication 2010/0224567 A1.

Iannicelli. 2010. *Method and Composition for Sorbing Toxic Substances/Accelerated Sedimentation of Sulfidized Red Mud.* USPTO Publication 2010/0224576 A1.

Iannicelli. 2010. *Method of Sorbing Discolored Organic Matter (DOC) by Sulfidized Red Mud (Sorbing DOC with Sulfidized Red Mud).* USPTO Publication 2010/0243568 A1.

Iannicelli. 2010. U.S. Patent No. 7,763,566. B2 Method and Composition for Sorbing Toxic Substances.

Iannicelli. 2010. U.S. Patent No. 7,807,058. B2 Method and Composition for Sorbing Toxic Substances/Filter.

MConshie, David. 2005. Cites from Virotec Web site: virotec.com/global.htm.

Ryle, Gerard. 2002. *The Great Red Mud Experiment That Went Radioactive.*

Tedder, D. William. 1984. *Bauxite Residue Fractionation with Magnetic Separators.* Chapter 33. AIME Bauxite Symposium.

Index

Note: f. indicates figure; t. indicates table.

L

M

T

U

V

W

X

Z